Audio/Video Professional's Field Manual

On-Line Updates

Additional updates relating to audio/video technology in general, and this book in particular, can be found at the *Standard Handbook of Video and Television Engineering* web site:

www.tvhandbook.com

The tvhandbook.com web site supports the professional audio/video community with news, updates, and product information relating to the broadcast, post production, consumer, and business/industrial applications of digital video.

Check the site regularly for news, updated chapters, and special events related to video engineering. The technologies encompassed by *Audio/Video Professional's Field Manual* are changing rapidly, with new developments each month. Changing market conditions and regulatory issues are adding to the rapid flow of news and information in this area.

Specific services found at **www.tvhandbook.com** include:

- **Audio/Video Technology News**. News reports and technical articles on the latest developments in digital radio and television, both in the U.S. and around the world. Check in at least once a month to see what's happening in the fast-moving area of digital broadcasting.
- **Technology Resource Center**. Check for the latest information on professional and broadcast audio/video systems. The Resource Center provides updates on implementation and standardization efforts, plus links to related web sites.
- **tvhandbook.com Update Port**. Updated material for *Audio/Video Professional's Field Manual* is posted on the site regularly. Material available includes updated sections and chapters in areas of rapidly advancing technologies.
- **tvhandbook.com Book Store**. Check to find the latest books on digital video and audio technologies. Direct links to authors and publishers are provided. You can also place secure orders from our on-line bookstore.

In addition to the resources outlined above, detailed information is available on other books in the McGraw-Hill Video/Audio Series.

www.tvhandbook.com

Audio/Video Professional's Field Manual

Jerry C. Whitaker

McGraw-Hill
New York Chicago San Francisco
Lisbon London Madrid Mexico City Milan
New Delhi San Juan Seoul Singapore
Sydney Toronto

Cataloging-in-Publication Data is on file with the Library of Congress.

McGraw-Hill

A Division of The ***McGraw-Hill*** *Companies*

1 2 3 4 5 6 7 8 9 0 DOC/DOC 0 9 8 7 6 5 4 3 2 1

ISBN 0-07-137209-1

The sponsoring editor for this book was Steve Chapman and the production supervisor was Pamela Pelton. The book was set in Times New Roman and Helvetica by Technical Press, Morgan Hill, CA.

Printed and bound by R. R. Donnelley & Sons Company.

McGraw-Hill books are available at special quantity discounts to use as premiums and sales promotions, or for use in corporate training programs. For more information, please write to the Director of Special Sales, Professional Publishing, McGraw-Hill, Two Penn Plaza, New York, NY 10121-2298. Or contact your local bookstore.

This book is printed on recycled, acid-free paper containing a minimum of 50% recycled, de-inked fiber.

Contents

Chapter 6: Reference Documents by Subject 459

Chapter 7: Index of Figures, Tables, and Subjects 531

Preface

The arrival of practical digital broadcasting systems for radio and television represent significant milestones in broadcasting. New concepts, technologies, and business models have reshaped the technical landscape, raising a host of new challenges and opportunities. The transition to digital broadcasting is progressing at a remarkable rate, already accomplishing far more that most thought possible.

Being involved with audio and video engineering every day, it is easy to view this unprecedented buildup as just another day's work. However, viewed from a perspective, the accomplishments have been monumental. There are countless examples of technologies unveiled in technical papers given at industry conferences two and three years ago that are now widely implemented—even commonplace.

At the same time, the conventional video systems of NTSC, PAL, and SECAM are still the bread-and-butter of television. Likewise, the conventional radio systems of AM and FM still reign supreme. The classic techniques and conventions of analog audio and video need not be discarded simply because digital technology offers another way of accomplishing the same task. It is certain, therefore, that the broadcast industry will exist in a hybrid mode of both analog and digital systems for many years to come.

With these issues in mind, the *Audio/Video Professional's Field Manual* was devised to provide easy access to a wealth of information for all technologies involved in audio and video engineering today. This book serves as a companion publication to McGraw-Hill's *Standard Handbook of Video and Television Engineering* and *Standard Handbook of Audio and Radio Engineering*, providing readers with data in a "just-the-facts" format. The Standard Handbooks provide detailed background and tutorial chapters covering all aspects of audio and video engineering—from human vision to loudspeakers—while in the *Audio/Video Professional's Field Manual*, a considerably different approach is taken. The key elements of this book include the following:

- **Engineering Fundamentals.** Chapter 1 covers the fundamentals of electronics engineering. Focusing on components and basic circuits, these sections are intended to provide readers with a firm engineering foundation, covering subjects that form the basis of applied electronics.

- **Reference Data, Tables, and Figures.** Chapters 2 and 3 form the core of the *Audio/Video Professional's Field Manual*. Contained within the 200+ pages of these chapters are hundreds of tables and figures relating to analog and digital audio/video engineering. The material included was selected because it provides quick answers to specific questions or needs. This section is not tutorial in nature; rather, it is intended as a reference resource.

- **Dictionary.** An extensive glossary—more than 100 pages in length—provides definitions of terms relating to electronics in general, and audio/video engineering in particular (Chapters 4 and 5).

- **Reference Documents.** Chapter 6 represents the most complete listing of reference documents ever assembled for audio and video engineering. It is provided as a way of helping readers find additional information on any aspect of analog and digital video technologies.

- **Extensive Index.** A unique index is provided in Chapter 7 that lists all of the figures and tables contained in this book, organized by subject matter. In addition, a conventional subject index is provided.

The *Audio/Video Professional's Field Manual* is intended for audio/video professionals involved in the design, installation, operation, and maintenance of radio and television stations, video production centers, and transmission facilities at broadcast stations, post production facilities, corporate/industrial centers, and other related facilities.

I trust that you will find this publication informative and useful as you continue to match the rapid pace of audio and video engineering.

Jerry C. Whitaker
October 2001

For

John and Christine Skeen

good neighbors, good friends

Audio/Video Professional's Field Manual

Chapter

1

The Principles of Electronics Reviewed

Jerry C. Whitaker and K. Blair Benson

1.1 Introduction

The atomic theory of matter specifies that each of the many chemical elements is composed of unique and identifiable particles called atoms. In ancient times only 10 were known in their pure, uncombined form; these were carbon, sulfur, copper, antimony, iron, tin, gold, silver, mercury, and lead. Of the several hundred now identified, less than 50 are found in an uncombined, or chemically free, form on earth.

Each atom consists of a compact nucleus of positively and negatively charged particles (protons and electrons, respectively). Additional electrons travel in well-defined orbits around the nucleus. The electron orbits are grouped in regions called *shells*, and the number of electrons in each orbit increases with the increase in orbit diameter in accordance with quantum-theory laws of physics. The diameter of the outer orbiting path of electrons in an atom is in the order of one-millionth (10^{-6}) millimeter, and the nucleus, one-millionth of that. These typical figures emphasize the minute size of the atom.

1.1a Magnetic Effects

The nucleus and the free electrons for an iron atom are shown in the schematic diagram in Figure 1.1. Note that the electrons are spinning in different directions. This rotation creates a magnetic field surrounding each electron. If the number of electrons with positive spins is equal to the number with negative spins, then the net field is zero and the atom exhibits no magnetic field.

In the diagram, although the electrons in the first, second, and fourth shells balance each other, in the third shell five electrons have clockwise positive spins, and one a counterclockwise negative spin, which gives the iron atom in this particular electron configuration a cumulative magnetic effect.

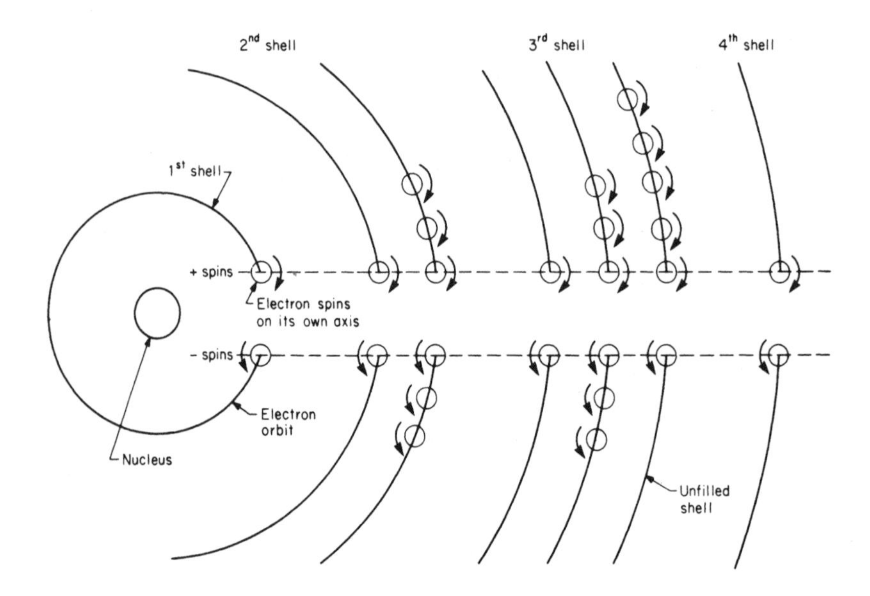

Figure 1.1 Schematic of the iron (Fe) atom.

The parallel alignment of electron spins over regions, known as *domains*, containing a large number of atoms. When a magnetic material is in a demagnetized state, the direction of magnetization in the domain is in a random order. Magnetization by an external field takes place by a change or displacement in the isolation of the domains, with the result that a large number of the atoms are aligned with their charged electrons in parallel.

1.1b Conductors and Insulators

In some elements, such as copper, the electrons in the outer shells of the atom are so weakly bound to the nucleus that they can be released by a small electrical force, or voltage. A voltage applied between two points on a length of a metallic conductor produces the flow of an electric current, and an electric field is established around the conductor. The conductivity is a constant for each metal that is unaffected by the current through or the intensity of any external electric field.

In some nonmetallic materials, the free electrons are so tightly bound by forces in the atom that, upon the application of an external voltage, they will not separate from their atom except by an electrical force strong enough to destroy the insulating properties of the material. However, the charges will realign within the structure of their atom. This condition occurs in the insulating material (*dielectric*) of a capacitor when a voltage is applied to the two conductors encasing the dielectric.

Semiconductors are electronic conducting materials wherein the conductivity is dependent primarily upon impurities in the material. In addition to negative mobile charges of electrons, positive mobile charges are present. These positive charges are called *holes* because each exists as an absence of electrons. Holes (+) and electrons (–), because they are oppositely charged, move in opposite directions in an electric field. The conductivity of semiconductors is highly sensitive to, and increases with, temperature.

1.1c Direct Current (dc)

Direct current is defined as a unidirectional current in which there are no significant changes in the current flow. In practice, the term frequently is used to identify a voltage source, in which case variations in the load can result in fluctuations in the current but not in the direction.

Direct current was used in the first systems to distribute electricity for household and industrial power. For safety reasons, and the voltage requirements of lamps and motors, distribution was at the low nominal voltage of 110. The losses in distribution circuits at this voltage seriously restricted the length of transmission lines and the size of the areas that could be covered. Consequently, only a relatively small area could be served by a single generating plant. It was not until the development of alternating-current systems and the voltage transformer that it was feasible to transport high levels of power at relatively low current over long distances for subsequent low-voltage distribution to consumers.

1.1d Alternating Current (ac)

Alternating current is defined as a current that reverses direction at a periodic rate. The average value of alternating current over a period of one cycle is equal to zero. The effective value of an alternating current in the supply of energy is measured in terms of the *root mean square* (rms) value. The rms is the square root of the square of all the values, positive and negative, during a complete cycle, usually a sine wave. Because rms values cannot be added directly, it is necessary to perform an rms addition as shown in the equation:

$$V_{total} = \sqrt{V_{rms1}^2 + V_{rms2}^2 + \ldots + V_{rmsn}^2} \tag{1.1}$$

As in the definition of direct current, in practice the term frequently is used to identify a voltage source.

The level of a sine-wave alternating current or voltage can be specified by two other methods of measurement in addition to rms. These are *average* and *peak*. A sine-wave signal and the rms and average levels are shown in Figure 1.2. The levels of complex, symmetrical ac signals are specified as the peak level from the axis, as shown in the figure.

1.2 Electronic Circuits

Electronic circuits are composed of elements such as resistors, capacitors, inductors, and voltage and current sources, all of which may be interconnected to permit the flow of electric cur-

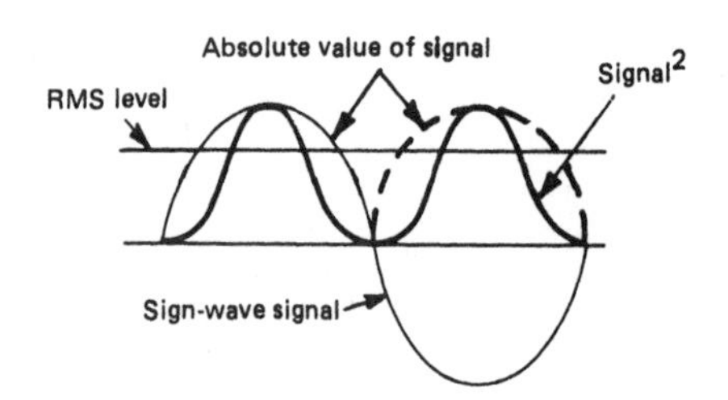

Figure 1.2 Root mean square (rms) measurements. The relationship of rms and average values is shown.

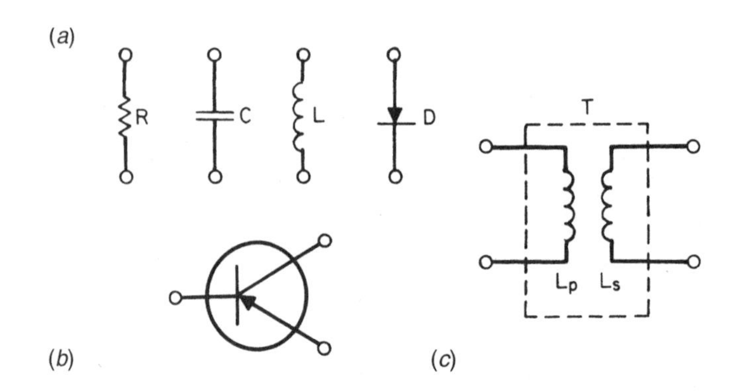

Figure 1.3 Schematic examples of circuit elements: (*a*) two-terminal element, (*b*) three-terminal element, (*c*) four-terminal element.

rents. An *element* is the smallest component into which circuits can be subdivided. The points on a circuit element where they are connected in a circuit are called *terminals*.

Elements can have two or more terminals, as shown in Figure 1.3. The resistor, capacitor, inductor, and diode shown in the Figure 1.3*a* are *two-terminal elements*; the transistor in Figure 1.3*b* is a three-terminal element; and the transformer in Figure 1.3*c* is a four-terminal element.

Circuit elements and components also are classified as to their function in a circuit. An element is considered *passive* if it absorbs energy and *active* if it increases the level of energy in a signal. An element that receives energy from either a passive or active element is called a *load*. In addition, either passive or active elements, or components, can serve as loads.

The basic relationship of current and voltage in a two-terminal circuit where the voltage is constant and there is only one source of voltage is given in Ohm's law. This states that the voltage V between the terminals of a conductor varies in accordance with the current I. The ratio of voltage, current, and resistance R is expressed in Ohm's law as follows:

$$E = I \times R \tag{1.2}$$

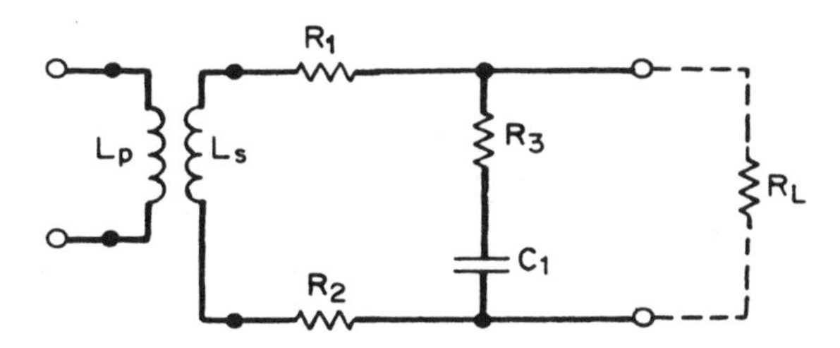

Figure 1.4 Circuit configuration composed of several elements and branches, and a closed loop (R_1, R, C_1, R_2, and L_s).

Using Ohm's law, the calculation for power in watts can be developed from $P = E \times I$ as follows:

$$P = \frac{E^2}{R} \text{ and } P = I^2 \times R \tag{1.3}$$

A circuit, consisting of a number of elements or components, usually amplifies or modifies a signal before delivering it to a load. The terminal to which a signal is applied is an *input port*, or *driving port.* The pair or group of terminals that delivers a signal to a load is the *output port.* An element or portion of a circuit between two terminals is a *branch.* The circuit shown in Figure 1.4 is made up of several elements and branches. R_1 is a branch, and R_1 and C_1 make up a two-element branch. The secondary of transformer T, a voltage source, and R_2 also constitute a branch. The point at which three or more branches join together is a *node.* A series connection of elements or branches, called a *path*, in which the end is connected back to the start is a *closed loop.*

1.2a Circuit Analysis

Relatively complex configurations of linear circuit elements, that is, where the signal gain or loss is constant over the signal amplitude range, can be analyzed by simplification into the equivalent circuits. After the restructuring of a circuit into an equivalent form, the current and voltage characteristics at various nodes can be calculated using network-analysis theorems, including Kirchoff's current and voltage laws, Thevenin's theorem, and Norton's theorem.

- **Kirchoff's current law** (KCL). The algebraic sum of the instantaneous currents entering a node (a common terminal of three or more branches) is zero. In other words, the currents from two branches entering a node add algebraically to the current leaving the node in a third branch.
- **Kirchoff's voltage law** (KVL). The algebraic sum of instantaneous voltages around a closed loop is zero.

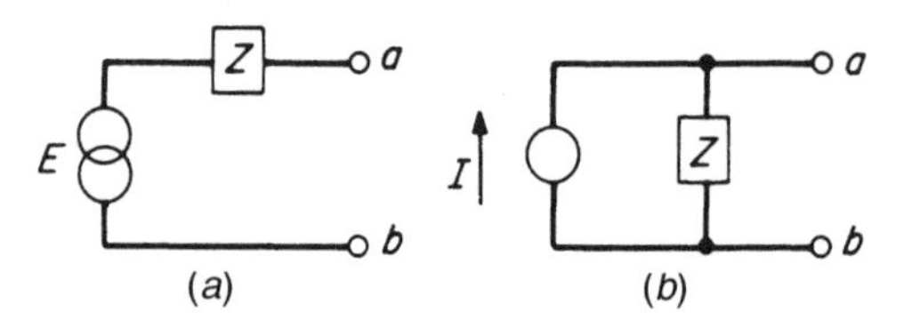

Figure 1.5 Equivalent circuits: (*a*) Thevenin's equivalent voltage source, (*b*) Norton's equivalent current source. (*After* [2]. *Used with permission.*)

- **Thevenin's theorem**. The behavior of a circuit at its terminals can be simulated by replacement with a voltage E from a dc source in series with an impedance Z (see Figure 1.5*a*).
- **Norton's theorem**. The behavior of a circuit at its terminals can be simulated by replacement with a dc source I in parallel with an impedance Z (see Figure 1.5*b*).

1.3 Static Electricity

The phenomenon of static electricity and related potential differences concerns configurations of conductors and insulators where no current flows and all electrical forces are unchanging; hence the term *static*. Nevertheless, static forces are present because of the number of excess electrons or protons in an object. A static charge can be induced by the application of a voltage to an object. A flow of current to or from the object can result from either a breakdown of the surrounding nonconducting material or by the connection of a conductor to the object.

Two basic laws regarding electrons and protons are:

- Like charges exert a repelling force on each other; electrons repel other electrons and protons repel other protons
- Opposite charges attract each other; electrons and protons are attracted to each other

Therefore, if two objects each contain exactly as many electrons as protons in each atom, there is no electrostatic force between the two. On the other hand, if one object is charged with an excess of protons (deficiency of electrons) and the other an excess of electrons, there will be a relatively weak attraction that diminishes rapidly with distance. An attraction also will occur between a neutral and a charged object.

Another fundamental law, developed by Faraday, governing static electricity is that all of the charge of any conductor not carrying a current lies in the surface of the conductor. Thus, any electric fields external to a completely enclosed metal box will not penetrate beyond the surface. Conversely, fields within the box will not exert any force on objects outside the box. The box need not be a solid surface; a conduction cage or grid will suffice. This type of isolation frequently is referred to as a *Faraday shield*.

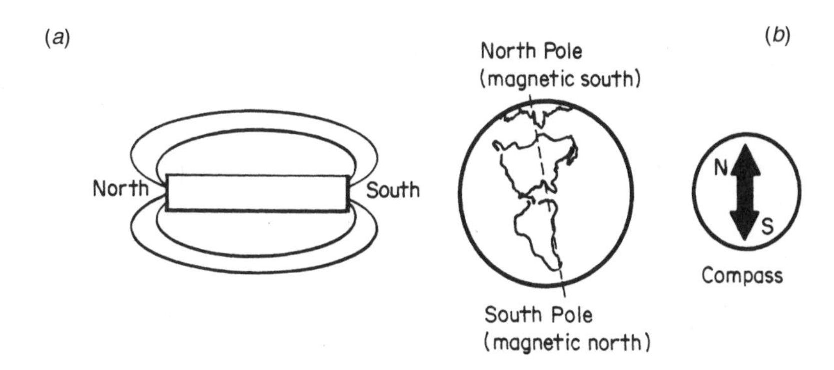

Figure 1.6 The properties of magnetism: (*a*) lines of force surrounding a bar magnet, (*b*) relation of compass poles to the earth's magnetic field.

1.4 Magnetism

The elemental magnetic particle is the spinning electron. In magnetic materials, such as iron, cobalt, and nickel, the electrons in the third shell of the atom (see Figure 1.1) are the source of magnetic properties. If the spins are arranged to be parallel, the atom and its associated domains or clusters of the material will exhibit a magnetic field. The magnetic field of a magnetized bar has lines of magnetic force that extend between the ends, one called the north pole and the other the south pole, as shown in Figure 1.6*a*. The lines of force of a magnetic field are called *magnetic flux lines*.

1.4a Electromagnetism

A current flowing in a conductor produces a magnetic field surrounding the wire as shown in Figure 1.7*a*. In a coil or solenoid, the direction of the magnetic field relative to the electron flow (– to +) is shown in Figure 1.7*b*. The attraction and repulsion between two iron-core electromagnetic solenoids driven by direct currents is similar to that of two permanent magnets described previously.

The process of magnetizing and demagnetizing an iron-core solenoid using a current being applied to a surrounding coil can be shown graphically as a plot of the magnetizing field strength and the resultant magnetization of the material, called a *hysteresis loop* (Figure 1.8). It will be found that the point where the field is reduced to zero, a small amount of magnetization, called *remnance*, remains.

1.4b Magnetic Shielding

In effect, the shielding of components and circuits from magnetic fields is accomplished by the introduction of a magnetic short circuit in the path between the field source and the area to be protected. The flux from a field can be redirected to flow in a partition or shield of magnetic material, rather than in the normal distribution pattern between north and south poles.

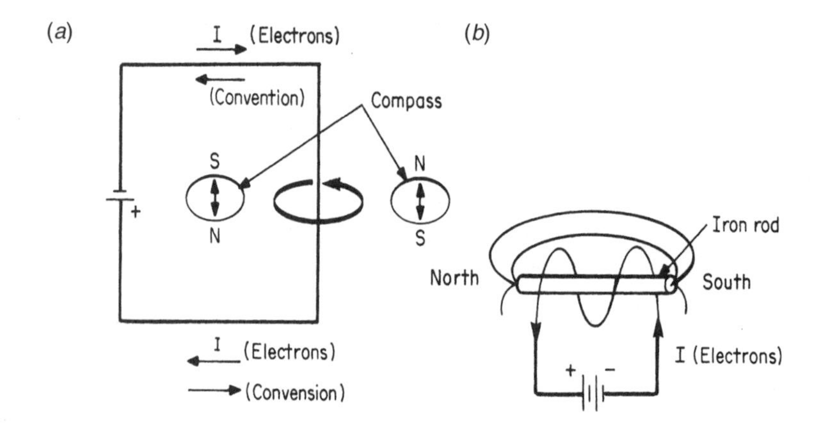

Figure 1.7 Magnetic field surrounding a current-carrying conductor: (*a*) Compass at right indicates the polarity and direction of a magnetic field circling a conductor carrying direct current. *I* indicates the direction of electron flow. Note that the convention for flow of electricity is from + to –, the reverse of the actual flow. (*b*) Direction of magnetic field for a coil or solenoid.

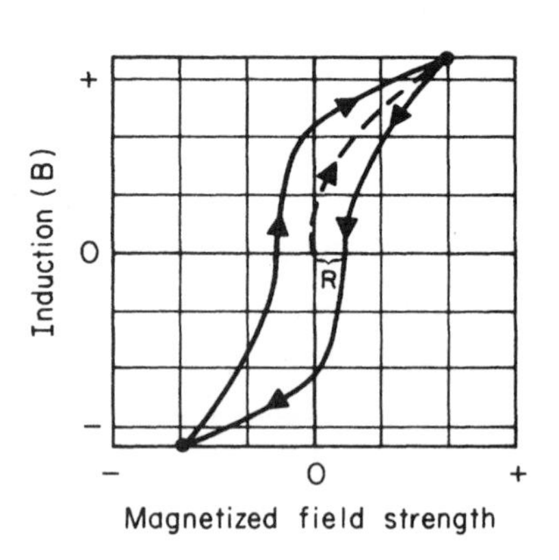

Figure 1.8 Graph of the magnetic hysteresis loop resulting from magnetization and demagnetization of iron. The dashed line is a plot of the induction from the initial magnetization. The solid line shows a reversal of the field and a return to the initial magnetization value. *R* is the remaining magnetization (remnance) when the field is reduced to zero.

The effectiveness of shielding depends primarily upon the thickness of the shield, the material, and the strength of the interfering field.

Some alloys are more effective than iron. However, many are less effective at high flux levels. Two or more layers of shielding, insulated to prevent circulating currents from magnetization of the shielding, are used in low-level audio, video, and data applications.

1.4c Electromagnetic-Radiation Spectrum

The usable spectrum of electromagnetic-radiation frequencies extends over a range from below 100 Hz for power distribution to 10^{20} for the shortest X-rays. Services using various frequency bands in the spectrum are shown in Figure 1.9. The lower frequencies are used pri-

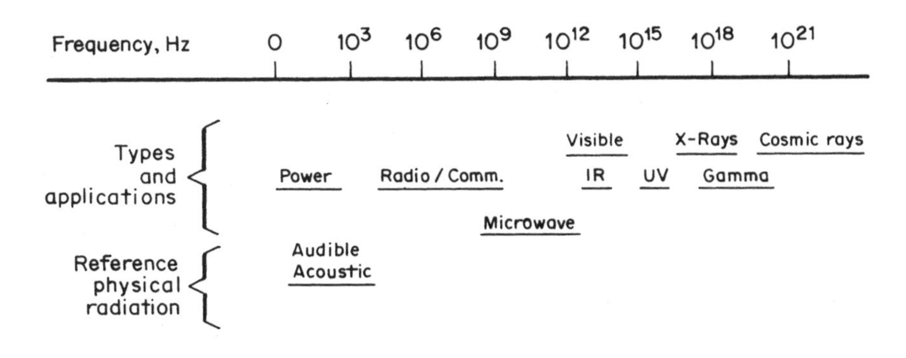

Figure 1.9 The electromagnetic spectrum. (*From* [1]. *Used with permission.*)

marily for terrestrial broadcasting and communications. The higher frequencies include visible and near-visible infrared and ultraviolet light, and X-rays.

Low-End Spectrum Frequencies (1 to 1000 Hz)

Electric power is transmitted by wire but not by radiation at 50 and 60 Hz, and in some limited areas, at 25 Hz. Aircraft use 400-Hz power in order to reduce the weight of iron in generators and transformers. The restricted bandwidth that would be available for communication channels is generally inadequate for voice or data transmission, although some use has been made of communication over power distribution circuits using modulated carrier frequencies. The sound-transmission frequencies noted in Figure 1.9 are acoustic rather than electromagnetic.

Low-End Radio Frequencies (1000 to 100 kHz)

These low frequencies are used for very long distance radio-telegraphic communication where extreme reliability is required and where high-power and long antennas can be erected.

Medium-Frequency Radio (20 kHz to 2 MHz)

The low-frequency portion of the band is used for around-the-clock communication services over moderately long distances and where adequate power is available to overcome the high level of atmospheric noise. The upper portion is used for AM radio, although the strong and quite variable *sky wave* occurring during the night results in substandard quality and severe fading at times. The greatest use is for AM broadcasting, in addition to fixed and mobile service, LORAN ship and aircraft navigation, and amateur radio communication.

High-Frequency Radio (2 to 30 MHz)

This band provides reliable medium-range coverage during daylight and, when the transmission path is in total darkness, worldwide long-distance service, although the reliability and signal quality of the latter is dependent to a large degree upon ionospheric conditions and related long-term variations in sun-spot activity affecting sky-wave propagation. The primary applications include broadcasting, fixed and mobile services, telemetering, and amateur transmissions.

Very High and Ultrahigh Frequencies (30 MHz to 3 GHz)

VHF and UHF bands, because of the greater channel bandwidth possible, can provide transmission of a large amount of information, either as television detail or data communication. Furthermore, the shorter wavelengths permit the use of highly directional parabolic or rnultielement antennas. Reliable long-distance communication is provided using high-power tropospheric scatter techniques. The multitude of uses include, in addition to television, fixed and mobile communication services, amateur radio, radio astronomy, satellite communication, telemetering, and radar.

Microwaves (3 to 300 GHz)

At these frequencies, many transmission characteristics are similar to those used for shorter optical waves, which limit the distances covered to line of sight. Typical uses include television relay, satellite, radar, and wide-band information services.

Infrared, Visible, and Ultraviolet Light

The portion of the spectrum visible to the eye covers the gamut of transmitted colors ranging from red, through yellow, green, cyan, and blue. It is bracketed by infrared on the low-frequency side and ultraviolet (UV) on the high side. Infrared signals are used in a variety of consumer and industrial equipments for remote controls and sensor circuits in security systems. The most common use of UV waves is for excitation of phosphors to produce visible illumination.

X-Rays

Medical and biological examination techniques and industrial and security inspection systems are the best-known applications of X-rays. X-rays in the higher-frequency range are classified as *hard X-rays* or *gamma rays*. Exposure to X-rays for long periods can result in serious irreversible damage to living cells or organisms.

1.5 Passive Circuit Components

Components used in electrical circuitry can be categorized into two broad classifications as passive or active. A voltage applied to a passive component results in the flow of current and the dissipation or storage of energy. Typical passive components are resistors, coils or inductors, and capacitors. For an example, the flow of current in a resistor results in radiation of heat; from a light bulb, the radiation of light as well as heat.

On the other hand, an active component either (1) increases the level of electric energy or (2) provides available electric energy as a voltage. As an example of (1), an amplifier produces an increase in energy as a higher voltage or power level, while for (2), batteries and generators serve as energy sources.

1.5a Resistors

Resistors are components that have a nearly 0° phase shift between voltage and current over a wide range of frequencies with the average value of resistance independent of the instantaneous value of voltage or current. Preferred values of ratings are given ANSI standards or corresponding ISO or MIL standards. Resistors are typically identified by their construction and by the resistance materials used. *Fixed resistors* have two or more terminals and are not adjustable. *Variable resistors* permit adjustment of resistance or voltage division by a control handle or with a tool.

Wire-Wound Resistor

The resistance element of most wire-wound resistors is resistance wire or ribbon wound as a single-layer helix over a ceramic or fiberglass core, which causes these resistors to have a residual series inductance that affects phase shift at high frequencies, particularly in large-size devices. Wire-wound resistors have low noise and are stable with temperature, with temperature coefficients normally between ±5 and 200 ppm/°C. Resistance values between 0.1 and 100,000 Ω with accuracies between 0.001 and 20 percent are available with power dissipation ratings between 1 and 250 W at 70°C. The resistance element is usually covered with a vitreous enamel, which can be molded in plastic. Special construction includes such items as enclosure in an aluminum casing for heatsink mounting or a special winding to reduce inductance. Resistor connections are made by self-leads or to terminals for other wires or printed circuit boards.

Metal Film Resistor

Metal film, or *cermet*, resistors have characteristics similar to wire-wound resistors except a much lower inductance. They are available as axial lead components in 1/8, 1/4, or 1/2 W ratings, in chip resistor form for high-density assemblies, or as resistor networks containing multiple resistors in one package suitable for printed circuit insertion, as well as in tubular form similar to high-power wire-wound resistors. Metal film resistors are essentially printed circuits using a thin layer of resistance alloy on a flat or tubular ceramic or other suitable insulating substrate. The shape and thickness of the conductor pattern determine the resistance value for each metal alloy used. Resistance is trimmed by cutting into part of the conductor pattern with an abrasive or a laser. Tin oxide is also used as a resistance material.

Carbon Film Resistor

Carbon film resistors are similar in construction and characteristics to axial lead metal film resistors. Because the carbon film is a granular material, random noise may be developed because of variations in the voltage drop between granules. This noise can be of sufficient level to affect the performance of circuits providing high grain when operating at low signal levels.

Carbon Composition Resistor

Carbon composition resistors contain a cylinder of carbon-based resistive material molded into a cylinder of high-temperature plastic, which also anchors the external leads. These resistors can have noise problems similar to carbon film resistors, but their use in electronic equipment for the last 50 years has demonstrated their outstanding reliability, unmatched by other components. These resistors are commonly available at values from 2.7 Ω with tolerances of 5, 10, and 20 percent in 1/8-, 1/4-, 1/2-, 1-, and 2-W sizes.

Control and Limiting Resistors

Resistors with a large negative temperature coefficient, *thermistors*, are often used to measure temperature, limit inrush current into motors or power supplies, or to compensate bias circuits. Resistors with a large positive temperature coefficient are used in circuits that have to match the coefficient of copper wire. Special resistors also include those that have a low resistance when cold and become a nearly open circuit when a critical temperature or current is exceeded to protect transformers or other devices.

Resistor Networks

A number of metal film or similar resistors are often packaged in a single module suitable for printed circuit mounting. These devices see applications in digital circuits, as well as in fixed attenuators or padding networks.

Adjustable Resistors

Cylindrical wire-wound power resistors can be made adjustable with a metal clamp in contact with one or more turns not covered with enamel along an axial stripe. *Potentiometers* are resistors with a movable arm that makes contact with a resistance element, which is connected to at least two other terminals at its ends. The resistance element can be circular or linear in shape, and often two or more sections are mechanically coupled or *ganged* for simultaneous control of two separate circuits. Resistance materials include all those described previously.

Trimmer potentiometers are similar in nature to conventional potentiometers except that adjustment requires a tool.

Most potentiometers have a *linear taper*, which means that resistance changes linearly with control motion when measured between the movable arm and the "low," or counterclockwise, terminal. Gain controls however, often have a *logarithmic taper* so that attenuation changes linearly in decibels (a logarithmic ratio). The resistance element of a potentiometer may also contain taps that permit the connection of other components as required in a specialized circuit.

Attenuators

Variable attenuators are adjustable resistor networks that show a calibrated increase in attenuation for each switched step. For measurement of audio, video, and RF equipment, these steps may be decades of 0.1, 1, and 10 dB. Circuits for unbalanced and balanced fixed attenuators are shown in Figure 1.10. Fixed attenuator networks can be cascaded and switched to provide step adjustment of attenuation inserted in a constant-impedance network.

Audio attenuators generally are designed for a circuit impedance of 150 Ω although other impedances can be used for specific applications. Video attenuators are generally designed to operate with unbalanced 75-Ω grounded-shield coaxial cable. RF attenuators are designed for use with 75- or 50-Ω coaxial cable.

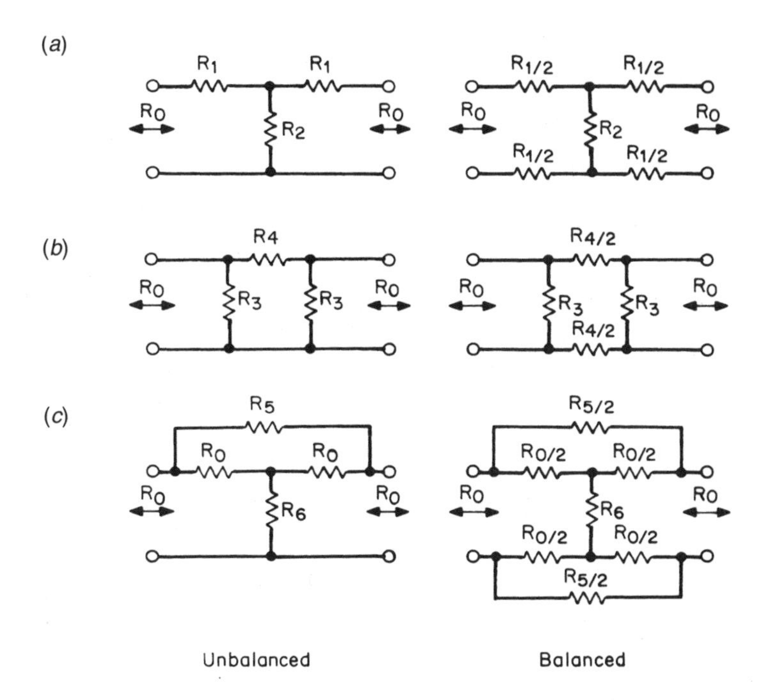

Figure 1.10 Unbalanced and balanced fixed attenuator networks for equal source and load resistance: (*a*) T configuration, (*b*) π configuration, (*c*) bridged-T configuration.

1.5b Capacitors

Capacitors are passive components in which current leads voltage by nearly 90° over a wide range of frequencies. Capacitors are rated by capacitance, voltage, materials, and construction.

A capacitor may have two voltage ratings. *Working voltage* is the normal operating voltage that should not be exceeded during operation. Use of the *test* or *forming voltage* stresses the capacitor and should occur only rarely in equipment operation. Good engineering practice is to use components only at a fraction of their maximum ratings.

Polarized Capacitors

Polarized capacitors can be used in only those applications where a positive sum of all dc and peak-ac voltages is applied to the positive capacitor terminal with respect to its negative terminal. These capacitors include all tantalum and most aluminum electrolytic capacitors. These devices are commonly used in power supplies or other electronic equipment where these restrictions can be met.

Losses in capacitors occur because an actual capacitor has various resistances. These losses are usually measured as the *dissipation factor* at a frequency of 120 Hz. Leakage resistance in parallel with the capacitor defines the time constant of discharge of a capacitor. This time constant can vary between a small fraction of a second to many hours depending on capacitor construction, materials, and other electrical leakage paths, including surface contamination.

The *equivalent series resistance* of a capacitor is largely the resistance of the conductors of the capacitor plates and the resistance of the physical and chemical system of the capacitor. When an alternating current is applied to the capacitor, the losses in the equivalent series resistance are the major causes of heat developed in the device. The same resistance also determines the maximum attenuation of a filter or bypass capacitor and the loss in a coupling capacitor connected to a load.

The *dielectric absorption* of a capacitor is the residual fraction of charge remaining in a capacitor after discharge. The residual voltage appearing at the capacitor terminals after discharge is of little concern in most applications but can seriously affect the performance of analog-to-digital (A/D) converters that must perform precision measurements of voltage stored in a sampling capacitor.

The *self-inductance* of a capacitor determines the high-frequency impedance of the device and its ability to bypass high-frequency currents. The self-inductance is determined largely by capacitor construction and tends to be highest in common metal foil devices.

Nonpolarized Capacitors

Nonpolarized capacitors are used in circuits where there is no direct voltage bias across the capacitor. They are also the capacitor of choice for most applications requiring capacity tolerances of 10 percent or less.

Film Capacitors

Plastic is a preferred dielectrical material for capacitors because it can be manufactured with minimal imperfections in thin films. A metal-foil capacitor is constructed by winding layers of metal, plastic, metal, and plastic into a cylinder and then making a connection to the two layers of metal. A *metallized foil* capacitor uses two layers, each of which has a very thin layer of metal evaporated on one surface, thereby obtaining a higher capacity per volume in exchange for a higher equivalent series resistance. Metallized foil capacitors are self-repairing in the sense that the energy stored in the capacitor is often sufficient to burn away the metal layer surrounding the void in the plastic film.

Depending on the dielectric material and construction, capacitance tolerances between 1 and 20 percent are common, as are voltage ratings from 50 to 400 V. Construction types include axial leaded capacitors with a plastic outer wrap, metal-encased units, and capacitors in a plastic box suitable for printed circuit board insertion.

Polystyrene has the lowest dielectric absorption of 0.02 percent, a temperature coefficient of –20 to –100 ppm/°C, a temperature range to 85°C, and extremely low leakage. Capacitors between 0.001 and 2 μF can be obtained with tolerances from 0.1 to 10 percent.

Polycarbonate has an upper temperature limit of 100°C, with capacitance changes of about 2 percent up to this temperature. Polypropylene has an upper temperature limit of 85°C. These capacitors are particularly well suited for applications where high inrush currents occur, such as switching power supplies. Polyester is the lowest-cost material with an upper temperature limit of 125°C. Teflon and other high-temperature materials are used in aerospace and other critical applications.

Foil Capacitors

Mica capacitors are made of multiple layers of silvered mica packaged in epoxy or other plastic. Available in tolerances of 1 to 20 percent in values from 10 to 10,000 pF, mica capacitors exhibit temperature coefficients as low as 100 ppm. Voltage ratings between 100 and 600 V are common. Mica capacitors are used mostly in high-frequency filter circuits where low loss and high stability are required.

Electrolytic Capacitors

Aluminum foil electrolytic capacitors can be made nonpolar through use of two cathode foils instead of anode and cathode foils in construction. With care in manufacturing, these capacitors can be produced with tolerance as tight as 10 percent at voltage ratings of 25 to 100 V peak. Typical values range from 1 to 1000 μF.

Ceramic Capacitors

Barium titanate and other ceramics have a high dielectric constant and a high breakdown voltage. The exact formulation determines capacitor size, temperature range, and variation of capacitance over that range (and consequently capacitor application). An alphanumeric code defines these factors, a few of which are given here.

- Ratings of Y5V capacitors range from 1000 pF to 6.8 μF at 25 to 100 V and vary + 22 to – 82 percent in capacitance from –30 to + 85°C.
- Ratings of Z5U capacitors range to 1.5 μF and vary +22 to –56 percent in capacitance from +10 to +85°C. These capacitors quite small in size and are used typically as bypass capacitors.
- X7R capacitors range from 470 pF to 1 μF and vary 15 percent in capacitance from –55 to + 125°C.
- Nonpolarized (NPO) rated capacitors range from 10 to 47,000 pF with a temperature coefficient of 0 to +30 ppm over a temperature range of –55 to +125°C.

Ceramic capacitors come in various shapes, the most common being the radial-lead disk. Multilayer monolithic construction results in small size, which exists both in radial-lead styles and as chip capacitors for direct surface mounting on a printed circuit board.

Polarized-Capacitor Construction

Polarized capacitors have a negative terminal—the cathode—and a positive terminal—the anode—and a liquid or gel between the two layers of conductors. The actual dielectric is a thin oxide film on the cathode, which has been chemically roughened for maximum surface area. The oxide is formed with a *forming voltage*, higher than the normal operating voltage, applied to the capacitor during manufacture. The direct current flowing through the capacitor forms the oxide and also heats the capacitor.

Whenever an electrolytic capacitor is not used for a long period of time, some of the oxide film is degraded. It is reformed when voltage is applied again with a leakage current that decreases with time. Applying an excessive voltage to the capacitor causes a severe increase in leakage current, which can cause the electrolyte to boil. The resulting steam may escape by

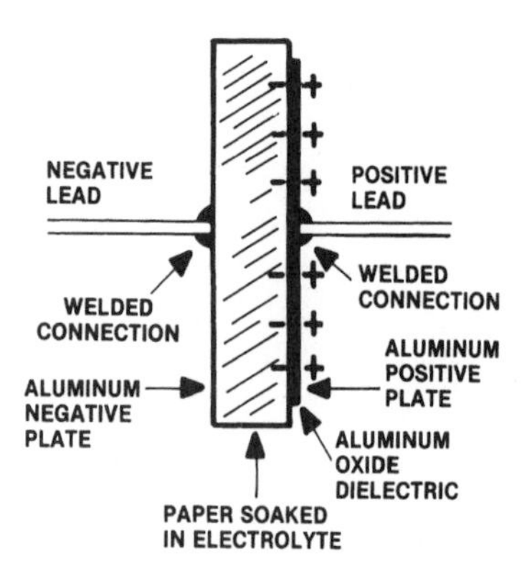

Figure 1.11 The basic construction of an aluminum electrolytic capacitor.

way of the rubber seal or may otherwise damage the capacitor. Application of a reverse voltage in excess of about 1.5 V will cause forming to begin on the unetched anode electrode. This can happen when pulse voltages superimposed on a dc voltage cause a momentary voltage reversal.

Aluminum Electrolytic Capacitors

Aluminum electrolytic capacitors use very pure aluminum foil as electrodes, which are wound into a cylinder with an interlayer paper or other porous material that contains the electrolyte. (See Figure 1.11.) Aluminum ribbon staked to the foil at the minimum inductance location is brought through the insulator to the anode terminal, while the cathode foil is similarly connected to the aluminum case and cathode terminal.

Electrolytic capacitors typically have voltage ratings from 6.3 to 450 V and rated capacitances from 0.47 μF to several hundreds of microfarads at the maximum voltage to several farads at 6.3 V. Capacitance tolerance may range from ±20 to +80/–20 percent. The operating temperature range is often rated from –25 to +85°C or wider. Leakage current of an electrolytic capacitor may be rated as low as 0.002 times the capacity times the voltage rating to more than 10 times as much.

Tantalum Electrolytic Capacitors

Tantalum electrolytic capacitors are the capacitors of choice for applications requiring small size, 0.33- to 100-μF range at 10 to 20 percent tolerance, low equivalent series resistance, and low leakage current. These devices are well suited where the less costly aluminum electrolytic capacitors have performance issues. Tantalum capacitors are packaged in hermetically sealed metal tubes or with axial leads in epoxy plastic, as illustrated in Figure 1.12.

1.5c Inductors and Transformers

Inductors are passive components in which voltage leads current by nearly 90° over a wide range of frequencies. Inductors are usually coils of wire wound in the form of a cylinder. The current through each turn of wire creates a magnetic field that passes through every turn of

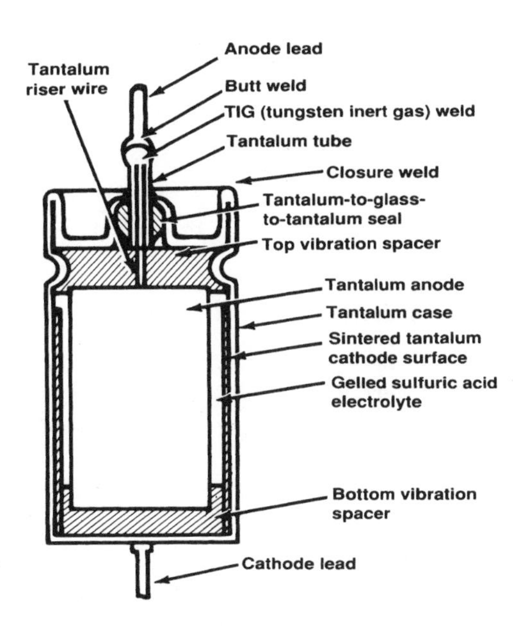

Figure 1.12 Basic construction of a tantalum capacitor.

wire in the coil. When the current changes, a voltage is induced in the wire and every other wire in the changing magnetic field. The voltage induced in the same wire that carries the changing current is determined by the inductance of the coil, and the voltage induced in the other wire is determined by the *mutual inductance* between the two coils. A transformer has at least two coils of wire closely coupled by the common magnetic core, which contains most of the magnetic field within the transformer.

Inductors and transformers vary widely in size, weighing less than 1 g or more than 1 ton, and have specifications ranging nearly as wide.

Losses in Inductors and Transformers

Inductors have resistive losses because of the resistance of the copper wire used to wind the coil. An additional loss occurs because the changing magnetic field causes *eddy currents* to flow in every conductive material in the magnetic field. Using thin magnetic laminations or powdered magnetic material reduces these currents.

Losses in inductors are measured by the Q, or quality, factor of the coil at a test frequency. Losses in transformers are sometimes given as a specific insertion loss in decibels. Losses in power transformers are given as core loss in watts when there is no load connected and as a regulation in percent, measured as the relative voltage drop for each secondary winding when a rated load is connected.

Transformer loss heats the transformer and raises its temperature. For this reason, transformers are rated in watts or volt-amperes and with a temperature code designating the maximum hotspot temperature allowable for continued safe long-term operation. For example, class A denotes 105°C safe operating temperature. The volt-ampere rating of a power transformer must be always larger than the dc power output from the rectifier circuit connected because volt-amperes, the product of the rms currents and rms voltages in the transformer, are larger by a factor of about 1.6 than the product of the dc voltages and currents.

Inductors also have capacitance between the wires of the coil, which causes the coil to have a self-resonance between the winding capacitance and the self-inductance of the coil. Circuits are normally designed so that this resonance is outside of the frequency range of interest. Transformers are similarly limited. They also have capacitance to the other winding(s), which causes stray coupling. An electrostatic shield between windings reduces this problem.

Air-Core Inductors

Air-core inductors are used primarily in radio frequency applications because of the need for values of inductance in the microhenry or lower range. The usual construction is a multilayer coil made self-supporting with adhesive-covered wire. An inner diameter of 2 times coil length and an outer diameter 2 times as large yields maximum Q, which is also proportional to coil weight.

Ferromagnetic Cores

Ferromagnetic materials have a permeability much higher than air or vacuum and cause a proportionally higher inductance of a coil that has all its magnetic flux in this material. Ferromagnetic materials in audio and power transformers or inductors usually are made of silicon steel laminations stamped in the forms of letters E or I (Figure 1.13). At higher frequencies, powdered ferric oxide is used. The continued magnetization and remagnetization of silicon steel and similar materials in opposite directions does not follow the same path in both directions but encloses an area in the magnetization curve and causes a hysteresis loss at each pass, or twice per ac cycle.

All ferromagnetic materials show the same behavior; only the numbers for permeability, core loss, saturation flux density, and other characteristics are different.

Shielding

Transformers and coils radiate magnetic fields that can induce voltages in other nearby circuits. Similarly, coils and transformers can develop voltages in their windings when subjected to magnetic fields from another transformer, motor, or power circuit. Steel mounting frames or chassis conduct these fields, offering less reluctance than air.

The simplest way to reduce the stray magnetic field from a power transformer is to wrap a copper strip as wide as the coil of wire around the transformer enclosing all three legs of the core. Shielding occurs by having a short circuit turn in the stray magnetic field outside of the core.

1.5d Diodes and Rectifiers

A diode is a passive electronic device that has a positive anode terminal and a negative cathode terminal and a nonlinear voltage-current characteristic. A rectifier is assembled from one or more diodes for the purpose of obtaining a direct current from an alternating current; this term also refers to large diodes used for this purpose. Many types of diodes exist.

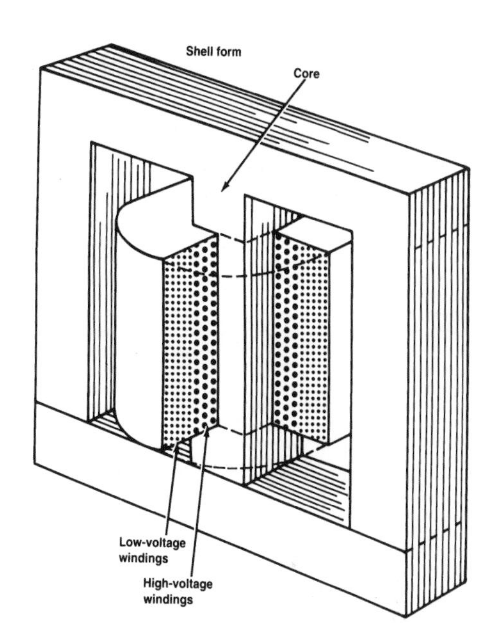

Figure 1.13 Physical construction of an E-shaped power transformer. The low- and high-voltage windings are stacked as shown.

Over the years, a great number of constructions and materials have been used as diodes and rectifiers. Rectification in electrolytes with dissimilar electrodes resulted in the *electrolytic rectifier*. The voltage-current characteristic of conduction from a heated cathode in vacuum or low-pressure noble gases or mercury vapor is the basis of vacuum tube diodes and rectifiers. Semiconductor materials such as germanium, silicon, selenium, copper-oxide, or gallium arsenide can be processed to form a *pn* junction that has a nonlinear diode characteristic. Although all these systems of rectification have seen use, the most widely used rectifier in electronic equipment is the silicon diode. The remainder of this section deals only with these and other silicon two-terminal devices.

The *pn* Junction

When biased in a reverse direction at a voltage well below breakdown, the diode reverse current is composed of two currents. One current is caused by leakage due to contamination and is proportional to voltage. The intrinsic diode reverse current is independent of voltage but doubles for every 10°C in temperature (approximately). The forward current of a silicon diode is approximately equal to the leakage current multiplied by *e* (= 2.718) raised to the power given by the ratio of forward voltage divided by 26 mV with the junction at room temperature. In practical rectifier calculations, the reverse current is considered to be important in only those cases where a capacitor must hold a charge for a time, and the forward voltage drop is assumed to be constant at 0.7 V, unless a wide range of currents must be considered.

All diode junctions have a *junction capacitance* that is approximately inversely proportional to the square of the applied reverse voltage. This capacitance rises further with applied forward voltage. When a rectifier carries current in a forward direction, the junction capacitance builds up a charge. When the voltage reverses across the junction, this charge must flow

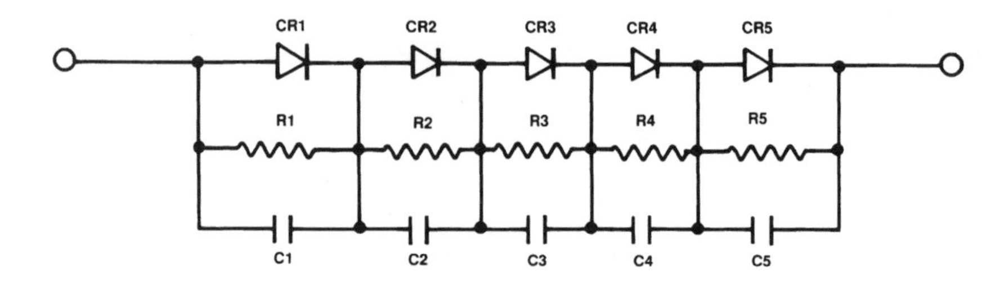

Figure 1.14 A high-voltage rectifier stack.

out of the junction, which now has a lower capacitance, giving rise to a current spike in the opposite direction of the forward current. After the *reverse-recovery time*, this spike ends, but interference may be radiated into low-level circuits. For this reason, rectifier diodes are sometimes bypassed with capacitors of about 0.1 μF located close to the diodes. Rectifiers used in high-voltage assemblies use bypass capacitors and high value resistors to reduce noise and equalize the voltage distribution across the individual diodes (Figure 1.14).

Tuning diodes have a controlled reverse capacitance that varies with applied direct tuning voltage. This capacitance may vary over a 2-to-1 to as high as a 10-to-1 range and is used to change the resonant frequency of tuned RF circuits. These diodes find application in radio and television receiver circuits.

Zener Diodes and Reverse Breakdown

When the reverse voltage on a diode is increased to a certain *critical voltage*, the reverse leakage current will increase rapidly or *avalanche*. This breakdown or *zener* voltage sets the upper voltage limit a rectifier can experience in normal operation because the peak reverse currents may become as high as the forward currents. Rectifier and other diodes have a rated peak reverse voltage, and some rectifier circuits may depend on this reverse breakdown to limit high-voltage spikes that may enter the equipment from the power line. It should also be noted that diode dissipation is very high during these periods.

The reverse breakdown voltage can be controlled in manufacture to a few percent and used to advantage in a class of devices known as *zener diodes*, used extensively in voltage-regulator circuits. It should be noted that the voltage-current curve of a *pn* junction may go through a region where a *negative resistance* occurs and voltage decreases a small amount while current increases. This condition can give rise to noise and oscillation, which can be minimized by connecting a ceramic capacitor of about 0.02 μF and an electrolytic capacitor of perhaps 100 μF in parallel with the zener diode. Voltage-regulator diodes are available in more than 100 types, covering voltages from 2.4 to 200 V with rated dissipation between 1/4 and 10 W (typical). The forward characteristics of a zener diode usually are not specified but are similar to those of a conventional diode.

Precision voltage or *bandgap* reference diodes make use of the difference in voltage between two diodes carrying a precise ratio of forward currents. Packaged as a two-terminal device including an operational amplifier, these devices produce stable reference voltages of 1.2, 2.5, 5, and 10 V, depending on type.

Current Regulators

The *current regulator* diode is a special class of device used in many small signal applications where constant current is needed. These diodes are *junction field-effect transistors* (FETs) with the gate connected to the source and effectively operated at zero-volt bias. Only two leads are brought out. Current-regulator diodes require a minimum voltage of a few volts for good regulation. Ratings from 0.22 to 4.7 mA are commonly available.

Varistor

Varistors are symmetrical nonlinear voltage-dependent resistors, behaving not unlike two zener diodes connected back to back. The current in a varistor is proportional to applied voltage raised to a power N. These devices are normally made of zinc oxide, which can be produced to have an N factor of 12 to 40. In circuits at normal operating voltages, varistors are nearly open circuits shunted by a capacitor of a few hundred to a few thousand picofarads. Upon application of a high voltage pulse, such as a lightning discharge, they conduct a large current, thereby absorbing the pulse energy in the bulk of the material with only a relatively small increase in voltage, thus protecting the circuit. (See Figure 1.15.) Varistors are available for operating voltages from 10 to 1000 V rms and can handle pulse energies from 0.1 to more than 100 J and maximum peak currents from 20 to 2000 A. Typical applications include protection of power supplies and power-switching circuits, and the protection of telephone and data-communication lines.

1.5e Indicators

Indicators are generally passive components that send a message to the operator of the equipment. This message is most commonly a silent visual indication that the equipment is operating in some particular mode, is ready to operate, or is not ready. Indicator lights of different colors illuminating a legend or having an adjacent legend are most commonly used. Alphanumeric codes and complete messages are often displayed on cathode ray tubes or on liquid crystal displays. These more complex displays are computer- or microprocessor-controlled.

Miniature light bulbs are incandescent devices operating at low voltage between 1 and 48 V, with currents from 0.01 to 4 A and total power requirements from 0.04 to more than 20 W, resulting in light output from 0.001 to more than 20 cd. The rated life of normally 10 to 50,000 h will typically decrease to one-tenth of the rating if the lamp voltage is increased to 20 percent above rated value. The resistance of the filament increases with temperature, varying by as much as a factor of 16 from cold to hot.

Solid-state lamps or *light-emitting diodes* (LED) are *pn*-junction lasers that generate light when diode current exceeds a critical threshold value. Visible red light is emitted from gallium arsenide phosphide junctions. Green or amber light is emitted from doped gallium phosphide junctions. The junctions have a forward voltage drop of 1.7 to 2.2 V at a normal operating current of 10 to 50 mA. Other visible colors are commercially available.

The LED is encased singly in round or rectangular plastic cases or assembled as multiples. A linear array of LEDs is often used in an arrangement similar to a thermometer to indicate

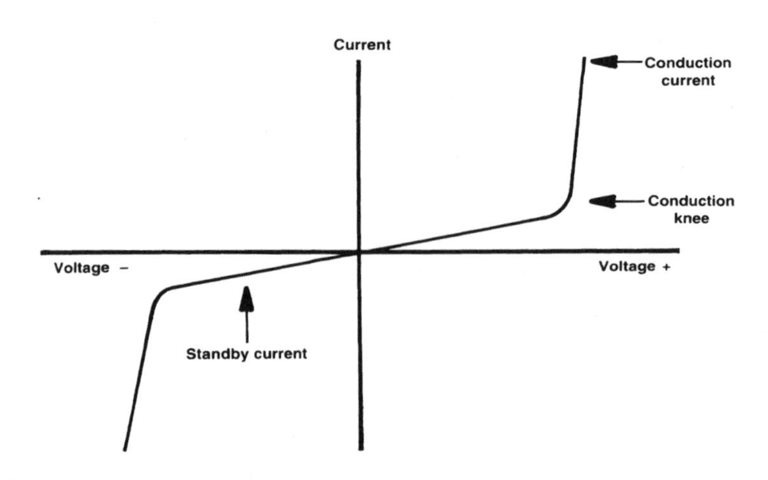

Figure 1.15 The current-vs.-voltage transfer curve for a varistor. Note the *conduction knee.*

volume or transmission level in audio or video circuits. An array, typically seven segments, can form the shapes of numerals and letters by selectively applying power to some or all segments. An array of 35 lamps in a 5 × 7 matrix can be connected to power to show the shape of letters, numerals, and punctuation marks. Semiconductor integrated circuits are available to achieve such functions with groups of these digit or indicator assemblies.

Light-emitting diodes have a typical operating life of about 50,000 h but have the disadvantage of relatively high current consumption, limited colors and shapes, and reduced visibility in bright light.

Electrons emitted from a heated cathode or a cold cathode can cause molecules of low-pressure gas, such as neon, to ionize and to emit light. Neon lamps require a current-limited supply of at least 90 V to emit orange light. The most frequent use of neon lamps is to indicate the presence of power line voltage. By means of a series resistor, the current is limited to a permissible value.

Emitted electrons can also strike a target connected to an anode terminal and coated with fluorescent phosphors. By directing the electron flow to "flood" different segments, alphanumeric displays can be produced similar to LED configurations with supply voltages found in battery-operated circuits.

When certain solutions of organic chemicals are exposed to an electric field, these crystal-like ions align themselves with the field and cause light of only one polarization to pass through the liquid. A second polarizer of light then causes the assembly to be a voltage-controlled attenuator of light. *Liquid crystal displays* come in many shapes, require low operating power, and can be backlit or used with external light only. These displays are found in many different types of systems, including test equipment, watches, computer terminals, and television sets.

1.6 Active Circuit Components

Active components can generate more alternating signal power into an output load resistance than the power absorbed at the input at the same frequency. Active components are the major building blocks in system assemblies such as amplifiers and oscillators.

1.6a Vacuum Tubes

Vacuum tubes are the active components that enabled the amplification and control of audio, radio frequency, and other signals and helped bring about the growth of the electronics industry from a laboratory curiosity early in the twentieth century to a high state of maturity in the 1960s. Since this time, the transistor and the integrated solid-state circuit have largely replaced vacuum tubes in most applications. The major uses of vacuum tubes today are as displays for television sets and computers, as generators of radio frequency power in selected applications, and the generation of X-rays for medical and industrial use.

A heated cathode coated with rare-earth oxides in a vacuum causes a cloud of electrons to exist near the cathode. A positive anode voltage with respect to the cathode causes some of these electrons to flow as a current to the anode. A grid of wires at a location between anode and cathode and biased at a control voltage with respect to the cathode causes a greater or lesser amount of anode current to flow. Other intervening grids also control the anode current and, if biased with a positive voltage, draw grid current from the total cathode current.

1.6b Bipolar Transistors

A bipolar transistor has two *pn* junctions that behave in a manner similar to that of the diode *pn* junctions described previously. These junctions are the *base-emitter junction* and the *base-collector junction*. In typical use, the first junction would normally have a forward bias, causing conduction, and the second junction would have a reverse bias. If the material of the base were very thick, the flow of electrons into the *p*-material base junction (of an *NPN* transistor) would go entirely into the base junction and no current would flow in the reverse-biased collector-base junction.

If, however, the base junction were quite thin, electrons would diffuse in the semiconductor crystal lattice into the base-collector junction, having been injected into the base material of the base-emitter junction. This diffusion occurs because an excess electron moving into one location will bump out an electron in the adjacent semiconductor molecule, which will bump its neighbor. Thus, a collector current will flow that is nearly as large as the injected emitter current.

The ratio of collector to emitter current is *alpha* or the common-base current gain of the transistor, normally a value a little less than 1.000. The portion of the emitter current not flowing into the collector will flow as a base current in the same direction as the collector current. The ratio of collector current to base current, or *beta*, is the conventional current gain of the transistor and may be as low as 5 in power transistors operating at maximum current levels to as high as 5000 in super-beta transistors operated in the region of maximum current gain.

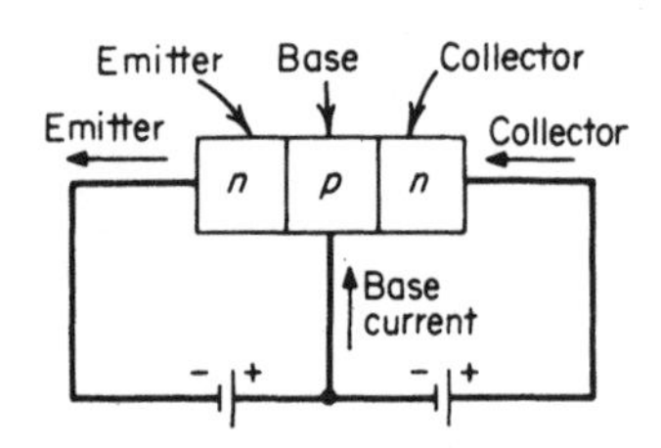

Figure 1.16 An NPN junction transistor. (*From* [1]. *Used with permission.*)

NPN and PNP Transistors

Bipolar transistors are identified by the sequence of semiconductor material going from emitter to collector. NPN transistors operate normally with a positive voltage on the collector with respect to the emitter, with *PNP* transistors requiring a negative voltage at the collector and the flow of current being internally mostly a flow of holes or absent excess electrons in the crystal lattice at locations of flow. (See Figure 1.16.)

Because the diffusion velocity of holes is slower than that *pn* electrons, PNP transistors have more junction capacitance and slower speed than NPN transistors of the same size. Holes and electrons in *pn* junctions are minority carriers of electric current as opposed to electrons, which are majority carriers and which can move freely in resistors or in the conductive channel of field-effect transistors. Consequently, bipolar transistors are known as minority carrier devices.

The most common transistor material is silicon, which permits transistor junction temperatures as high as 200°C. The normal base-emitter voltage is about 0.7 V, and collector-emitter voltage ratings of up to hundreds of volts are available. At room temperature these transistors may dissipate from tens of milliwatts to hundreds of watts with proper heat removal.

Transistors made of gallium arsenide and similar materials are also available for use in microwave and high speed circuits, taking advantage of the high diffusion speeds and low capacitances characteristic of such materials.

Transistor Impedance and Gain

Transistor impedances and gain are normally referred to the *common-emitter* connection, which also results in the highest gain. It is useful to treat transistor parameters first as if the transistor were an ideal device and then to examine degradations resulting from nonideal behavior.

If we assume that the transistor has a fixed current gain, then the collector current is equal to the base current multiplied by the current gain of the transistor, and the emitter current is the sum of both of these currents. Because the collector-base junction is reverse-biased, the output impedance of the ideal transistor is very high.

Actual bipolar transistors suffer degradations from this ideal model. Each transistor terminal may be thought of having a resistor connected in series, although these resistors are actually distributed rather than lumped components. These resistors cause the transistor to have lower gain than predicted and to have a saturation voltage in both input and output circuits. In addition, actual transistors have resistances connected between terminals that cause further reductions in available gain, particularly at low currents and with high load resistances.

In addition to resistances, actual transistors also exhibit stray capacitance between terminals, causing further deviation from the ideal case. These capacitances are—in part—the result of the physical construction of the devices and also the finite diffusion velocities in silicon. The following effects result:

- Transistor current gain decreases with increasing frequency, with the transistor reaching unity current gain at a specific *transition frequency.*
- A feedback current exists from collector to base through the base-collector capacitance.
- Storage of energy in the output capacitance similar to energy storage in a rectifier diode. This stored energy limits the turn-off speed of transistors, a critical factor in video-amplifier and switching-circuit design.

Transistor Configurations

Table 1.1 summarizes the most common transistor operating modes. For stages using a single device, the common-emitter arrangement is by far the most common. Power output stages of push-pull amplifiers make use of the common-collector or emitter-follower connection. Here, the collector is directly connected to the supply voltage, and the load is connected to the emitter terminal with signal and bias voltage applied to the base terminal. The voltage gain of such a circuit is a little less than 1.000, and the load impedance at the emitter is reflected to the base circuit as if it were increased by the current gain of the transistor.

At high frequencies, the base of a transistor is often grounded for high-frequency signals, which are fed to the emitter of the transistor. With this arrangement, the input impedance of the transistor is low, which is easily matched to radio frequency transmission lines, assisted in part by the minimal capacitive feedback within the transistor.

So far in this discussion, transistor analysis has dealt primarily with the small signal behavior. For operation under large signal conditions, other limitations must be observed. When handling low-frequency signals, a transistor can be viewed as a variable-controlled resistor between the supply voltage and the load impedance. The quiescent operating point in the absence of ac signals is usually chosen so that the maximum signal excursions in both positive and negative directions can be handled without limiting resulting from near-zero voltage across the transistor at maximum output current or near-zero current through the transistor at maximum output voltage. This is most critical in class B push-pull amplifiers where first one transistor stage conducts current to the load during part of one cycle and then the other stage conducts during the other part. Similar considerations also apply for distortion reduction considerations.

Limiting conditions also constitute the maximum capabilities of transistors under worst-case conditions of supply voltage, load impedance, drive signal, and temperature consistent with safe operation. In no case should the maximum voltage across a transistor ever be exceeded.

Switching and Inductive-Load Ratings

When using transistors for driving relays, deflection yokes of cathode ray tubes, or any other inductive or resonant load, current in the inductor will tend to flow in the same direction, even if interrupted by the transistor. The resultant voltage spike caused by the collapse of the magnetic field can destroy the switching device unless it is designed to handle the energy of these

Table 1.1 Basic Amplifier Configurations of Bipolar Transistors

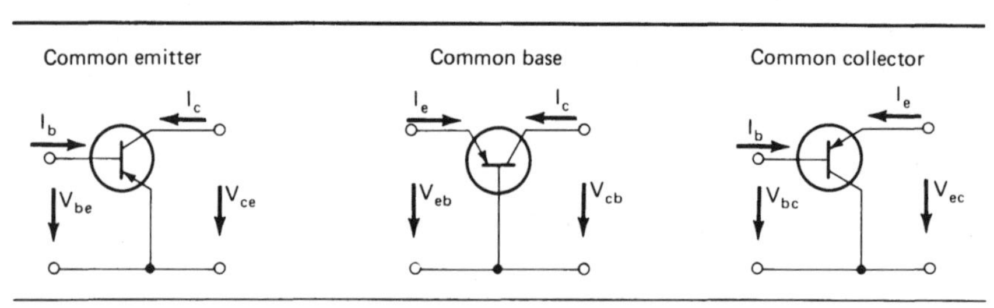

Characteristics of basic configurations	Common emitter	Common base	Common collector
Input impedance Z_1	Medium	Low	High
	Z_{1e}	$Z_{1b} \approx \frac{Z_{1e}}{h_{fe}}$	$Z_{1c} \approx h_{fe} R_L$
Output impedance Z_2	High	Very high	Low
	Z_{2e}	$Z_{2b} \approx Z_{2e} h_{fe}$	$Z_{2c} \approx \frac{Z_{1e} + R_g}{h_{fe}}$
Small-signal current gain	High	< 1	High
	h_{fe}	$h_{fb} \approx \frac{h_{fe}}{h_{fe} + 1}$	$\gamma \approx h_{fe} + 1$
Voltage gain	High	High	< 1
Power gain	Very high	High	Medium
Cutoff frequency	Low	High	Low
	f_{hfe}	$f_{hfb} \approx h_{fe} f_{hfc}$	$f_{hfc} \approx f_{hfe}$

voltage excursions. The manufacturers of power semiconductors have special transistor types and application information relating to inductive switching circuits. In many cases, the use of protection diodes are sufficient.

Transistors are often used to switch currents into a resistive load. The various junction capacitances are voltage-dependent in the same manner as the capacitance of tuning diodes that have maximum capacitance at forward voltages, becoming less at zero voltage and lowest at reverse voltages. These capacitances and the various resistances combine into the switching delay times for turn-on and turn-off functions. If the transistor is prevented from being saturated when turned on, shorter delay times will occur for nonsaturated switching than for saturated switching. These delay times are of importance in the design of switching amplifiers or D/A converters.

Noise

Every resistor creates noise with equal and constant energy for each hertz of bandwidth, regardless of frequency. A useful number to remember is that a 1000-Ω resistor at room temperature has an open-circuit output noise voltage of 4 nanovolts per *root-hertz*. This converts to 40 nV in a 100-Hz bandwidth or 400 μV in a 10-kHz bandwidth.

Bipolar transistors also create noise in their input and output circuits, and every resistor in the circuit also contributes its own noise energy. The noise of a transistor is effectively created in its input junction, and all transistor noise ratings are referred to it.

In an ideal bipolar transistor, the voltage noise at the base is created by an equivalent resistor that has a value of twice the transistor input conductance at its emitter terminal, and the current noise is created by a resistor that has the value of twice the transistor input conductance at its input terminal. This means that the current noise energy is less at the base terminal of a common-emitter stage by the current gain of the transistor when compared to the current noise at the input of a grounded-base stage.

The highest signal-to-noise ratio in an amplifier can be achieved when the resistance of the signal source is equal to the ratio of amplifier input noise voltage and input noise current, and the reactive impedances have been tuned to zero. Audio frequency amplifiers usually cannot be tuned, and minimum noise may be achieved by matching transformers or by bias current adjustment of the input transistor. With low source impedances, the optimum may not be reached economically, and the equipment must then be designed to have an acceptable input noise voltage.

Practical transistors are not ideal from the standpoint of noise performance. All transistors show a voltage and current noise energy that increases inversely with frequency. At a corner frequency this noise will become independent of frequency. Very low noise transistors may have a corner frequency as low as a few hertz, and ordinary high-frequency devices may have a corner frequency well above the audio frequency range. Transistor noise may also be degraded by operating a transistor at more than a few percent of its maximum current rating. Poor transistor design or manufacturing techniques can result in transistors that exhibit "popcorn" noise, so named after the audible characteristics of a random low-level switching effect.

1.6c Field-Effect Transistors

Field-effect transistors (FETs) have a conducting channel terminated by source and drain electrodes and a gate terminal that effectively widens or narrows the channel by the electric field between the gate and each portion of the channel. No gate current is required for steady-state control.

Current flow in the channel is by majority carriers only, analogous to current flow in a resistor. The onset of conduction is not limited by diffusion speeds but by the electric field accelerating the charged electrons.

The input impedance of an FET is a capacitance. Because of this, electrostatic charges during handling may reach high voltages that are capable of breaking down gate insulation.

FETs for audio and video frequency applications use silicon as the semiconducting material. Field-effect transistors are made both in *p*-channel and *n*-channel configurations. An *n*-channel FET has a positive drain voltage with respect to the source voltage, and a positive increase in gate voltage causes an increase in channel current. Reverse polarities exist for *p*-channel devices (Figure 1.17).

An *n*-channel FET has a drain voltage that is normally positive, and a positive increase in gate-to-source voltage increases drain current and transconductance. In single-gate field-effect transistors, drain and source terminals may often be interchanged without affecting circuit performance; however, power handling and other factors may be different. Such an interchange is not possible when two FETs are interconnected internally to form a dual-gate

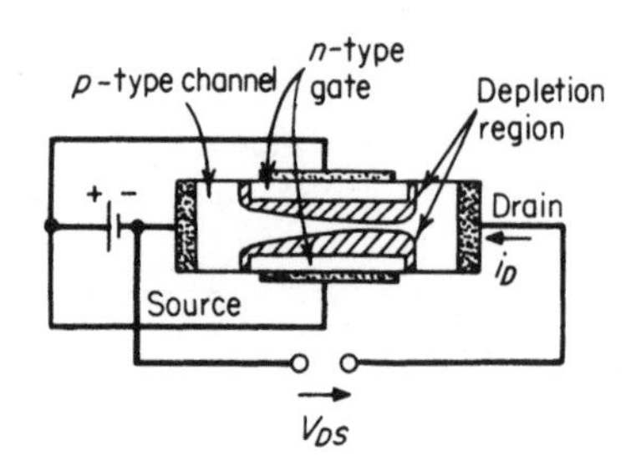

Figure 1.17 Construction of a *p*-channel junction field-effect transistor. (*From* [1]. *Used with permission.*)

cascode-connected FET, or matched pairs, or when channel conductance is controlled by gates on two sides of the channel as in insulated-gate FETs.

FET Impedance and Gain

The input impedance of a field-effect transistor is very high at both audio and video frequencies and is primarily capacitive. The input capacitance consists of the gate-source capacitance in parallel with the gate-drain capacitance multiplied by the stage gain + 1, assuming the FET has its source at ac-ground potential.

The output impedance of a common-source FET is also primarily capacitive as long as the drain voltage is above a critical value, which, for a junction-gate FET, is equal to the sum of the *pinch-off voltage* and gate-bias voltage. When the pinch-off voltage is applied between the gate and source terminals, the drain current is nearly shut off (the channel is pinched off). Actual FETs have a high drain resistance in parallel with this capacitance. At low drain voltages near zero volts, the drain impedance of an ideal FET is a resistor reciprocal in value to the transconductance of the FET in series with the residual end resistances between the source and drain terminals and the conducting FET channel. This permits an FET to be used as a variable resistor in circuits controlling analog signals.

At drain voltages between zero and the critical voltage, the drain current will increase with both increasing drain voltage and increasing gate voltage. This factor will cause increased saturation voltages in power amplifier circuits when compared to circuits with bipolar transistors.

Table 1.2 summarizes the basic FET amplifier configurations.

1.6d Integrated Circuits

An integrated circuit (IC) is a combination of circuit elements that are interconnected and formed on and in a continuous substrate material. Usually, an integrated circuit is monolithic and formed by steps that produce semiconductor elements along with resistors and capacitors. A hybrid integrated circuit contains silicon chips along with circuit elements partially formed on the substrate.

The circuit elements formed in integrated circuits are more closely matched to each other than separately selected components, and these elements are in intimate thermal contact with each other. The circuit configurations used in integrated circuits take advantage of this matching and thermal coupling.

Digital Integrated Circuits

The basis of digital circuits is the logic gate that produces a high (or 1) or low (or 0) logic-level output with the proper combination of logic-level inputs. A number of these gates are combined to form a digital circuit that is part of the hardware of computers or controllers of equipment or other circuits. A digital circuit may be extremely complex, containing up to more than 1,000,000 gates.

Bipolar and field-effect transistors are the active elements of digital integrated circuits, divided into families such as transistor-transistor logic (TTL), high-speed complementary metal-oxide-gate semiconductor (HCMOS), and many others. Special families include memories, microprocessors, and interface circuits between transmission lines and logic circuits. Thousands of digital integrated circuit types in tens of families have been produced.

Linear Integrated Circuits

Linear integrated circuits are designed to process linear signals in their entirety or in part, as opposed to digital circuits that process logic signals only. Major classes of linear integrated circuits include operational amplifiers, voltage regulators, digital-to-analog and analog-to-digital circuits, circuits for consumer electronic equipment and communications equipment, power control circuits, and others not as easily classified.

Operational Amplifiers

An operational amplifier has a pair of differential input terminals that have very high gain to the output for differential signals of opposite phase at each input and relatively low gain for *common-mode* signals that have the same phase at each input (see Figure 1.18). An external feedback network between the output and the minus (–) input and ground or signal, sets the circuit gain, with the plus (+) input at signal or ground level. Most operational amplifiers require a positive and a negative power supply voltage. One to eight operational amplifiers may be contained on one substrate mounted in a plastic, ceramic, or hermetically sealed metal-can package. Operational amplifiers may require external capacitors for circuit stability or may be internally compensated. Input stages may be field-effect transistors for high input impedance or bipolar transistors for low-offset voltage and low-voltage noise. Available types of operational amplifiers number in the hundreds. Precision operational amplifiers generally have more tightly controlled specifications than general-purpose types.

Current and Voltage Ratings

The input-bias current of an operational amplifier is the average current drawn by each of the two inputs, + and –, from the input and feedback circuits. Any difference in dc resistance between the circuits seen by the two inputs multiplied by the input-bias current will be amplified by the circuit gain and become an *output-offset* voltage.

The *input-offset current* is the difference in bias current drawn by the two inputs, which when multiplied by the sum of the total dc resistance in the input and feedback circuits and the circuit gain, becomes an additional output-offset voltage.

The *input-offset voltage* is the internal difference in bias voltage within the operational amplifier, which when multiplied by the circuit gain, becomes an additional output-offset voltage.

Table 1.2 Basic Field-Effect Transistor Amplifier Configurations and Operating Characteristics

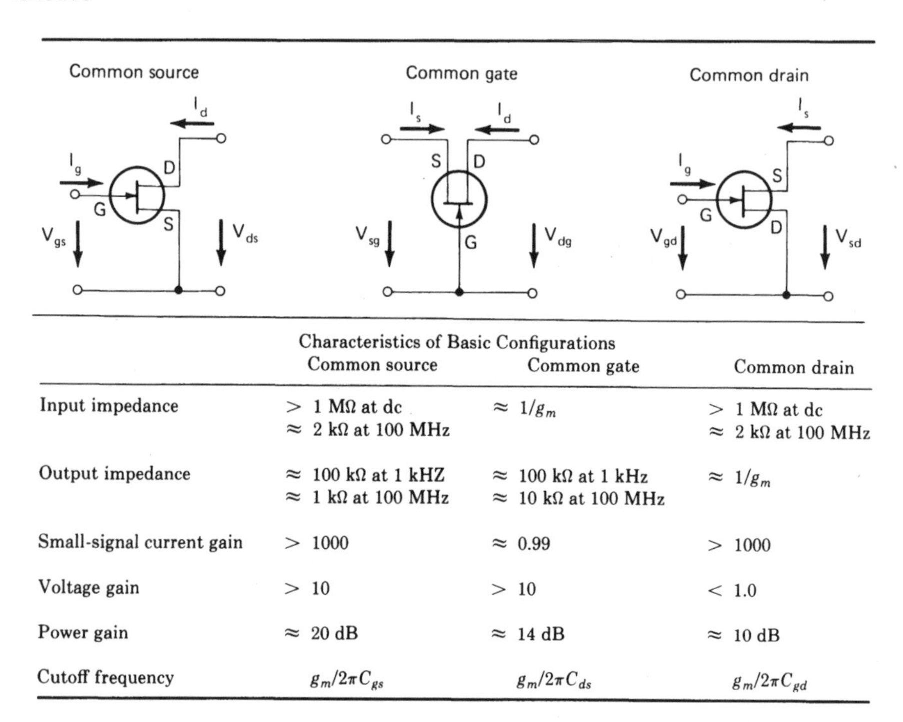

Characteristics of Basic Configurations

	Common source	Common gate	Common drain
Input impedance	$>$ 1 MΩ at dc $\approx$ 2 kΩ at 100 MHz	$\approx 1/g_m$	$>$ 1 MΩ at dc $\approx$ 2 kΩ at 100 MHz
Output impedance	$\approx$ 100 kΩ at 1 kHZ $\approx$ 1 kΩ at 100 MHz	$\approx$ 100 kΩ at 1 kHz $\approx$ 10 kΩ at 100 MHz	$\approx 1/g_m$
Small-signal current gain	$>$ 1000	$\approx$ 0.99	$>$ 1000
Voltage gain	$>$ 10	$>$ 10	$<$ 1.0
Power gain	$\approx$ 20 dB	$\approx$ 14 dB	$\approx$ 10 dB
Cutoff frequency	$g_m/2\pi C_{gs}$	$g_m/2\pi C_{ds}$	$g_m/2\pi C_{gd}$

If the normal input voltage is zero, the open-circuit output voltage is the sum of the three offset voltages.

1.7 Analog and Digital Circuits

Amplifiers are the functional building blocks of audio and video systems, and each of these building blocks typically contains several amplifier stages coupled together. An amplifier may contain its own power supply or require one or more external sources of power. The active component of each amplifier stage is usually a transistor or an FET. Other amplifying components, such as vacuum tubes, can also be used in amplifier circuits if the operating power and/or frequency of the application demands it.

1.7a Single-Stage Transistor/FET Amplifier

The single-stage amplifier can best be described using a single transistor or FET connected as a common-emitter or common-source amplifier, using an *npn* transistor (Figure 1.19*a*) or an *n*-channel FET (Figure 1.19*b*) and treating *pnp* transistors or *p*-channel FET circuits by simply reversing the current flow and the polarity of the voltages.

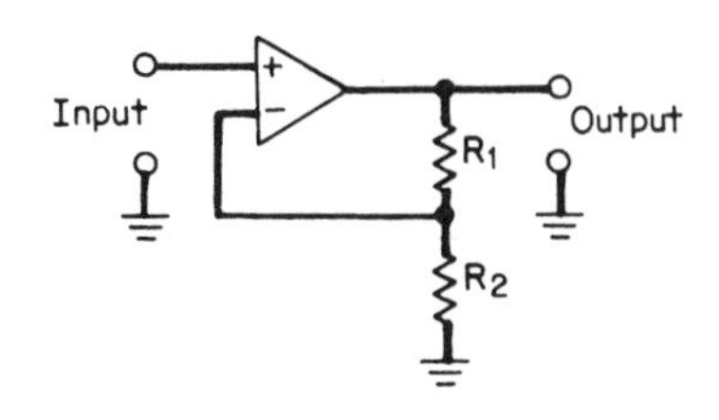

Figure 1.18 Operational amplifier with unbalanced input and output signals and a fixed level of feedback to set the voltage gain V_g, which is equal to the quantity (1 + *R*)/*R*.

At zero frequency (dc) and at low frequencies, the transistor or FET amplifier stage requires an input voltage E_1 equal to the sum of the input voltages of the device (the transistor V_{be} or FET V_{gs}) and the voltage across the resistance R_e or R_s between the common node (ground) and the emitter or source terminal. The input current I_1 to the amplifier stage is equal to the sum of the current through the external resistor connected between ground and the base or gate and the base current I_b or gate current I_g drawn by the device. In most FET circuits, the gate current may be so small that it can be neglected, while in transistor circuits the base current I_b is equal to the collector current I_c divided by the current gain beta of the transistor. The input resistance R_1 to the amplifier stage is equal to the ratio of input voltage E_1 to input current I_1.

The input voltage and the input resistance of an amplifier stage increases as the value of the emitter or source resistor becomes larger.

The output voltage E_2 of the amplifier stage, operating without any external load, is equal to the difference of supply voltage $V+$ and the product of collector or drain load resistor R_1 and collector current I_c or drain current I_d. An external load will cause the device to draw an additional current I_2, which increases the device output current.

As long as the collector-to-emitter voltage is larger than the saturation voltage of the transistor, collector current will be nearly independent of supply voltage. Similarly, the drain current of an FET will be nearly independent of drain-to-source voltage as long as this voltage is greater than an equivalent saturation voltage. This saturation voltage is approximately equal to the difference between gate-to-source voltage and *pinch-off voltage*, the latter being the bias voltage that causes nearly zero drain current. In some FET data sheets, the pinch-off voltage is referred to as the *threshold voltage*. At lower supply voltages, the collector or drain current will become less until it reaches zero, when the drain-to-source voltage is zero or the collector-to-emitter voltage has a very small reverse value.

The output resistance R_2 of a transistor or FET amplifier stage is—in effect—the parallel combination of the collector or drain load resistance and the series connection of two resistors, consisting of R_e or R_s, and the ratio of collector-to-emitter voltage and collector current or the equivalent drain-to-source voltage and drain current. In actual devices, an additional resistor, the relatively large output resistance of the device, is connected in parallel with the output resistance of the amplifier stage.

The collector current of a single-stage transistor amplifier is equal to the base current multiplied by the current gain of the transistor. Because the current gain of a transistor may be specified as tightly as a two-to-one range at one value of collector current, or it may have just a minimum value, knowledge of the input current is usually not quite sufficient to specify the output current of a transistor.

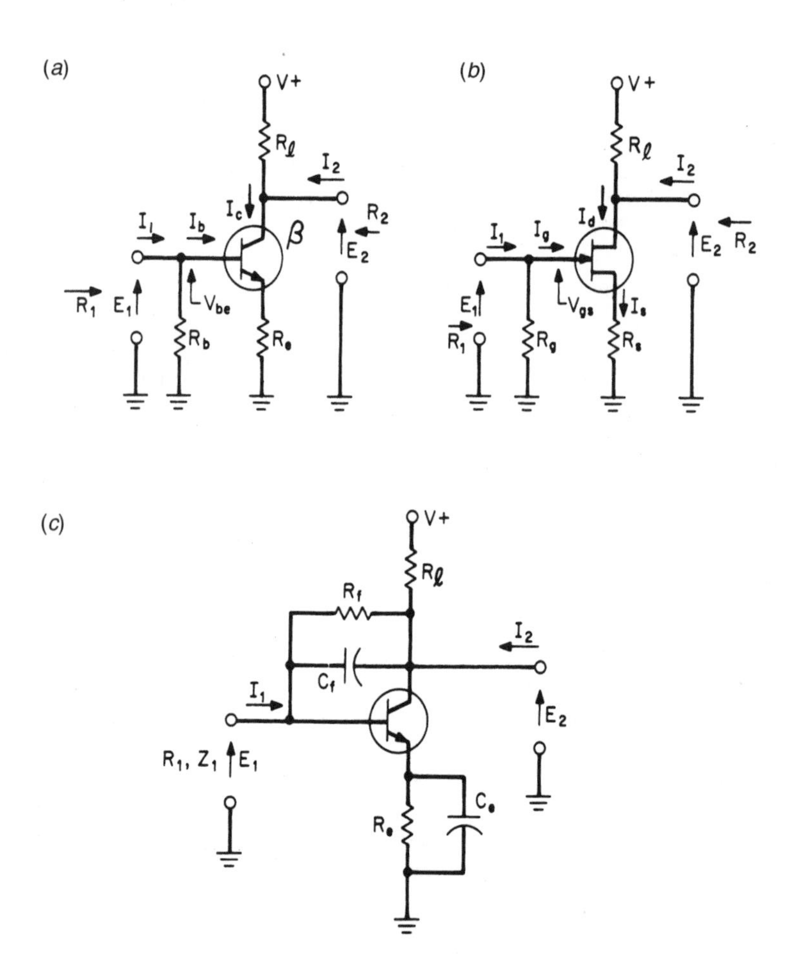

Figure 1.19 Single-stage amplifier circuits: (*a*) common-emitter *NPN*, (*b*) common-source *n*-channel FET, (*c*) single-stage with current and voltage feedback.

Impedance and Gain

The input impedance is the ratio of input voltage to input current, and the output impedance is the ratio of output voltage to output current. As the input current increases, the output current into the external output load resistor will increase by the current *amplification factor* of the stage. The output voltage will decrease because the increased current flows from the collector or drain voltage supply source into the collector or drain of the device. Therefore, the voltage amplification is a negative number having the magnitude of the ratio of output voltage change to input voltage change.

The magnitude of voltage amplification is often calculated as the product of transconductance G_m of the device and the load resistance value. This can be done as long as the emitter or source resistance is zero or the resistor is bypassed with a capacitor that effectively acts as a short circuit for all signal changes of interest but allows the desired bias currents to flow through the resistor. In a bipolar transistor, the transconductance is approximately equal to the emitter current multiplied by 39, which is the charge of a single electron divided by the product of *Boltzmann's constant* and absolute temperature in degrees Kelvin. In a field-effect transistor, this value will be less and usually proportional to the input-bias voltage, with reference to the pinch-off voltage.

The power gain of the device is the ratio of output power to input power, often expressed in decibels. Voltage gain or current gain can be stated in decibels but must be so marked.

The resistor in series with the emitter or source causes negative feedback of most of the output current, which reduces the voltage gain of the single amplifier stage and raises its input impedance (Figure 1.19*c*). When this resistor R_e is bypassed with a capacitor C_e, the amplification factor will be high at high frequencies and will be reduced by approximately 3 dB at the frequency where the impedance of capacitor C_e is equal to the emitter or source input impedance of the device, which in turn is approximately equal to the inverse of the transconductance G_m of the device (Figure 1.20*a*). The gain of the stage will be approximately 3 dB higher than the dc gain at the frequency where the impedance of the capacitor is equal to the emitter or source resistor. These simplifications hold in cases where the product of transconductance and resistance values are much larger than 1.

A portion of the output voltage may also be fed back to the input, which is the base or gate terminal. This resistor R_f will lower the input impedance of the single amplifier stage, reduce current amplification, reduce output impedance of the stage, and act as a supply voltage source for the base or gate. This method is used when the source of input signals, and internal resistance R_s, is coupled with a capacitor to the base or gate and a group of devices with a spread of current gains, transconductances, or pinch-off voltages must operate with similar amplification in the same circuit. If the feedback element is also a capacitor C_f, high-frequency current amplification of the stage will be reduced by approximately 3 dB when the impedance of the capacitor is equal to the feedback resistor R_f and voltage gain of the stage is high (Figure 1.20*b*). At still higher frequencies, amplification will decrease at the rate of 6 dB per octave of frequency. It should be noted that the base-collector or gate-drain capacitance of the device has the same effect of limiting high-frequency amplification of the stage; however, this capacitance becomes larger as the collector-base or drain-gate voltage decreases.

Feedback of the output voltage through an impedance lowers the input impedance of an amplifier stage. Voltage amplification of the stage will be affected only as this lowered input impedance loads the source of input voltage. If the source of input voltage has a finite source impedance and the amplifier stage has very high voltage amplification and reversed phase, the effective amplification for this stage will approach the ratio of feedback impedance to source impedance and also have reversed phase.

Common-Base or Common-Gate Connection

For the common-base or common-gate case, voltage amplification is the same as in the common-emitter or common-source connection; however, the input impedance is approximately the inverse of the transconductance of the device. (See Figure 1.21*a*.) As a benefit, high-frequency amplification will be less affected because of the relatively lower emitter-collector or source-drain capacitance and the relatively low input impedance. This is the reason why the cascade connection (Figure 1.21*b*) of a common-emitter amplifier stage driving a common-base amplifier stage exhibits nearly the dc amplification of a common-emitter stage with the wide bandwidth of a common-base stage. Another advantage of the common-base or com-

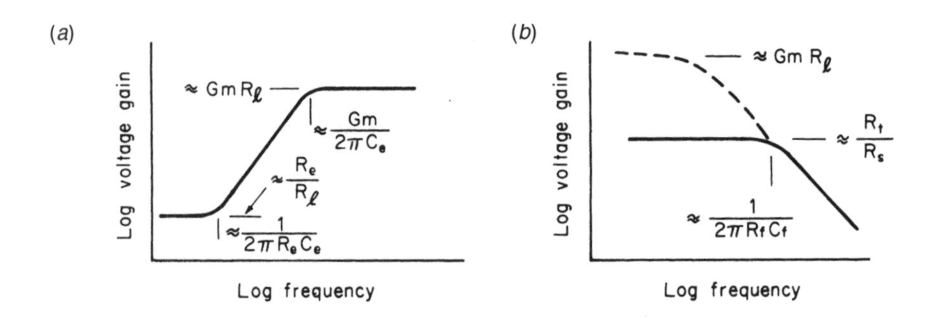

Figure 1.20 Feedback amplifier voltage gains: (*a*) current feedback, (*b*) voltage feedback.

mon-gate amplifier stage is stable amplification at very high frequencies (VHF) and ease of matching to RF transmission-line impedances, usually 50 to 75 Ω.

Common-Collector or Common-Drain Connection

The voltage gain of a transistor or FET is slightly below 1.0 for the common-collector or common-drain configuration. However, the input impedance of a transistor so connected will be equal to the value of the load impedance multiplied by the current gain of the device plus the inverse of the transconductance of the device (Figure 1.21*c*). Similarly, the output impedance of the stage will be the impedance of the source of signals divided by the current gain of the transistor plus the inverse of the transconductance of the device.

When identical resistors are connected between the collector or drain and the supply voltage and the emitter or source and ground, an increase in base or gate voltage will result in an increase of emitter or source voltage that is nearly equal to the decrease in collector or drain voltage. This type of connection is known as the *split-load phase inverter*, useful for driving push-pull amplifiers, although the output impedances at the two output terminals are unequal (Figure 1.21*d*).

The current gain of a transistor decreases at high frequencies as the emitter-base capacitance shunts a portion of the transconductance, thereby reducing current gain until it reaches a value of 1 at the transition frequency of the transistor (Figure 1.22). From this figure it can be seen that the output impedance of an emitter-follower or common-collector stage will increase with frequency, having the effect of an inductive source impedance when the input source to the stage is resistive. If the source impedance is inductive, as it might be with cascaded-emitter followers, the output impedance of such a combination can be a negative value at certain high frequencies and be a possible cause of amplifier oscillation. Similar considerations also apply to common-drain FET stages.

Bias and Large Signals

When large signals have to be handled by a single-stage amplifier, distortion of the signals introduced by the amplifier itself must be considered. Although feedback can reduce distortion, it is necessary to ensure that each stage of amplification operates in a region where normal signals will not cause the amplifier stage to operate with nearly zero voltage drop across

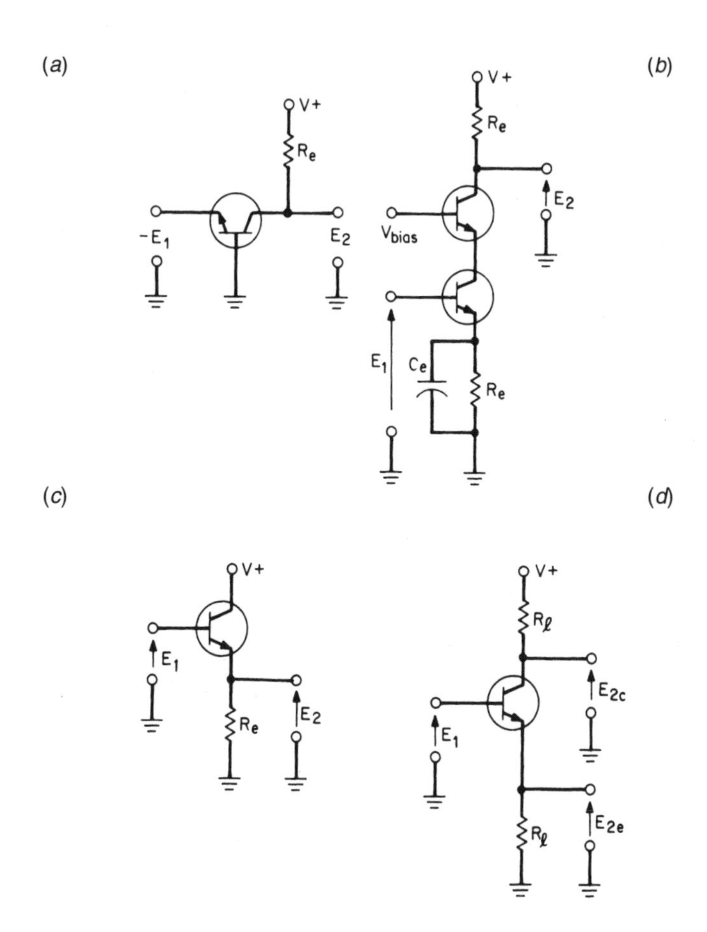

Figure 1.21 Transistor amplifier circuits: (*a*) common-base NPN, (*b*) cascode NPN, (*c*) common-collector NPN emitter follower, (*d*) split-load phase inverter.

the device or to operate the device with nearly zero current during any portion of the cycle of the signal. Although described primarily with respect to a single-device-amplifier stage, the same holds true for any amplifier stage with multiple devices, except that here at least one device must be able to control current flow in the load without being saturated (nearly zero voltage drop) or cut off (nearly zero current).

If the single-device-amplifier load consists of the collector or drain load resistor only, the best operating point should be chosen so that in the absence of a signal, one-half of the supply voltage appears as a quiescent voltage across the load resistor R_l. If an additional resistive load R_l is connected to the output through a coupling capacitor C_c (Figure 1.23*a*), the maximum peak load current I_l in one direction is equal to the difference between quiescent current I_I of the stage and the current that would flow if the collector resistor and the external load resistor were connected in series across the supply voltage. In the other direction, the maxi-

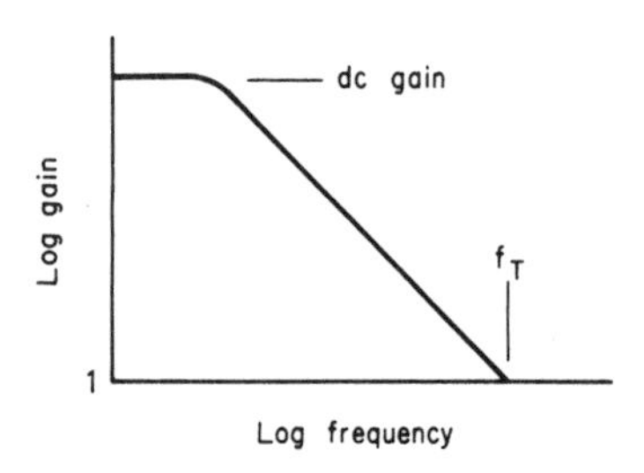

Figure 1.22 Amplitude-frequency response of a common-emitter or common-source amplifier.

mum load current is limited by the quiescent voltage across the device divided by the load resistance. The quiescent current flows in the absence of an alternating signal and is caused by bias voltage or current only. Because most audio frequency signals have positive and negative peak excursions of equal probability, it is advisable to have the two peak currents be equal. This can be accomplished by increasing the quiescent current as the external load resistance decreases. Video signals, on the other hand, are typically unidirectional in nature.

When several devices contribute current into an external load resistor (Figure 1.23*b*), one useful strategy is to set bias currents so that the sum of all transconductances remains as constant as practical, which means a design for minimum distortion. This operating point for one device is near one-quarter the peak device current for push-pull FET stages and at a lesser value for bipolar push-pull amplifiers.

When the load resistance is coupled to the single-device-amplifier stage with a transformer (Figure 1.23*c*), the optimum bias current should be nearly equal to the peak current that would flow through the load impedance at the transformer with a voltage drop equal to the supply voltage.

1.7b Digital Circuits

Digital signals differ from analog in that only two steady-state levels are used for the storage, processing, and/or transmission of information. The concept of requiring no more than two levels is not new, since two-level, on-off control circuits are a basic component in all electrical systems. The adaptation the key-and-sounder clicks of Morse's telegraphy to continuous-wave (cw) radio transmission utilizes a two-level (on or oft) transmission, but with one additional parameter: the short and long coding of dots and dashes. These, in turn, are arranged in a variety of sequences to represent letters of the alphabet and numbers (alphanumeric characters). Another common application of two-level coding is the synchronizing signal in the analog transmission of television pictures. Their purpose is to supply information necessary for the timing of picture-scanning circuits and the on-off control of the analog picture signal. Thus, no information is transmitted in the amplitude range between blanking and sync tips; the only reasons for using the large sync-signal amplitude relative to the video signal (40 percent of the peak-white video signal) are to: (1) assure stable synchronization in the presence of noise or other spurious signals, and (2) to permit the use of relatively simple circuits to separate the sync signal from the wideband video signal.

Digitizing an analog signal for processing or transmission requires the conversion of the signals, which are varying in level over a contiguous range of values, to a meaningful coding

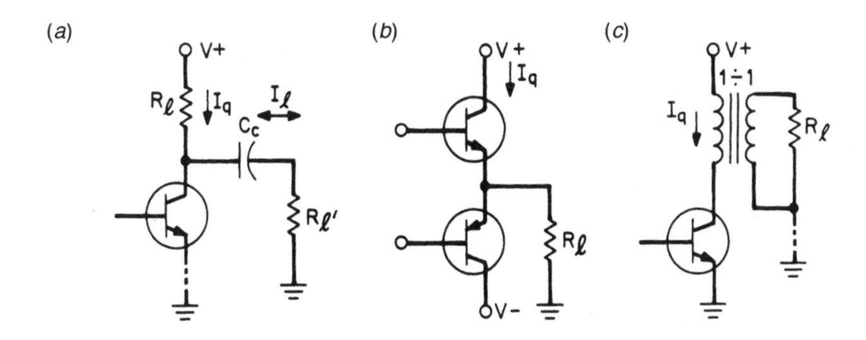

Figure 1.23 Output load-coupling circuits: (*a*) ac-coupled, (*b*) series-parallel ac, push-pull half-bridge, (*c*) single-ended transformer-coupled.

of only two different levels. Then for viewing or listening, these discrete levels must be decoded into the original analog signal, again a continuously varying range of levels.

Binary Coding

The definition of a digital transmission format requires specification of the following parameters:

- The type of information corresponding to each of the binary levels
- The frequency or rate at which the information is transmitted as a bilevel signal

The digital coding of signals for television and audio systems uses a scheme of binary numbers in which only two digits, 0 and 1, are used. This is called a *base*, or *radix*, of 2. It is of interest that systems of other bases are used for some more complex mathematical applications, the principal ones being *octal* (8) and *hexadecimal* (16). Table 1.3 compares the decimal, binary, and octal counting systems. Note that numbers in the decimal system are equal to the number of items counted, if used for a tabulation.

Combinational Logic

When the inputs to a logic circuit have only one meaning for each, the circuit is said to be *combinational*. These devices tend to have names reflecting the function they will perform, such as AND, OR, exclusive OR, latch, flip-flop, counter, and gate. Logic circuits are usually documented through the use of schematic diagrams. For simple devices, the *shape* of the symbol tells the function it performs, while the presence of small bubbles at the points of connection tell whether that point is high or low when the function is being performed. More complicated functions are shown as rectangular boxes. Figure 1.24 shows a collection of common logic symbols.

The *clocking input* to memory devices and counters is indicated by a small triangle at (usually) the inside left edge of the box. If the device is a *transparent latch*, the output follows the input while the clock input is active, and the output is "frozen" when the clock becomes inactive. A flip-flop, on the other hand, is an edge-triggered device. The output is allowed to change only upon a transition of the clock input from low to high (no bubble) or high to low (bubble present).

Three types of flip-flops are shown in Figure 1.24:

- A *T* (*toggle*) flip-flop, which will reverse its output state when clocked while the *T* input is active.
- A *D* flip-flop, which will allow the output to assume the state of the *D* input when clocked.
- A *J-K* flip-flop. If both *J* and *K* inputs are inactive, the output does not change when clocked. If both are active, the output will toggle as in *T*. If *J* and *K* are different, the output will assume the state of the *J* input when clocked, similar to the *D* case.

The origin of the designation *J-K* flip-flop is an interesting bit of history: Early flip-flops were designed using a dual-triode vacuum tube mounted on a plug-in module. The pins of the module were named for the letters of the alphabet. The particular pins that carried the logic inputs to the device were pins *J* and *K*.

Flip-flops, latches, and counters are often supplied with additional inputs used to force the output to a known state. An active *set* input will force the output into the active state, while a *reset* input will force the output into the inactive state. Counters also have inputs to force the output states; there are two types:

- *Asynchronous,* in which the function (preset or clear) is performed immediately
- *Synchronous,* in which the action occurs on the next clock transition

Usually, if both preset and clear are applied at once, the clear function outranks the preset function. Figure 1.25 shows some common logic stages and their truth tables. These gates and a few simple rules of boolean algebra, the basics of which are shown in Table 1.4, facilitate the design of very complex circuits.

1.7c Logic Device Families

Resistor-transistor-logic (RTL) is mostly of historic interest only. It used a 3.6-V positive power supply, and was essentially incompatible with the logic families that came later. The packages were round with a circular array of wires (not pins) for circuit board mounting. Inputs were applied to the base of a transistor, and the transistor was turned on directly by the input signal if it was high. An open input could usually be considered as an *"off"* or "0."

Diode-Transistor Logic (DTL)

RTL was followed by the popular DTL, mounted in a DIP (dual in-line package). It had 14 or 16 stiff pins arranged in two parallel rows 0.3 in apart with the pins on 0.1-in centers. For simple devices, such as a two-input NAND gate, four gates were packaged into one DIP. The stiff pins made possible the use of sockets. An internal resistor attached to the positive 5.0-V supply turned on the input transistor. Input signals were applied through diodes such that if an input signal were low, it pulled down the resistor's current, and the transistor turned off. It is important to remember that a disconnected DTL or TTL input is a logic high. The DTL output circuit was pulled low by a transistor and pulled up to +5 V by an internal resistor. As a result, fall times were faster than rise times.

Table 1.3 Comparison of Counting in the Decimal, Binary, and Octal Systems

Decimal	Binary	Octal
0	0	0
1	1	1
2	10	2
3	11	3
4	100	4
5	101	5
6	110	6
7	111	7
8	1000	10
9	1001	11
10	1010	12
11	1011	13
12	1100	14
13	1101	15
14	1110	16
15	1111	17

Transistor-Transistor Logic (TTL)

TTL, like DTL, supplies its own turn-on current but uses a transistor instead of a resistor. The inputs do not use diodes but instead use multiple emitters on an input transistor. The output is pulled down by one transistor and pulled up by another. There are a considerable number of family variations on this basic design. For example, the 7400 device (a two-input NAND gate) has the following common variations:

- 7400—the prototype
- 74L00—a low-power version, but with relatively slow switching speed
- 74S00—(Schottky) fast but power-hungry
- 74LS00—low power and relatively slow speed
- 74AS00—advanced Schottky
- 74ALS00—similar to LS, but with improved performance
- 74F00—*F* for fast

All variants can be used in the presence of the others, but doing so complicates the design rules that determine how many inputs can be driven by one output. The dividing line between an input high and an input low in this example is about 1.8 V. A high output is guaranteed to be 2.4 V or greater, while an output low will be 0.8 V or less.

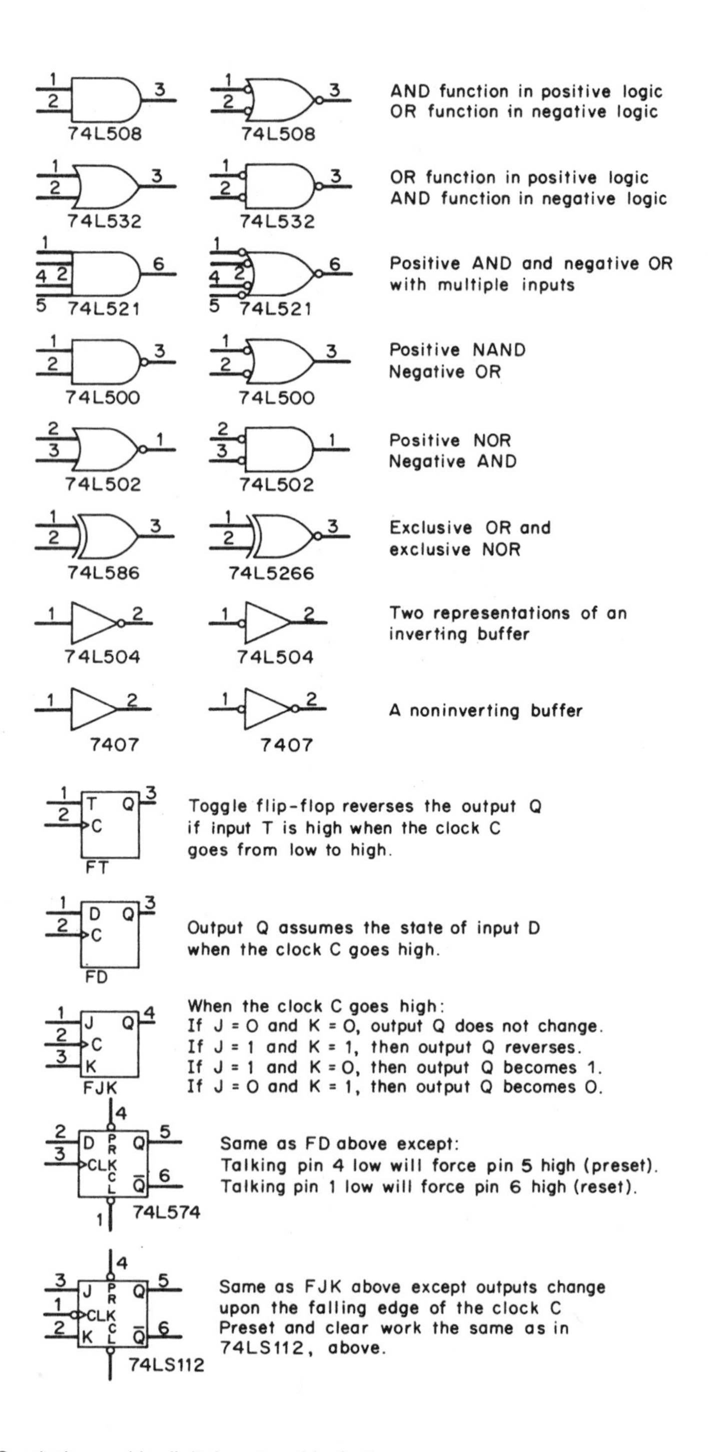

Figure 1.24 Symbols used in digital system block diagrams.

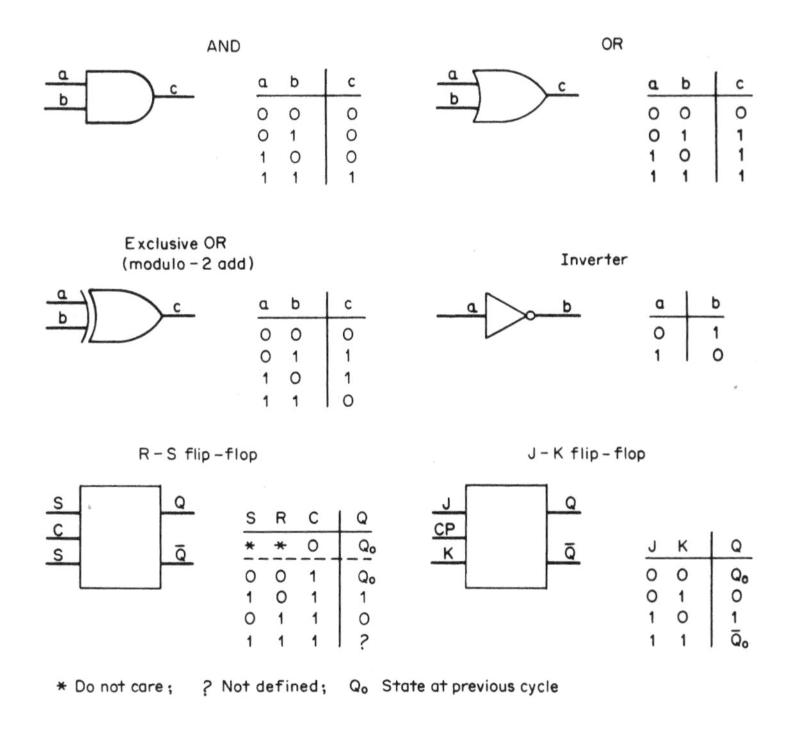

Figure 1.25 Basic logic circuits and truth tables.

NMOS and PMOS

Metal-oxide semiconductor (MOS) logic devices use field-effect transistors as the switching elements. The initial letter tells whether the device uses *n*- or *p*-type dopant on the silicon. At low frequencies, MOS devices are very frugal in power consumption. Early MOSs were fairly slow, but smaller conductor sizes have reduced on-chip capacitance and therefore charging time.

Complementary MOS (CMOS)

A very popular logic family, CMOS devices uses both *p*- and *n*-type transistors. At direct current, input currents are almost zero. Output current rises with frequency since the output circuit must charge and discharge the capacitance of the inputs it is driving. Early CMOS devices were fairly slow when powered with a 5-V supply, but performance improved when powered at 10 or 15 V. Modern microscopic geometry produces CMOS parts that challenge TTL speeds while using less power.

The input decision level of a CMOS device is nominally midway between the positive supply and ground. The logic state of an open input is indeterminate. It can and will wander around depending on which of the two input transistors is leaking the most. Unused inputs *must* be returned either to ground or the supply rail. CMOS outputs, unlike TTL, are very close to ground when low and very close to the supply rail when high. CMOS can drive TTL

Table 1.4 Fundamental Rules of Boolean Algebra

Rules and relations	$0 + 0 = 0$
	$0 \bullet 0 = 0$
	$1 + 1 = 1$
	$1 \bullet 1 = 1$
	$0 \bullet 1 = 0$
	$0 + 1 = 1$
	$\overline{0} = 1$
	$\overline{1} = 0$
	$a \bullet a = a$
	$a + a = a$
	$\overline{a} \bullet a = 0$
	$\overline{a} + a = 1$
	$0 + a = a$
	$0 \bullet a = 0$
	$1 + a = 1$
	$1 \bullet a = a$
	$a + a \bullet b = a$
	$a \bullet (a + b) = a$
	$a + a \bullet b = a + b$
Commutative law	$a + b = b + a$
	$a \bullet b = b \bullet a$
Associative law	$(a + b) + c = a + (b + c)$
	$(a \bullet b) \bullet c = a \bullet (b \bullet c)$
Distributive law	$a \bullet (b + c) = a \bullet b + a \bullet c$
DeMorgan's rules	$\overline{(a \bullet b)} = \overline{a} + \overline{b}$
	$\overline{(a + b)} = \overline{a} \bullet \overline{b}$

† • denotes AND, + denotes OR.

inputs, however, in a 5-V environment, the CMOS decision level of 2.5 V is too close to the TTL guaranteed output high for reliable operation. The solution is an external pull-up resistor between the output pin of the TTL part and the supply rail.

Early CMOS devices had their own numbering system (beginning at 4000) that was totally different from the one used for TTL parts. Improvements in speed and other performance metrics spawned subfamilies that tend toward a return to the use of the 7400 convention; for example, 74HC00 is a high-speed CMOS part.

Emitter-Coupled Logic (ECL)

ECL has almost nothing in common with the families previously discussed. Inputs and outputs are push-pull. The supply voltage is negative with respect to ground at –5.2 V. Certain advantages accrue from this configuration:

- Because of the push-pull input-output, inverters are not needed. To invert, simply reverse the two connections.

- The differential-amplifier construction of ECL input and output stages causes the total current through the device to be almost constant.
- The output voltage swing is small and, from a crosstalk standpoint, is opposed by the complementary output.
- Driving a balanced transmission line does not require a line-driver because an ECL output (with some resistors) *is* a line-driver.

Because the transistors in ECL are never saturated (maximum conduction state), they operate at maximum speed. Early ECL was power-hungry, but new ECL gate-array products are available that will toggle well into the gigahertz range without running hot.

Scaling of Digital Circuit Packages

The term *small-scale integration* (SSI) includes those packages containing, for example, a collection of four gates, a 4-bit counter, a 4-bit adder, and any other item of less than about 100 gate equivalents. *Large-scale integration (LSI)* describes more complex circuitry, such as an asynchronous bit-serial transmitter-receiver, or a DMA (direct memory access) controller, involving a few thousand gate equivalents. *Very large scale integration* (VLSI) represents tens of thousands of gate equivalents or more, such as a microprocessor or a graphics controllers. LSI and VLSI devices are typically packaged in a larger version of the DIP package, usually with the two rows spaced 0.6 in or more, and having 24 to 68 pins or more.

Many devices are available in dual in-line packages designed to be soldered to the surface of a circuit board rather than using holes in the circuit board. The pin spacing is 0.05 in or less. The *leadless chip carrier* is another surface-mount device with contact spacing of 0.05 in or less, and an equal number of contacts along each edge of the square. Well over 100 contacts can be accommodated in such packages. Sockets are available for these packages, but once the package is installed, a special tool is required to extract it. Yet another large-scale package is called the *pin-grid array,* with pins protruding from the bottom surface of a flat, square package in a row-and-column "bed-of-nails" array. The pin spacing is 0.1 in or less. For this device, more than 200 pins may be incorporated. Extraction tools are available for these packages as well.

1.7d Representation of Numbers and Numerals

A single bit, terminal, or flip-flop in a binary system can have only two states. When a single bit is used to describe numerals, by convention those two numerals are 0 and 1. A group of bits, however, can describe a larger range of numbers. Conventional groupings are identified in the following sections.

Nibble

A nibble is a group of 4 bits. It is customary to show the binary representation with the *least significant bit* (LSB) on the right. The LSB has a decimal value of 1 or 0. The next most significant bit has a value of 2 or 0, and the next, 4 or 0, and the *most significant bit* (MSB), 8 or 0. The nibble can describe any value from binary 0000 (= 0 decimal) and 1111 (= 8 + 4 + 2 + 1 = 15 decimal), inclusive. The 16 characters used to signify the 16 values of a nibble are the ordinary numerals 0 through 9, followed by the letters of the alphabet A through F. The 4-bit "digit" is a hexadecimal representation.

Octal, an earlier numbering scheme, used groupings of 3 bits to describe the numerals 0 through 7. Used extensively by the Digital Equipment Corporation, it has fallen out of use, but is still included in some figures for reference.

Byte

A byte is a collection of 8 bits, or 2 nibbles. It can represent numbers (a *number* is a collection of numerals) in two ways:

- Two hexadecimal digits, the least significant representing the number of 1s, and the most significant the number of 16s. The total range of values is 0 through 255 (FF).
- Two *decimal* digits, the least significant representing the number of 1s, and limited to the range of numerals 0 through 9, and the most significant representing the number of 10s, again limited to the range 0 through 9.

The use of 4 bits to represent decimal numbers is called *binary-coded decimal* (BCD). The use of a byte to store two numerals is called *packed* BCD. The least significant nibble is limited to the range of 0 through 9, as is the upper nibble, thus representing 00 through 90. The maximum value of the byte is 99.

Word

A *word,* usually a multiple of 8 bits, is the largest array of bits that can be handled by a system in one action of its logic. In most personal computers, a word is 16 or 32 bits. Larger workstations use words of 32 and 64 bits in length. In all cases, the written and electronically mapped representation of the numeric value of the word is either hexadecimal or packed BCD.

Negative Numbers

When a byte or word is used to describe a *signed* number (one that may be less than zero), it is customary for the most significant bit to represent the sign of the number, 0 meaning positive and 1 negative. This representation is known as *two's complement.* To negate (make negative) a number, simply show the number in binary, make all the zeros into 1s, and all the 1s into zeros, and then *add* 1.

Floating Point

In engineering work, the range of numerical values is tremendous and can easily overflow the range of values offered by 64-bit (and smaller) systems. Where the accuracy of a computation can be tolerably expressed as a *percentage* of the input values and the result, *floating-point* calculation is used. One or two bytes are used to express the *characteristic* (a power of 10 by which to multiply everything), and the rest are used to express the *mantissa* (that fractional power of 10 to be multiplied by). This is commonly referred to as *engineering notation.* (See Table 1.5.)

Compare

A *comparison* involves negating one of the two numbers being compared, then adding them and testing the result. If the test shows zero, the two numbers are equal. If not, the test reveals which of the two is greater than or less than the other, and the appropriate bits in the status register are set.

Jump

The orderly progression of the program counter may be interrupted and instructions fetched from a new location in memory, usually based upon a test or a comparison. For example, "If the result is zero, jump to location *X* and begin execution there; if the result is positive, jump to *Y* and begin execution there; else keep on counting." This ability is probably the most powerful asset of a computer because it permits logic-based branching of a program.

1.7e Errors in Digital Systems

When a digital signal is transmitted through a noisy path, errors can occur. Early methods to deal with this problem included generating one or more digital words, using check sums, cyclic redundancy checks, and similar error-coding schemes, and appending the result at the end of a block of transmitted data. Upon reception, the same arithmetic was used to generate the same results, which were compared to the data appended to the transmission. If they were identical, it was unlikely that an error had occurred. If they differed, an error was assumed to have occurred, and a retransmission was requested. Such methods, thus, performed *only* error detection. In the case of many digital transmission systems, however, retransmission is not possible and methods must be employed that not only *detect* but *correct* errors.

Error Detection and Correction

Given a string of 8-bit bytes, additional bytes can be generated using Galois field arithmetic and appended to the end of the string. The length of the string and the appended bytes must be 256 or less, since 8 bits can have no more than 256 different states. If 2 bytes are generated, upon reconstruction 2 *syndrome* (symptom) bytes are generated. If they are zero, there was likely no error. If they are nonzero, then after arithmetic processing, 1 byte "points" to the location of the damaged byte in the string, while the other contains the 8-bit error pattern. The error pattern is used in a bit-wise exclusive OR function upon the offending byte, thus reversing the damaged bits and correcting the byte. With 4 check bytes, 2 flawed bytes can be pinpointed and corrected; with 6, 3 can be treated; and so on. If the number of bytes in the string is significantly less than 256, for example, 64, the error-detection function becomes more robust because, if the error pointer points to a nonexistent byte, it may be assumed that the error-detection system itself made a mistake.

Errors in digital-video recorders, for example, fall into two classes: *random* errors brought on by thermal random noise in the reproduce circuitry, and *dropouts* (long strings of lost signal resulting from tape imperfections. The error detection and correction system of digital-video recorders is designed to cope with both types of errors. Figure 1.26 shows how data can be arranged in rows and columns, with separate check bytes generated for each row and each column in a two-dimensional array. The data is recorded (and reproduced) in row order. In the

Table 1.5 Number and Letter Representations

Decimal	Hexadecimal	Octal	Binary
0	0	0	0000
1	1	1	0001
2	2	2	0010
3	3	3	0011
4	4	4	0100
5	5	5	0101
6	6	6	0110
7	7	7	0111
8	8	10	1000
9	9	11	1001
10	A	12	1010
11	B	13	1011
12	C	14	1100
13	D	15	1101
14	E	16	1110
15	F	17	1111
81	51	121	01010001, 16 × 5 + 1
250	FA	372	11111010, 16 × 15 + 10
+127	7F	177	01111111, (signed)
–1	FF	377	11111111, (signed)
–128	80	200	10000000, (signed)

example given in the figure, it can be seen that a long interruption of signal will disrupt every tenth byte. The row corrector cannot cope with this, but it is likely that the column corrector can because it "sees" the burst error as being spread out over a large number of columns.

The column corrector, if taken alone, can correct $N/2$ errors, where N is the number of check bytes. Given knowledge of which *rows* are uncorrectable by the row corrector, then N errors can be corrected. Generally, the row (or "inner") corrector acts on errors caused by random noise, while the column (or "outer") corrector takes care of burst errors.

Generally, error detection and correction schemes have the following characteristics:

- Up to a threshold error rate, all errors are corrected.
- If the error rate is greater than the above first threshold, the system will flag the blocks of data it is unable to correct. This allows other circuits to attempt to *conceal* the error.
- Above an even higher error rate, the system will occasionally fail and either stop producing output data entirely, or simply pass along the data, correcting what it can and letting the rest pass through.

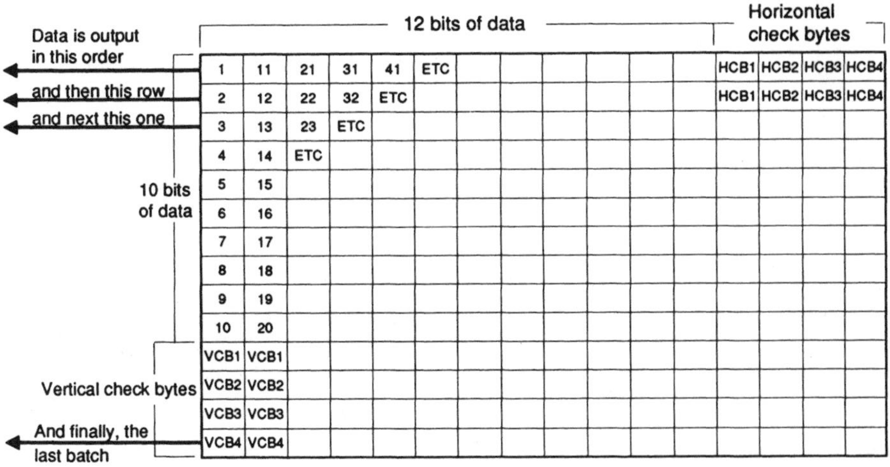

1	11	21	31	41	ETC							HCB1	HCB2	HCB3	HCB4
2	12	22	32	ETC								HCB1	HCB2	HCB3	HCB4
3	13	23	ETC												
4	14	ETC													
5	15														
6	16														
7	17														
8	18														
9	19														
10	20														
VCB1	VCB1														
VCB2	VCB2														
VCB3	VCB3														
VCB4	VCB4														

Figure 1.26 An example of row and column two-dimensional error-detection coding.

Error Concealment

When the error-correction system is overloaded and error-ridden samples are identified, it is typical practice to calculate an estimation of the bad sample. In video applications, samples that are visually nearby and that are not corrupted can be used to calculate an estimate of the damaged sample. The estimate is then substituted for the unusable sample. In the recording or transmission process, the video data samples are scrambled in a way that maximizes the chance that a damaged sample will be surrounded by good ones.

In the case of audio, the samples can be scrambled such that failure of the correction system is most likely to result in every *alternate* sample being in error. Replacement of a damaged audio sample can then consist of summing the previous (good) sample and the following (good) sample and dividing by 2. If the error rate becomes unreasonable, then the last good sample is simply repeated, or "held."

Video error concealment is roughly 10 times more effective than audio concealment, due in large part to differences in the way the eye and ear interpret and process input information.

1.8 References

1. Fink, Donald G., and Don Christiansen (eds.): *Electronic Engineers' Handbook*, McGraw-Hill, New York, N.Y., 1982.

1.9 Bibliography

Benson, K. Blair, and Jerry C. Whitaker: *Television and Audio Handbook for Technicians and Engineers*, McGraw-Hill, New York, N.Y., 1990.

Benson, K. Blair: *Audio Engineering Handbook*, McGraw-Hill, New York, N.Y., 1988.

Rhode, U., J. Whitaker, and T. Bucher: *Communications Receivers*, 2nd ed., McGraw-Hill, New York, N.Y., 1996.

Whitaker, Jerry C., and K. Blair Benson (eds): *Standard Handbook of Video and Television Engineering*, McGraw-Hill, New York, N.Y., 2000.

Benson, K. Blair (ed.): *Audio Engineering Handbook*, McGraw-Hill, New York, N.Y., 1988.

Boyer, Robert, JeanLuc Grimaldi, Jacques Oyaux, and Jacques Vallee: "Serial Interface Within the Digital Studio," *1. Soc. Motion Pict. Telev.*, November 1984.

Busby, E. Stanley: "Digital Fundamentals," in *Television and Audio Handbook for Technicians and Engineers*, K. Blair Benson and Jerry C. Whitaker (eds.), McGraw-Hill, New York, N.Y., 1990.

EBU: Publication Tech 3247.E, Technical Centre of the EBU, Brussels, 1985.

Fink, Donald (ed.): *Electronics Engineers' Handbook*, McGraw-Hill, New York, N.Y., 1982.

SMPTE: "Bit-Parallel Digital Interface for Component Video Signals." SMPTE RP 125, SMPTE, White Plains, N.Y., 1984.

Texas Instruments: 2-µm *CMOS Standard Cell Data Book*, Chapter 8, Texas Instruments, Dallas, Texas, 1986.

Whitaker, Jerry C., and K. Blair Benson (eds.): *Standard Handbook of Video and Television Engineering*, third ed., McGraw-Hill, New York, N.Y., 2000.

Chapter

2

Reference Data, Tables, and Figures

This section contains a compilation of figures, tables, and related data of importance to video engineers. The material is divided into 24 elements, as outlined below.

2.1 The Physical Nature of Sound

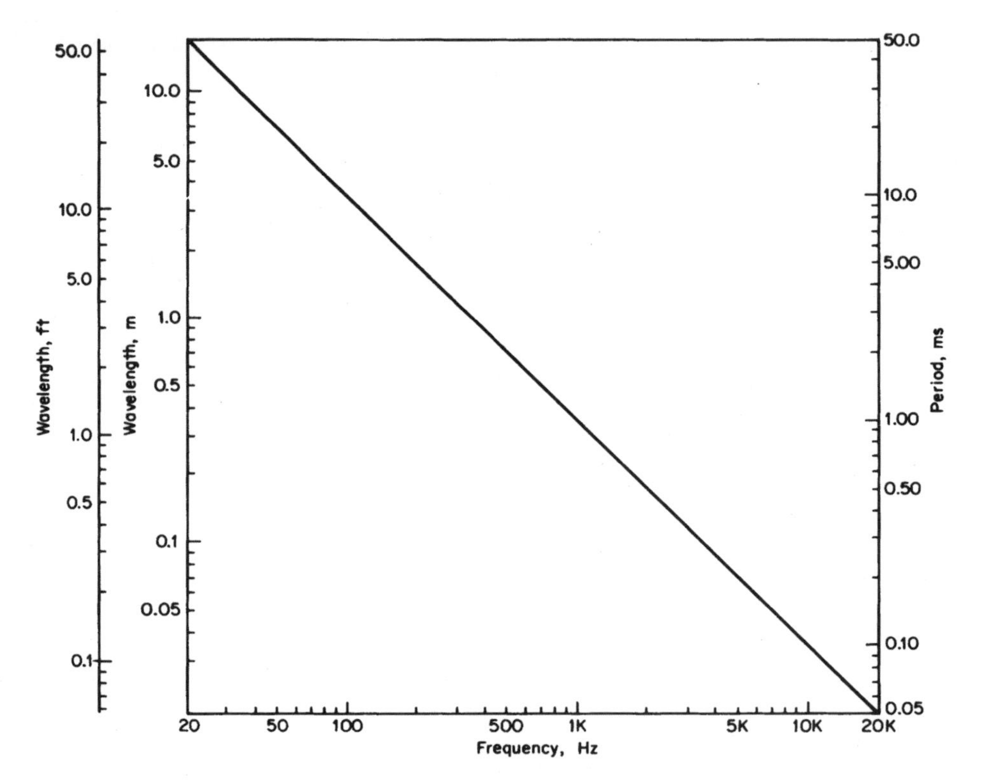

Figure 2.1.1 Relationships between wavelength, period, and frequency for sound waves in air.

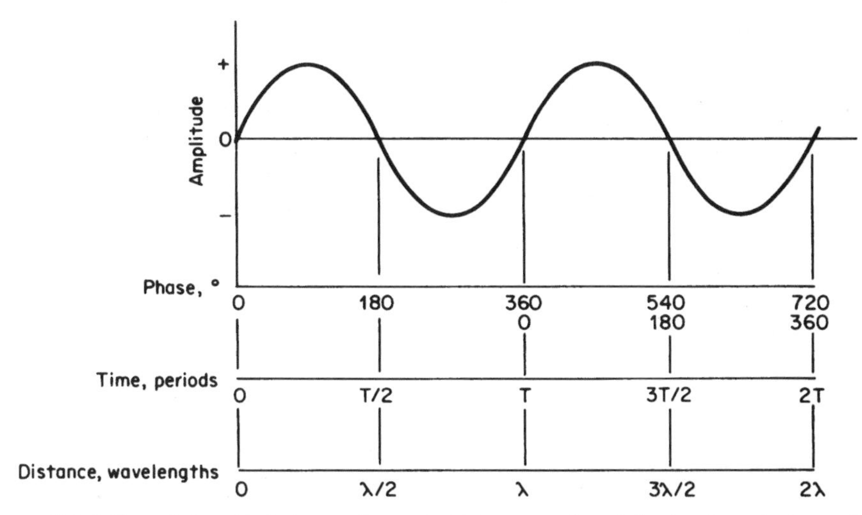

Figure 2.1.2 The relationship between the period *T* and wavelength λ of a sinusoidal waveform and the phase expressed in degrees. Although it is normal to consider each repetitive cycle as an independent 360°, it is sometimes necessary to sum successive cycles starting from a reference point in one of them.

Table 2.1.1 Typical Sound Pressure Levels and Intensities for Various Sound Sources*

Sound source	Sound pressure level, dB	Intensity, W/m²	Listener reaction
	160		Immediate damage
Jet engine at 10 m	150	10^3	
	140		Painful feeling
	130		
SST takeoff at 500 m	120	1	Discomfort
Amplified rock music	110		
Chain saw at 1 m	100		fff
Power mower at 1.5 m	90	10^{-3}	
75-piece orchestra at 7 m	80		f
City traffic at 15 m	70		
Normal speech at 1 m	60	10^{-6}	p
Suburban residence	50		
Library	40		ppp
Empty auditorium	30	10^{-9}	
Recording studio	20		
Breathing	10		
	0†	10^{-12}	Inaudible

* The relationships illustrated in this table are necessarily approximate because the conditions of measurement are not defined. Typical levels should, however, be within about 10 dB of the stated values.

† O-dB sound pressure level (SPL) represents a reference sound pressure of 0.0002 μbar, or 0.00002 N/m².

Table 2.1.2 Various Power and Amplitude Ratios and their Decibel Equivalents*

Sound or electrical power ratio	Decibels	Sound pressure, voltage, or current ratio	Decibels
1	0	1	0
2	3.0	2	6.0
3	4.8	3	9.5
4	6.0	4	12.0
5	7.0	5	14.0
6	7.8	6	15.6
7	8.5	7	16.9
8	9.0	8	18.1
9	9.5	9	19.1
10	10.0	10	20.0
100	20.0	100	40.0
1,000	30.0	1,000	60.0
10,000	40.0	10,000	80.0
100,000	50.0	100,000	100.0
1,000,000	60.0	1,000,000	120.0

* Other values can be calculated precisely by using Eqs. (1.6) and (1.7) or estimated by using this table and the following rules:

Power ratios that are multiples of 10 are converted into their decibel equivalents by multiplying the appropriate exponent by 10. For example, a power ratio of 1000 is 10^3, and this translates into $3 \times 10 = 30$ dB. Since power is proportional to the square of amplitude, the exponent of 10 must be doubled to arrive at the decibel equivalent of an amplitude ratio.

Intermediate values can be estimated by combining values in this table by means of the rule that the multiplication of power or amplitude ratios is equivalent to adding level differences in decibels. For example, increasing a sound level by 27 dB requires increasing the power by a ratio of 500 (20 dB is a ratio of 100, and 7 dB is a ratio of 5; the product of the ratios is 500). The corresponding increase in sound pressure or electrical signal amplitude is a factor of just over 20 (20 dB is a ratio of 10, and 7 dB falls between 6.0 and 9.5 and is therefore a ratio of something in excess of 2); the calculated value is 22.4. Reversing the process, if the output from a power amplifier is increased from 40 to 800 W, a ratio of 20, the sound pressure level would be expected to increase by 13 dB (a power ratio of 10 is 10 dB, a ratio of 2 is 3 dB, and the sum is 13 dB). The corresponding voltage increase measured at the output of the amplifier would be a factor of between 4 and 5 (by calculation, 4.5).

Figure 2.1.3 Behavior of a point monopole sound source in full space (4π) and in close proximity to reflecting surfaces that constrain the sound radiation to progressively smaller solid angeles. (*After: Olson, Harry F.:* Acoustical Engineering, *Van Nostrand, New York, N.Y., 1957.*)

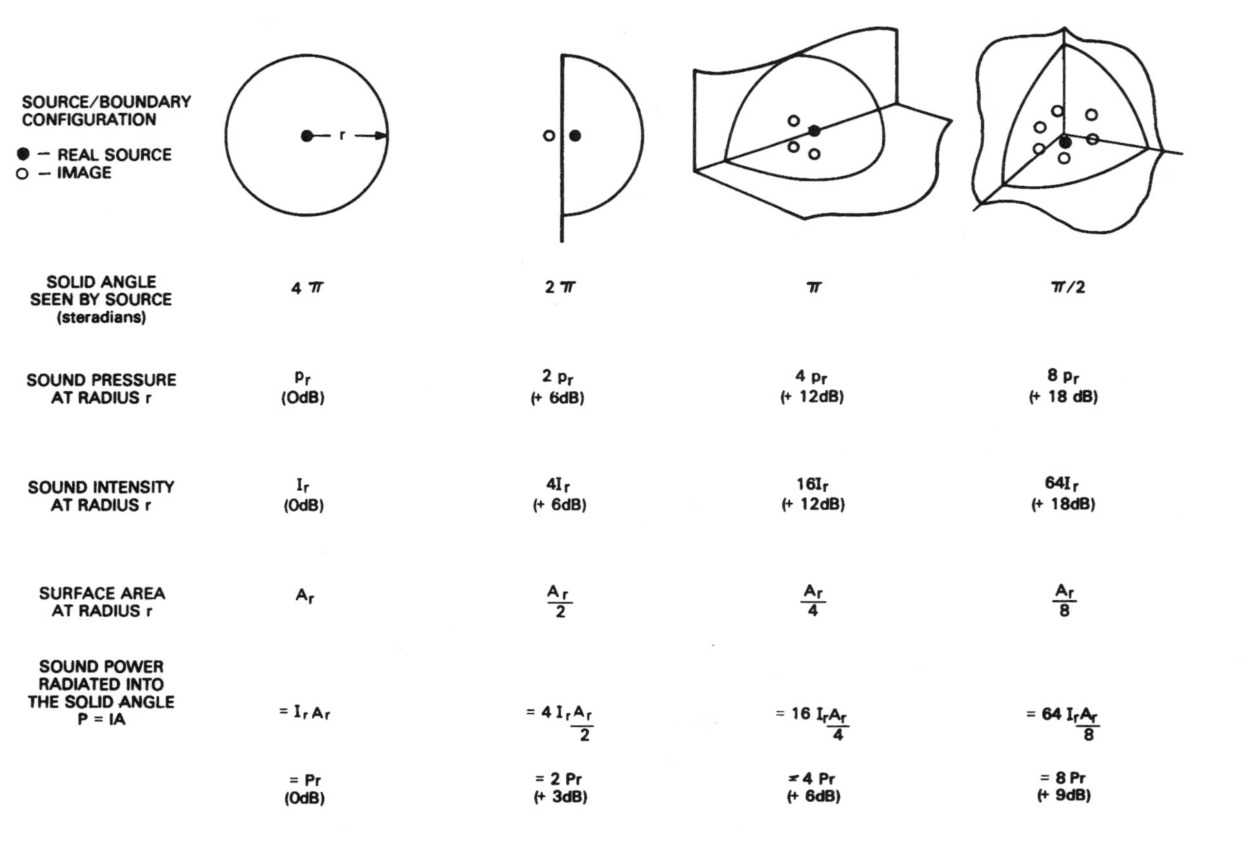

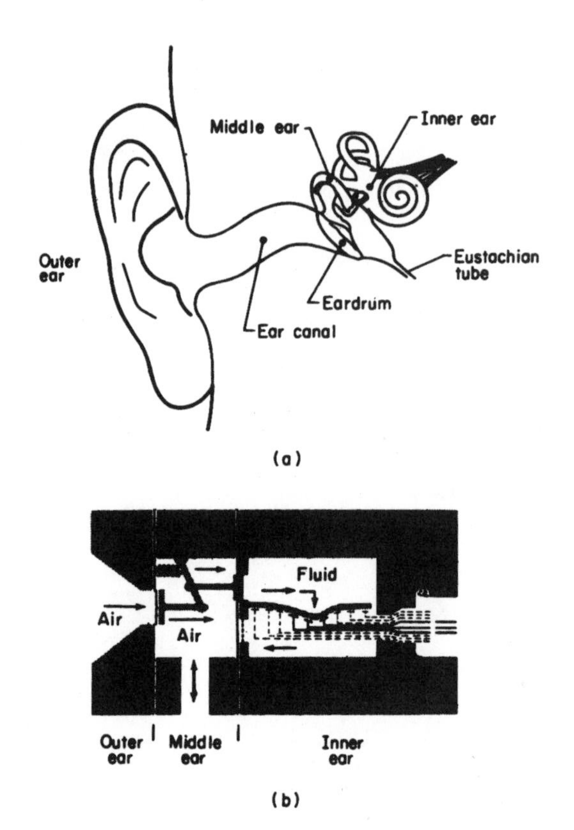

Figure 2.1.4 The human ear: (*a*) cross-sectional view showing the major anatomical elements, (*b*) a simplified functional representation.

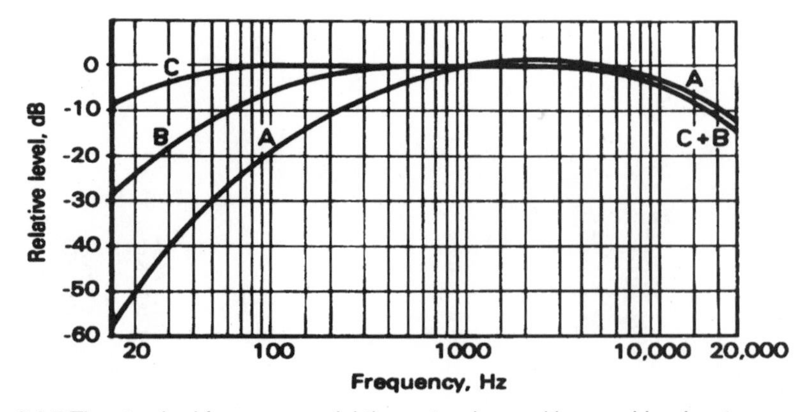

Figure 2.1.5 The standard frequency-weighting networks used in sound-level meters.

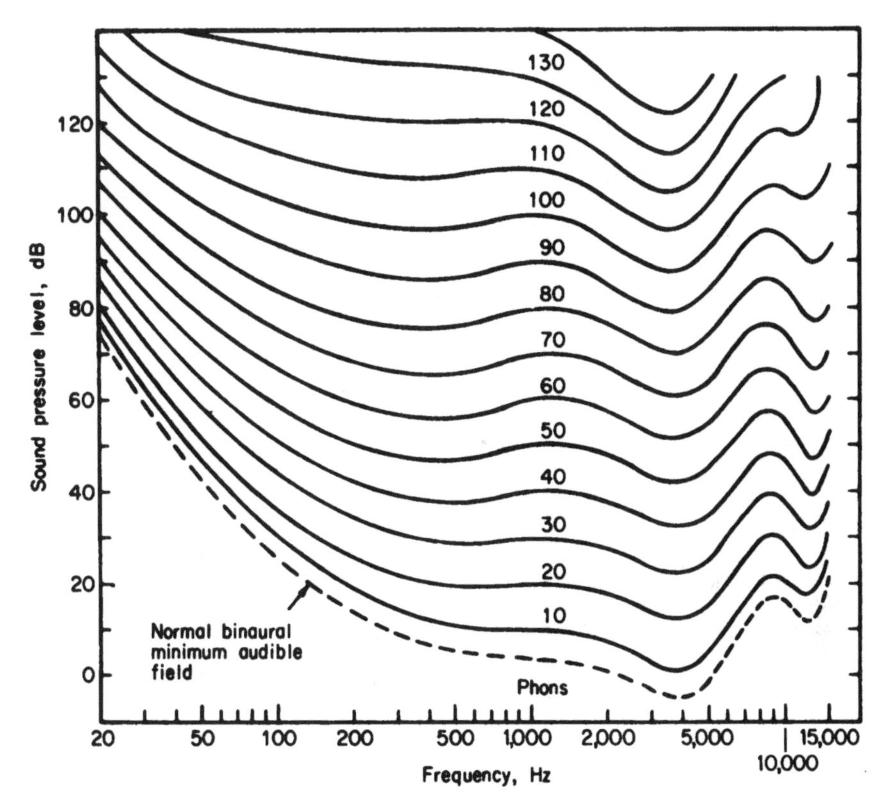

Figure 2.1.6 Contours of equal loudness showing the sound pressure level required for pure tones at different frequencies to sound as loud as a reference tone of 1000 Hz. (*From ISO Recommendation R226.*)

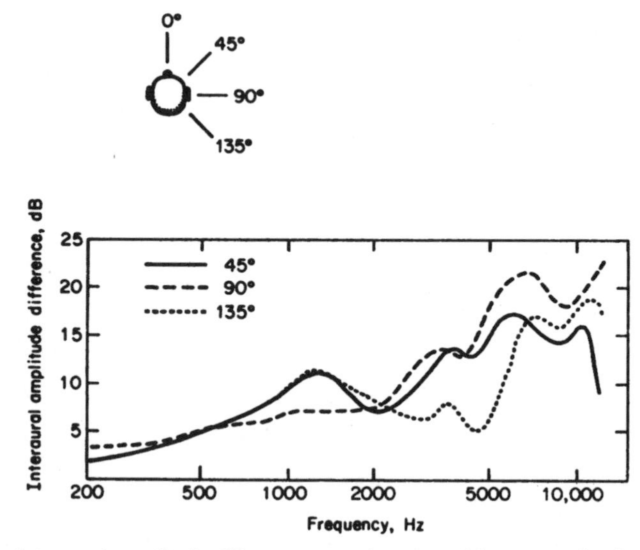

Figure 2.1.7 The interaural amplitude difference as a function of frequency for three angles of incidence. (*After: Shaw, E. A. G., and M. M. Vaillancourt: "Transformation of Sound-Pressure Level from the Free Field to the Eardrum Presented in Numerical Form,"* J. Acoust. Soc. Am., *vol. 78, pp. 1120–1123, 1985.*)

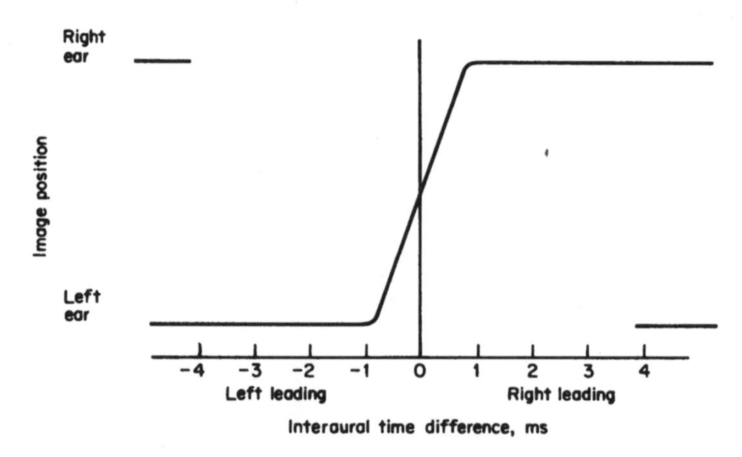

Figure 2.1.8 Perceived positions of the dominant auditory images resulting from impulsive signals (clicks) presented through headphones when the interaural time difference is varied.

2.2 The Audio Spectra

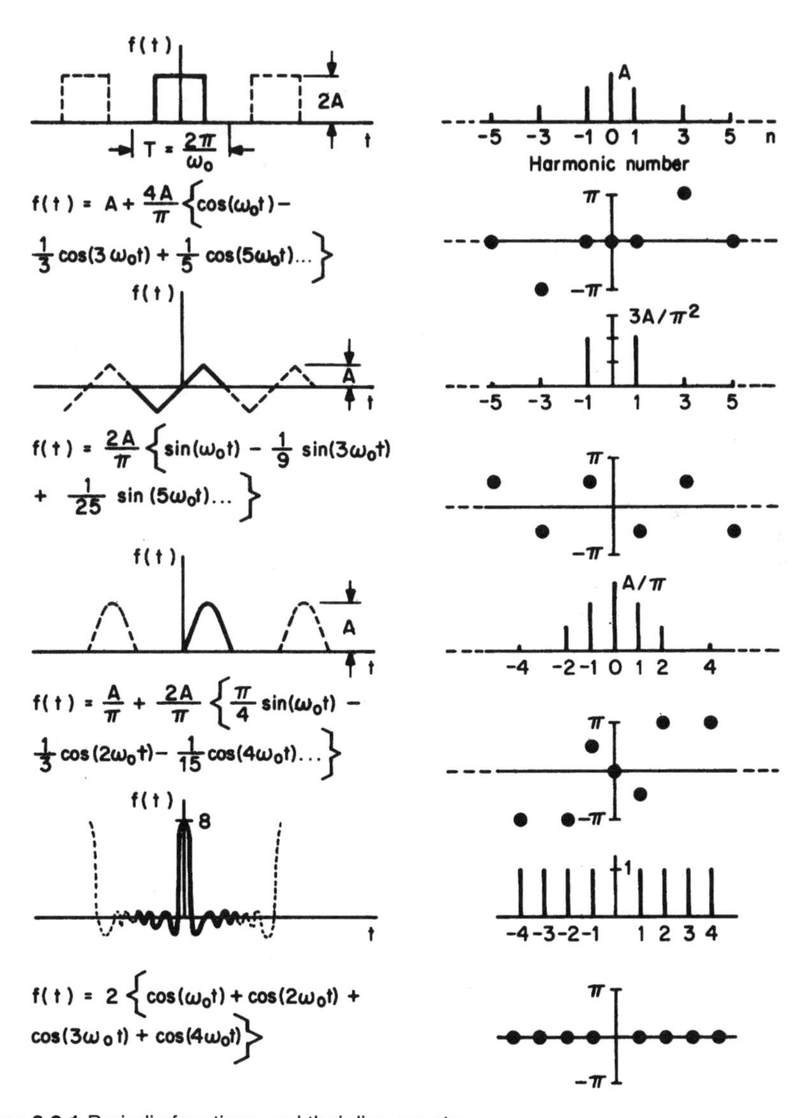

Figure 2.2.1 Periodic functions and their line spectra.

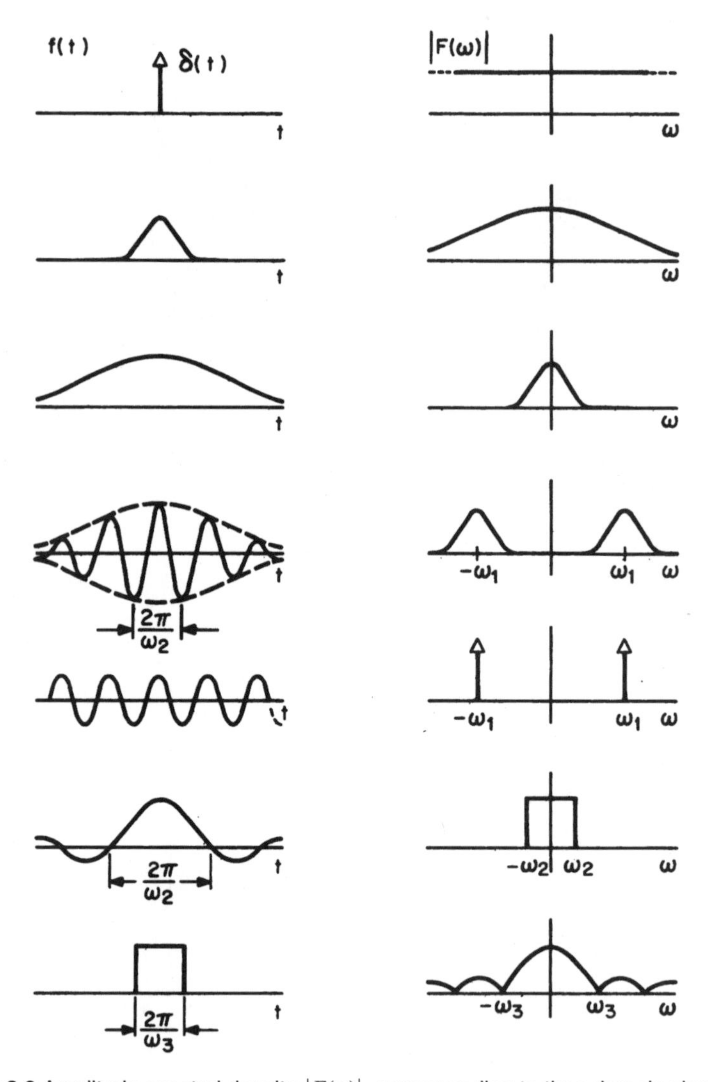

Figure 2.2.2 Amplitude spectral density $|F(\omega)|$ corresponding to time-domain signals $f(t)$.

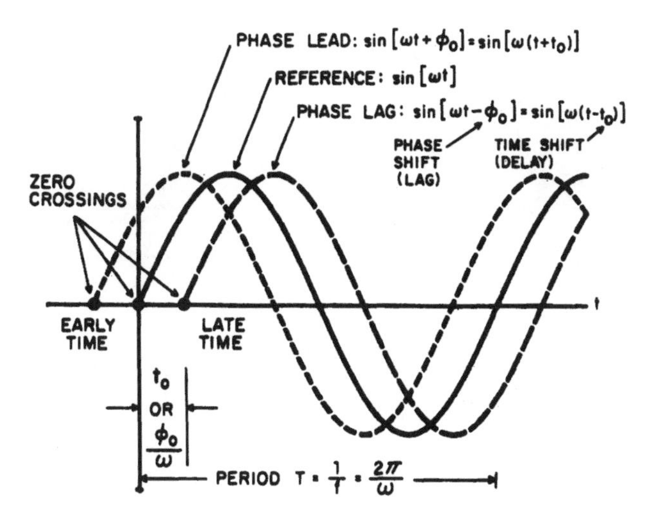

Figure 2.2.3 Standard conventions for sinusoids having steady-state phase shifts relative to a reference sinusoid (solid curve). The leading sinusoid (short dashes) has its first zero crossing at an earlier time than the reference, whereas the lagging sinusoid (long dashes) has its first zero crossing at a later time.

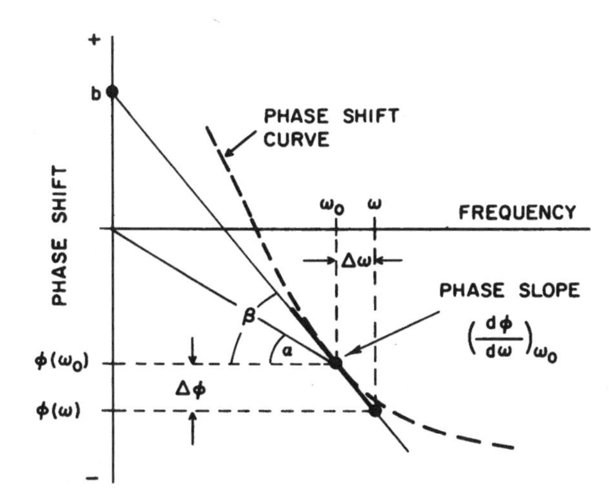

Figure 2.2.4 Arbitrary phase-shift-versus-frequency characteristic (long dashes). For frequencies ω near ω_0 the slope of the phase shift curve is nearly constant and can be approximated by the derivative $d\phi/d\omega$ evaluated at ω_0. The numerical value of the phase slope (or derivative) indicates how the phase shift varies near ω_0.

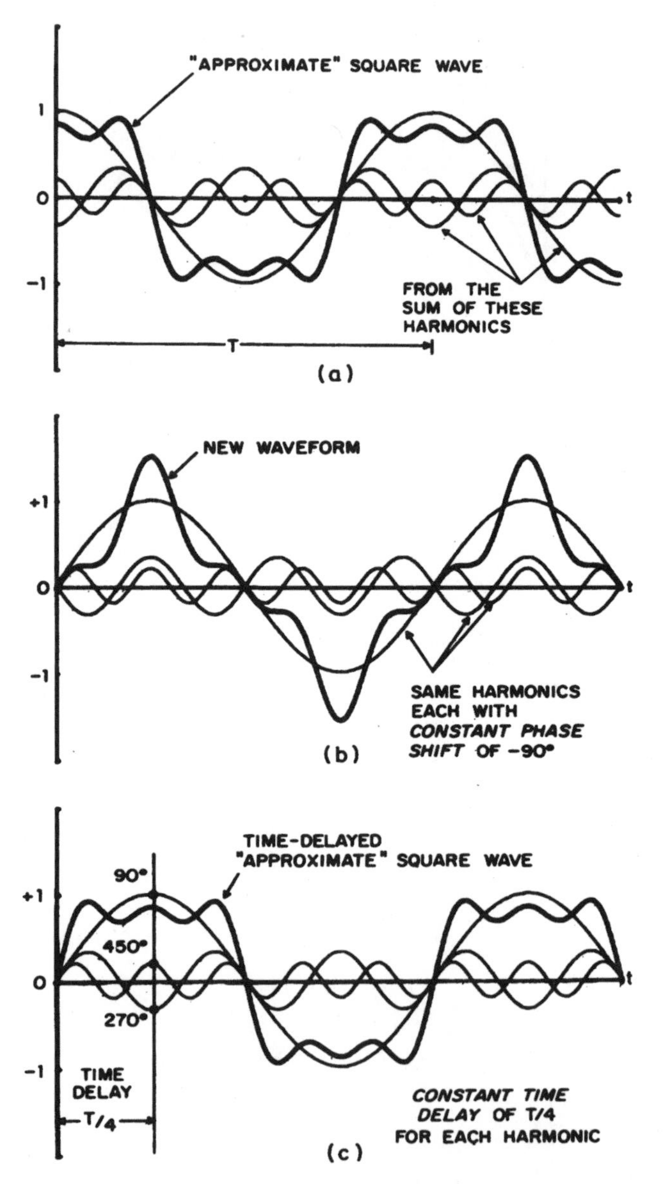

Figure 2.2.5 Characteristics of various waveforms: (*a*) an approximate square wave constructed from the first three (nonzero) Fourier harmonics, (*b*) constant phase shift for each harmonic yields a new waveform that is linearly distorted, (*c*) constant time delay for each harmonic uniformly delays the square wave while preserving its shape.

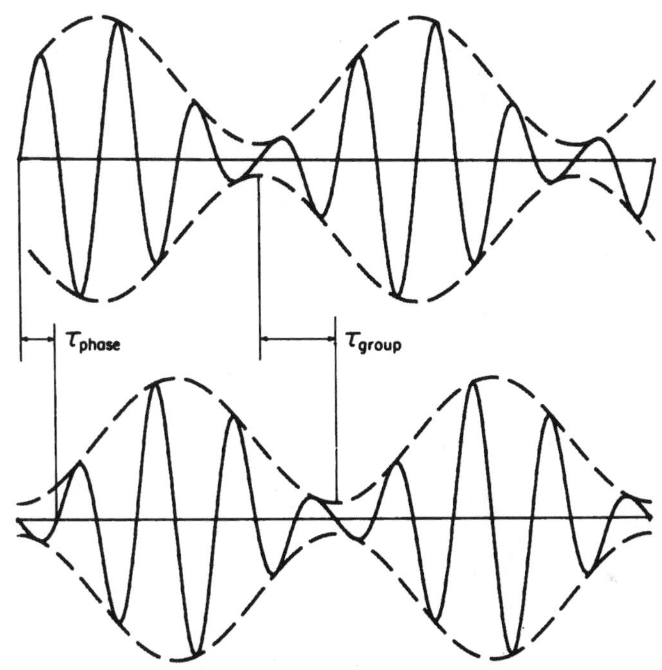

Figure 2.2.6 The difference between phase delay τ_{phase} and group delay τ_{group} is illustrated by comparing these two amplitude-modulated waveforms. The lower waveform has positive phase delay and positive group delay relative to the upper waveform. Because the envelope of the high-frequency oscillation is delayed by an amount of time τ_{group} group delay is sometimes referred to as *envelope delay.*

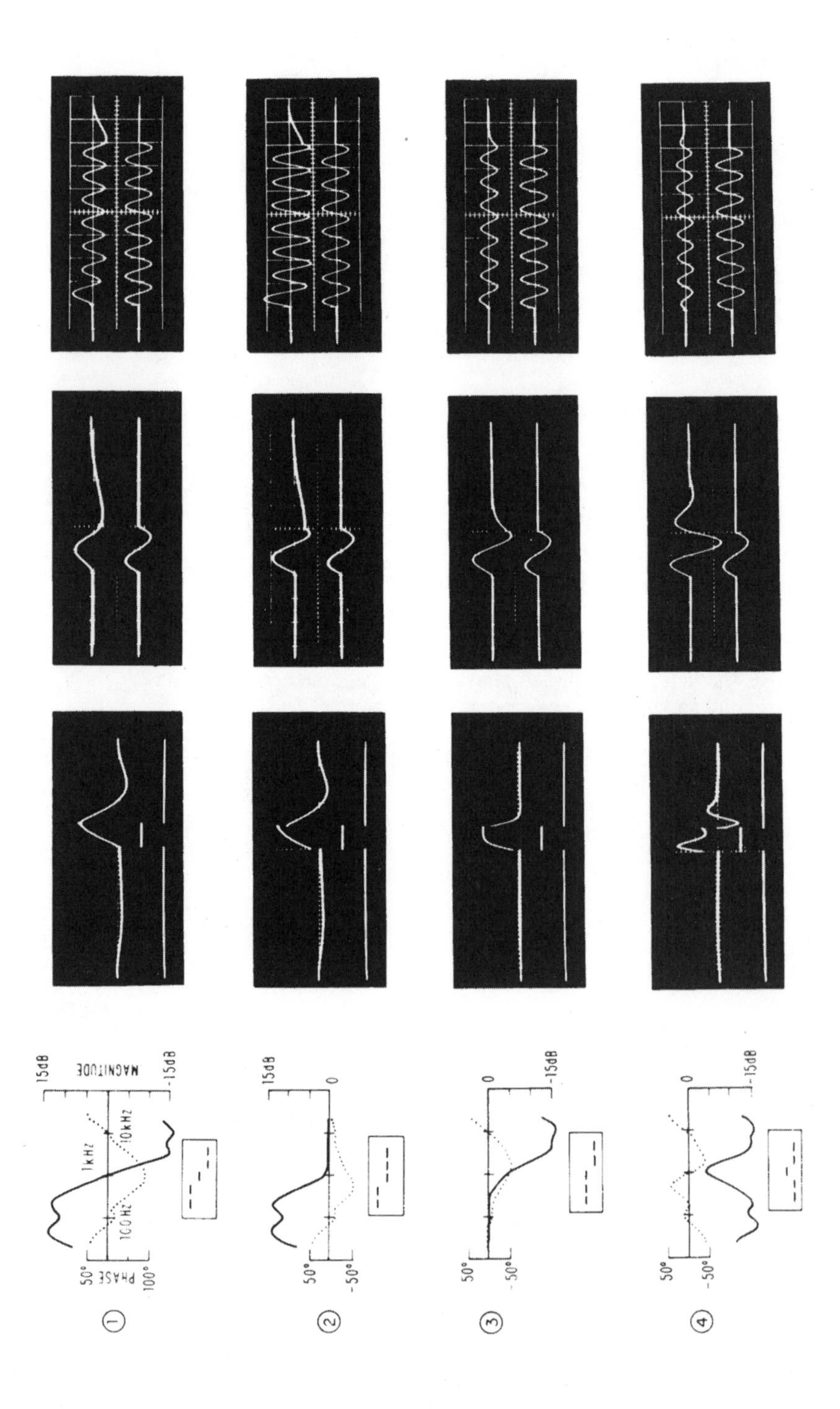
15dB
MAGNITUDE
-15dB
1kHz
10kHz
100Hz
50°
PHASE
100°
0
-50°
1
2
3
4

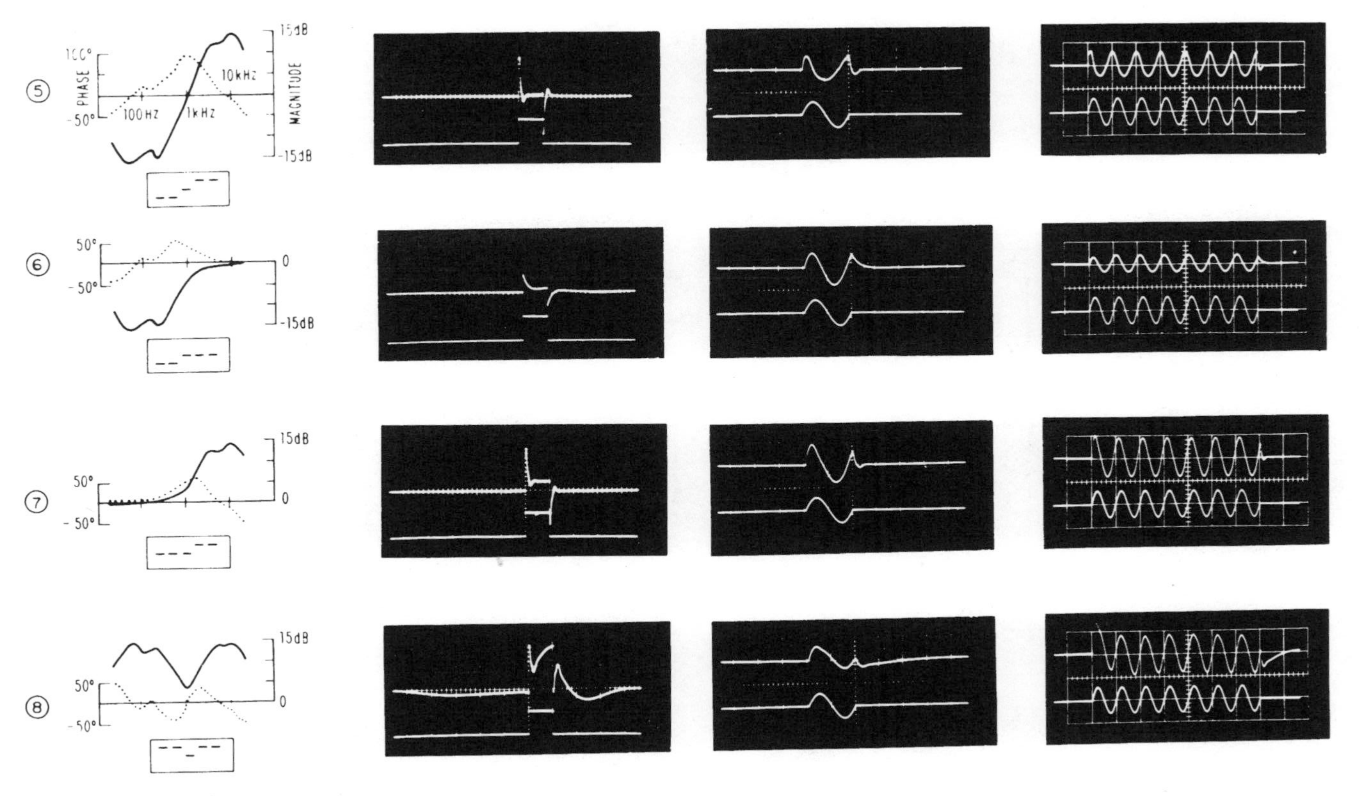

Figure 2.2.7 Comparison of steady-state magnitude and phase response measurements (of a five-band graphic equalizer) with responses to various transient signals. System output and input signals are shown in the upper and lower traces, respectively.

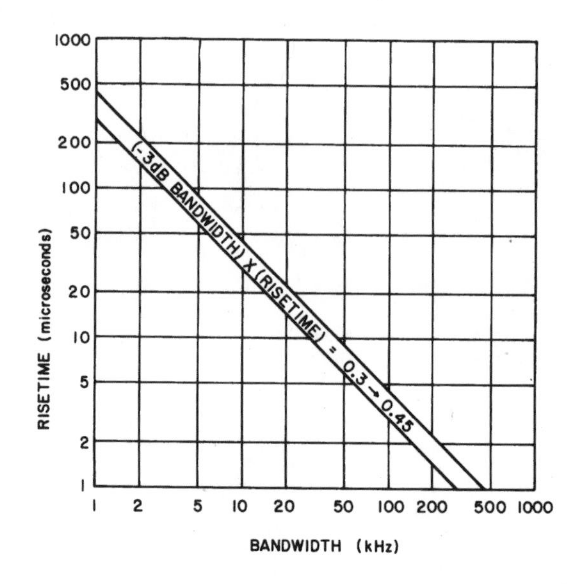

Figure 2.2.8 Relationship between rise time and bandwidth of practical linear systems. For a given bandwidth, the rise time will lie within the tolerance strip shown; conversely, the bandwidth requirements for a specific rise time also can be found.

2.3 Architectural Acoustic Principles and Design Techniques

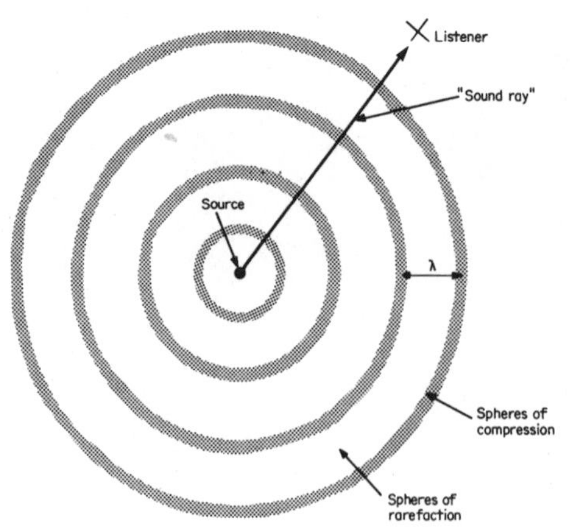

Figure 2.3.1 Simple sound source.

Table 2.3.1 Limits of Frequency Passbands

Band number	Nominal center frequency	One-third-octave passband	Octave passband
14	25	22.4–28.2	
15	31.5	28.2–35.5	22.4–44.7
16	40	35.5–44.7	
17	50	44.7–56.2	
18	63	56.2–70.8	44.7–89.1
19	80	70.8–89.1	
20	100	89.1–112	
21	125	112–141	89.1–178
22	160	141–178	
23	200	178–224	
24	250	224–282	178–355
25	315	282–355	
26	400	355–447	
27	500	447–562	355–708
28	630	562–708	
29	800	708–891	
30	1,000	891–1,120	708–1410
31	1,250	1,120–1,410	
32	1,600	1,410–1,780	
33	2,000	1,780–2,240	1,410–2,820
34	2,500	2,240–2,820	
35	3,150	2,820–3,550	
36	4,000	3,550–4,470	2,820–5,620
37	5,000	4,470–5,620	
38	6,300	5,620–7,080	
39	8,000	7,080–8,910	5,620–11,200
40	10,000	8,910–11,200	

Table 2.3.2 Typical Absorption Coefficients

Material	Sound absorption coefficient						NRC number
	125 Hz	250 Hz	500 Hz	1000 Hz	2000 Hz	4000 Hz	
Concrete masonry units, painted	0.08	0.05	0.05	0.07	0.08	0.08	0.06
Gypsum wallboard, ½ in thick, studs spaced 24 in on center	0.27	0.10	0.05	0.04	0.07	0.08	0.07
Typical window glass	0.30	0.22	0.17	0.13	0.07	0.03	0.15
Plaster on lath	0.15	0.10	0.06	0.05	0.05	0.03	0.07
Light fabric, flat against concrete wall	0.08	0.06	0.10	0.16	0.25	0.32	0.14
Thick drapery, draped to half area	0.15	0.36	0.55	0.70	0.73	0.75	0.59
Linoleum on concrete	0.02	0.03	0.03	0.03	0.03	0.03	0.03
Typical wood floor	0.15	0.12	0.10	0.06	0.06	0.06	0.09
Thin carpet on concrete	0.03	0.06	0.10	0.20	0.43	0.63	0.20
Thick carpet with underpadding	0.08	0.28	0.38	0.40	0.48	0.70	0.39
Typical ½-in-thick mineral-fiber acoustic ceiling tile	0.45	0.50	0.53	0.69	0.85	0.93	0.64
Typical ¾-in-thick glass-fiber acoustic ceiling tile	0.44	0.65	0.90	0.92	0.94	0.97	0.85

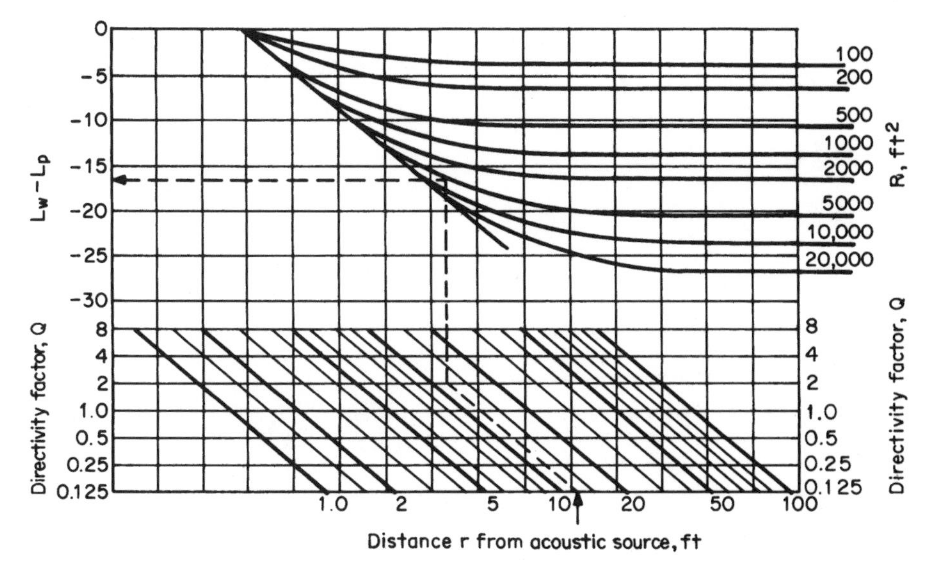

Figure 2.3.2 Relationship between distance from source, directivity factor, room constant, and $L_w - L_p$.

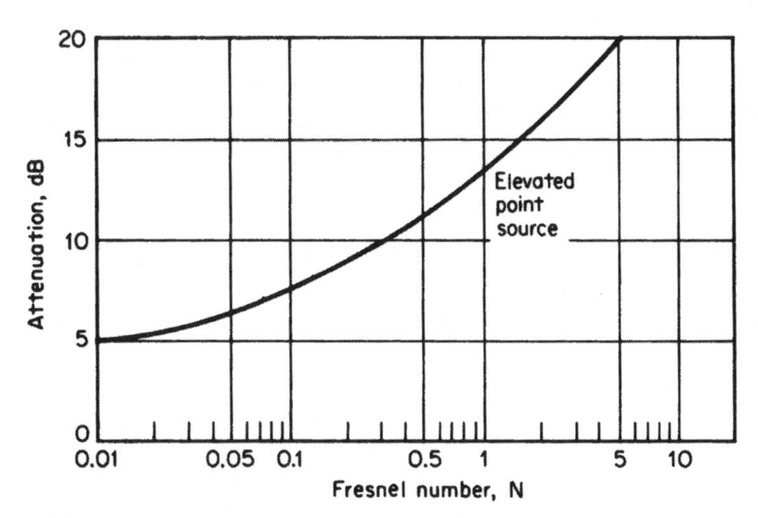

Figure 2.3.3 Noise reduction of a barrier for different Fresnel numbers.

Table 2.3.3 Typical STC and Transmission-Loss Data

Material	Transmission loss, dB						STC rating
	125 Hz	250 Hz	500 Hz	1000 Hz	2000 Hz	4000 Hz	
Gypsum wallboard, ½ in thick	14	20	24	30	30	27	27
Two layers in gypsum wallboard, both ½ in thick	19	26	30	30	29	36	31
Flat concrete panel, medium weight, 6 in thick	37	43	51	59	67	73	55
One layer of ½-in-thick gypsum wallboard on each side of 2- by 4-in wood studs (16 in off center) with 2-in-thick glass-fiber batt in the cavity	20	28	33	43	43	40	38
Same as above, but with two layers of ½-in-thick gypsum wallboard on each side	24	37	44	49	50	50	46
Same as above, but with staggered studs	34	43	49	54	54	52	51
Same as above, but with double row of 2- by 4-in studs spaced 1 in apart on separate plates, using type X (fire-rated) gypsum wallboard, and two layers of 3-in glass-fiber batt in the cavity	45	54	63	66	66	64	63
Same as above, but with bracing across cavity at third points of studs	40	45	56	62	57	60	57
4-in face brick, mortared together	31	33	39	47	55	61	45
Two layers of mortared 4-in face brick separated by 2-in air space, with metal ties	36	36	46	54	61	66	50
Same as above, but air space filled with concrete grout	41	47	56	62	66	70	59
6-in-thick three-cell dense concrete masonry units, mortared together	36	38	42	49	53	60	48
2-in-thick hollow-core door, ungasketed	13	19	23	18	17	21	19
2-in-thick solid wood door with airtight gasketing and drop seal	29	31	31	31	39	43	35
Typical window glass, ⅛ in thick, single plate	15	23	26	30	32	30	29
Typical thermal glazing window (3/16-in glass, ⅝-in air space, 3/16-in glass)	22	21	29	34	30	32	30
½-in-thick laminated glass	34	35	36	37	40	51	39
Composite window (½-in laminated glass, 5-in air space, ¼-in glass)	33	54	60	57	55	63	55
Typical ½-in-thick mineral-fiber acoustic ceiling tile	6	10	12	16	21	21	17

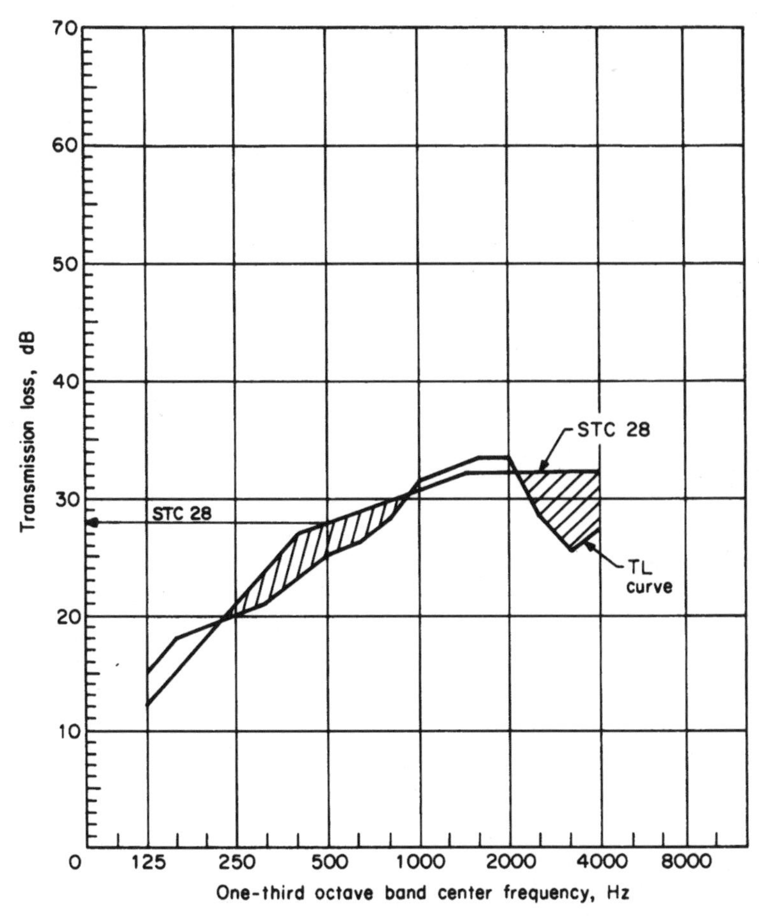

Figure 2.3.4 Sound transmission class (STC) for 1/2-in-thick gypsum wallboard. When a TL data point falls below the STC curve, the difference is termed a *deficiency.* The sum of all deficiencies must not be greater than 32 dB. No single deficiency shall be greater than 8 dB.

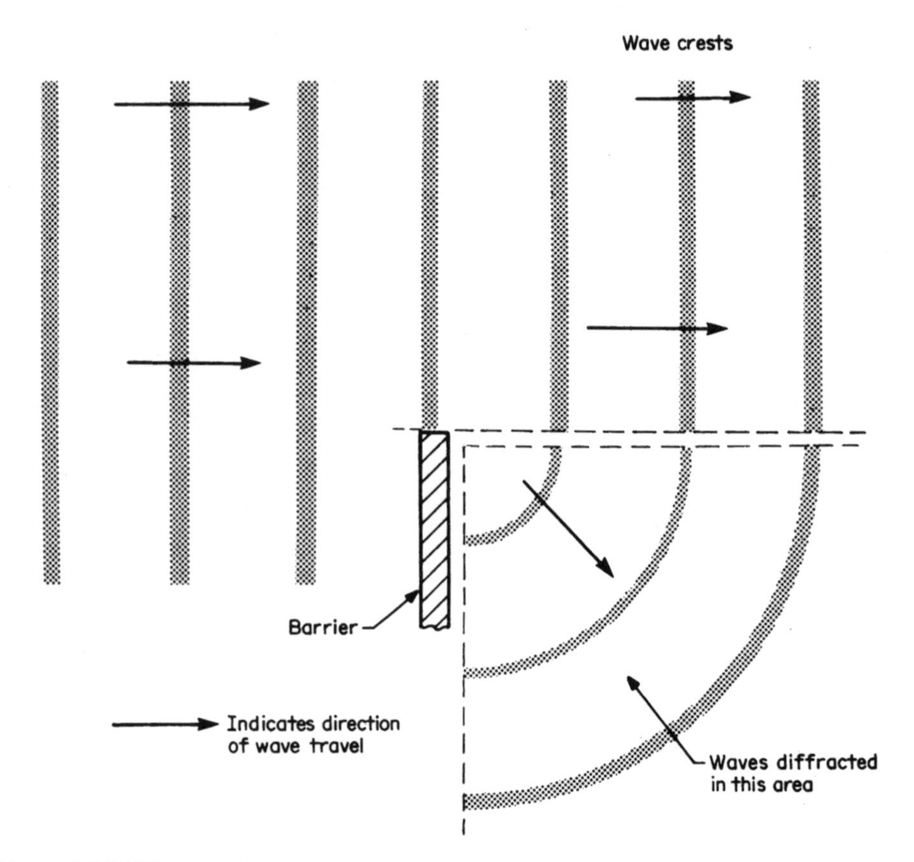

Figure 2.3.5 Diffraction of sound by a solid barrier.

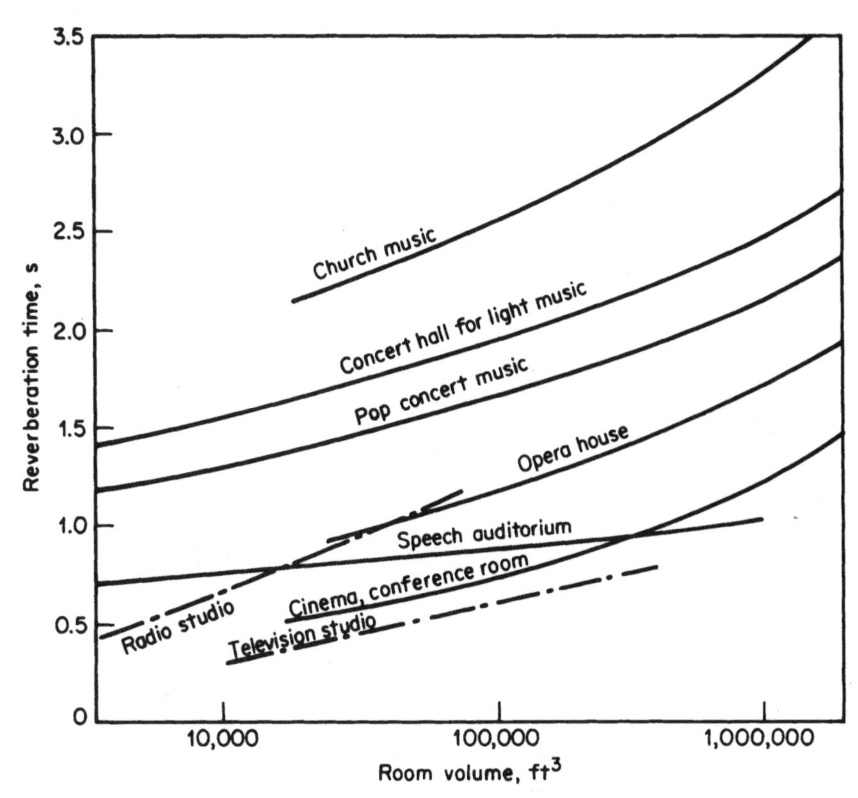

Figure 2.3.6 Typical reverberation times for different sizes of rooms and auditoria according to usage.

Table 2.3.4 Design Goals for Mechanical-System Noise Levels.

Type of area	NC range
Residences	
Single-family homes	20–30
Apartments and condominiums	25–35
Hotels and motels	
Guest rooms	30–40
Meeting or banquet rooms	30–35
Corridors and lobbies	35–45
Offices	
Boardrooms	20–30
Executive offices	30–40
Open-plan areas	35–40
Public circulation	35–45
Hospitals	
Private rooms	25–35
Operating rooms	30–40
Wards	30–40
Laboratories	30–40
Public circulation	35–45
Churches	
Sanctuaries	20–30
Public circulation	30–40
Schools and universities	
Libraries	30–40
Lecture rooms and classrooms	20–35
Laboratories	35–45
Cafeterias and recreation halls	35–45
Public buildings	
Libraries and museums	30–40
Courtrooms	25–35
Post offices and banks	30–40
Auditoria, theaters, and studios	
Concert halls	15–25
Recording studios	15–25
Multipurpose halls	20–30
TV studios	20–30
Movie theaters	30–35

Table 2.3.5 Limits of Frequency Passbands

Band number	Nominal center frequency	One-third-octave passband	Octave passband
14	25	22.4–28.2	
15	31.5	28.2–35.5	22.4–44.7
16	40	35.5–44.7	
17	50	44.7–56.2	
18	63	56.2–70.8	44.7–89.1
19	80	70.8–89.1	
20	100	89.1–112	
21	125	112–141	89.1–178
22	160	141–178	
23	200	178–224	
24	250	224–282	178–355
25	315	282–355	
26	400	355–447	
27	500	447–562	355–708
28	630	562–708	
29	800	708–891	
30	1,000	891–1,120	708–1410
31	1,250	1,120–1,410	
32	1,600	1,410–1,780	
33	2,000	1,780–2,240	1,410–2,820
34	2,500	2,240–2,820	
35	3,150	2,820–3,550	
36	4,000	3,550–4,470	2,820–5,620
37	5,000	4,470–5,620	
38	6,300	5,620–7,080	
39	8,000	7,080–8,910	5,620–11,200
40	10,000	8,910–11,200	

Table 2.3.6 Approximate Duct Attenuation

Duct type	Duct diameter, in	Approximate duct attenuation, dB/ft							
		63 Hz	125 Hz	250 Hz	500 Hz	1 kHz	2 kHz	4 kHz	8 kHz
Bare, rectangular or square	5–15	0.3	0.3	0.2	0.1	0.1	0.1	0.1	0.1
	16–45	0.3	0.2	0.1	0.1	0.1	0.1	0.1	0.1
	46–90	0.2	0.1	0.1	0.1	0.1	0.1	0.1	0.1
Bare, oval or round	5–15	0.15	0.15	0.1	0.05	0.05	0.05	0.05	0.05
	16–45	0.15	0.1	0.05	0.05	0.05	0.05	0.05	0.05
	46–90	0.1	0.05	0.05	0.05	0.05	0.05	0.05	0.05
1-in-thick lining, all duct shapes	5–15	0.4	0.5	1.2	2.5	5.0	4.5	3.0	1.5
	16–30	0.4	0.3	0.7	1.3	2.5	2.0	1.5	0.7
	31–45	0.4	0.3	0.3	0.8	1.4	1.2	0.9	0.5
	46–60	0.3	0.3	0.2	0.6	1.0	0.8	0.7	0.4
	61–75	0.3	0.3	0.2	0.4	0.8	0.6	0.5	0.3
	76–90	0.3	0.3	0.2	0.4	0.7	0.5	0.4	0.2
2-in-thick lining, all duct shapes	5–15	0.6	1.0	2.3	4.0	5.0	4.5	3.0	1.5
	16–30	0.4	0.4	1.0	2.0	2.5	2.1	1.5	0.7
	31–45	0.4	0.4	0.7	1.4	1.4	1.2	0.9	0.5
	46–60	0.3	0.3	0.5	0.9	1.0	0.8	0.7	0.4
	61–75	0.3	0.3	0.4	0.8	0.8	0.6	0.5	0.3
	76–90	0.3	0.3	0.3	0.7	0.7	0.5	0.4	0.2

Table 2.3.7 Approximate Elbow Acoustic Attenuation

90° elbows and turns	Duct diameter, in	Approximate elbow attenuation, dB/elbow							
		63 Hz	125 Hz	250 Hz	500 Hz	1 kHz	2 kHz	4 kHz	8 kHz
Unlined round elbow or square turn	5–30	0	0	0	0	1	2	3	3
	31–45	0	0	0	1	2	3	3	3
	46–90	0	0	1	2	3	3	3	3
Lined round elbow or square turn with turning vanes	5–30	0	0	0	1	2	3	4	5
	31–45	0	0	1	2	3	4	5	5
	46–90	0	1	2	3	4	5	5	6
Lined square turn without turning vanes	5–30	0	0	1	2	3	5	7	9
	31–45	0	1	2	3	5	7	9	11
	46–90	1	2	3	5	7	9	11	11

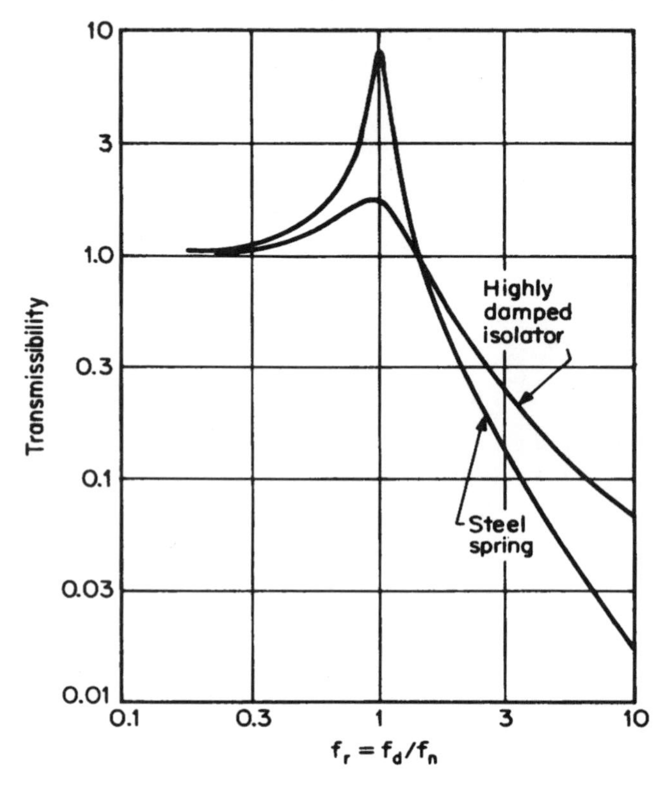

Figure 2.3.7 Transmissibility of a steel spring and a highly damped isolator.

2.4 Light, Vision, and Photometry

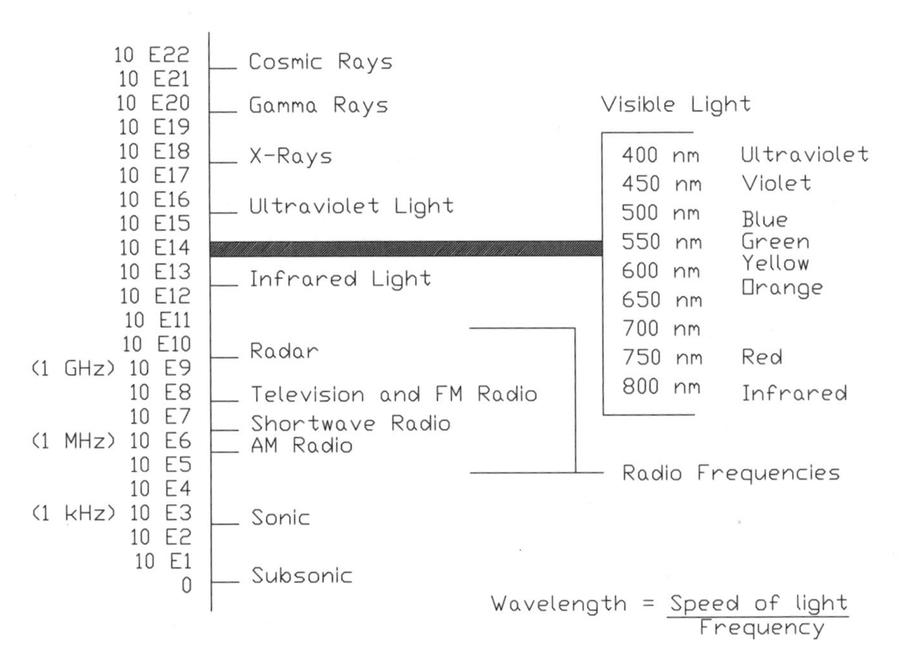

Figure 2.4.1 The electromagnetic spectrum.

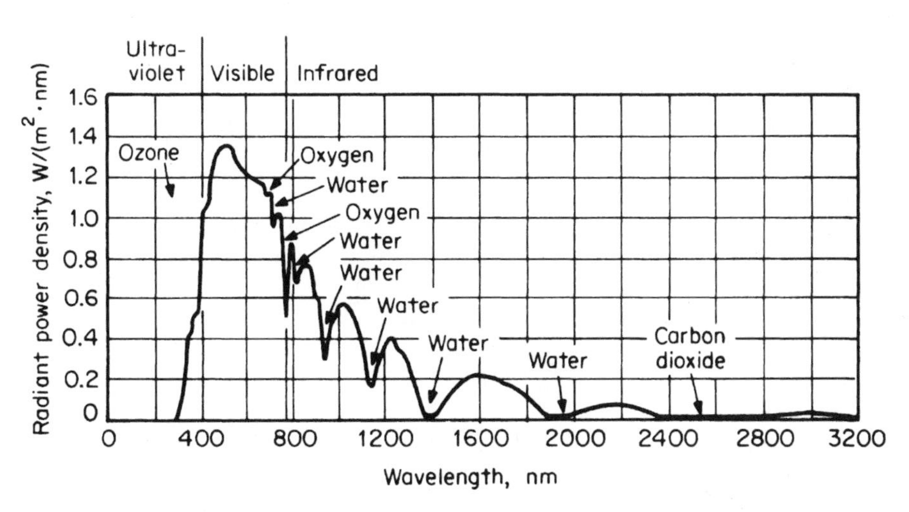

Figure 2.4.2 Spectral distribution of solar radiant power density at sea level, showing the ozone, oxygen, and carbon dioxide absorption bands. (*After*: IES Lighting Handbook, *Illuminating Engineering Society of North America, New York, N.Y., 1981.*)

Table 2.4.1 Psychophysical and Psychological Characteristics of Color

Psychophysical Properties	Psychological Properties
Dominant wavelength	Hue
Excitation purity	Saturation
Luminance	Brightness
Luminous transmittance	Lightness
Luminous reflectance	Lightness

Table 2.4.2 Typical Luminance Values (*After:* IES Lighting Handbook, *Illuminating Engineering Society of North America, New York, N.Y., 1981.*)

Parameter	Luminance, ft· L
Sun at zenith	4.28×10^8
Perfectly reflecting, diffusing surface in sunlight	9.29×10^3
Moon, clear sky	2.23×10^3
Overcast sky	$9–20 \times 10^2$
Clear sky	$6–17.5 \times 10^2$
Motion-picture screen	10

Table 2.4.3 Conversion Factors for Illuminance Units *(After: Fink, D. G.:* Television Engineering Handbook, *McGraw-Hill, New York, N.Y., 1957.*)

Parameter	Lux	Phot	Footcandle
Lux (meter-candle); lumens/m^2	1.00	1×10^{-4}	9.290×10^{-2}
Phot; lumens/cm^2	1×10^4	1.00	9.290×10^2
Footcandle; lumens/ft^2	1.076×10	1.076×10^{-3}	1.00
Multiply the quantity expressed in units *X* by the conversion factor to obtain the quantity in units of *Y*.			

Table 2.4.4 Relative Luminosity Values for Photopic and Scotopic Vision

Wavelength, nm	Photopic Vision	Scotopic Vision
390	0.00012	0.0022
400	0.0004	0.0093
410	0.0012	0.0348
420	0.0040	0.0966
430	0.0116	0.1998
440	0.023	0.3281
450	0.038	0.4550
460	0.060	0.5670
470	0.091	0.6760
480	0.139	0.7930
490	0.208	0.9040
500	0.323	0.9820
510	0.503	0.9970
520	0.710	0.9350
530	0.862	0.8110
540	0.954	0.6500
550	0.995	0.4810
560	0.995	0.3288
570	0.952	0.2076
580	0.870	0.1212
590	0.757	0.0655
600	0.631	0.0332
610	0.503	0.0159
620	0.381	0.0074
630	0.265	0.0033
640	0.175	0.0015
650	0.107	0.0007
660	0.061	0.0003
670	0.032	0.0001
680	0.017	0.0001
690	0.0082	
700	0.0041	
710	0.0021	
720	0.00105	
730	0.00052	
740	0.00025	
750	0.00012	
760	0.00006	

Table 2.4.5 Conversion Factors for Luminance and Retinal Illuminance Units (*After: Fink, D. G.:* Television Engineering Handbook, *McGraw-Hill, New York, N.Y., 1957.*)

Multiply Quantity Expressed in Units of X by Conversion Factor to Obtain Quantity in Units of Y

X \ Y	Candelas per square centimeter	Candelas per square meter	Candelas per square inch	Candelas per square foot	Lamberts	Millilamberts	Footlamberts	Trolands†‡
Candelas per square centimeter	1	1×10^4	6.452	9.290×10^2	3.142	3.142×10^3	2.919×10^3	7.854×10^3
Candelas per square meter (nit)§	1×10^{-4}	1	6.452×10^{-4}	9.290×10^{-2}	3.142×10^{-4}	3.142×10^{-1}	2.919×10^{-1}	7.854×10^{-1}
Candelas per square inch	1.550×10^{-1}	1.550×10^3	1	1.440×10^2	4.869×10^{-1}	4.869×10^2	4.524×10^2	1.217×10^3
Candelas per square foot	1.076×10^{-3}	1.076×10	6.944×10^{-3}	1	3.382×10^{-3}	3.382	3.142	8.454
Lamberts	3.183×10^{-1}	3.183×10^3	2.054	2.957×10^2	1	1×10^3	9.290×10^2	2.5×10^3
Millilamberts	3.183×10^{-4}	3.183	2.054×10^{-3}	2.957×10^{-1}	1×10^{-3}	1	9.290×10^{-1}	2.500
Footlamberts	3.426×10^{-4}	3.426	2.210×10^{-3}	3.183×10^{-1}	1.076×10^{-3}	1.076	1	2.691
Trolands‡	1.273×10^{-4}	1.273	8.213×10^{-4}	1.183×10^{-1}	4.000×10^{-4}	4.000×10^{-1}	3.716×10^{-1}	1

†In converting luminance to trolands it is necessary to multiply the conversion factor by the square of the pupil diameter in millimeters.
‡In converting trolands to luminance it is necessary to divide the conversion factor by the square of the pupil diameter in millimeters.

§As recommended at Session XII in 1951 of the International Commission on Illumination, one nit equals one candela per square meter.

Table 2.4.6 Luminous Outputs and Efficiencies

Type and wattage of lamp	Initial lumens output	Initial lumens per watt
Incandescent:		
40-W	465	11.7
60-W	835	13.9
100-W	1,630	16.3
200-W	3,650	18.3
300-W	5,900	19.6
500-W	9,950	19.9
1,000-W	21,500	21.5
1,000-W, spot	22,500	22.5
5,000-W, studio lighting	164,000	32.7
10,000-W, studio lighting	325,000	32.7
Fluorescent:†		
15-W	600	40
20-W	860	43
30-W	1,380	46
40-W	2,100	43
100-W	4,000	40

†Hot-cathode type, 4500 K white.

Table 2.4.7 Wavelengths of Various Radiations (in nanometers)

Spectrum region or line	Wavelength	Spectrum region or line	Wavelength
Ultraviolet	200–400	Peak of $\bar{x}$ distribution coefficient	600
Visible region (CIE limits)	380–780	Peak of $\bar{z}$ distribution coefficient	445
Visible region (practical limits)	400–700	Peak of $\bar{r}$ distribution coefficient	605
		Peak of $\bar{g}$ distribution coefficient	545
Violet region	400–450	Peak of $\bar{b}$ distribution coefficient	445
Blue region	450–490	Fraunhofer lines:	
Green region	490–550	*A*	766.1
Yellow region	550–590	*B* (oxygen)	687.0
Orange region	590–630	*C* (hydrogen)	656.28
Red region	630–700	D_1 (sodium)	589.59
Near infrared	700–3,000	D_3 (helium)	587.56
CIE red spectral primary	700	*E* (iron)	526.96
CIE green spectral primary	564.1	*F* (hydrogen)	486.14
		G (iron, calcium)	430.79
CIE blue spectral primary	435.8	*H*(calcium)	396.84
Peak of luminosity function ($\bar{y}$)	555	*K* (calcium)	393.36

Table 2.4.8 Light Conversion Factors

(Exact Conversions Are Shown in ***Boldface*** *Type. Repeating Decimals Are* *Underlined.)*

A. Luminance units. The SI unit of luminance is the candela per square meter (cd/m²)†

	Candelas per square meter (cd/m²)	Candelas per square foot (cd/ft²)	Candelas per square inch (cd/in²)	Apostilbs (asb)	Stilbs (sb)	Lamberts (L)	Footlamberts (fL)
1 candela per square meter =	**1**	**0.092 903 04**	**6.451 6** $\times$ **10**$^{-4}$	π = 3.141 592 65	**0.000 1**	**(0.000 1)** π = 3.141 592 65 $\times 10^{-4}$	0.291 863 51
1 candela per square foot =	10.763 910 4	**1**	**1/144** = 0.006 944 44	33.815 821 8	1.076 391 04 $\times 10^{-3}$	3.381 582 18 $\times 10^{-3}$	π = 3.141 592 65
1 candela per square inch =	1 550.003 1	**144**	**1**	4 869.478 4	0.155 000 31	0.486 947 84	452.389 342
1 apostilb =	**1/**π = 0.318 309 89	0.029 571 96	2.053 608 06 $\times 10^{-4}$	**1**	3.183 098 86 $\times 10^{-5}$	**0.000 1**	**0.092 903 04**
1 stilb =	**10 000**	**929.030 4**	**6.451 6**	31 415.926 5	**1**	π = 3.141 592 65	2 918.635
1 lambert =	**10 000/**π = 3 183.098 86	295.719 561	2.053 608 06	**10 000**	**1/**π = 0.318 309 89	**1**	**929.030 4**
1 footlambert =	3.426 259 1	**1/**π = 0.318 309 89	2.210 485 32 $\times 10^{-3}$	10.763 910 4	3.426 259 1 $\times 10^{-4}$	1.076 391 03 $\times 10^{-3}$	**1**

B. Illuminance units. The SI unit of illuminance is the lux (lux)‡

	Luxes (lx)	Phots (ph)	Footcandles (fc)	Lumens per square inch (lm/in²)
1 lux =	**1**	**0.000 1**	**0.092 903 04**	**6.451 6** $\times$ **10**$^{-4}$
1 phot =	**10 000**	**1**	**929.030 4**	**6.451 6**
1 footcandle =	10.763 910 4	1.076 391 04 $\times 10^{-3}$	**1**	**1/144** = 0.006 944 44
1 lumen per square inch =	1 550.003 1	0.155 000 31	**144**	**1**

†*Note:* 1 nit (nt) = 1 candela per square meter (cd/m²).
1 stilb (sb) = 1 candela per square centimeter (cd/cm²).
‡*Note:* 1 lux (lux) = **1** lumen per square meter (lm/m²)
1 phot (ph) = **1** lumen per square centimeter (lm/cm²)
1 footcandle (fc) = **1** lumen per square foot (lm/ft²)

Source: Fink and Beaty, *Standard Handbook for Electrical Engineers*, 11th ed., McGraw-Hill, New York, 1978.

2.5 Color Vision, Representation, and Reproduction

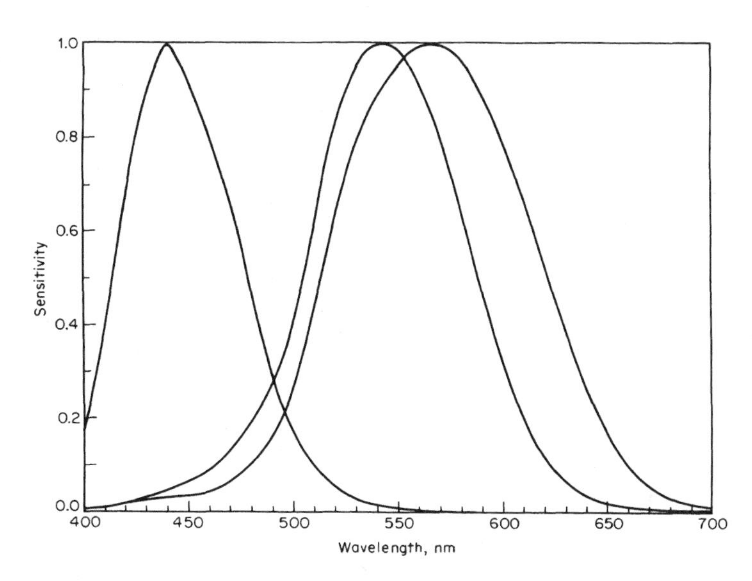

Figure 2.5.1 Spectral sensitivities of the three types of cones in the human retina. The curves have been normalized so that each is unity at its peak. (*After: Smith, V. C., and J. Pokorny: "Spectral Sensitivity of the Foveal Cone Pigments Between 400 and 500 nm,"* Vision Res., *vol. 15, pp. 161–171, 1975; and Boynton, R.M.:* Human Color Vision, *Holt, New York, N.Y., p. 404, 1979.*)

Table 2.5.1 Perceptual Terms and their Psychophysical Correlates

Perceptual (Subjective)	Psychophysical (Objective)
Hue	Dominant wavelength
Saturation	Excitation purity
Brightness	Luminance
Lightness	Luminous reflectance or luminous transmittance

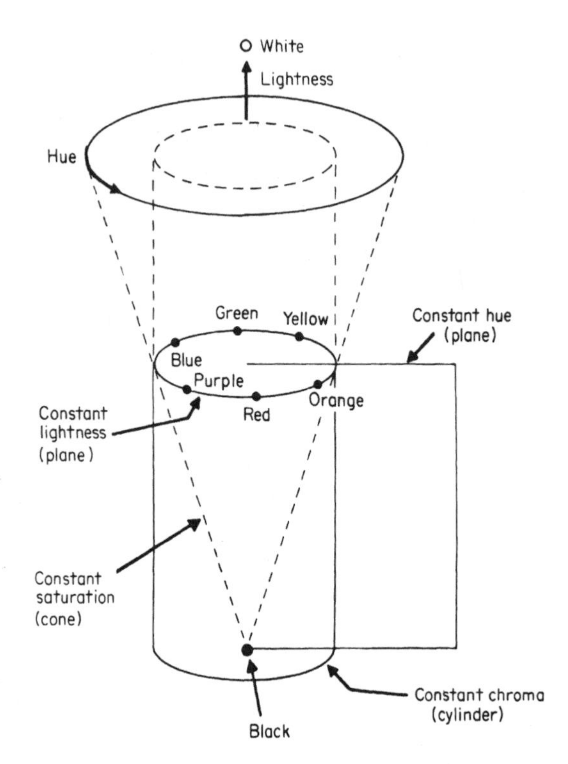

Figure 2.5.2 A geometrical model of perceptual color space for reflecting objects.

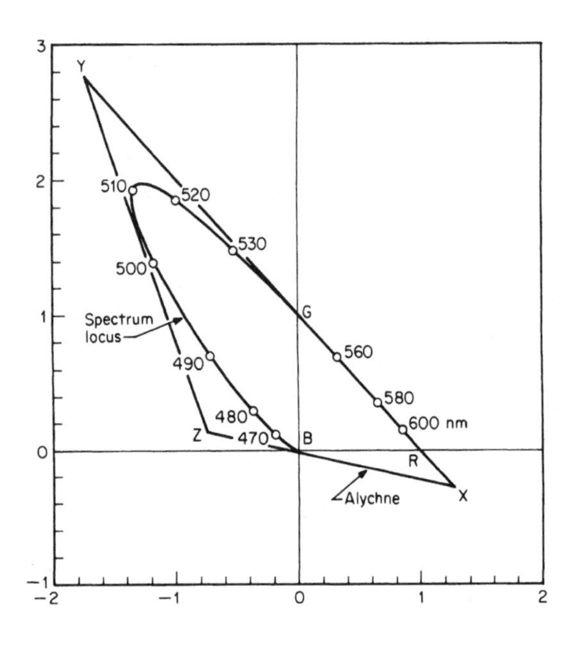

Figure 2.5.3 The spectrum locus and alychne of the CIE 1931 Standard Observer plotted in a chromaticity diagram based on matching stimuli of wavelengths 700.0, 546.1, and 435.8 nm. The locations of the CIE primary stimuli *X*, *Y*, and *Z* are shown.

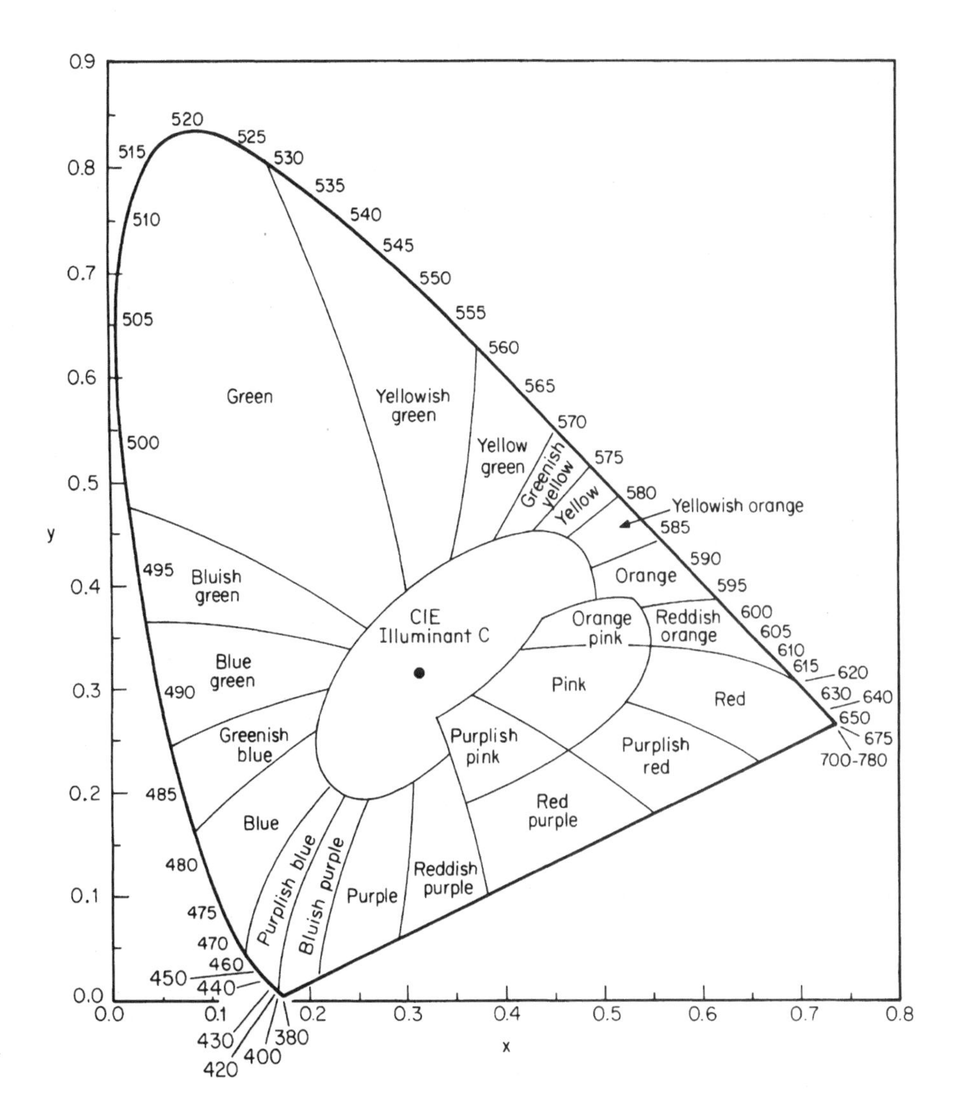

Figure 2.5.4 The CIE 1931 chromaticity diagram divided into various color names derived from observations of self-luminous areas against a dark background. (*After: Kelly, K. L.: "Color Designations for Lights,"* J. Opt. Soc. Am., *vol. 33, pp. 627–632, 1943.*)

Table 2.5.2 CIE Colorimetric Data (1931 Standard Observer)

Wave-length (mm)	Trichromatic Coefficients		Distribution Coefficients, Equal-energy Stimulus			Energy Distributions for Standard Illuminants			
	r	g	$\bar{r}$	$\bar{g}$	$\bar{b}$	E_A	E_B	E_C	E_{D65}
380	0.0272	–0.0115	0.0000	0.0000	0.0012	9.80	22.40	33.00	49.98
390	0.0263	–0.0114	0.0001	0.0000	0.0036	12.09	31.30	47.40	54.65
400	0.0247	–0.0112	0.0003	0.0001	0.0121	14.71	41.30	63.30	82.75
410	0.0225	–0.0109	0.0008	–0.0004	0.0371	17.68	52.10	80.60	91.49
420	0.0181	–0.0094	0.0021	–0.0011	0.1154	20.99	63.20	98.10	93.43
430	0.0088	–0.0048	0.0022	–0.0012	0.2477	24.67	73.10	112.40	86.68
440	–0.0084	0.0048	–0.0026	0.0015	0.3123	28.70	80.80	121.50	104.86
450	–0.0390	0.0218	0.0121	0.0068	0.3167	33.09	85.40	124.00	117.01
460	0.0909	0.0517	–0.0261	0.0149	0.2982	37.81	88.30	123.10	117.81
470	–0.1821	0.1175	–0.0393	0.0254	0.2299	42.87	92.00	123.80	114.86
480	–0.3667	0.2906	–0.0494	0.0391	0.1449	48.24	95.20	123.90	115.92
490	–0.7150	0.6996	–0.0581	0.0569	0.0826	53.91	96.50	120.70	108.81
500	1.1685	1.3905	0.0717	0.0854	0.0478	59.86	94.20	112.10	109.35
510	–1.3371	1.9318	–0.0890	0.1286	0.0270	66.06	90.70	102.30	107.80
520	–0.9830	1.8534	–0.0926	0.1747	0.0122	72.50	89.50	96.90	104.79
530	–0.5159	1.4761	0.0710	0.2032	0.0055	79.13	92.20	98.00	107.69
540	0.1707	1.1628	0.0315	0.2147	0.0015	85.95	96.90	102.10	104.41
550	0.0974	0.9051	0.0228	0.2118	–0.0006	92.91	101.00	105.20	104.05
560	0.3164	0.6881	0.0906	0.1970	–0.0013	100.00	102.80	105.30	100.00
570	0.4973	0.5067	0.1677	0.1709	–0.0014	107.18	102.60	102.30	96.33
580	0.6449	0.3579	0.2543	0.1361	0.0011	114.44	101.00	97.80	95.79
590	0.7617	0.2402	0.3093	0.0975	–0.0008	121.73	99.20	93.20	88.69
600	0.8475	0,1537	0.3443	0.0625	–0.0005	129.04	98.00	89.70	90.01
610	0.9059	0.0494	0.3397	0.0356	0.0003	136.35	98.50	88.40	89.60
620	0.9425	0.0580	0.2971	0.0183	–0.0002	143.62	99.70	88.10	87.70
630	0.9649	0.0354	0.2268	0.0083	–0.0001	150.84	101.00	88.00	83.29
640	0.9797	0.0205	0.1597	0.0033	0.000	157.98	102.20	87.80	83.70
650	0.9888	0.0113	0.1017	0.0012	0.0000	165.03	103.90	88.20	80.03
660	0.9940	0.0061	0.0593	0.0004	0.0000	171.96	105.00	87.90	80.21
670	0.9966	0.0035	0.0315	0.0001	0.0000	178.77	104.90	86.30	82.28
680	0.9984	0.0016	0.0169	0.0000	0.0000	185.43	103.90	84.00	78.28
690	0.9996	0.0004	0.0082	0.0000	0.0000	191.93	101.60	80.20	69.72
700	1.0000	0.0000	0.0041	0.0000	0.0000	198.26	99.10	76.30	71.61
710	1.0000	0.0000	0.0021	0.0000	0.0000	204.41	96.20	72.40	74.15
720	1.0000	0.0000	0.0011	0.0000	0.0000	210.36	92.90	68.30	61.60
730	1.0000	0.0000	0.0005	0.0000	0.0000	216.12	89.40	64.40	69.89
740	1.0000	0.0000	0.0003	0.0000	0.0000	221.67	86.90	61.50	75.09
750	1.0000	0.0000	0.0001	0.0000	0.0000	227.00	85.20	59.20	63.59
760	1.0000	0.0000	0.0001	0.0000	0.0000	232.12	84.70	58.10	46.42
770	1.0000	0.0000	0.0000	0.0000	0.0000	237.01	85.40	58.20	66.81
780	1.0000	0.0000	0.0000	0.0000	0.0000	241.68	87.00	59.10	63.38

Table 2.5.2 (continued)

Wave-length (mm)	Trichromatic Coefficients		Distribution Coefficients, Equal-energy Stimulus			Distribution Coefficients Weighted by Illuminant C		
	x	y	$\bar{x}$	$\bar{y}$	$\bar{z}$	$E_C\bar{x}$	$E_C\bar{y}$	$E_C\bar{z}$
380	0.1741	0.0050	0.0014	0.0000	0.0065	0.0036	0.0000	0.0164
390	0.1738	0.0049	0.0042	0.0001	0.0201	0.0183	0.0004	0.0870
400	0.1733	0,0048	0.0143	0.0004	0.0679	0.0841	0.0021	0.3992
410	0.1726	0.0048	0.0435	0.0012	0.2074	0.3180	0.0087	1.5159
420	0.1714	0.0051	0.1344	0.0040	0.6456	1.2623	0.0378	6.0646
430	0.1689	0.0069	0.2839	0.0116	1.3856	2.9913	0.1225	14.6019
440	0.1644	0.0109	0.3483	0.0230	1.7471	3.9741	0.2613	19.9357
450	0.1566	0.0177	0.3362	0.0380	1.7721	3.9191	0.4432	20.6551
460	0.1440	0.0297	0.2908	0.0600	1.6692	3.3668	0.6920	19.3235
470	0.1241	0.0578	0.1954	0.0910	1.2876	2.2878	1.0605	15.0550
480	0.0913	0.1327	0.0956	0.1390	0.8130	1.1038	1.6129	9.4220
490	0.0454	0.2950	0.0320	0.2080	0.4652	0.3639	2.3591	5.2789
500	0.0082	0.5384	0.0049	0.3230	0.2720	0.0511	3.4077	2.8717
510	0.0139	0.7502	0.0093	0.5030	0.1582	0.0898	4.8412	1.5181
520	0.0743	0.8338	0.0633	0.7100	0.0782	0.5752	6.4491	0.7140
530	0.1547	0.8059	0.1655	0.8620	0.0422	1.5206	7.9357	0.3871
540	0.2296	0.7543	0.2904	0.9540	0.0203	2.7858	9.1470	0.1956
550	0.3016	0.6923	0.4334	0.9950	0.0087	4.2833	9.8343	0.0860
560	0.3731	0.6245	0.5945	0.9950	0.0039	5.8782	9.8387	0.0381
570	0.4441	0.5547	0.7621	0.9520	0.0021	7.3230	9.1476	0.0202
580	0.5125	0.4866	0.9163	0.8700	0.0017	8.4141	7.9897	0.0147
590	0.5752	0.4242	1.0263	0.7570	0.0011	8.9878	6.6283	0.0101
600	0.6270	0.3725	1.0622	0.6310	0.0008	8.9536	5.3157	0.0067
610	0.6658	0.3340	1.0026	0.5030	0.0003	8.3294	4.1788	0.0029
620	0.6915	0.3083	0.8544	0.3810	0.0002	7.0604	3.1485	0.0012
630	0.7079	0.2920	0.6424	0.2650	0.0000	5,3212	2.1948	0.0000
640	0,7190	0.2809	0.4479	0.1750	0.0000	3.6882	1.4411	0.0000
650	0.7260	0.2740	0.2835	0.1070	0.0000	2.3531	0.8876	0.0000
660	0.7300	0.2700	0.1649	0,0610	0.0000	1.3589	0.5028	0.0000
670	0.7320	0.2680	0.0874	0.0320	0.0000	0.7113	0.2606	0.0000
680	0.7334	0.2666	0.0468	0.0170	0.0000	0.3657	0.1329	0.0000
690	0.7344	0.2656	0.0227	0.0082	0.0000	0.1721	0.0621	0.0000
700	0.7347	0.2653	0.0114	0.0041	0.0000	0.0806	0.0290	0.0000
710	0.7347	0,2653	0.0058	0.0021	0.0000	0.0398	0.0143	0.0000
720	0.7347	0.2653	0.0029	0.0010	0.0000	0.0183	0.0064	0.0000
730	0.7347	0.2653	0.0014	0.0005	0.0000	0.0085	0.0030	0.0000
740	0.7347	0.2653	0.0007	0.0003	0.0000	0.0040	0.0017	0.0000
750	0.7347	0.2653	0.0003	0.0001	0.0000	0.0017	0.0006	0.0000
760	0.7347	0.2653	0.0002	0.0001	0.0000	0.0008	0.0003	0.0000
770	0.7347	0.2653	0.0001	0.0000	0.0000	0.0003	0.0000	0.0000
780	0.7347	0.2653	0.0000	0.0000	0.0000	0.0000	0.000	0.0000

Table 2.5.3 Chromaticity Coordinates (Trichromatic Coefficients) of Standard Illuminants, Other Light Sources, and Primaries

Designation	x	y	z
Illuminant A (2854 K)	0.4476	0.4075	0.1450
Illuminant B (4870 K)	0.3484	0.3516	0.3000
Illuminant C (6770 K)	0.3101	0.3162	0.3738
Illuminant D6500 (6500 K)	0.3127	0.3291	0.3582
Illuminant E (equal-energy white)	0.3333	0.3333	0.3333
Planckian radiator (black body):			
2000 K	0.5266	0.4133	0.0601
2848 K	0.4475	0.4075	0.1450
3500 K	0.4052	0.3907	0.2041
5000 K	0.3450	0.3516	0.3034
7500 K	0.3003	0.3103	0.3894
10,000 K	0.2806	0.2883	0.4311
Sunlight	0.336	0.350	0.314
Average daylight	0.313	0.328	0.359
North-sky daylight	0.277	0.293	0.430
Zenith sky	0.263	0.278	0.459
White-flame carbon arc	0.315	0.332	0.353
Fluorescent lamp (4500 K)	0.539	0.363	0.278
Fluorescent lamp (3500 K)	0.404	0.396	0.200
CIE standard red primary (700 nm)	0.7347	0.2653	0.0000
CIE standard green primary (546.1 nm)	0.2738	0.7174	0.0088
CIE standard blue primary (435.6 nm)	0.1666	0.0089	0.8245
NTSC primaries:			
Red	0.67	0.33	0.00
Green	0.21	0.71	0.08
Blue	0.14	0.08	0.78

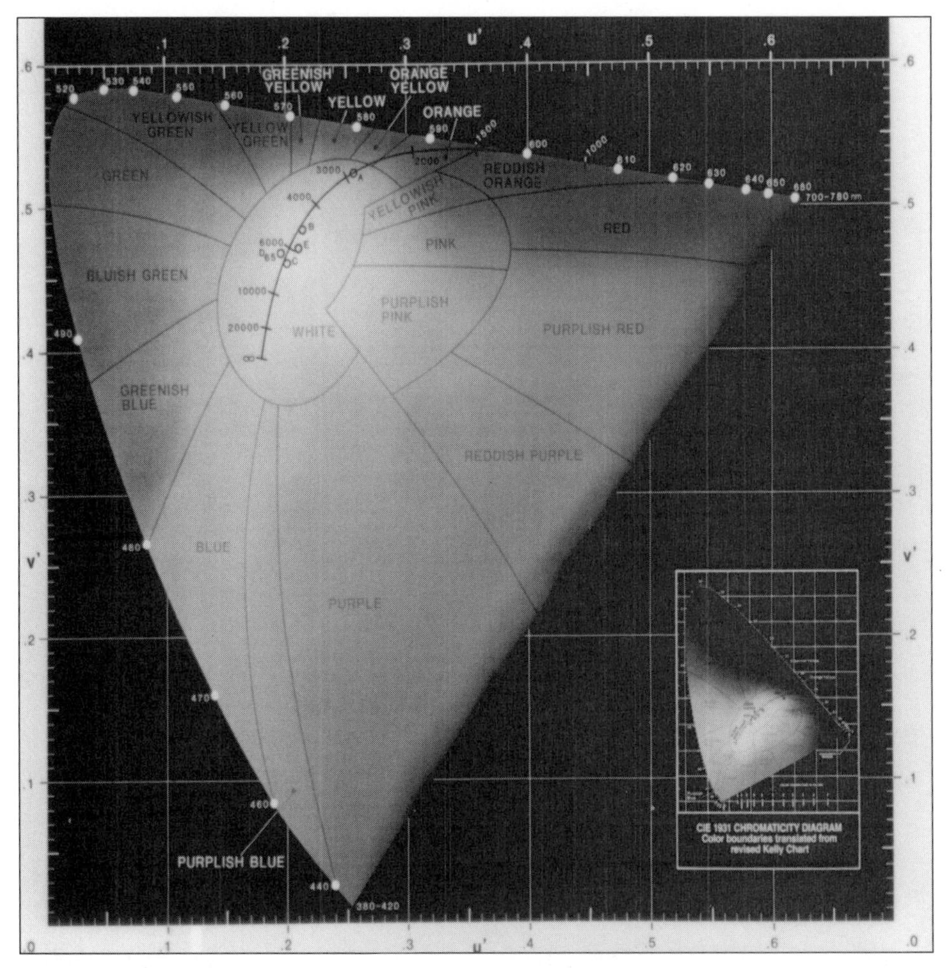

Figure 2.5.5 The 1976 CIE UCS diagram. The *u′* , *v′* chromaticity coordinates for any real color are located within the bounds the horse-shoe-shaped spectrum locus and the line of purples that joins the spectrum ends. (*Courtesy of Photo Research.*)

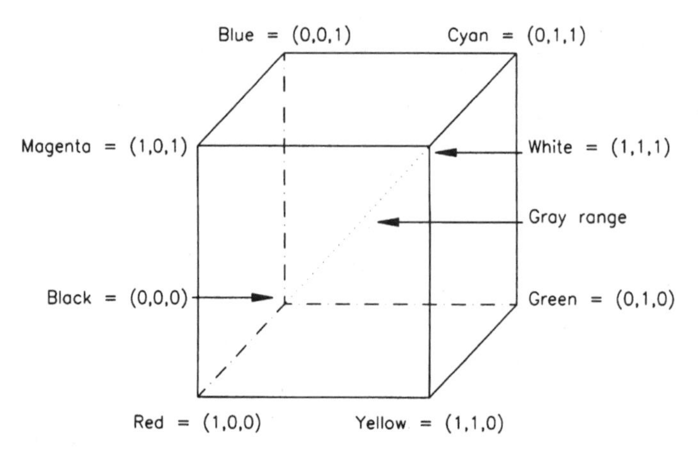

Figure 2.5.6 The RGB color model cube. (*After: Foley, James D., et al.:* Computer Graphics: Principles and Practice, *2nd ed., Addison-Wesley, Reading, Mass., pp. 584–592, 1991.*)

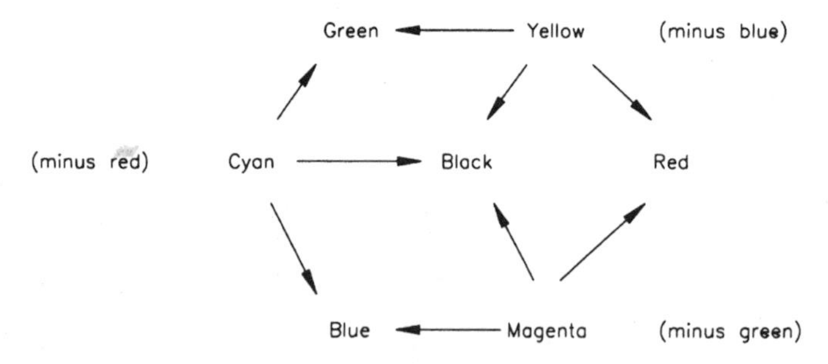

Figure 2.5.7 The CMY color model primaries and their mixtures. (*After: Foley, James D., et al.:* Computer Graphics: Principles and Practice, *2nd ed., Addison-Wesley, Reading, Mass., pp. 584–592, 1991.*)

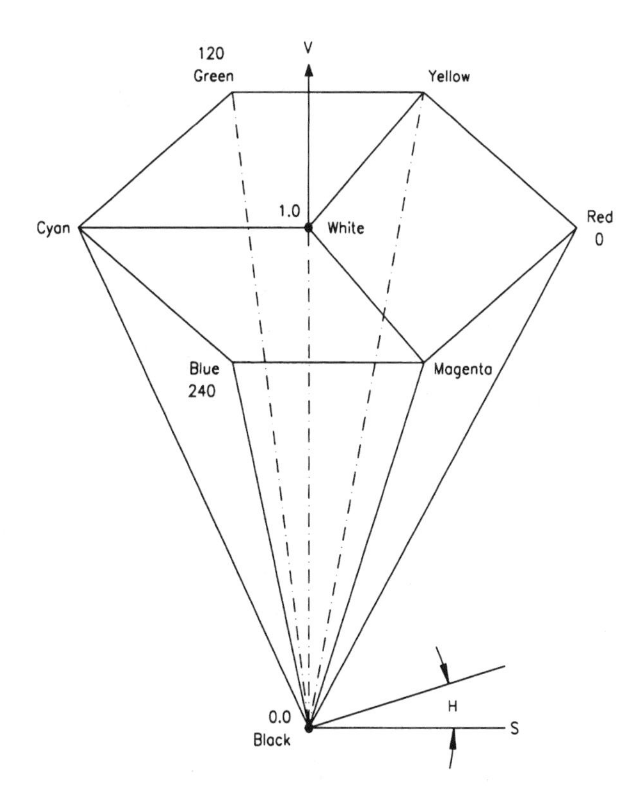

Figure 2.5.8 The single-hexcone HSV color model. The $V = 1$ plane contains the RGB model $R = 1$, $G = 1$, and $B = 1$ planes in the regions illustrated. (*After: Smith, A. R.: "Color Gamut Transform Pairs,"* SIGGRAPH 78, *12–19, 1978.*)

Table 2.5.4 Video and Sync Levels in IRE Units

Signal Level	IRE Level
Reference white	100
Color burst sine wave peak	+20 to –20
Reference black	7.5
Blanking	0
Sync level	–40

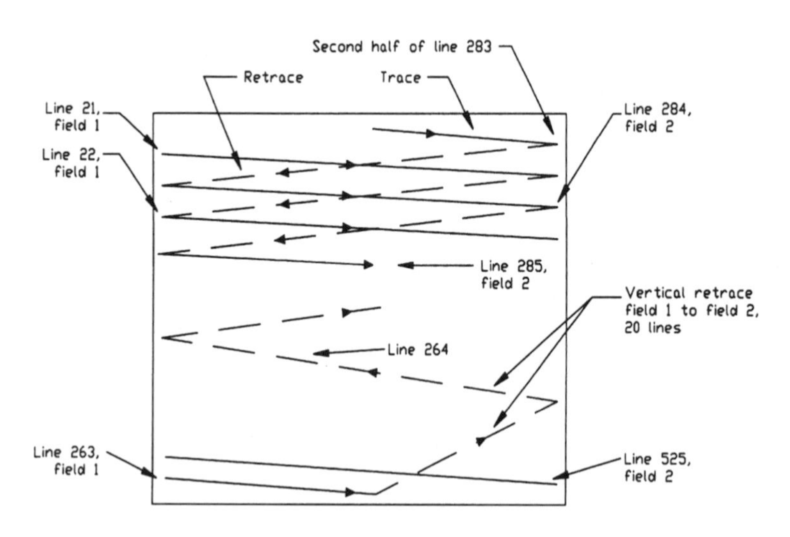

Figure 2.5.9 The interlace scanning pattern (raster) of the television image.

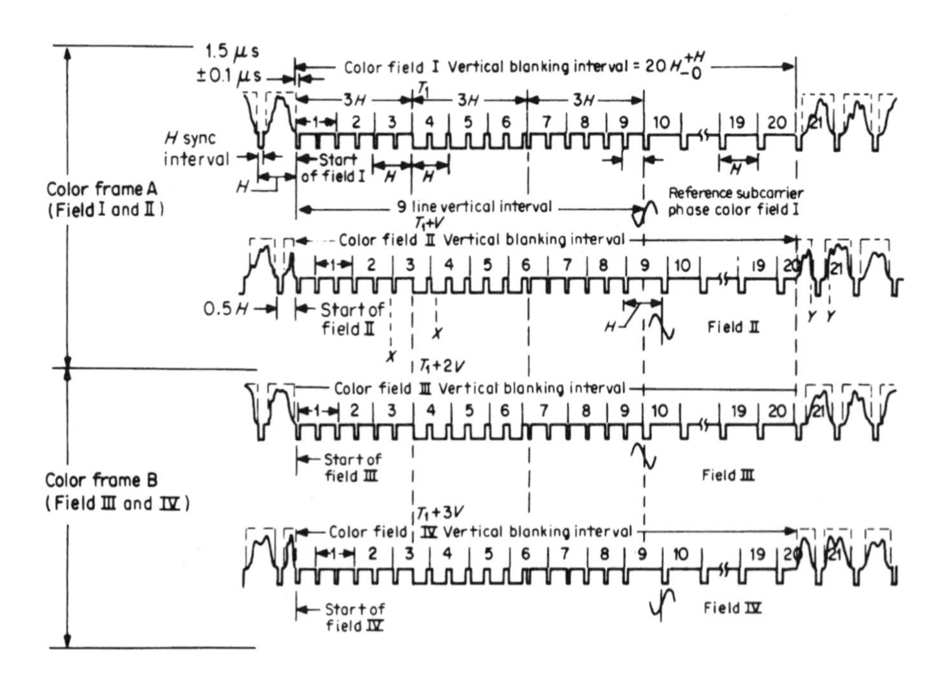

Figure 2.5.10 Principle components of the NTSC color television waveform. (*Source: Electronic Industries Association.*)

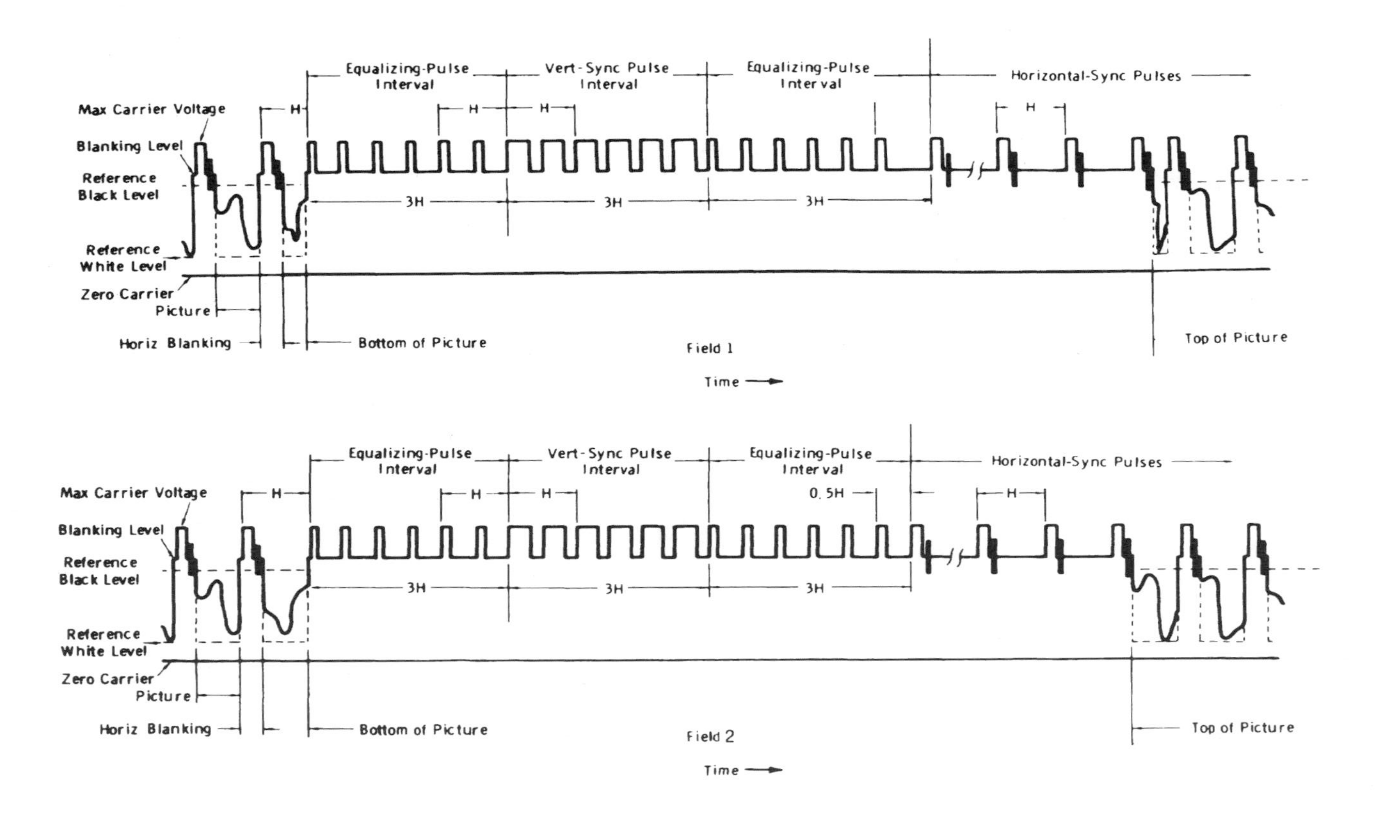

Figure 2.5.11 Detail of picture elements of the NTSC color television waveform. (*Source: Electronic Industries Association.*)

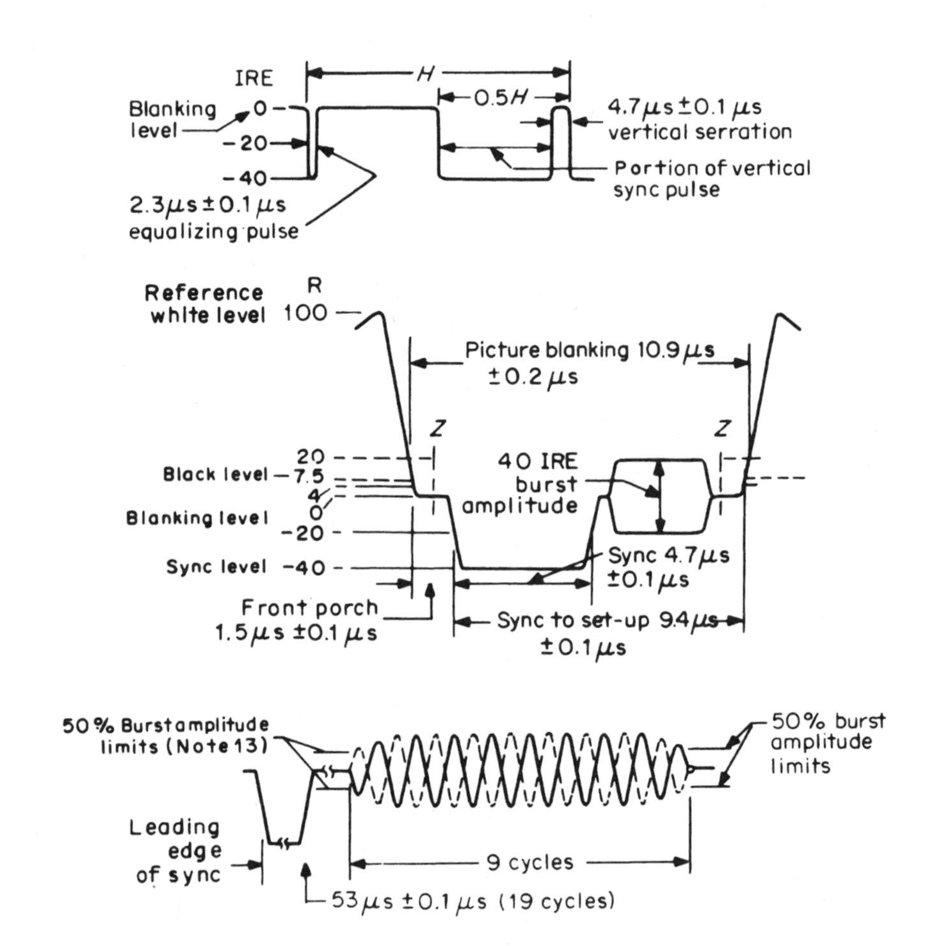

Figure 2.5.12 Detail of sync and color subcarrier pulse widths for the NTSC system. (*Source: Electronic Industries Association.*)

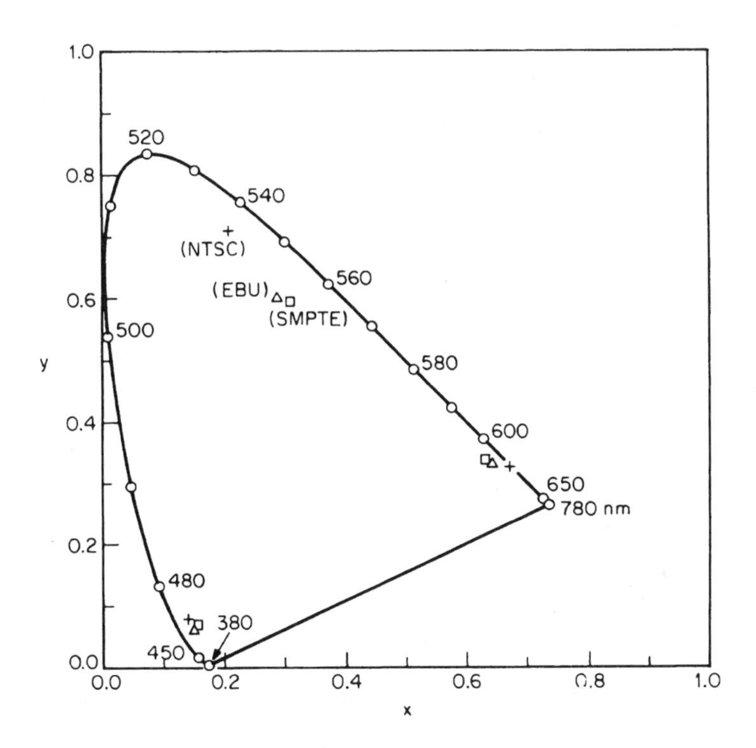

Figure 2.5.13 CIE 1931 chromaticity diagram showing three sets of phosphors used in color television displays.

Table 2.5.5 Relative Flicker Threshold for Various Luminances

(The luminances tabulated have only relative significance; they are based on a value of 180 ft·L for a flicker frequency of 60 Hz, which is typical performance under NTSC.)

Flicker Frequency (Hz)	System	Frames/sec	Flicker Threshold Luminance (ft·L)
48	Motion pictures	24	20
50	Television scanning	25	29
60	Television scanning	30	180

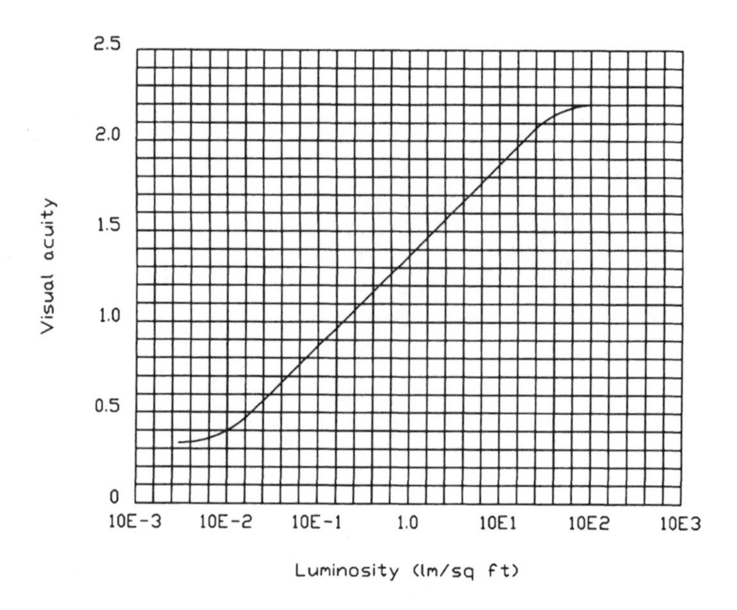

Figure 2.5.14 Visual acuity (the ability to resolve details of an image) as a function of the luminosity to which the eye is adapted. (*After R. J. Lythgoe.*)

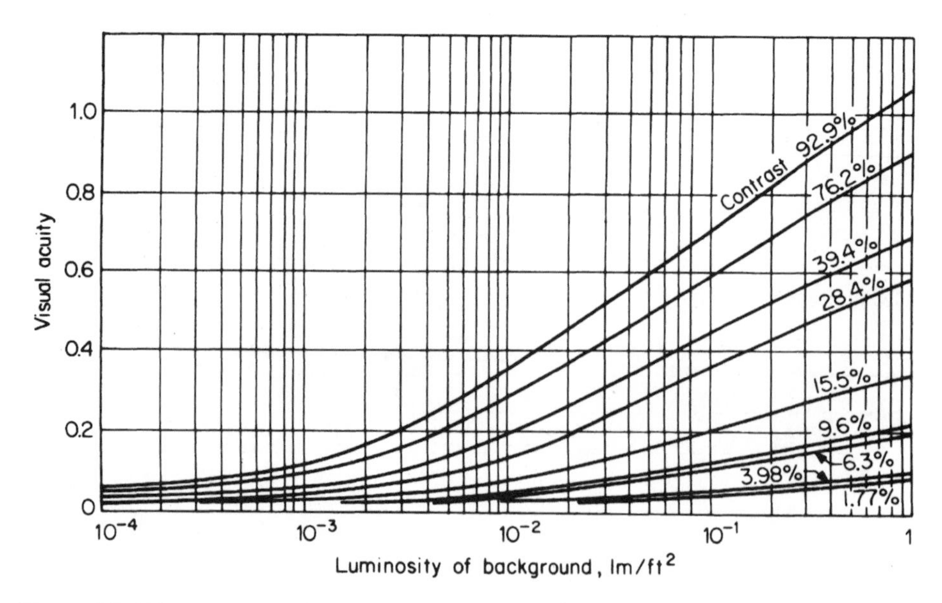

Figure 2.5.15 Visual acuity of the human eye as a function of luminosity and contrast (experimental data). (*After J. P. Conner and R. E. Ganoung.*)

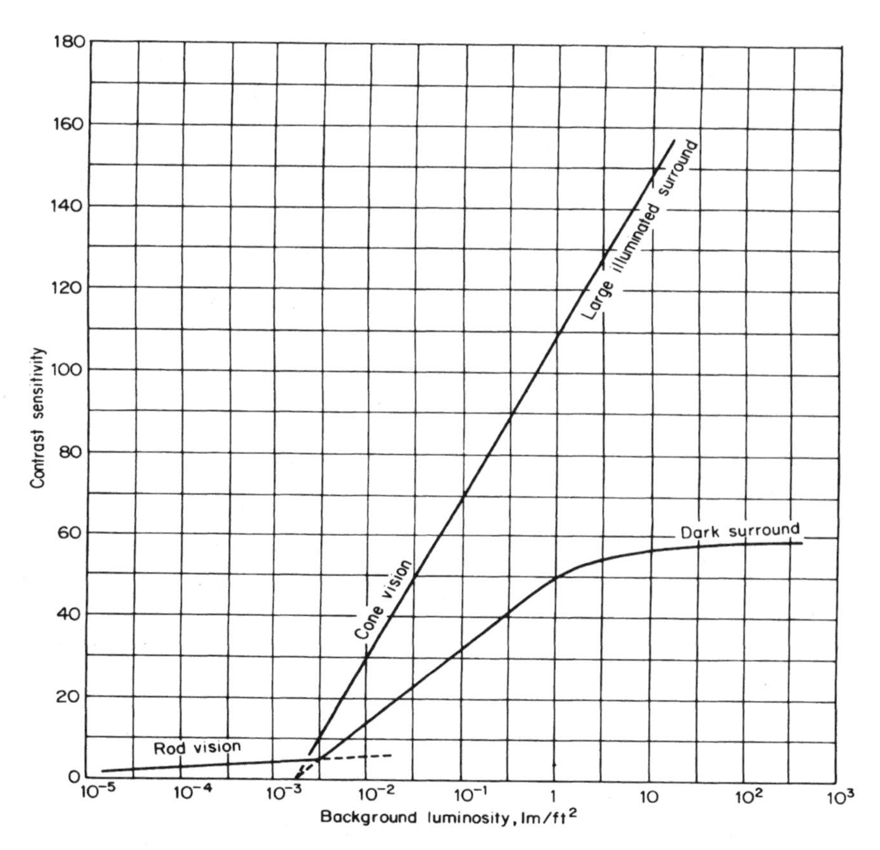

Figure 2.5.16 Contrast sensitivity as a function of background luminosity. (*After P. H. Moon.*)

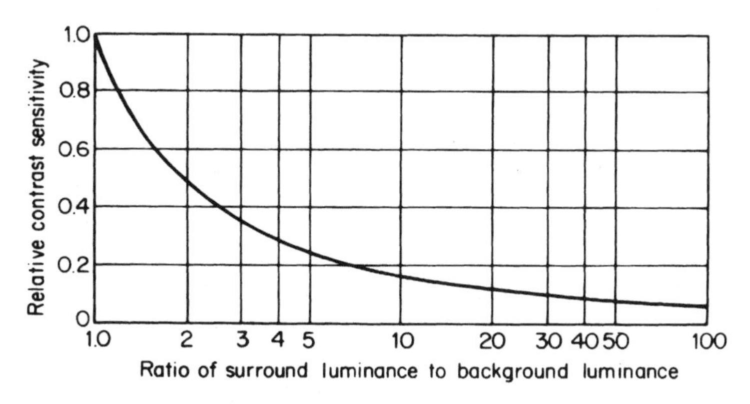

Figure 2.5.17 The effects of surround luminance on contrast sensitivity of the human eye. (*After P. H. Moon.*)

Table 2.5.6 Differing Colorimetry Standards used in Video Production (*From: Robin, Michael, and Michel Poulin:* Digital Television Fundamentals, *McGraw-Hill, New York, N.Y., 1997. Used with permission.*)

Standards	Color primaries: Primary	X	Y	Transfer characteristics (gamma)	Matrix coefficients
ITU-R BT.709	Red	640	330	$V = 1.099Lc^{0.45} - 0.099$	$E'_Y = 0.7152\,E'_G + 0.0722\,E'_B + 0.2126\,E'_R$
SMPTE 295M	Green	300	600	for $1 \geq Lc \geq 0.018$	$E'_{Pb} = -0.386\,E'_G + 0.500\,E'_B - 0.115\,E'_R$
	Blue	150	060	$V = 4.5Lc$	$E'_{Pr} = -0.454\,E'_G - 0.046\,E'_B + 0.500\,E'_R$
	White (D65)	312.7	329	for $0.018 > Lc \geq 0$	
ITU Rec. 624-system B, G		640	330	Gamma = 2.8	$E'_Y = 0.587\,E'_G + 0.114\,E'_B + 0.299\,E'_R$
		290	600		$E'_{Pb} = -0.331\,E'_G + 0.500\,E'_B - 0.169\,E'_R$
		150	060		$E'_{Pr} = -0.419\,E'_G - 0.081\,E'_B + 0.500\,E'_R$
		313	329		
SMPTE 170M (NTSC-1995)		630	340	$V = 1.099Lc^{0.45} - 0.099$	$E'_Y = 0.587\,E'_G + 0.114\,E'_B + 0.299\,E'_R$
		310	595	for $1 \geq Lc \geq 0.018$	$E'_{Pb} = 0.331\,E'_G + 0.500\,E'_B + 0.169\,E'_R$
		155	070	$V = 4.5Lc$	$E'_{Pr} = -0.419\,E'_G + 0.081\,E'_B + 0.500\,E'_R$
		312.7	329	for $0.018 > Lc \geq 0$	
ITU Rec. 624-system M		670	330	Gamma = 2.2	$E'_Y = 0.59\,E'_G + 0.11\,E'_B + 0.30\,E'_R$
NTSC-1953		210	710		$E'_{Pb} = -0.331\,E'_G + 0.500\,E'_B - 0.169\,E'_R$
(FCC)		140	080		$E'_{Pr} = -0.421\,E'_G + 0.079\,E'_B + 0.500\,E'_R$
		310	316		
SMPTE 240M		630	340	$V = 1.1115Lc^{0.45} - 0.1115$	$E'_Y = 0.701\,E'_G + 0.087\,E'_B + 0.212\,E'_R$
SMPTE 260M		310	595	for $Lc \geq 0.0228$	$E'_{Pb} = -0.384\,E'_G + 0.500\,E'_B - 0.116\,E'_R$
		155	070	$V = 4.0Lc$	$E'_{Pr} = -0.445\,E'_G + 0.055\,E'_B - 0.500\,E'_R$
		312.7	329	for $0.0228 > Lc$	

NOTE: The SMPTE 274M standards (1920×1080, Scanning and Interface for Multiple Picture Rates) specifies two reference primary chromaticities and white values. The ITU BT-709 standard is recommended for future generations of signal-generation equipment and the SMPTE 240M standard is reserved for interim implementations of currently used equipment.

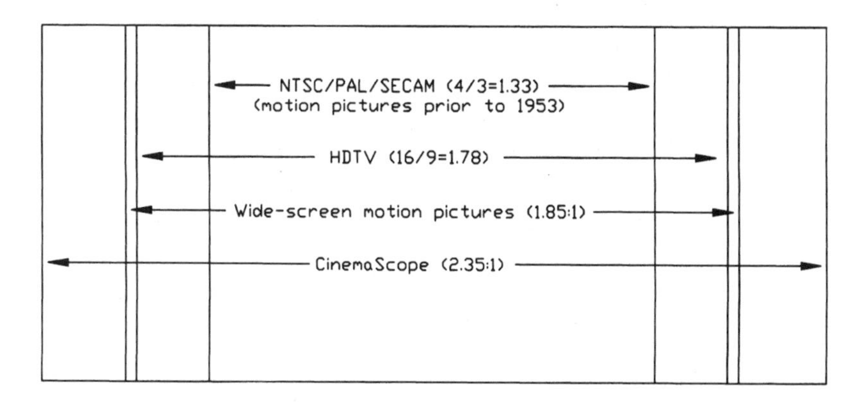

Figure 2.5.18 Comparison of the aspect ratios of television and motion pictures.

2.6 Optical Components and Systems

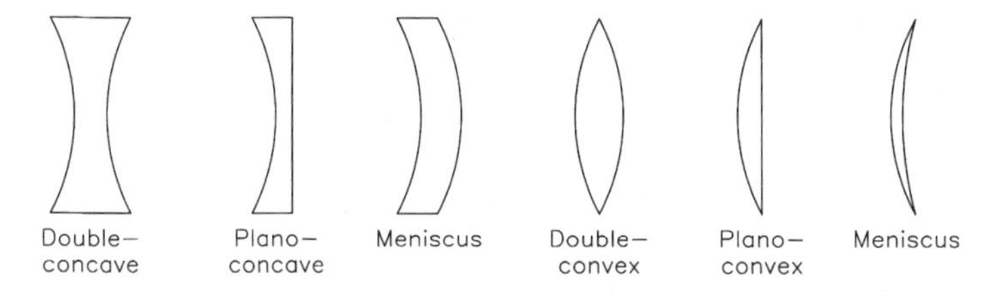

Figure 2.6.1 Various forms of simple converging and diverging lenses. (*After: Fink, D. G. (ed.):* Television Engineering Handbook, *McGraw-Hill, New York, N.Y., 1957.*)

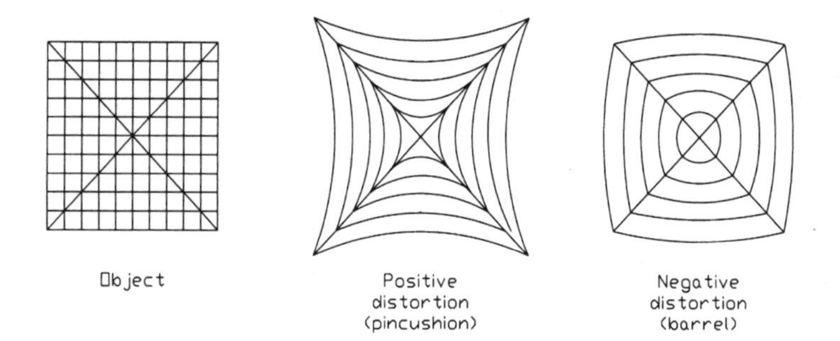

Figure 2.6.2 Pincushion and barrel distortion in the image of a lens system. (*After: Fink, D. G. (ed.):* Television Engineering Handbook, *McGraw-Hill, New York, N.Y., 1957.*)

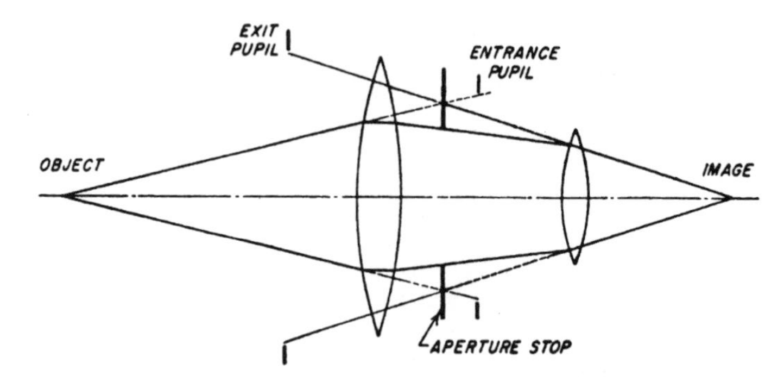

Figure 2.6.3 Aperture stop, entrance pupil, and exit pupil for a lens system. (*After: Fink, D. G. (ed.):* Television Engineering Handbook, *McGraw-Hill, New York, N.Y., 1957.*)

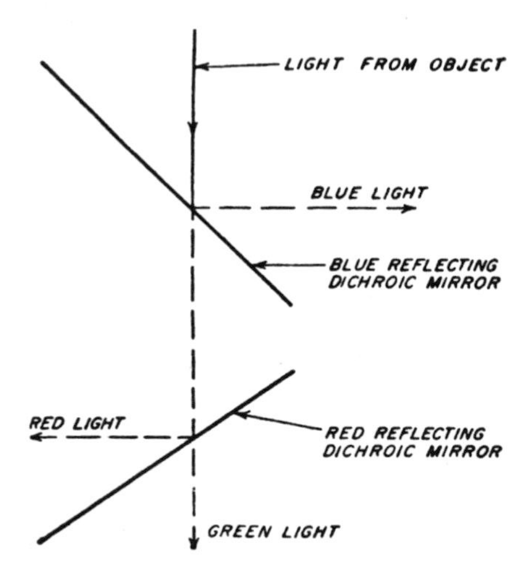

Figure 2.6.4 Arrangement of a dichroic mirror beam-splitting system. (*After: Fink, D. G. (ed.):* Television Engineering Handbook, *McGraw-Hill, New York, N.Y., 1957.*)

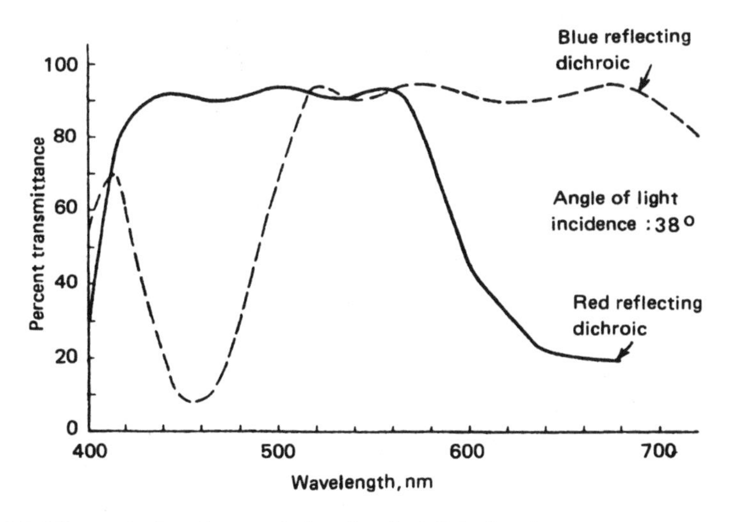

Figure 2.6.5 Transmission characteristics of typical dichroic mirrors.

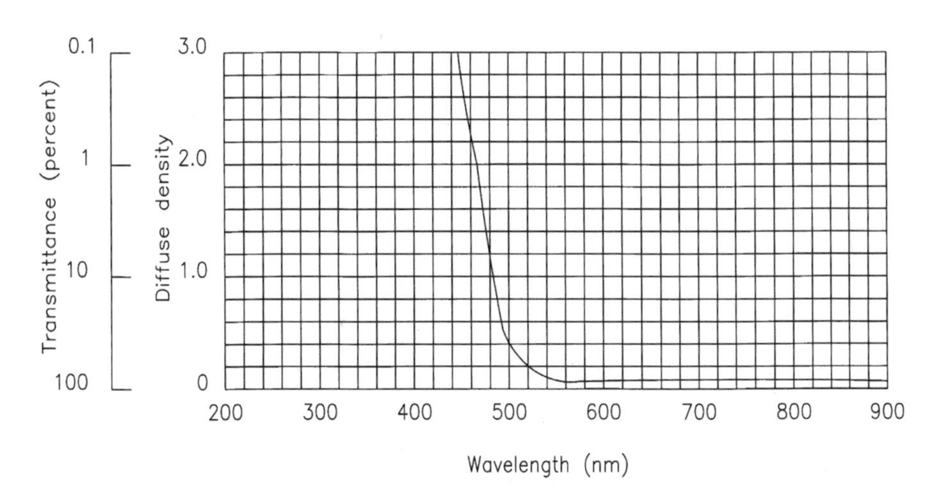

Figure 2.6.6 Response curve of a spectral trim filter (yellow). (*Source: Eastman Kodak Company.*)

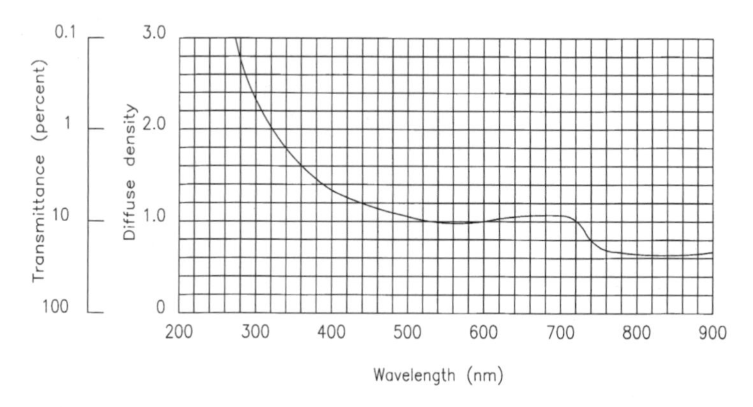

Figure 2.6.7 Response curve of a neutral density filter (visible spectrum). (*Source: Eastman Kodak Company.*)

2.7 Digital Coding of Audio/Video Signals

Table 2.7.1 Binary Values of Amplitude Levels for 8-Bit Words (*From: Benson, K. B., and D. G. Fink: "Digital Operations in Video Systems,"* HDTV: Advanced Television for the 1990s, *McGraw-Hill, New York, pp. 4.1–4.8, 1990. Used with permission.*)

Amplitude	Binary Level	Amplitude	Binary Level	Amplitude	Binary Level
0	00000000	120	01111000	240	11110000
1	00000001	121	01111001	241	11110001
2	00000010	122	01111010	242	11110010
3	00000011	123	01111011	243	11110011
4	00000100	124	01111100	244	11110100
5	00000101	125	01111101	245	11110101
6	00000110	126	01111110	246	11110110
7	00000111	127	01111111	247	11110111
8	00001000	128	10000000	248	11111000
9	00001001	129	10000001	249	11111001
10	00001010	130	10000010	250	11111010
11	00001011	131	10000011	251	11111011
12	00001100	132	10000100	252	11111100
13	00001101	133	10000101	253	11111101
14	00001110	134	10000110	254	11111110
15	00001111	135	10000111	255	11111111

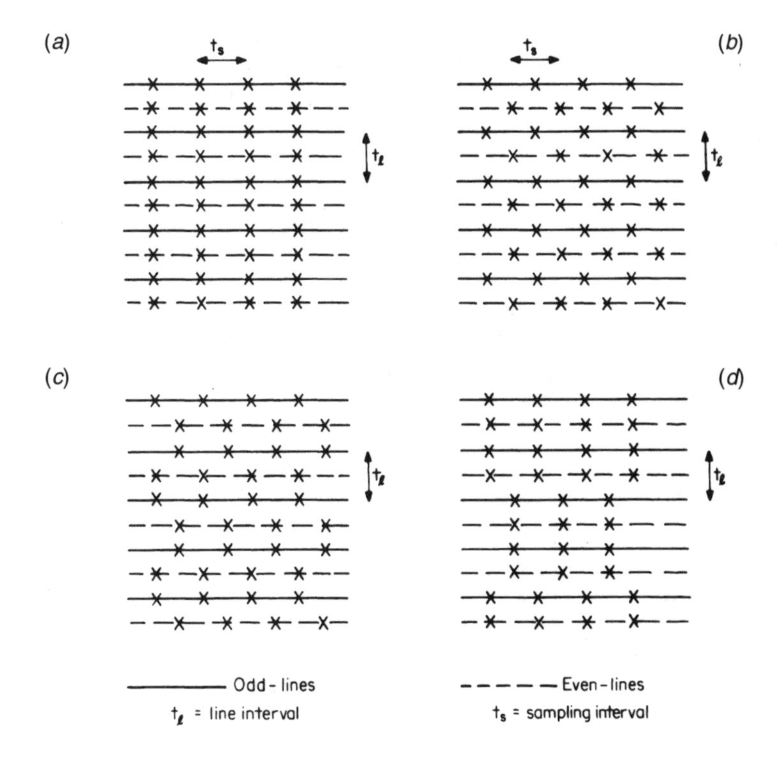

Figure 2.7.1 Rectangular sampling patterns: (*a*) rectangular field-aligned; (*b*) rectangular field-offset; (*c*) checkerboard; (*d*) field-aligned, double checkerboard.

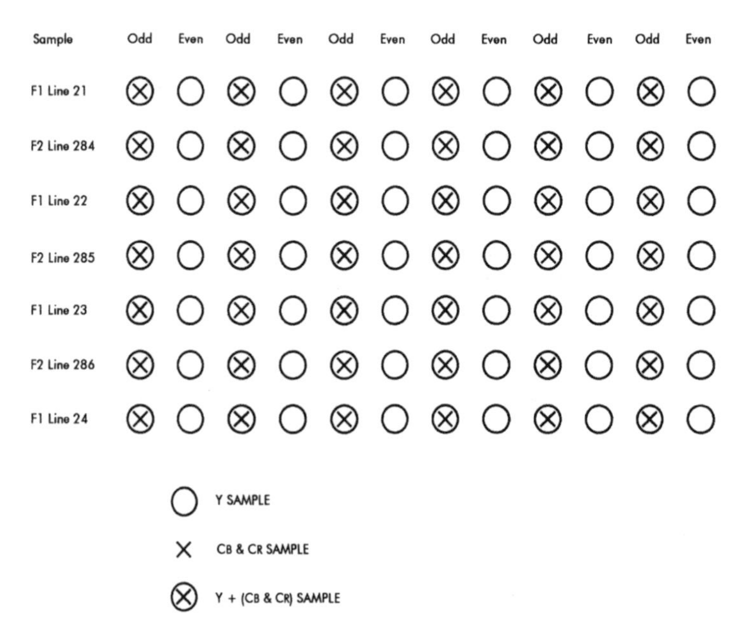

Figure 2.7.2 Details of 525/60 scanning standard line and field repetitive 4:2:2 orthogonal sampling structure showing the position of cosited $Y/C_B/C_R$ samples and isolated Y samples. (*From: Robin, Michael, and Michel Poulin:* Digital Television Fundamentals, *McGraw-Hill, New York, N.Y., 1997. Used with permission.*)

Table 2.7.2 Representative Sampling of Converter Operating Parameters

Converter Type	Sampling Rate	Resolution	S/Nq[1]	Max. Input Frequency	Power Consumption
A/D	400 Ms/s[2]	8 bits	43 dB	1 GHz	3 W
	200 Ms/s	10 bits	58 dB	400 MHz	2 W
	120 Ms/s	12 bits	70 dB	350 MHz	1 W
	70 Ms/s	14 bits	75 dB	300 MHz	1.3 W
D/A	**Sampling Rate**	**Resolution**	**Dynamic Range**		**Power Consumption**
	500 Ms/s[3]	10 bits	80 dB		250 mW
	300 Ms/s	12 bits	85 dB		300 mW
	200 Ms/s	14 bits	88 dB		350 mW

Notes:
[1] signal-to-quantization noise, [2] megasamples per second, [3] settling rime in megasamples per second

Table 2.7.3 Summary of 4:2:2 Encoding Parameters for the 525/60 Scanning Standard (*From: Robin, Michael, and Michel Poulin:* Digital Television Fundamentals, *McGraw-Hill, New York, N.Y., 1997. Used with permission.*)

Scanning standard	525/60
Coded signals	$E'_Y = 0.587\,E'_G + 0.11\,E'_B + 0.299\,E'_R$ $E'_{CB} = 0.564(E'_B - E'_Y)$ $E'_{CR} = 0.713(E'_R - E'_Y)$
Number of samples per total line	Y: 858 C_B: 429 C_R: 429 Total: 1716
Number of samples per digital active line	Y: 720 C_B: 360 C_R: 360 Total: 1440
Sampling structure	Orthogonal Line, field, and frame repetitive C_B and C_R samples cosited with odd Y samples in each line
Sampling frequency	Y: $858 \times f_H = 13.5$ MHz C_B and C_R: $429 \times f_H = 6.75$ MHz
Coding	Uniformly quantized PCM
Quantizing resolution	8 or 10 bits per sample for the luminance and each color-difference signal

Table 2.7.4 Summary of 4:2:2 Encoding Parameters for the 625/50 Scanning Standard (*From: Robin, Michael, and Michel Poulin:* Digital Television Fundamentals, *McGraw-Hill, New York, N.Y., 1997. Used with permission.*)

Scanning standard	625/50
Coded signals	$E'_Y = 0.587\,E'_G + 0.11\,E'_B + 0.299\,E'_R$ $E'_{CB} = 0.564(E'_B - E'_Y)$ $E'_{CR} = 0.713(E'_R - E'_Y)$
Number of samples per total line	Y: 864 C_B: 432 C_R: 432 Total: 1728
Number of samples per digital active line	Y: 720 C_B: 360 C_R: 360 Total: 1440
Sampling structure	Orthogonal Line, field, and frame repetitive C_B and C_R samples cosited with odd Y samples in each line
Sampling frequency	Y: $864 \times f_H = 13.5$ MHz C_B and C_R: $432 \times f_H = 6.75$ MHz
Coding	Uniformly quantized PCM
Quantizing resolution	8 or 10 bits per sample for the luminance and each color-difference signal

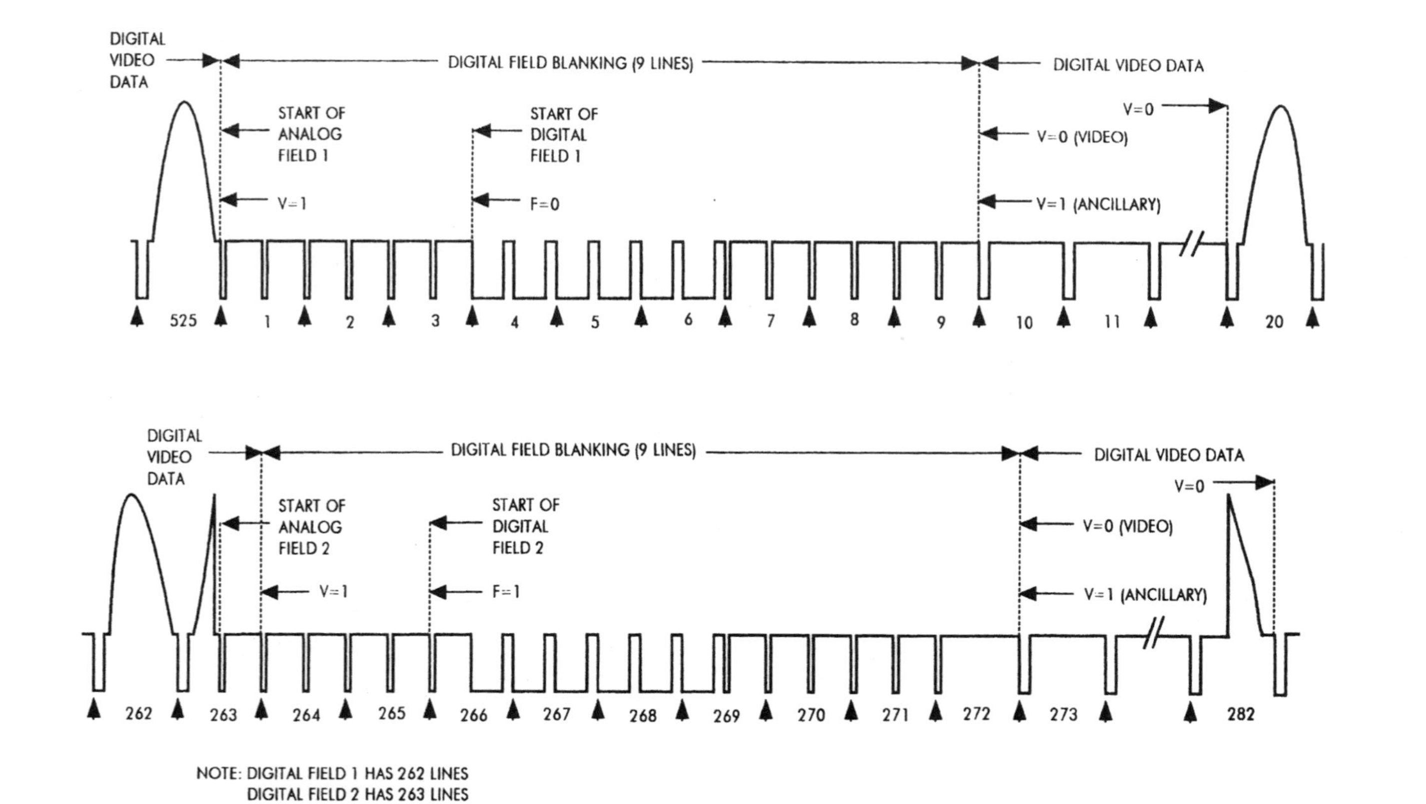

Figure 2.7.3 Relationship between 4:2:2 digital and 525/60 analog fields. (*From: Robin, Michael, and Michel Poulin:* Digital Television Fundamentals, *McGraw-Hill, New York, N.Y., 1997. Used with permission.*)

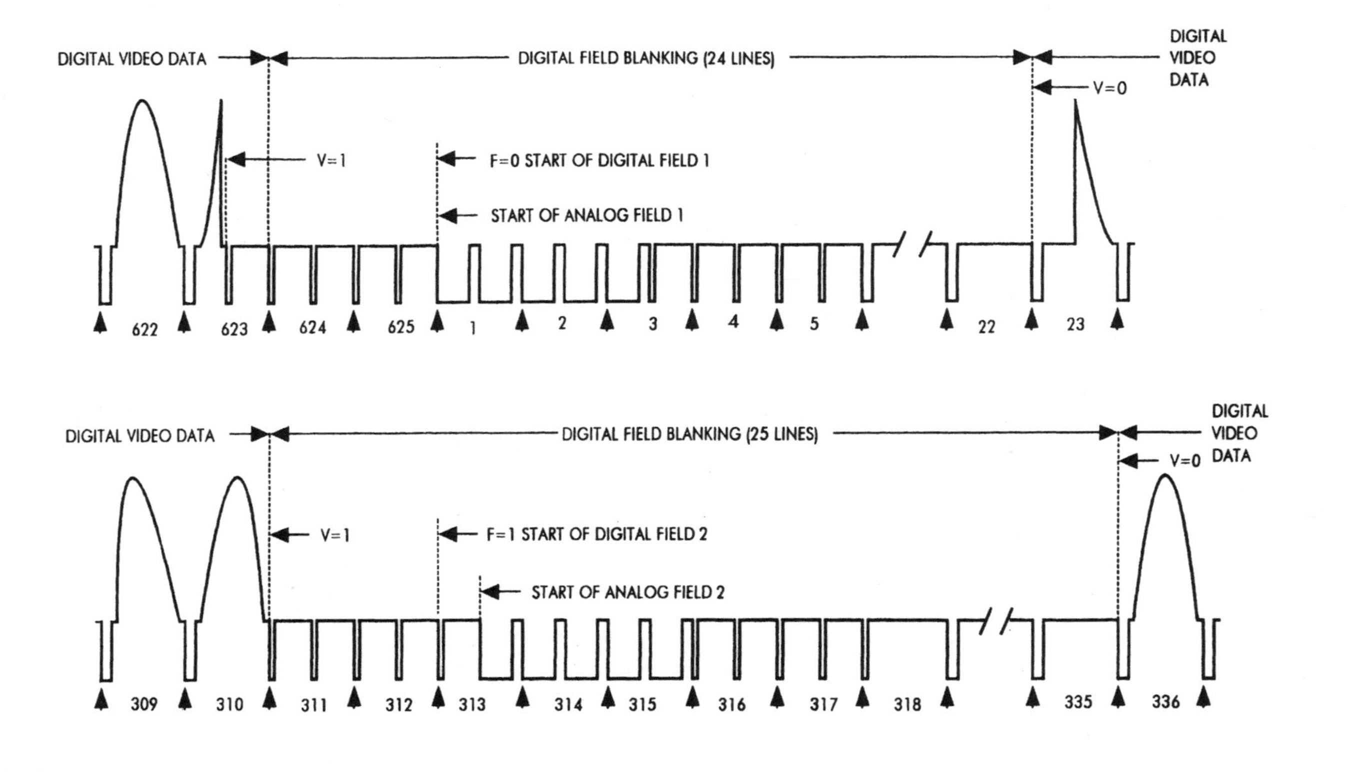

Figure 2.7.4 Relationship between 4:2:2 digital and 625/50 analog fields. (*From: Robin, Michael, and Michel Poulin:* Digital Television Fundamentals, *McGraw-Hill, New York, N.Y., 1997. Used with permission.*)

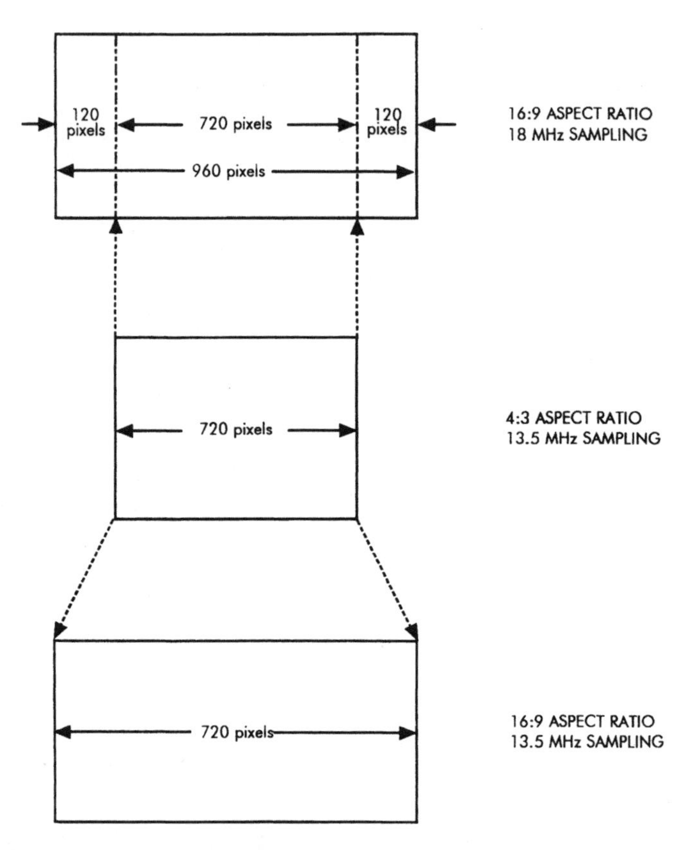

Figure 2.7.5 Graphic comparison between two 16:9 aspect ratio formats and the 4:2:2 4:3 format emphasizing the number of samples/active line. (*From: Robin, Michael, and Michel Poulin:* Digital Television Fundamentals, *McGraw-Hill, New York, N.Y., 1997. Used with permission.*)

Table 2.7.5 $4f_{sc}$ NTSC Ancillary Data Space (*From: Robin, Michael, and Michel Poulin:* Digital Television Fundamentals, *McGraw-Hill, New York, N.Y., 1997. Used with permission.*)

Horizontal ancillary data space (HANC)	55 Words/active line × 485 active lines/frame = 26,675 words/frame 26,675 Words/frame × 29.97 frames/s = 799,449.75 words/s 799,449.75 Words/s × 10 bits/word = 7.9944975 Mbps
Vertical ancillary data space (VANC)	Broad pulses: 376 Words/broad pulse × 6 broad pulses/field = 2256 words/field 2256 words/field × 59.94 fields/s = 135,224.6 words/s 135,224.6 words/s × 10 bits/word = 1.352246 Mbps
	Equalizing pulses: 21 Words/eq. pulse × 6 eq. pulses/field = 126 words/field 126 Words/field × 59.94 fields/s = 7552.44 words/s 7552.4 Words/s × 10 bits/word = 0.0755244 Mbps
	Sync tip of blanked lines: 55 Words/line × 20 blanked lines/field = 1100 words/field 1100 Words/field × 59.94 fields/s = 65,934 words/s 65,934 Words/s × 10 bits/word = 0.65934 Mbps
Total ancillary data space	7.9944975 Mbps (HANC) + 2.08711 Mbps (VANC) ≈ 10.0816075 Mbps Data formatting and exclusions may reduce this value by 10% to 20%
Total bit rate	910 Words/total line × 525 lines/frame × 29.97 frames/s × 10 bits/word = 143.1816 Mbps
Essential bit rate	143.1816 Mbps − 10.0816975 Mbps ≈ 133 Mbps

Table 2.7.6 $4f_{sc}$ PAL Ancillary Data Space (*From: Robin, Michael, and Michel Poulin:* Digital Television Fundamentals, *McGraw-Hill, New York, N.Y., 1997. Used with permission.*)

Horizontal ancillary data space (HANC)	64 Words/active line × 575 active lines/frame = 36,800 words/frame 36,800 words/frame × 25 frames/s = 920,000 words/s 920,000 Words/s × 10 bits/word = 9.2 Mbps
Vertical ancillary data space (VANC)	Broad pulses: 466 Words/broad pulse × 5 broad pulses /field = 2330 words/field 2330 Words/field × 50 fields/s = 116,500 words/s 116,500 Words/s × 10 bits/word = 1.165 Mbps
	Equalizing pulses: 23 Words/eq. pulse × 5 Eq. pulses/field = 115 word s/field 115 Words/field × 50 fields/s = 5750 words/s 5750 Words/s × 10 bits/word = 0.0575 Mbps
	Sync tip of blanked lines: 64 Words/line × 25 blanked lines/field = 1600 words/field 1600 Words/field × 50 fields/s = 80,000 words/s 80,000 Words/s × 10 bits/word = 0.8 Mbps
Total ancillary data space	9.2 Mbps (HANC) + 2.0225 Mbps (VANC) = 11.2225 Mbps Data formatting and exclusions may reduce this value by 10% to 20%
Total bit rate	1135 Words/total line × 625 lines/frame × 25 frames/s × 10 bits/word = 177.34375 Mbps
Essential bit rate	177.34375 Mbps − 11.2225 Mbps ≈ 166.1 Mbps

Table 2.7.7 4:2:2 525/60 Ancillary Data Space (*From: Robin, Michael, and Michel Poulin:* Digital Television Fundamentals, *McGraw-Hill, New York, N.Y., 1997. Used with permission.*)

Horizontal ancillary data space (HANC)	268 Words/line × 525 lines/frame = 140,700 words/frame 140,700 Words/frame × 29.97 frames/s = 4.216779 Mwords/s 4.216779 Mwords/s × 10 bits/word = 42.16779 Mbps
Vertical ancillary data space (VANC)	1440 Words/line × 38 vertical-interval lines = 54,720 words/frame 54,720 Words/frame × 29.97 frames/s = 1.6399584 Mwords/s 1.6399584 Words/s × 8 bits/word = 13.1196672 Mbps
Total ancillary data space	42.16779 Mbps (HANC) + 13.1196672 Mbps (VANC) ≈ 55.3 Mbps Data formatting and exclusions may reduce this value by 10% to 20%
Total bit rate	1716 Words/total line × 525 lines/frame × 29.97 frames/s × 10 bits/word ≈ 270 Mbps
Essential bit rate	270 Mbps − 55.3 mbps = 214.7 Mbps

Table 2.7.8 4:2:2 625/50 Ancillary Data Space (*From: Robin, Michael, and Michel Poulin:* Digital Television Fundamentals, *McGraw-Hill, New York, N.Y., 1997. Used with permission.*)

Horizontal ancillary data space (HANC)	280 Words/line × 625 lines/frame = 175,000 words/frame 175,000 Words/frame × 25 frames/s = 4.375 Mwords/s 4.375 Mwords/s × 10 bits/word = 43.75 Mbps
Vertical ancillary data space (VANC)	1440 Words/line × 48 vertical-interval lines = 69,120 words/frame 69,120 Words/frame×25 frames/s = 1.728 Mwords/s 1.728 Mwords/s × 8 bits/word = 13.824 Mbps
Total ancillary data space	43.75 Mbps (HANC) + 13.824 Mbps (VANC) = 57.574 Mbps Data formatting and exclusions may reduce this value by 10% to 20%
Total bit rate	1728 Words/total line × 625 lines/frame × 25 frames/s × 10 bits/word = 270 Mbps
Essential bit rate	270 Mbps − 57.574 Mbps = 212.426 Mbps

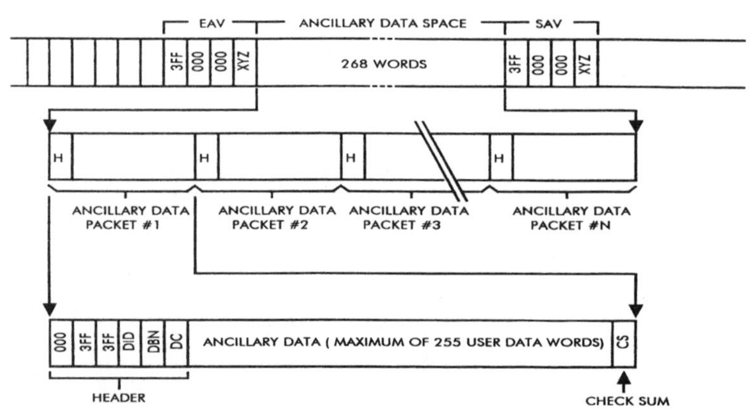

Figure 2.7.6 Ancillary packet structure for the component digital interface. (*From: Robin, Michael, and Michel Poulin:* Digital Television Fundamentals, *McGraw-Hill, New York, N.Y., 1997. Used with permission.*)

2.8 Microphone Devices and Systems

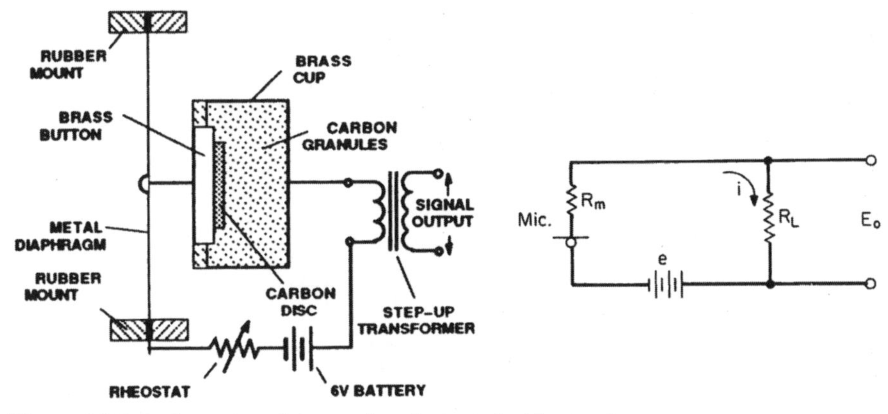

Figure 2.8.1 Carbon microphone and equivalent electric circuit.

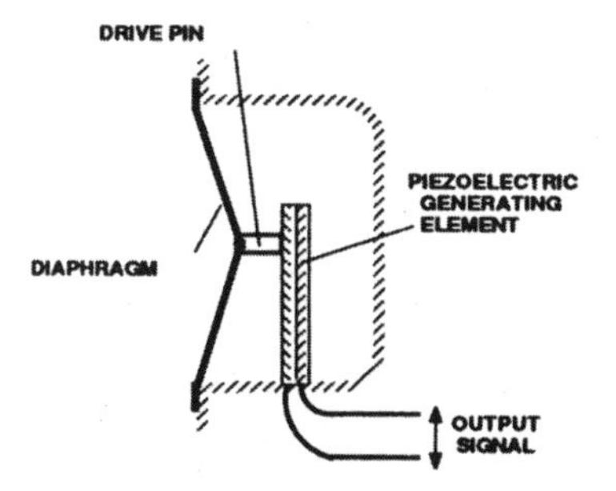

Figure 2.8.2 Typical construction of a ceramic microphone.

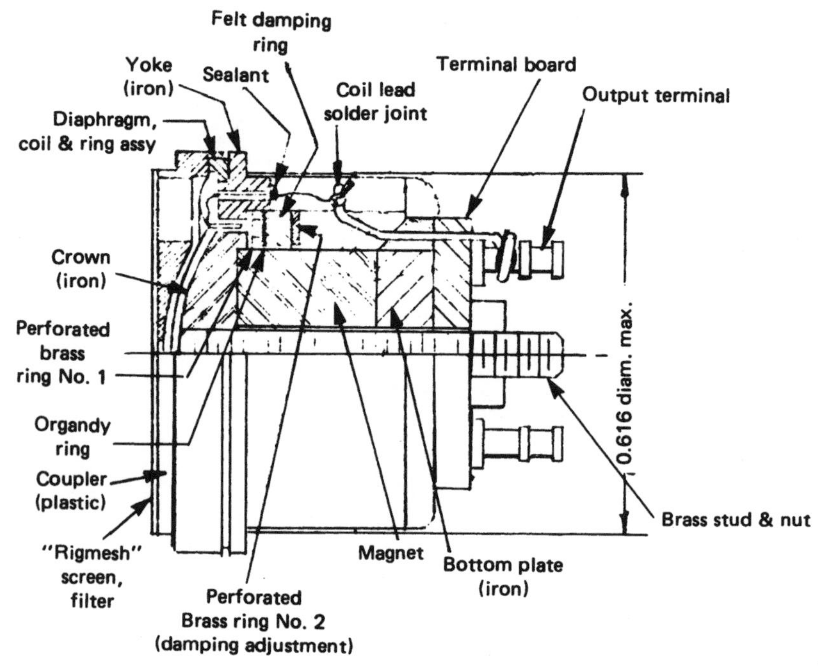

Figure 2.8.3 Dynamic moving-coil pressure-microphone cartridge.

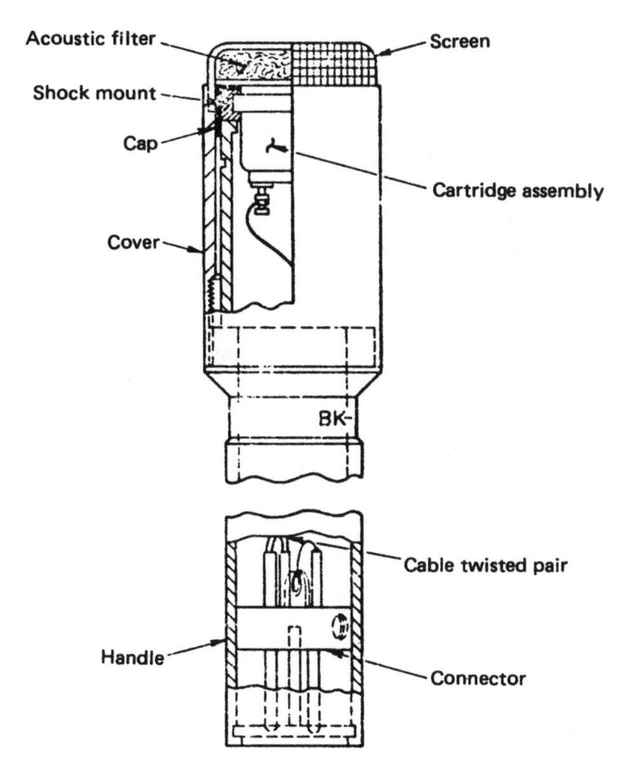

Figure 2.8.4 Dynamic moving-coil microphone.

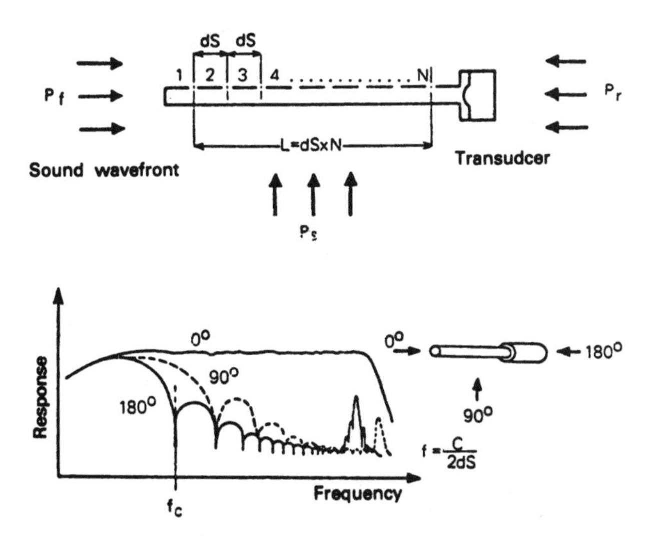

Figure 2.8.5 Operating principles of the line microphone. (*Courtesy of Sony.*)

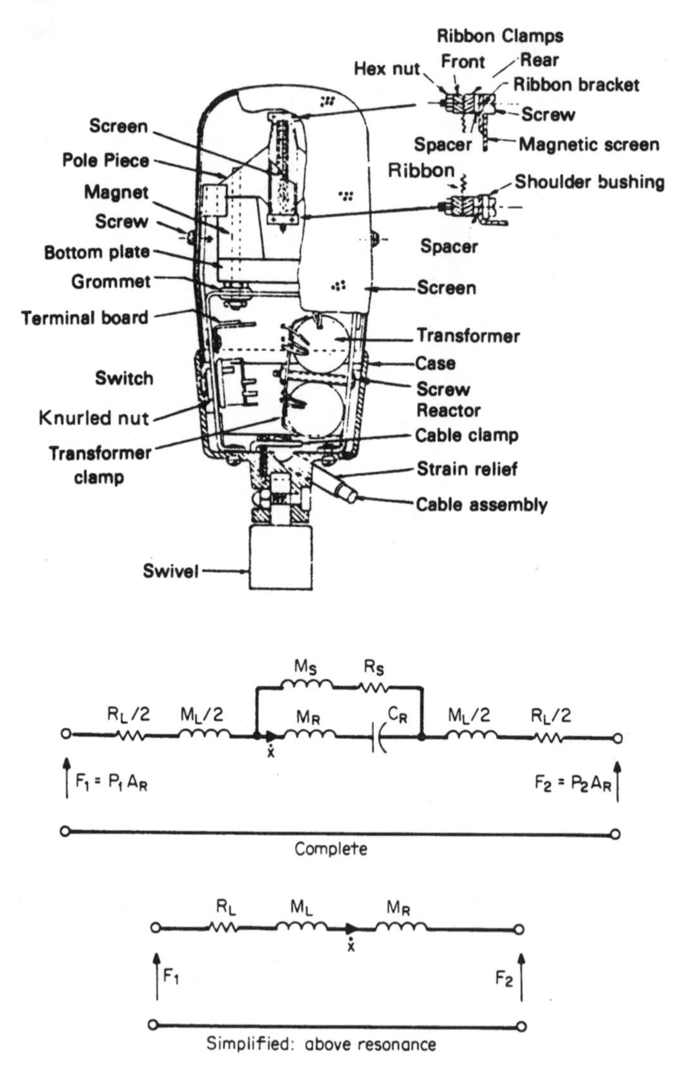

Figure 2.8.6 Classic ribbon velocity microphone (RCA type BK-11A) and mechanical networks.

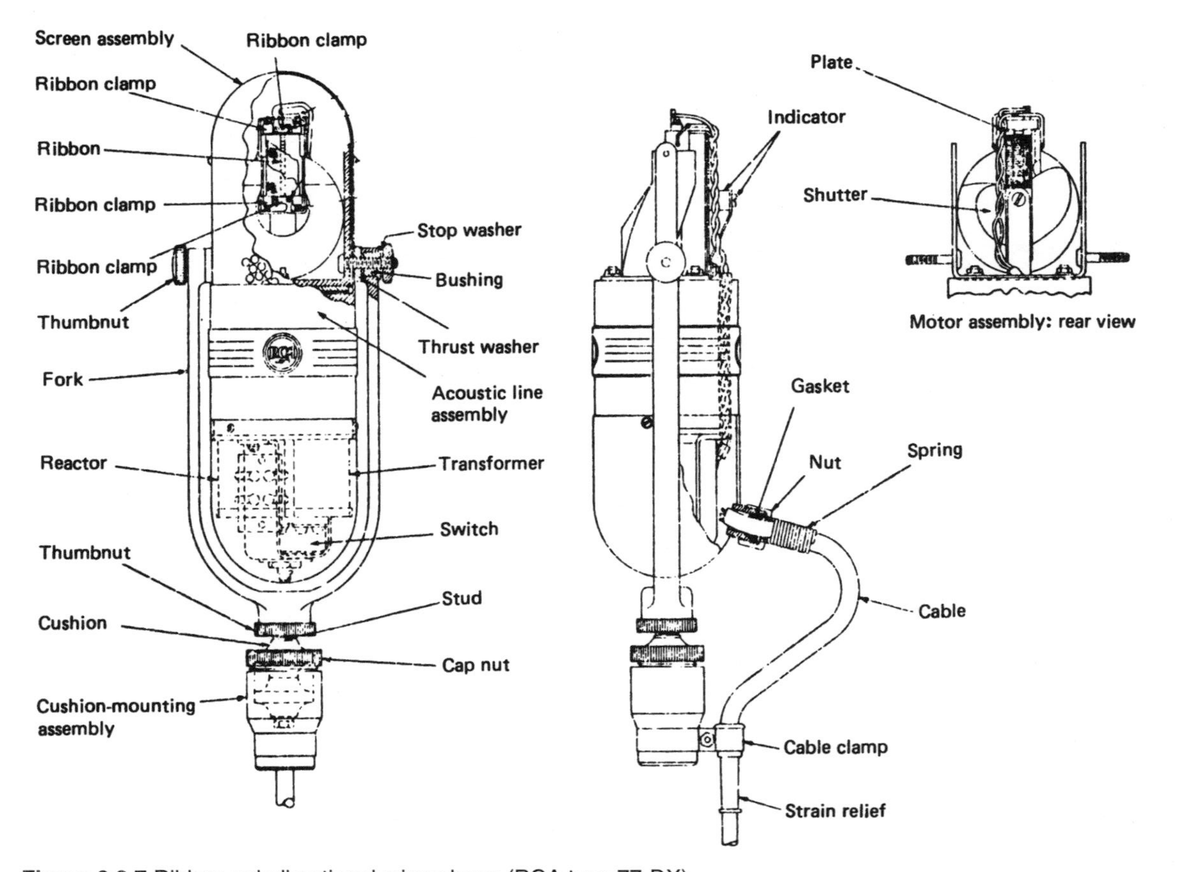

Figure 2.8.7 Ribbon polydirectional microphone (RCA type 77-DX).

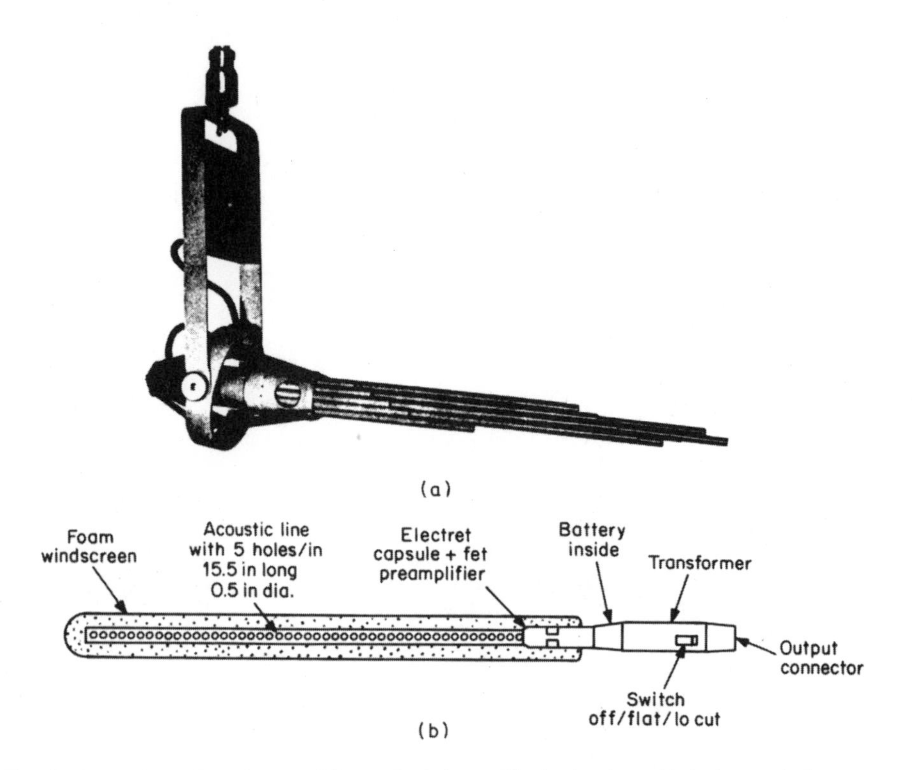

Figure 2.8.8 Line microphones: (*a*) bundled pipes, (*b*) single pipe with holes and electret condenser.

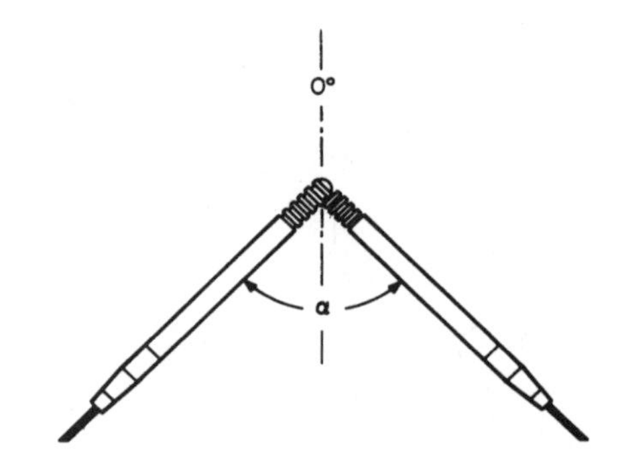

Figure 2.8.9 Coincident *XY* microphone pair.

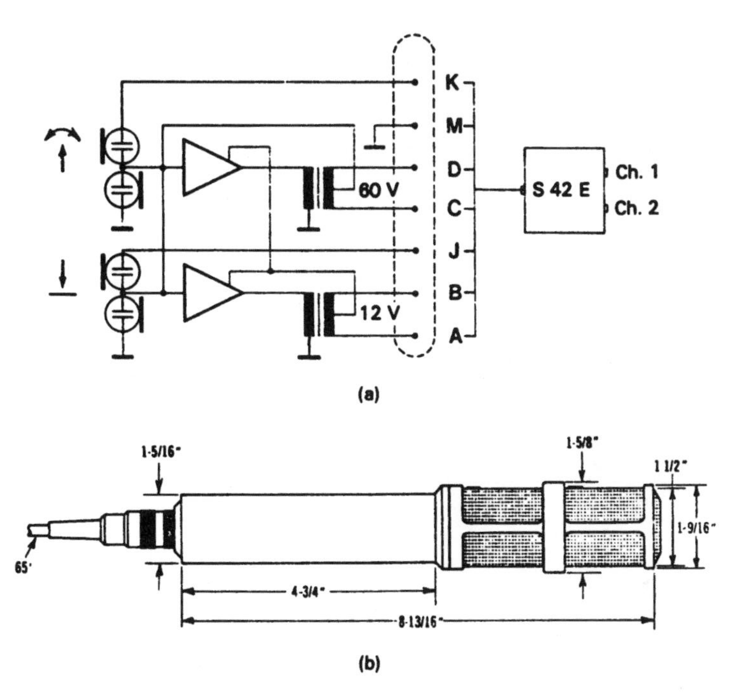

Figure 2.8.10 Stereophonic condenser microphone: (*a*) wiring diagram, (*b*) shell construction.

Figure 2.8.11 MS conversion to *XY*.

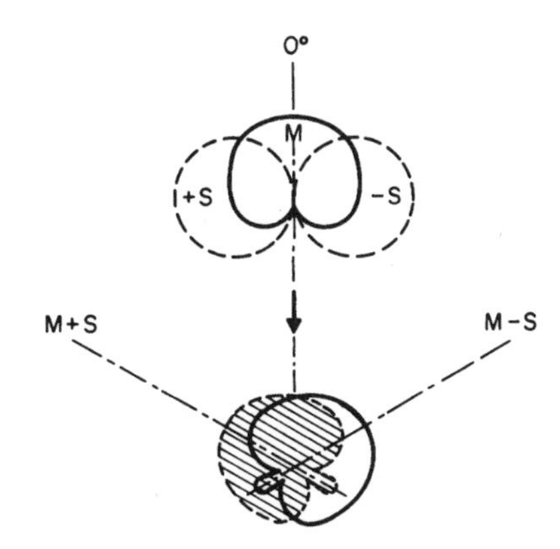

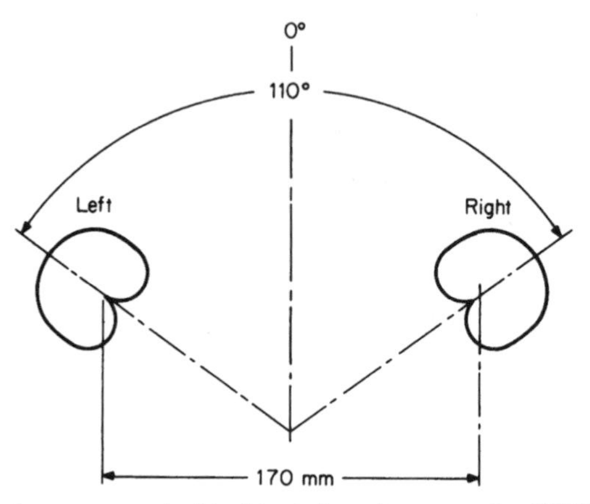

Figure 2.8.12 Two-microphone near-coincident technique knows as the ORTF configuration.

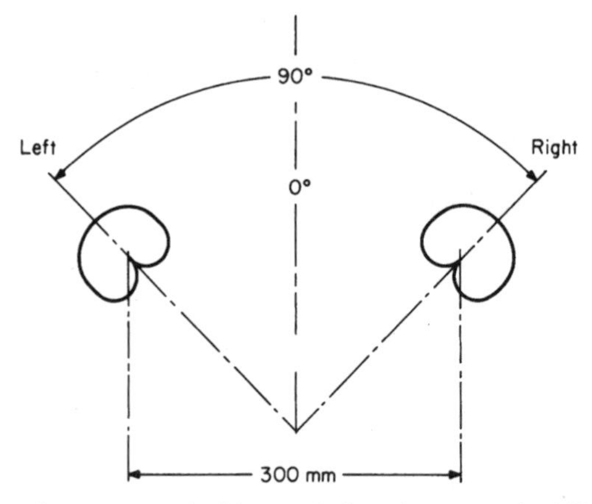

Figure 2.8.13 Two-microphone near-coincident technique knows as the NOS configuration.

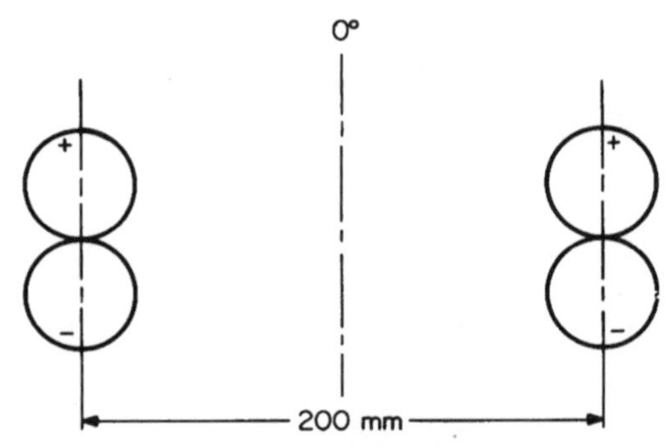

Figure 2.8.14 Two-microphone near-coincident technique knows as the Faulkner configuration.

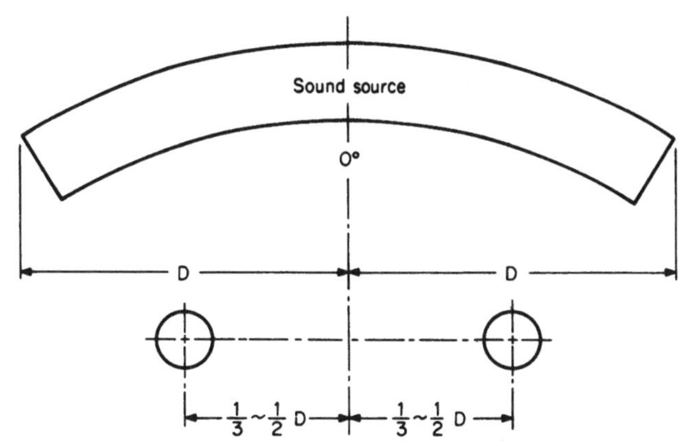

Figure 2.8.15 Spaced omnidirectional pair.

2.9 Sound Reproduction Devices and Systems

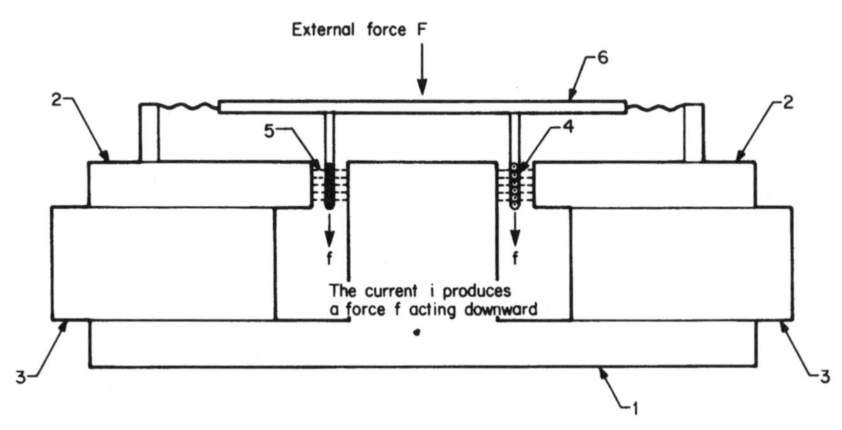

Figure 2.9.1 Simplified form of a moving-coil transducer consisting of a voice coil cutting a magnetic field of a flux density *B*. 1, 2 = pole pieces; 3 = permanent magnet; 4 = voice coil; 5 = magnetic flux; 6 = diaphragm.

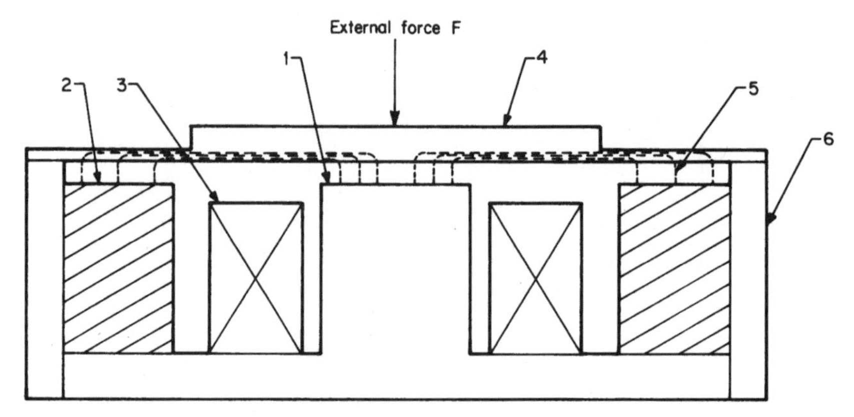

Figure 2.9.2 Simplified form of an electromagnetic transducer. 1 = pole piece; 2 = permanent magnet; 3 = drive coil; 4 = diaphragm; 5 = magnet flux; 6 = frame.

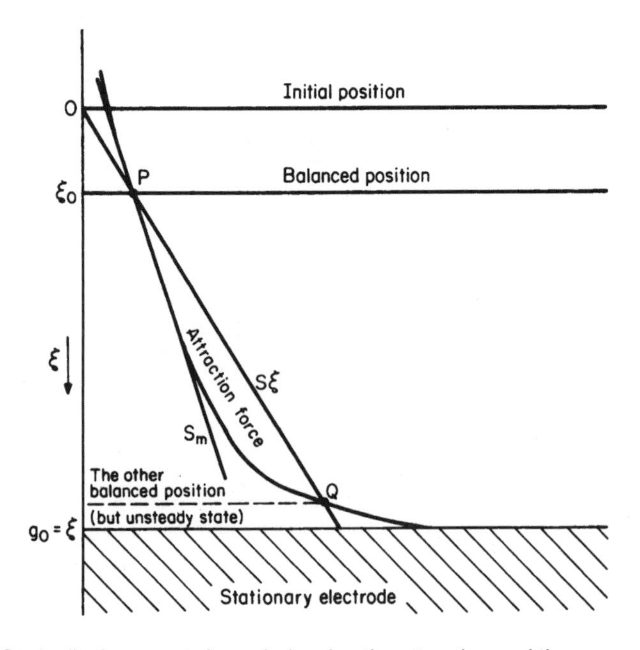

Figure 2.9.3 Static displacement shows balancing the attraction and the recover force.

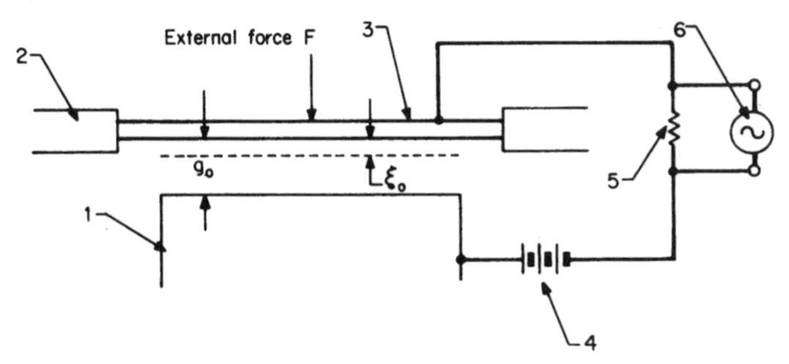

Figure 2.9.4 Cross-sectional view of an electroacoustic transducer. 1 = back electrode; 2 = clamping ring; 3 = diaphragm with electrode; 4 = polarizing power supply; 5 = polarizing electrical resistance; 6 = signal source.

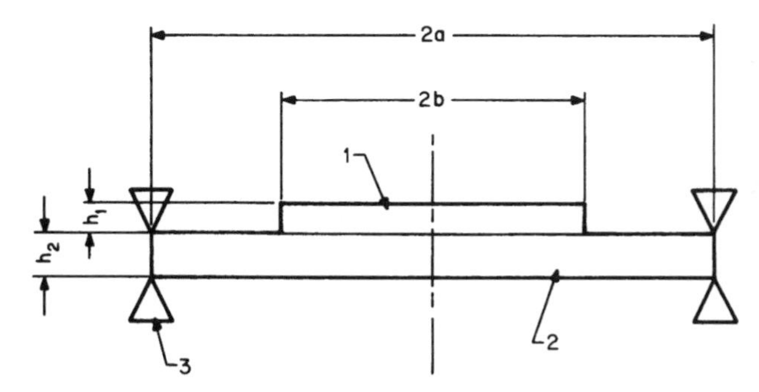

Figure 2.9.5 Simplified form of a monomorphic piezoelectric transducer. 1 = piezoelectric element (E_1, P_1 μ_1); 2 = metal plate (E_2, P_2 μ_2); 3 = supporting ring.

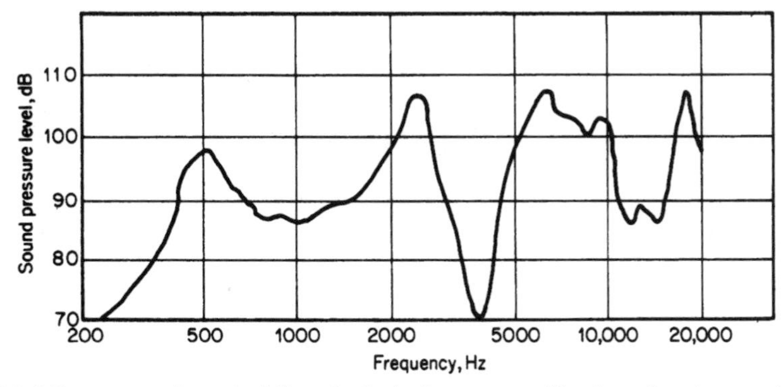

Figure 2.9.6 Frequency characteristics of a typical monomorphic piezoelectric transducer.

Table 2.9.1 Three Speaker Control Systems

	Resistance control	Mass control	Stiffness control
Z_m approximation	r	ωm	s/ω
$\lvert v/F\rvert$	$1/r$	$1/\omega m$	ω/s
Characteristics	$\lvert 1/z\rvert$ (0.1, 1, 10)	$\lvert\omega/z\rvert$ (1)	$\lvert 1/\omega z\rvert$ (1)
Applications	Horn speaker	Direct radiant-type speaker	Headphone

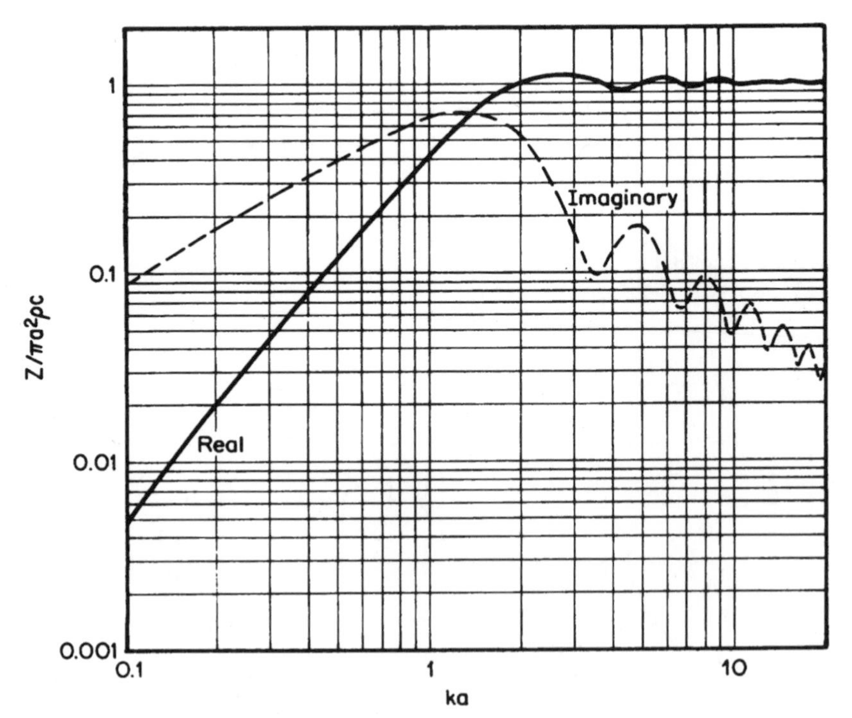

Figure 2.9.7 Radiation impedance for a rigid circular diaphragm in an infinite baffle as a function of $ka = 2\pi a/\lambda$.

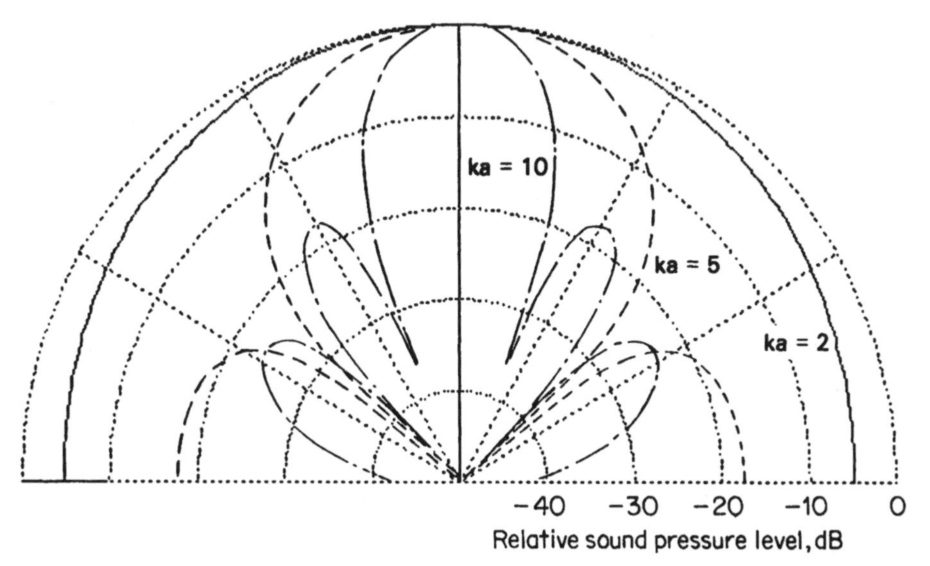

Figure 2.9.8 Directional characteristics of a circular diaphragm.

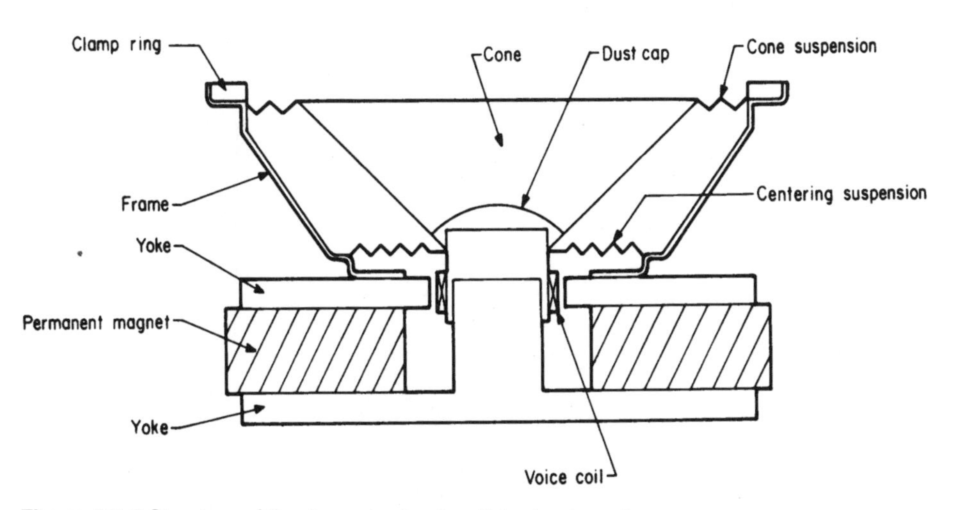

Figure 2.9.9 Structure of the dynamic direct-radiator loudspeaker.

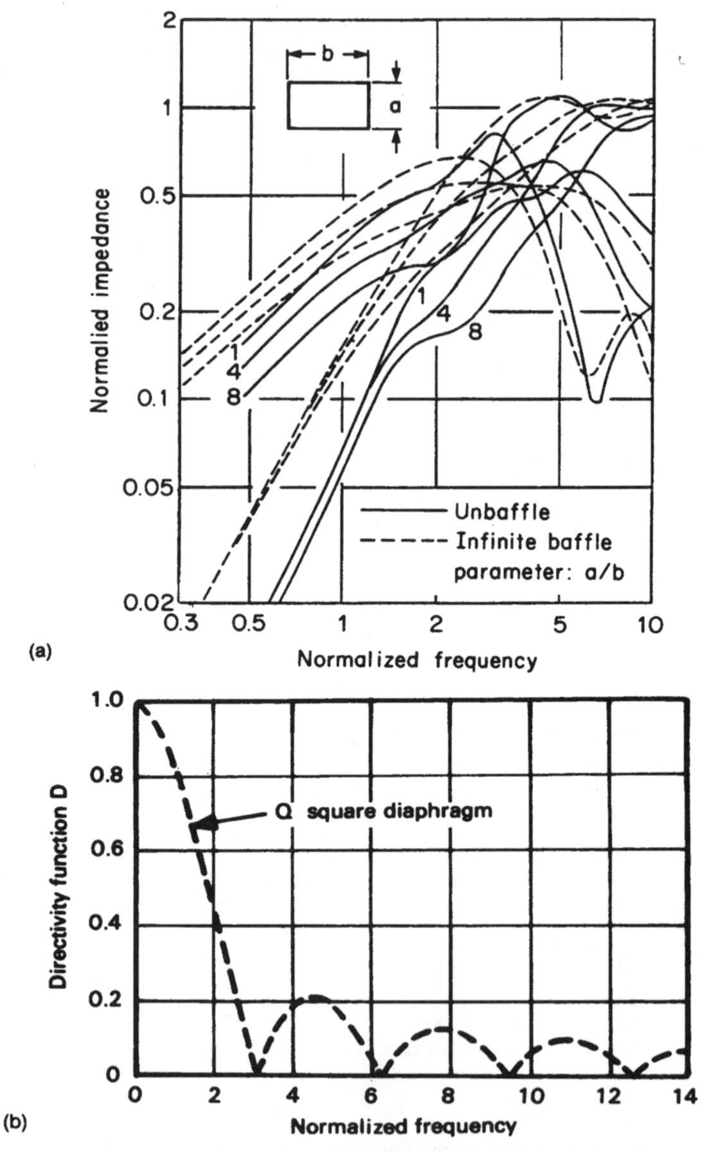

Figure 2.9.10 Diaphragm characteristics: (*a*) radiation impedance for a rigid rectangular diaphragm, (*b*) directivity function for a rigid square diaphragm. Note that in (*a*) solid lines, which have been calculated by using the finite element method (FEM), are instructive for practical designs.

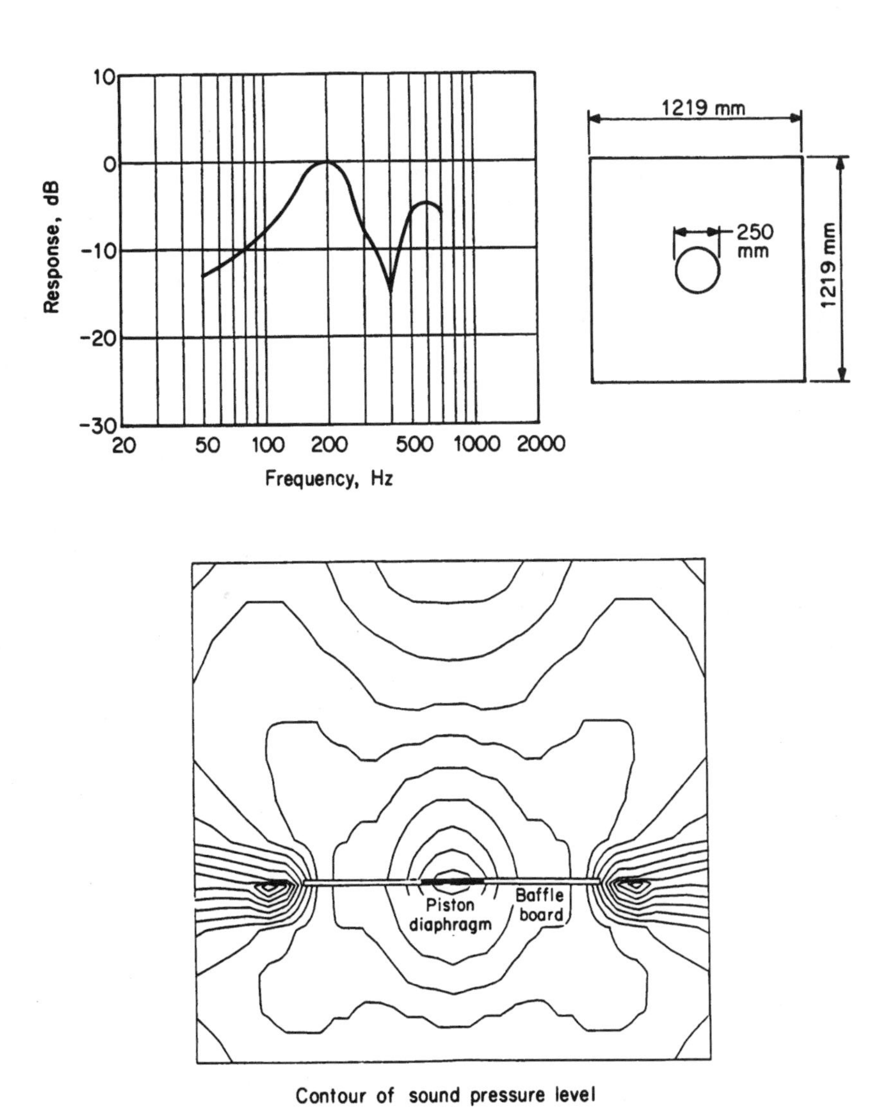

Figure 2.9.11 Pressure-response-frequency characteristics for a direct radiator installed in the center of a finite baffle, estimated by FEM.

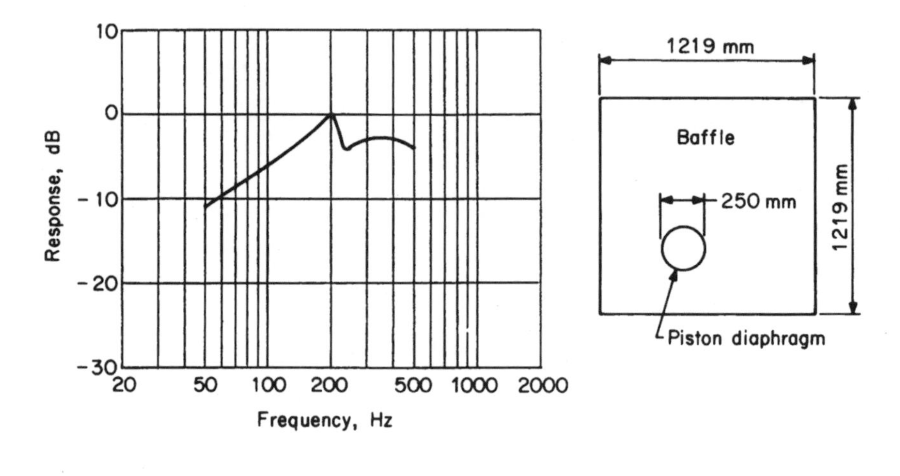

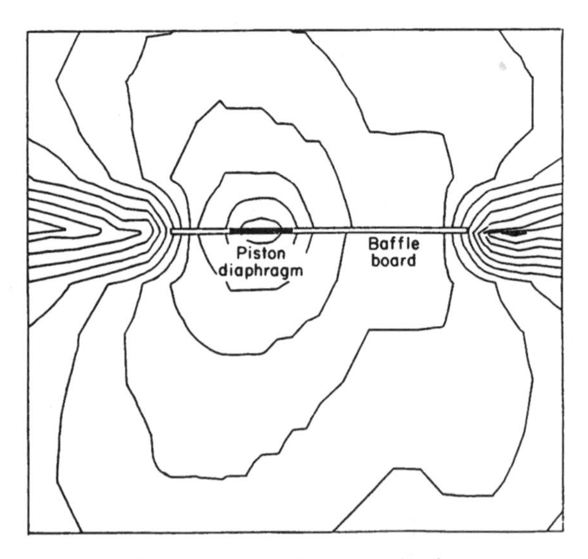

Contour of sound pressure level

Figure 2.9.12 Pressure-response-frequency characteristics of a direct radiator installed off center, estimated by FEM.

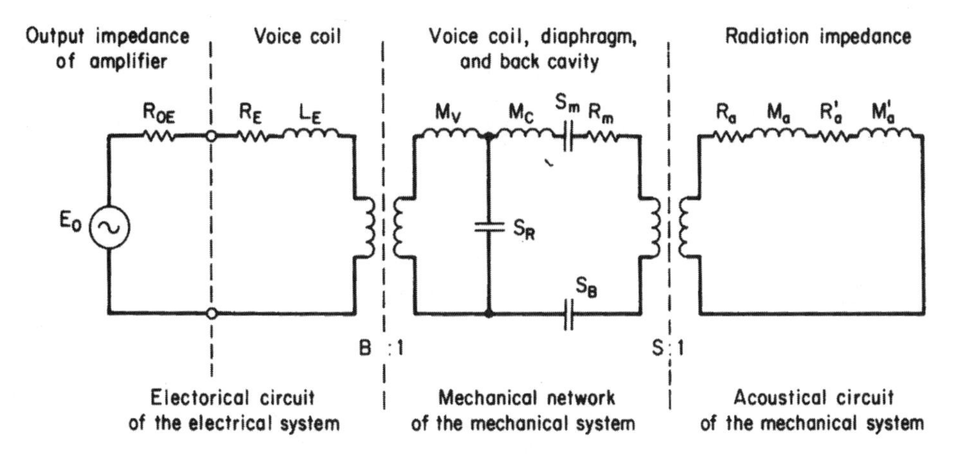

Figure 2.9.13 Electromechanical equivalent circuit. R_{OE} = output impedance of amplifier, Ω; R_E = resistance of voice coil, Ω; L_E = inductance of voice coil, H; M_V = mass of voice coil, kg; S_R = stiffness between cone and voice coil, N/m; M_C = mass of cone, kg; S_B = stiffness of back cavity, N/m; R_a, R_a' = radiation resistance of diaphragm, mechanical ohms; M_a, M_a' = radiation mass of diaphragm, kg; Bl = force factor; S = area of diaphragm, m^2.

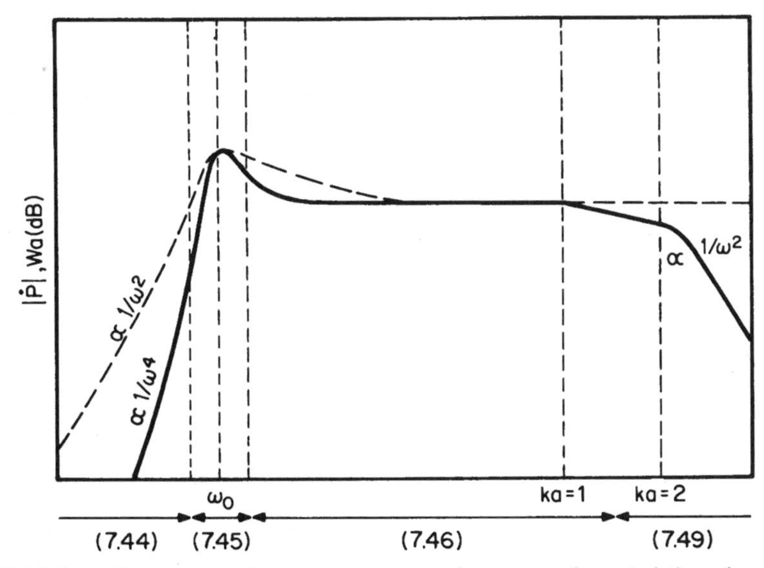

Figure 2.9.14 Acoustic power and pressure-response-frequency characteristics of a piston source in an infinite-plane baffle.

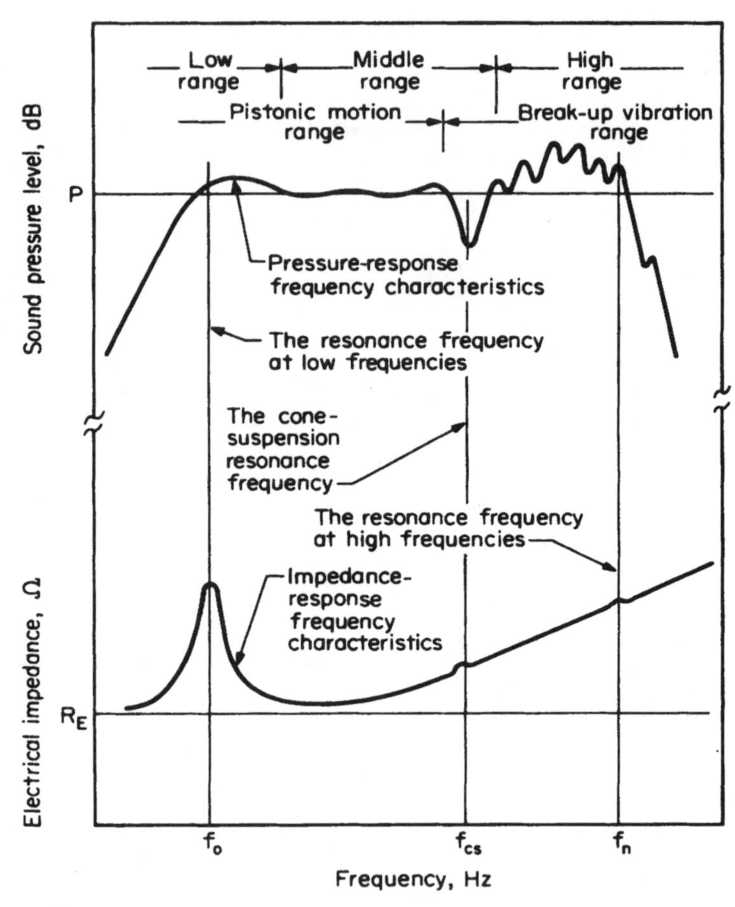

Figure 2.9.15 Frequency characteristics of the dynamic direct-radiator loudspeaker.

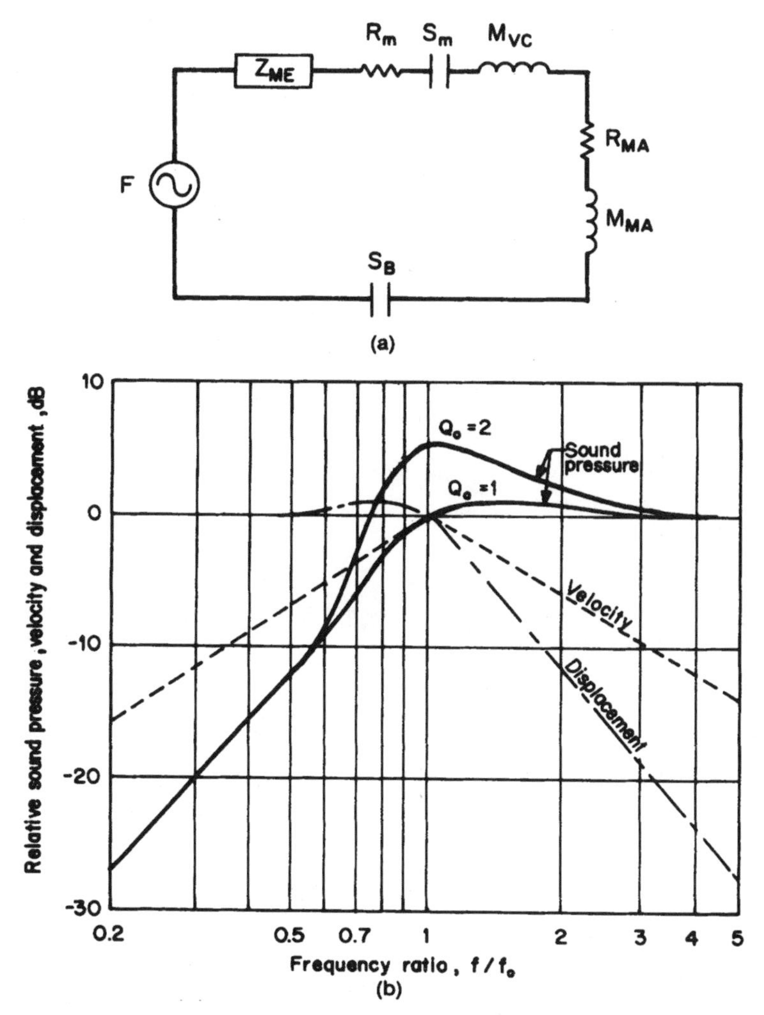

Figure 2.9.16 Loudspeaker characteristics: (*a*) mechanical equivalent circuit at a low-frequency range, (*b*) frequency characteristics of sound pressure, velocity, and displacement. Z_{ME} = motional impedance, mechanical ohms; R_m = resistance of vibrating system, mechanical ohms; S_m = stiffness of vibrating system, N/m; M_{VC} = mass of vibrating system, kg; R_{MA} = resistance of radiating system, mechanical ohms; M_{MA} = mass of radiating system, kg; S_B = stiffness of back cavity, N/m.

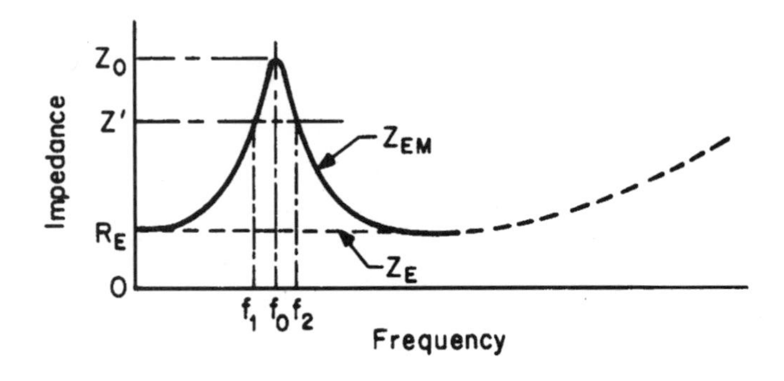

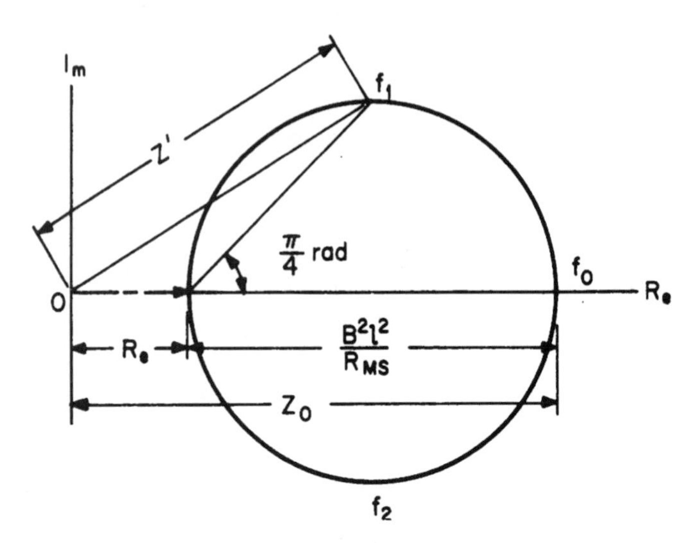

Figure 2.9.17 Loudspeaker voice-coil impedance and impedance locus. R_E = resistance of voice coil, Ω; $Z = Z_0/\sqrt{2}$, Ω; B = magnetic-flux density in the gap, Wb/m_2; l = length of wire on voice-coil winding, m; R_{MS} = resistance of vibrating system, mechanical ohms; f_0 = resonance at low-frequency range, Hz; f = frequency at –3 dB, Hz.

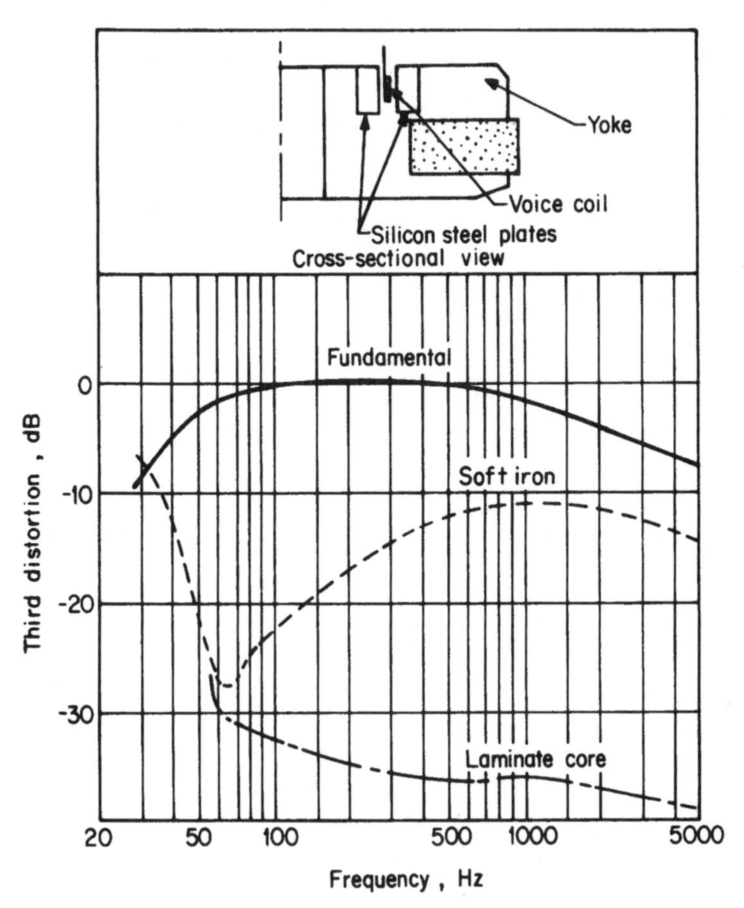

Figure 2.9.18 Comparison of the third-harmonic distortion between soft iron and silicon plates (solid line = fundamental current level, dashed line = soft-iron-type yoke, dash-dot line = laminate-core-type yoke).

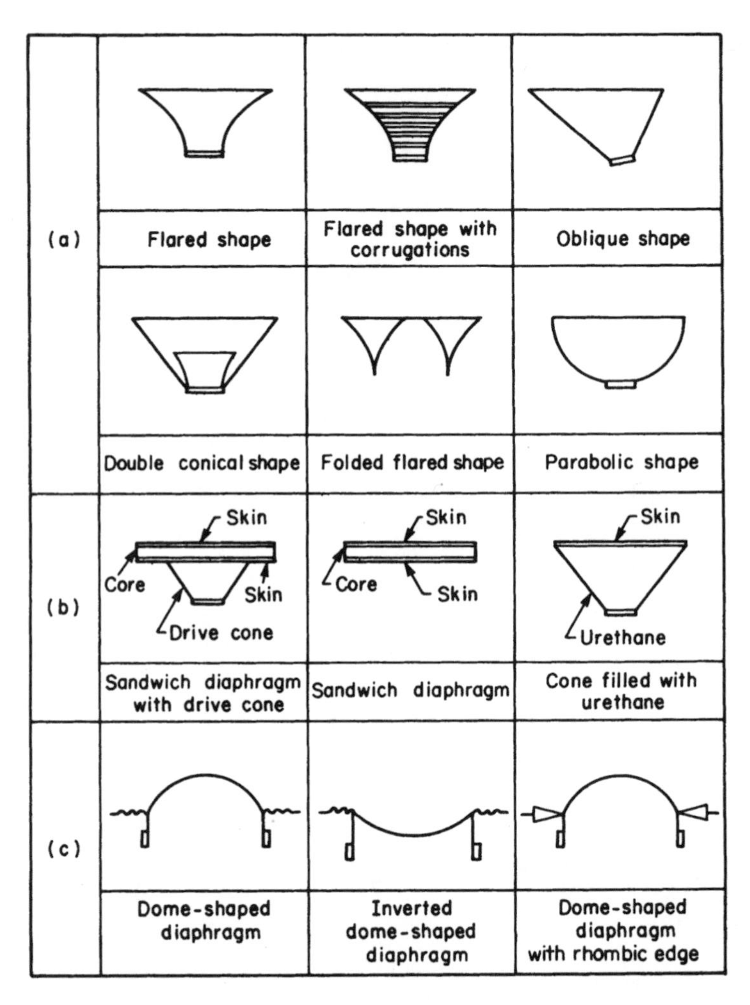

Figure 2.9.19 Sectional views of various diaphragm shapes: (*a*) diaphragm of the cone type extends to the reproducing band by changing the shape of the curved surface, (*b*) diaphragm of the plane type removes the cavity effect by using a flat radiation surface, (*c*) diaphragm of the dome type improves bending elasticity by forming thin plates into a domelike shape.

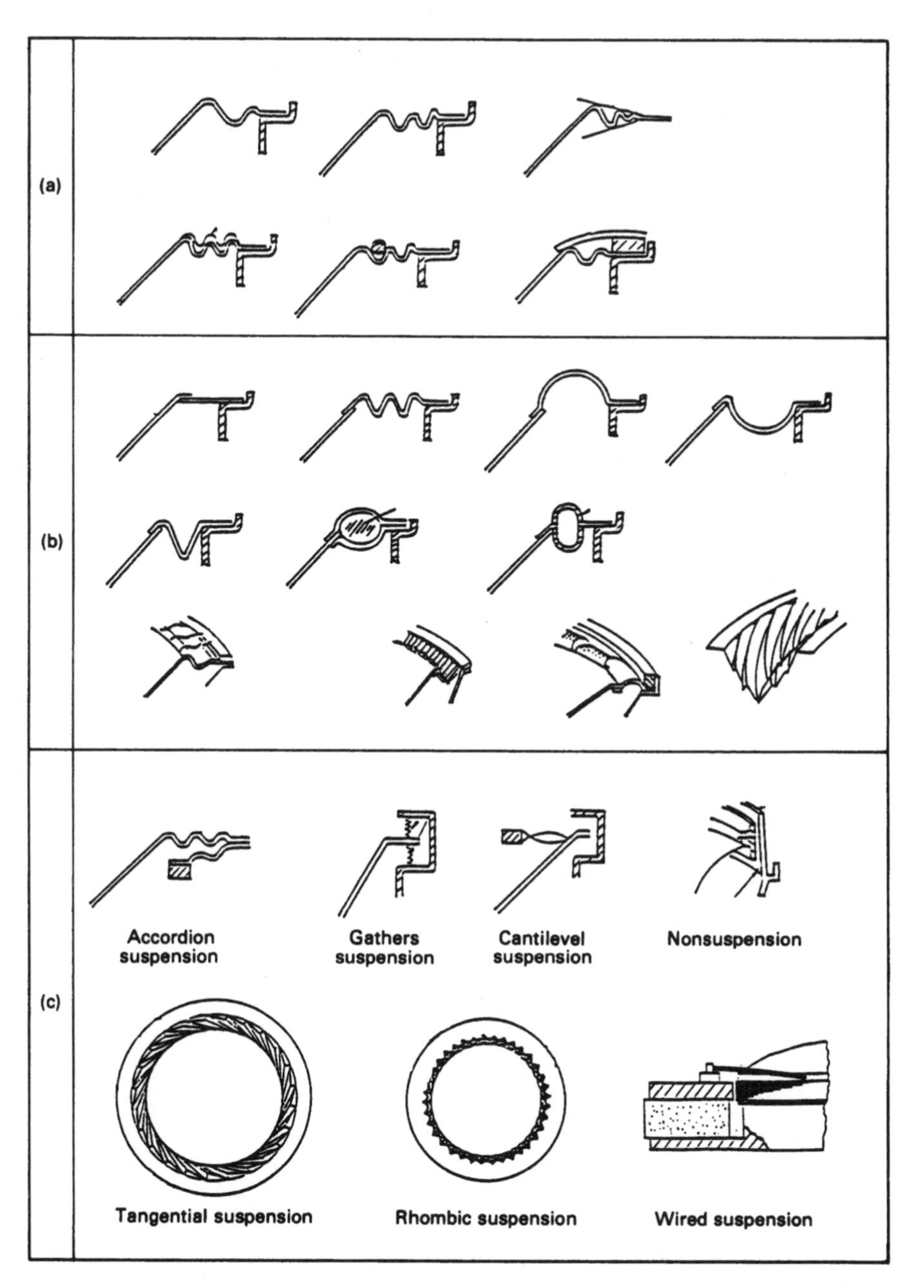

Figure 2.9.20 Sectional views of cone suspension systems: (*a*) the thinned edge of a diaphragm fulfills the function of the cone suspension, (*b*) material different from that of a diaphragm is used to fulfill the function of cone suspension, (*c*) exceptional cone suspensions.

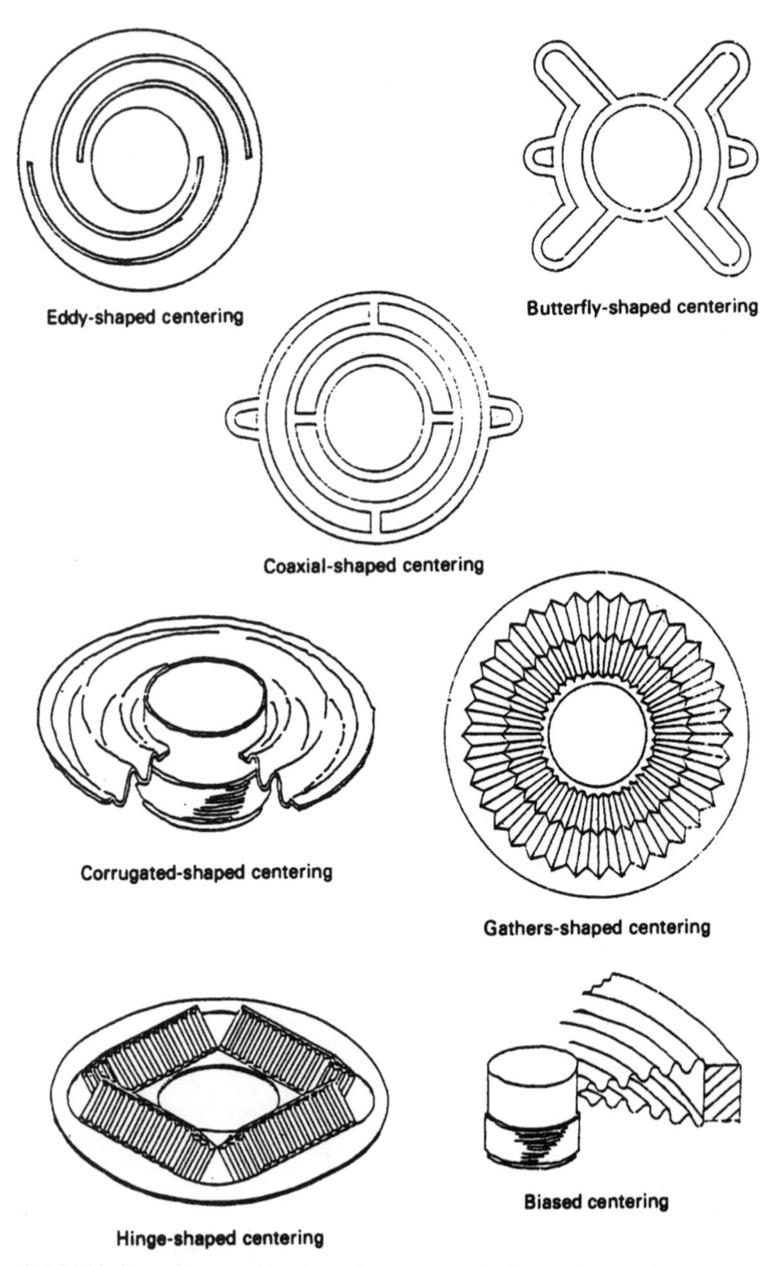

Figure 2.9.21 Various shapes of loudspeaker cone centering systems.

2.10 Electron Optics and Deflection

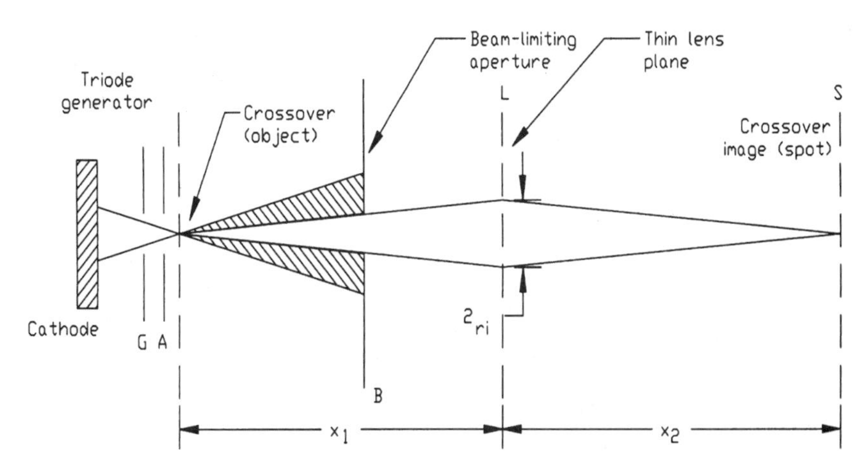

Figure 2.10.1 Electron beam shaping process. (*After: Moss, Hilary:* Narrow Angle Electron Guns and Cathode Ray Tubes, *Academic, New York, N.Y., 1968.*)

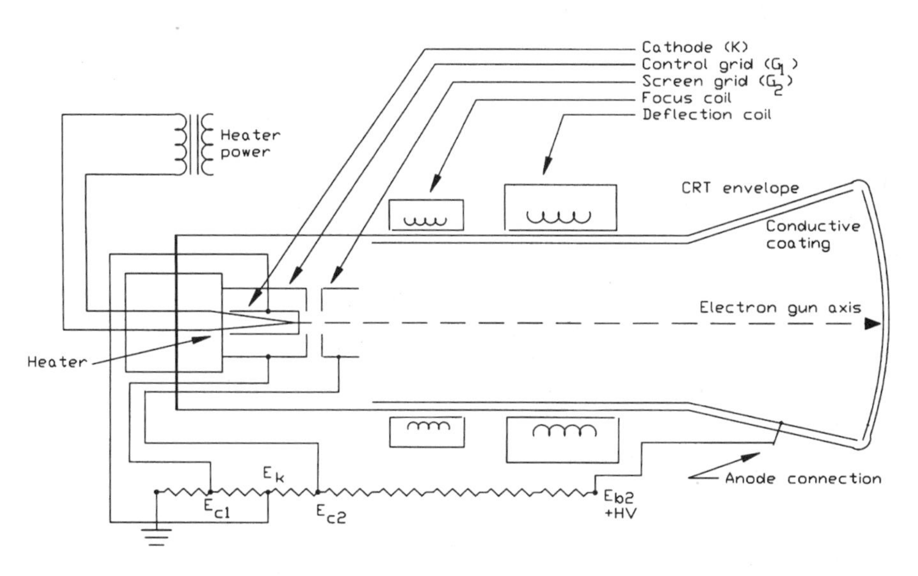

Figure 2.10.2 Generalized schematic of a CRT gun structure using electromagnetic focus and deflection. (*After:* Cathode Ray Tube Displays, *MIT Radiation Laboratory Series, vol. 22, McGraw-Hill, New York, N.Y., 1953.*)

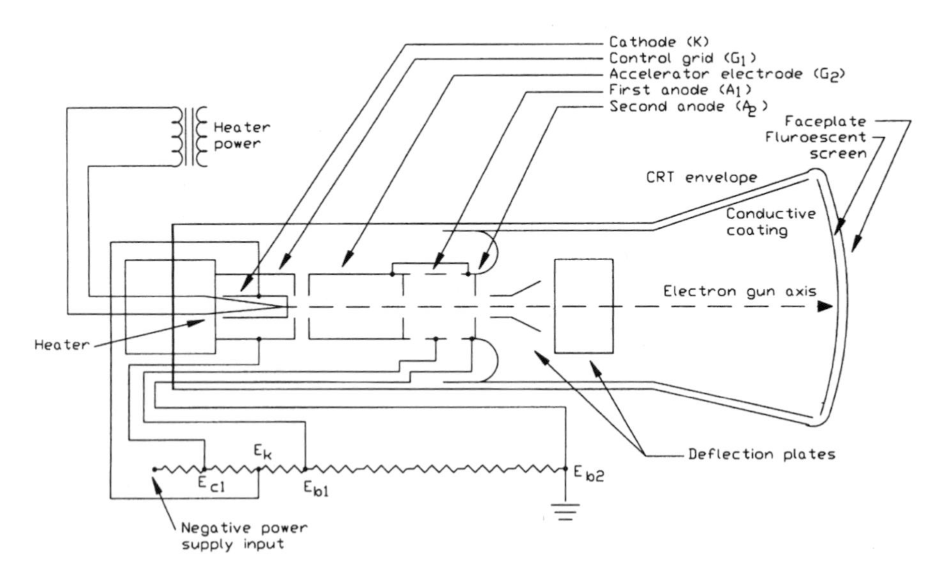

Figure 2.10.3 Generalized schematic of a CRT with electrostatic focus and deflection. An Einzell focusing lens is depicted. (*After:* Cathode Ray Tube Displays, *MIT Radiation Laboratory Series, vol. 22, McGraw-Hill, New York, N.Y., 1953.*)

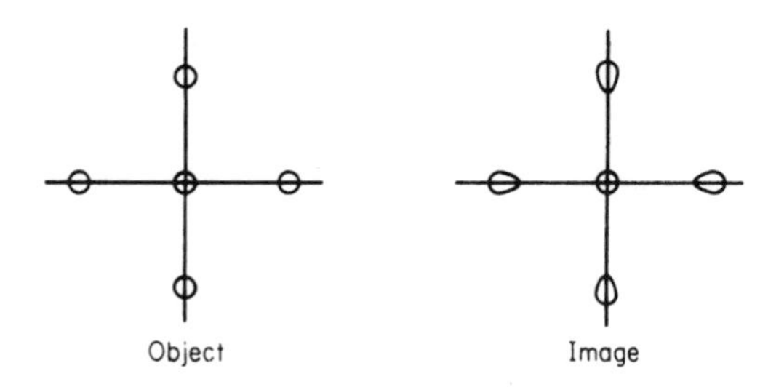

Figure 2.10.4 Illustration of coma. (*After: Spangenberg, K. R.:* Vacuum Tubes, *McGraw-Hill, New York, N.Y., 1948.*)

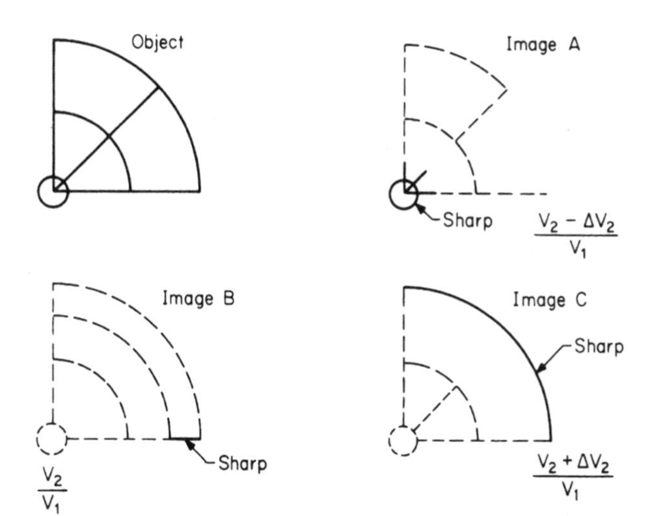

Figure 2.10.5 Illustration of astigmatism. (*After: Spangenberg, K. R.:* Vacuum Tubes, *McGraw-Hill, New York, N.Y., 1948.*)

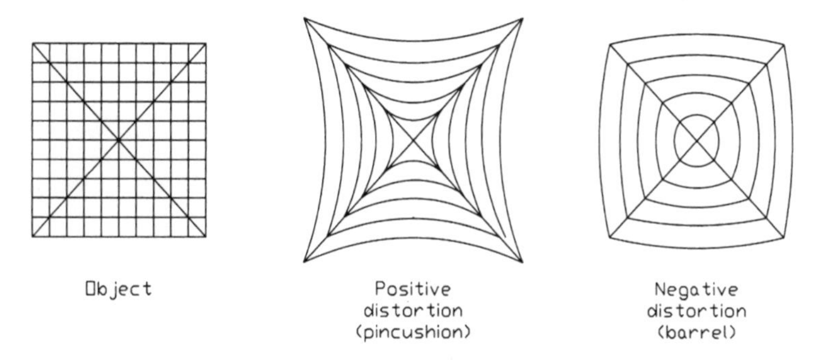

Figure 2.10.6 Illustration of pincusion and barrel distortion. (*After: Spangenberg, K. R.:* Vacuum Tubes, *McGraw-Hill, New York, N.Y., 1948.*)

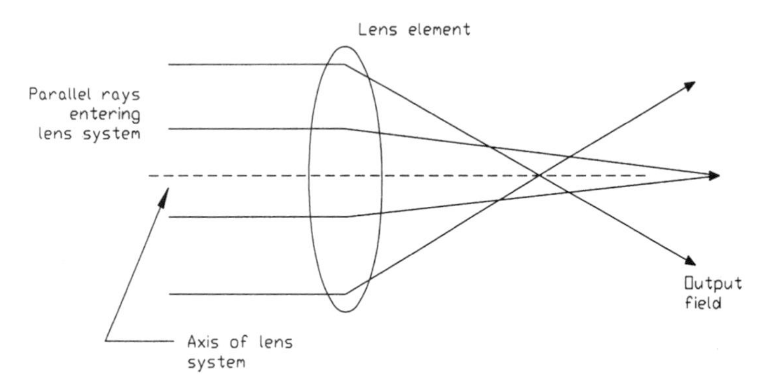

Figure 2.10.7 Illustration of spherical aberration. (*After: Zworykin, V. K., and G. Morton:* Television, *2d ed., Wiley, New York, N.Y., 1954.*)

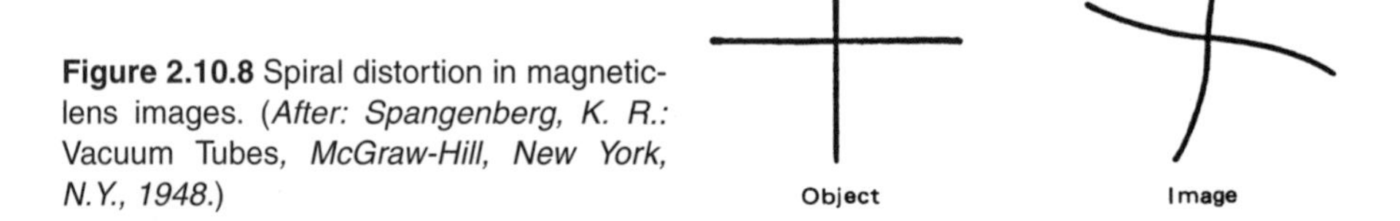

Figure 2.10.8 Spiral distortion in magnetic-lens images. (*After: Spangenberg, K. R.:* Vacuum Tubes, *McGraw-Hill, New York, N.Y., 1948.*)

Table 2.10.1 Comparison of Common Electromagnetic and Electrostatic Deflection CRTs

Parameter	Magnetic Deflection	Electrostatic Deflection
Deflection settling time	10 μs	< 1 μs to one spot diameter
Small-signal bandwidth	2 MHz	5 MHz
Video bandwidth	15–30 MHz	25 MHz
Linear writing speed	1 μs/cm	25 cm/μs
Resolution	1000 TV lines or more	17 lines/cm
Luminance	300 nits	150–500 nits
Spot size	0.25 mm (53 cm CRT)	0.25–0.38 mm
Accelerator voltage	10 kV	28 kV
Phosphor	Various	P-31 and others
Power consumption	250 W	130-140 W
Off-axis deflection	55°	20°
Physical length	Up to 30% shorter than electrostatic deflection	
Typical applications	Video display	Waveform display
1 nit = 1 candela per square meter (cd/m^2)		

2.11 Video Cameras

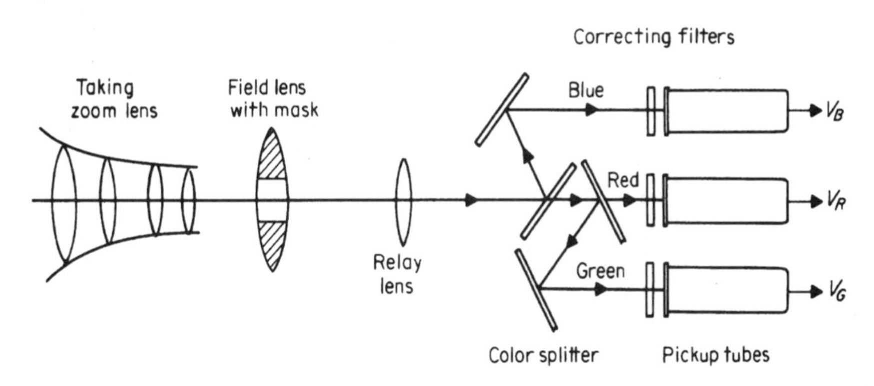

Figure 2.11.1 Color separation by relay lens and dichroic filters. (*From: Fink, D. G., and D. Christiansen (eds.):* Electronics Engineers' Handbook*, 2nd ed., McGraw-Hill, New York, N.Y., pp. 20–30, 1982. Used with permission.*)

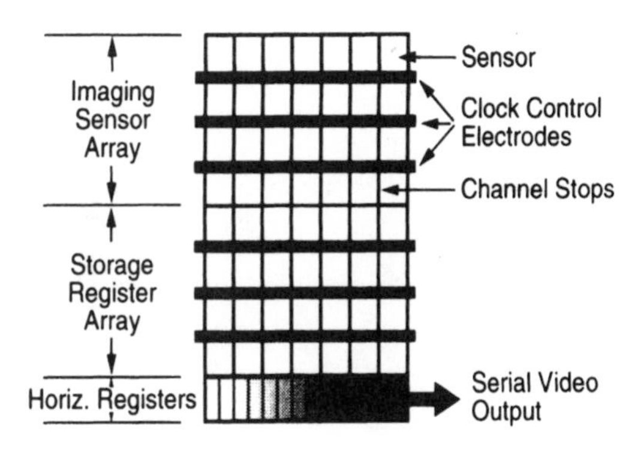

Figure 2.11.2 Simplified schematic of the frame transfer CCD structure. (*From: Thorpe, Laurence J.: "Television Cameras," in* Electronic Engineers' Handbook*, 4th ed., Donald Christiansen (ed.), McGraw-Hill, New York, N.Y., pp. 24.58–24.74, 1997. Used with permission.*)

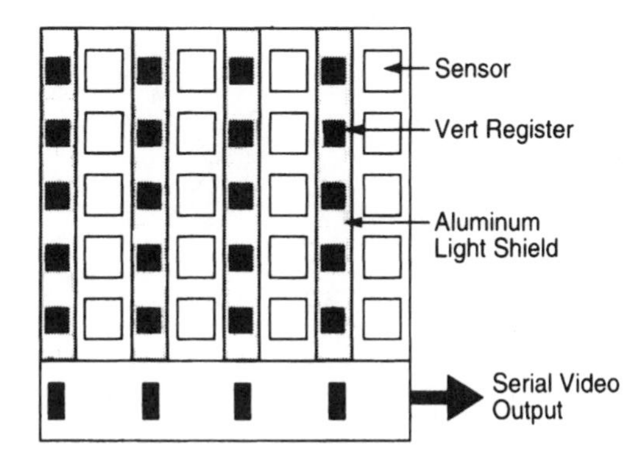

Figure 2.11.3 Simplified schematic of the interline transfer structure. (*From: Thorpe, Laurence J.: "Television Cameras," in* Electronic Engineers' Handbook*, 4th ed., Donald Christiansen (ed.), McGraw-Hill, New York, N.Y., pp. 24.58–24.74, 1997. Used with permission.*)

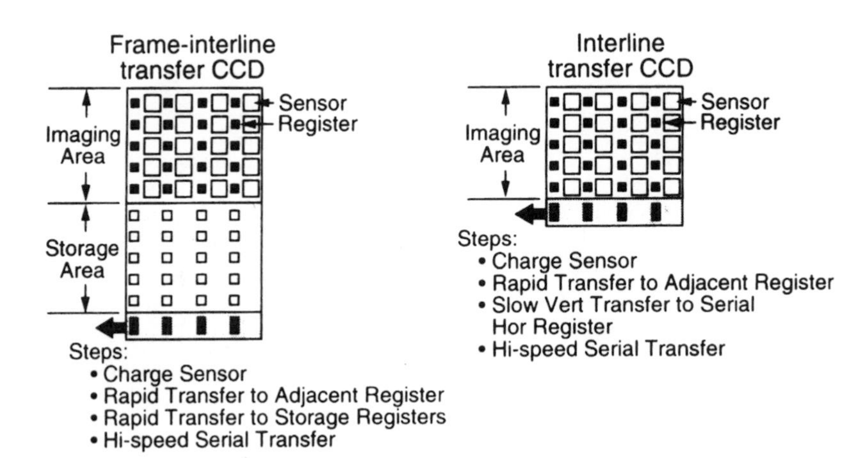

Figure 2.11.4 Comparison of features of a frame interline transfer CCD and an interline transfer CCD. (*From: Thorpe, Laurence J.: "Television Cameras," in* Electronic Engineers' Handbook, *4th ed., Donald Christiansen (ed.), McGraw-Hill, New York, N.Y., pp. 24.58–24.74, 1997. Used with permission.*)

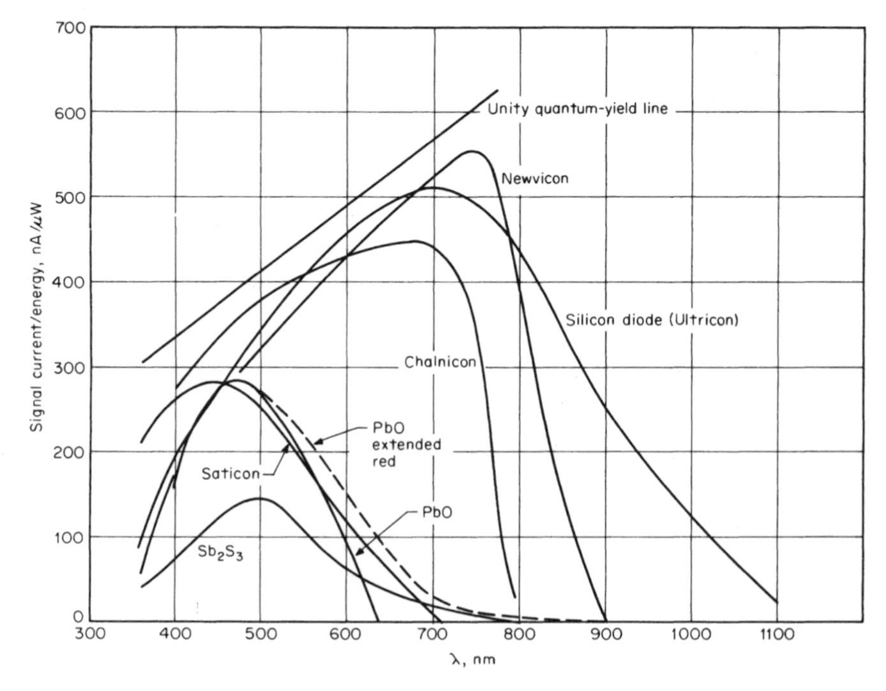

Figure 2.11.5 Absolute spectral response curves of various camera tube photoconductors.

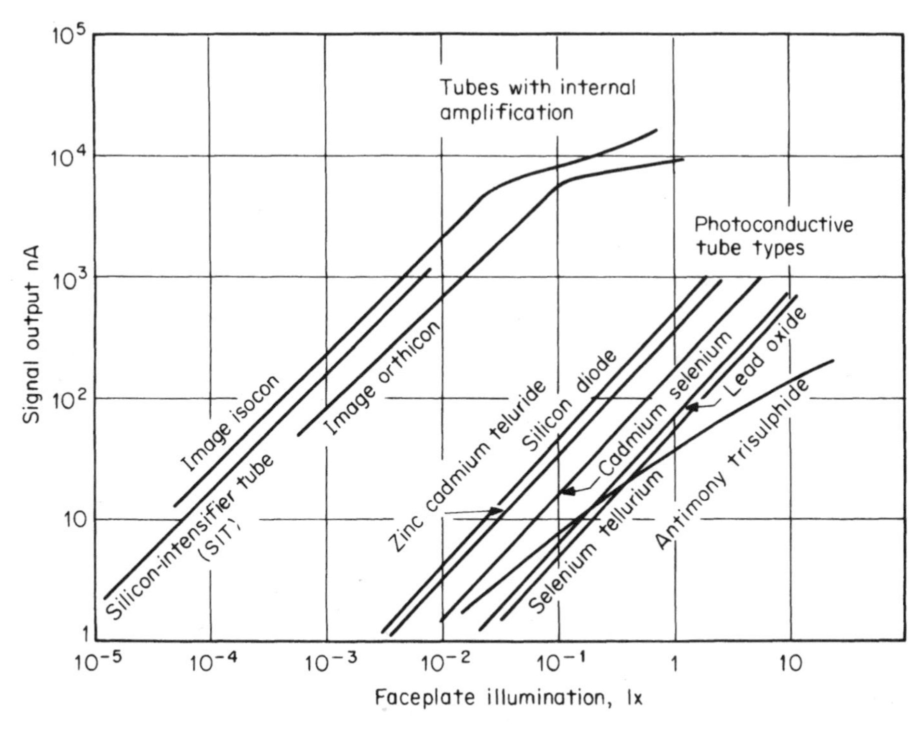

Figure 2.11.6 Light-transfer characteristics of typical camera tubes.

Table 2.11.1 Different Terminology Used in the Video and Film Industries for Comparable Imaging parameters (*After: Thorpe, Laurence J.: "HDTV and Film—Digitization and Extended Dynamic Range,"* 133rd SMPTE Technical Conference, *Paper no. 133-100, SMPTE, White Plains, N.Y., October 1991.*)

Video	Film
Sensitivity	Exposure index (EI) Speed
Resolution	Resolving power
Colorimetry	Sharpness Modulation transfer curves
Gray scale	Color reproduction
Dynamic range	Exposure latitude
Noise	Diffuse rms granularity

2.12 Monochrome and Color Image Display

Table 2.12.1 Computer Bus Data Speed Performance and Characteristics (*From: Robin, Michael, and Michel Poulin:* Digital Television Fundamentals, *McGraw-Hill, New York, N.Y., 1997. Used with permission.*)

	Theoretical bus speed, MBps	Bus format, bits	Bus frequency, MHz	CPU frequency, MHz
Main system buses				
▪ ISA	5	8/16	8	8–66
▪ EISA	33	32	8	8–130
▪ MCA	45	16/32	10	10–40
▪ Nu-bus	20	32	20–40	20–130
Local buses				
▪ VL	130	32	40	40–130
	260	64	50	50–130
▪ PCI	132	32	33	33–130
	264	64	33	33–130
	568	64	66	66–130
Over-the-top buses				
▪ Movie-2	242*	32/64	27	40–166
▪ HDV	432*	32/64	27	40–166
Switched bus				
▪ RACEway Interlink	1000	32/64	33–66	33–130

*8-bit and 10-bit video data words can be used.

Table 2.12.2 Examples of IBM-Type Computer Graphics Display Format Specifications (*From: Robin, Michael, and Michel Poulin:* Digital Television Fundamentals, *McGraw-Hill, New York, N.Y., 1997. Used with permission.*)

Standard	VGA	SVGA	SVGA	SVGA	SVGA	SVGA
Resolution, H×V	640×480	640×480	800×600	800×600	1024×768	1280×1024
Color depth	16/256K	16/256K	16/256K	16/256K	16/256K	16
V-active, lines	480	480	600	600	768	1024
V-total, lines	525	525	666	628	806	1068
Hor. freq., kHz	31.469	37.8	48.077	37.879	56.476	76.02
H-active, μs	25.422	20.317	16	20.00	13.653	10.119
Vert. freq., Hz	59.94	72.2	72.188	60.316	70.069	71.18
Pixel clock, MHz	25.2	31.5	50	40	75	126.5
Video bandwidth, MHz	12.6	15.75	25	20	37.5	63.24

Table 2.12.3 Examples of Mac-Type Computer Graphics Display Format Specifications (*From: Robin, Michael, and Michel Poulin:* Digital Television Fundamentals*, McGraw-Hill, New York, N.Y., 1997. Used with permission.*)

Standard	Mac II, 13", 15"	Mac II, 16", 17"	Mac II, 19"	Mac II, two-page
Resolution, H×V	640×480	832×624	1024×768	1152×870
Color depth	16/256/16M	16/256/16M	16/256	16/256
V-active, lines	480	624	768	870
V-total, lines	525	667	813	91
Hor. freq., kHz	35.000	49.724	60.24	68.681
H-active, μs	21.164	14.524	12.76	11.520
Vert. freq., Hz	66.67	74.55	75.00	75.06
Pixel clock, MHz	30.240	57.283	80	100.000
Video bandwidth, MHz	15.12	28.7	40	50

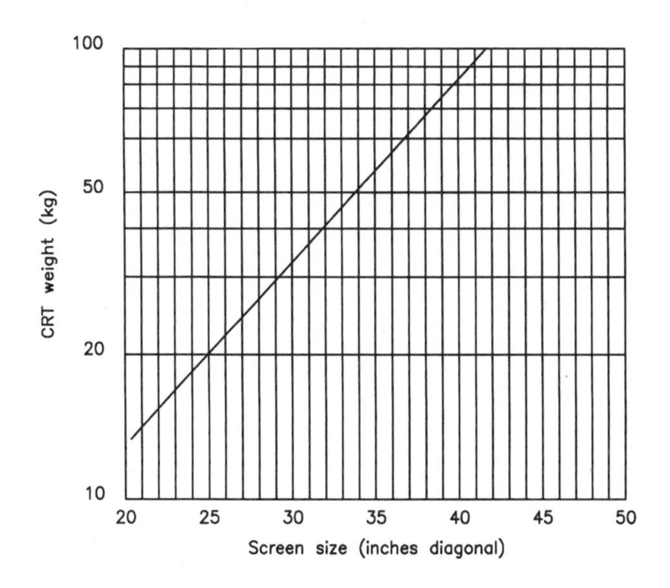

Figure 2.12.1 Relationship between screen size and CRT weight. (*After: Mitsuhashi, Tetsuo: "HDTV and Large Screen Display,"* Large-Screen Projection Displays II*, William P. Bleha, Jr. (ed.), Proc. SPIE 1255, SPIE, Bellingham, Wash., pp. 2-12, 1990.*)

Table 2.12.4 Color CRT Diagonal Dimension vs. Weight

Diagonal visible (in)	Weight (kg)	Weight (lb.)
19	12	26
25	23	51
30	40	88

Table 2.12.5 Comparative Resolution of Shadow-Mask Designs *(After: Benson, K. B., and D. G. Fink:* HDTV: Advanced Television for the 1990s*, McGraw-Hill, New York, 1990.*)

19V (48 cm) NGB tube type	Mask Material (mm)[1]	Vertical Pitch (mm)[1]	Center Hole Diameter (mm)[1]	Screen Vertical Pitch (mm)	N_t, Trios in Screen	N_r = sq. rt. (NT/1.33)
Conventional	0.15	0.56	0.27	0.60	400,000	500 lines
Monitor	0.15	0.40	0.19	0.43	800,000	775 lines
High resolution	0.13	0.30	0.15	0.32	1,400,000	1025 lines
1 Flat shadow mask						

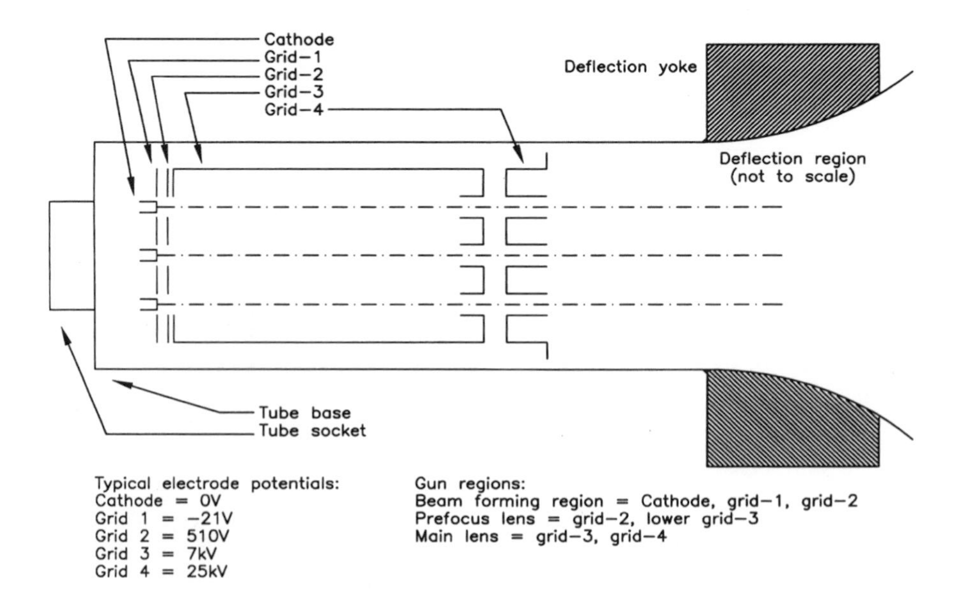

Figure 2.12.2 Simplified mechanical structure of a bipotential color electron gun.

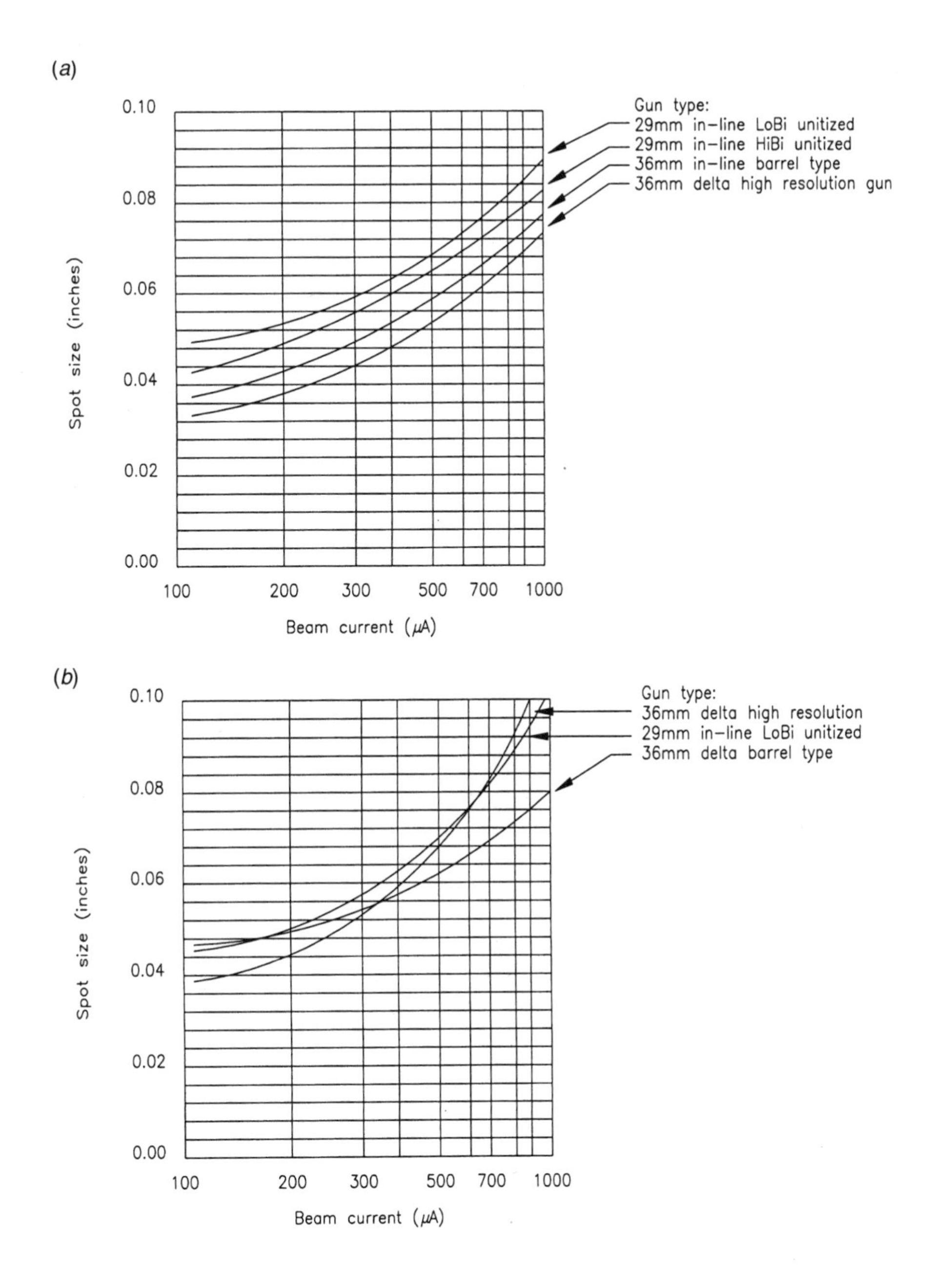

Figure 2.12.3 Spot-size comparison of high-resolution electron guns vs. commercial television guns: (*a*) 13-in vertical display, (*b*) 19-in vertical display.

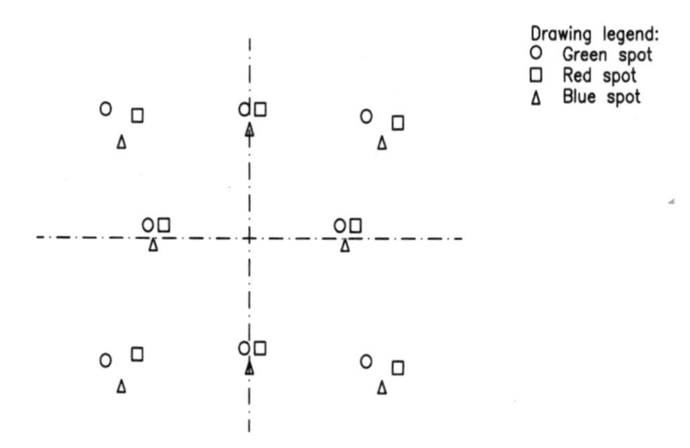

Figure 2.12.4 Astigmatism errors in a color CRT. (*After: Hutter, Rudolph G. E.: "The Deflection of Electron Beams,"* Advances in Image Pickup and Display, *B. Kazen (ed.), vol. 1, Academic Press, New York, pp. 212-215, 1974.*)

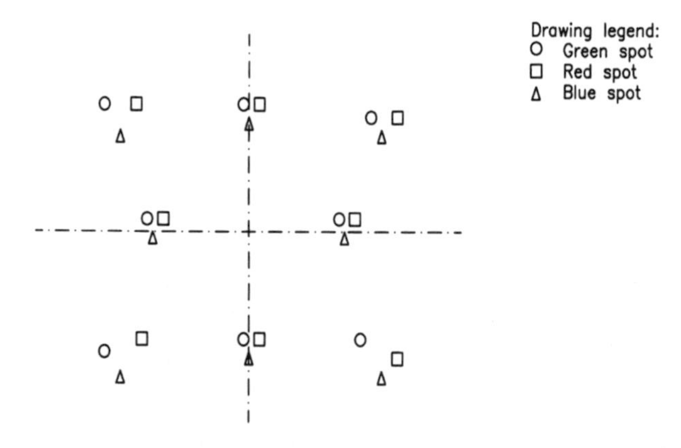

Figure 2.12.5 Astigmatism and coma errors in a color CRT. (*After: Hutter, Rudolph G. E.: "The Deflection of Electron Beams,"* Advances in Image Pickup and Display, *B. Kazen (ed.), vol. 1, Academic Press, New York, pp. 212-215, 1974.*)

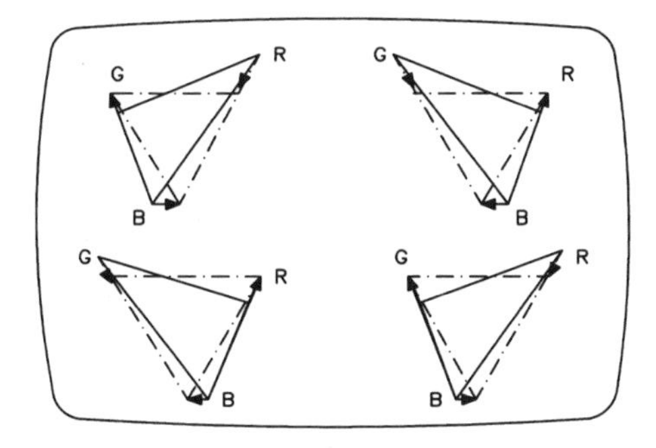

Figure 2.12.6 Misconvergence in the four corners of the raster in a color CRT. (*After: Hutter, Rudolph G. E.: "The Deflection of Electron Beams,"* Advances in Image Pickup and Display, *B. Kazen (ed.), vol. 1, Academic Press, New York, pp. 212-215, 1974.*)

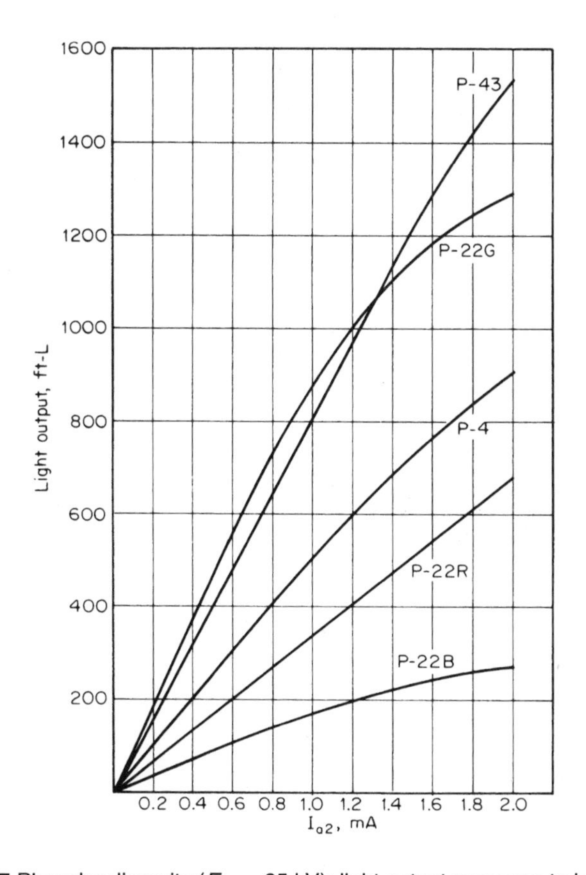

Figure 2.12.7 Phosphor linearity (E_{a2} = 25 kV); light output response to increasing current.

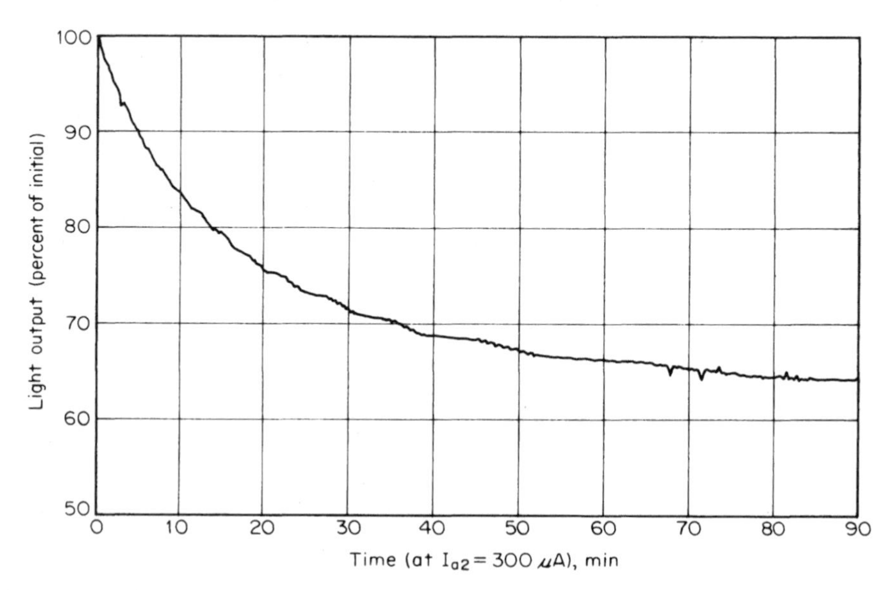

Figure 2.12.8 Loss in phosphor efficiency as screen heats at high-current operation (rare-earth green).

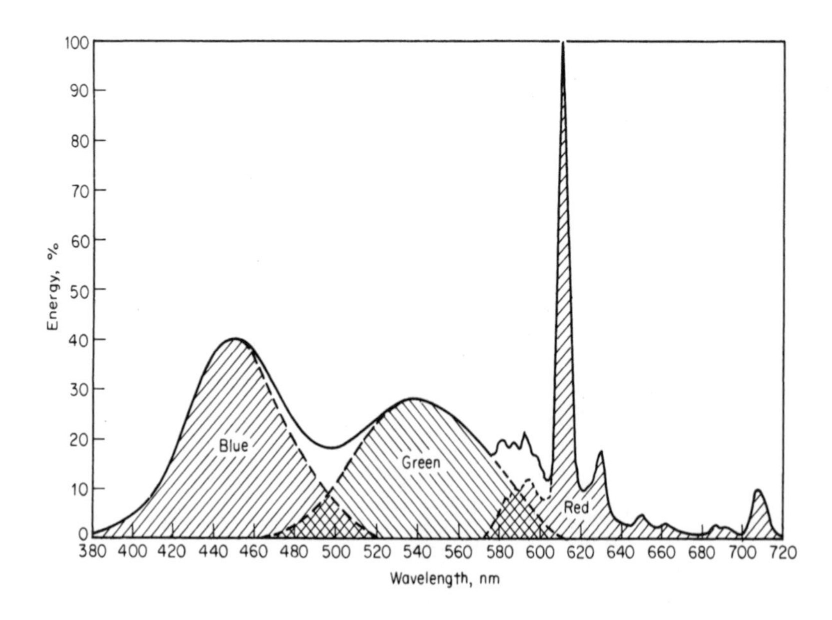

Figure 2.12.9 Typical spectral energy distribution (SED) color primaries at equal current density.

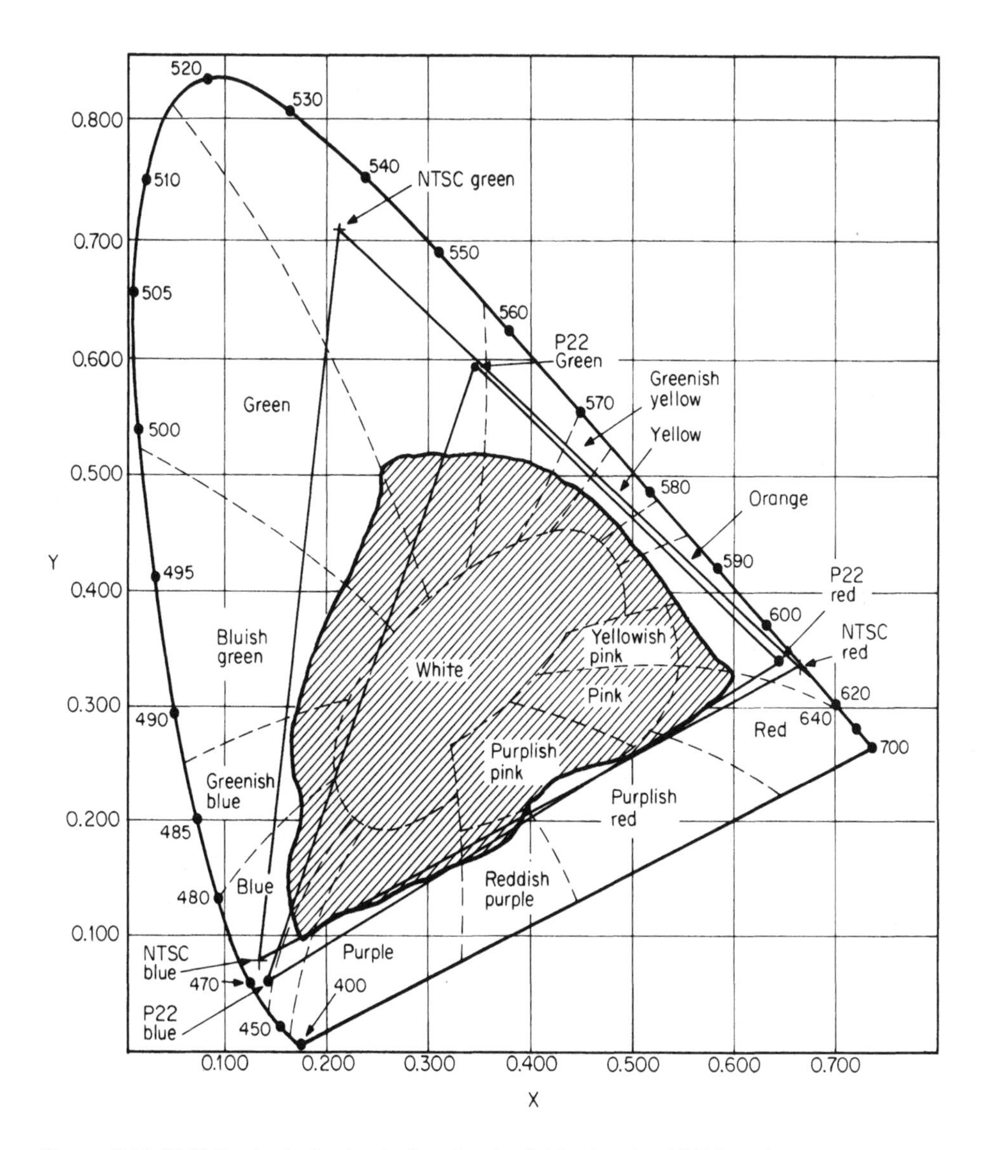

Figure 2.12.10 Kelly chart of color designation for lights showing NTSC and common commercial primary phosphors and locus of dyes, paints, and pigments.

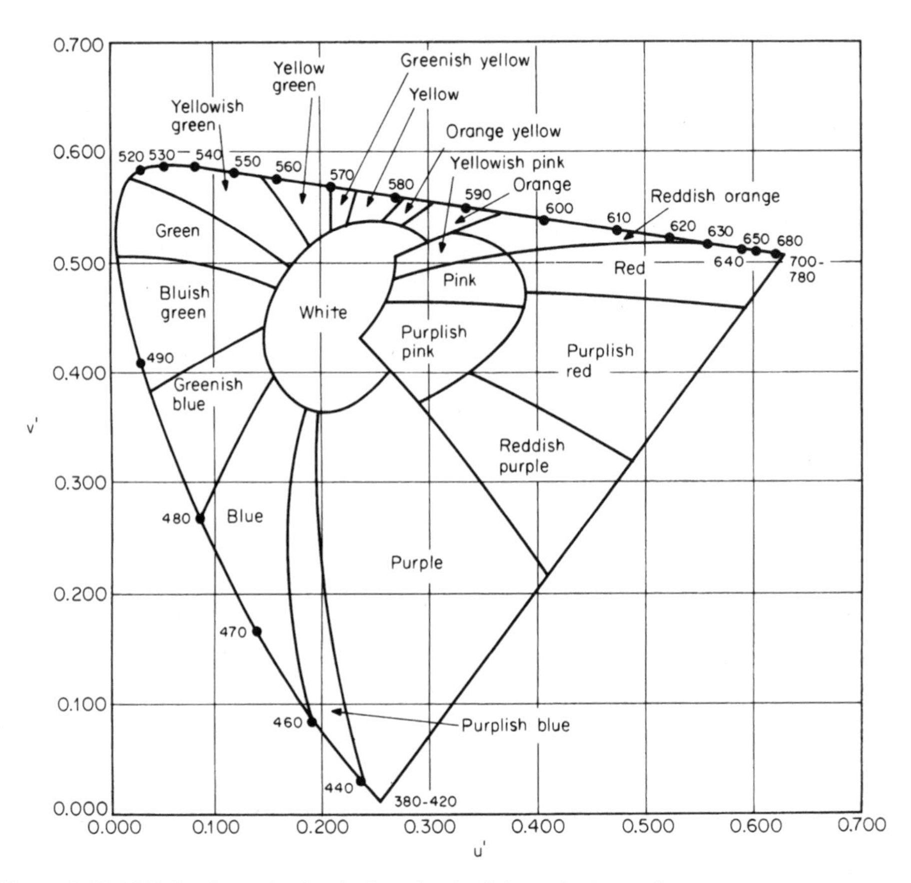

Figure 2.12.11 Kelly chart of color designation for lights, *u*′, *v*′ coordinates.

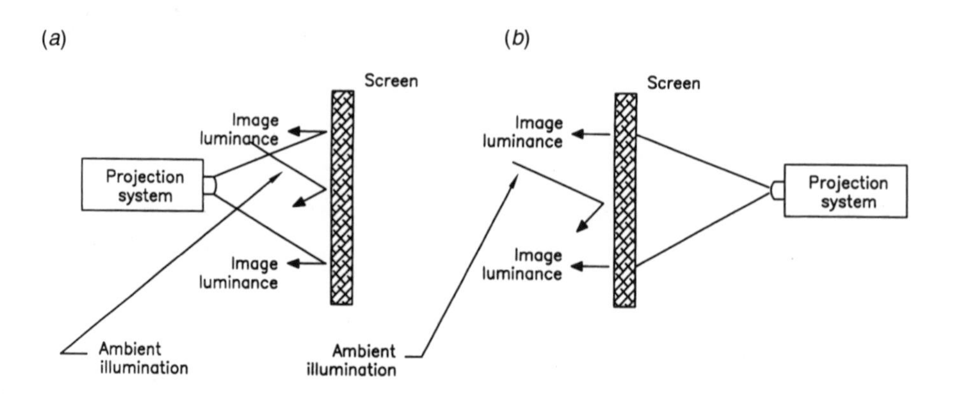

Figure 2.12.12 Projection-system screen characteristics: (*a*) front projection, (*b*) rear projection.

Table 2.12.6 Typical Characteristics of Common Phosphors

EIA no.	Worldwide designation	Use	Composition	Relative efficiency	Typical CIE coordinates				Decay
					x	y	u'	v'	
P-4	WW	Black and white television	$ZnS:Ag$ + $Zn_{(1-x)}Cd_xS:Cu,Al$	100	0.270	0.300	0.178	0.446	Medium short
P-1	GJ	Projection green	$ZnSiO_4:Mn$	130	0.218	0.712	0.079	0.577	Medium
P-43	GY	Projection green	$Gd_2O_2S:Tb$	155	0.333	0.556	0.148	0.556	Medium
P-22R	X	Red direct-view projection	$Y_2O_3:Eu$ $Y_2O_2S:Eu$	65	0.640 0.625	0.340 0.340	0.441 0.429	0.528 0.525	Medium short
P-22G	X	Green direct view	$Zn_{(1-x)}Cd_xS:Cu,Al$ $ZnS:Cu,Al$	180	0.340 0.285	0.595 0.600	0.144 0.119	0.566 0.561	Medium
P-22B	X	Blue direct-view projection	$ZnS:Ag$	25	0.150	0.065	0.172	0.168	Medium short

†Values are nominal; they may change with measurement methods and source of phosphor.

Table 2.12.7 Performance Levels of Video and Theater Displays

Display System	Luminous Output (Brightness), nits (ft-L)	Contrast Ratio at Ambient Illumination (fc)	Resolution (TVL)
Television receiver	200–400, 60–120	30:1 at 5	275
Theater (film projector)	34–69, 10–20[1]	100:1 at 0.1[2]	1000 and up

[1] U. S. standard (PH-22.124-1961)

[2] Limited by lens flare.

Table 2.12.8 Screen Gain Characteristics for Various Materials *(After: Luxenberg, H., and R. Kuehn:* Display Systems Engineering*, McGraw-Hill, New York, 1968.)*

Screen Type	Gain
Lambertian (flat-white paint, magnesium oxide)	1.0
White semigloss	1.5
White pearlescent	1.5–2.5
Aluminized	1–12
Lenticular	1.5–2
Beaded	1.5–3
Ektalite (Kodak)	10–15
Scotch-light (3M)	Up to 200

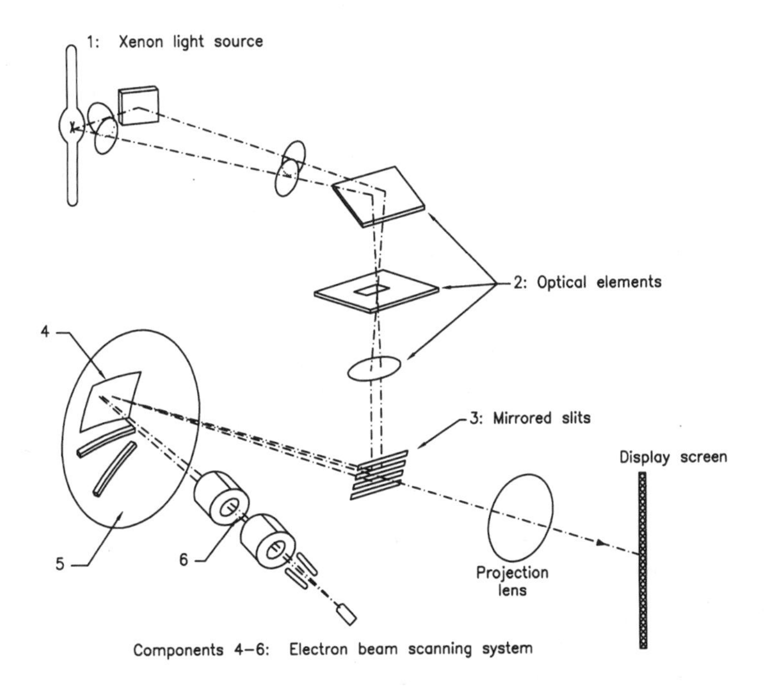

Figure 2.12.13 Mechanical configuration of the Eidophor projector optical system.

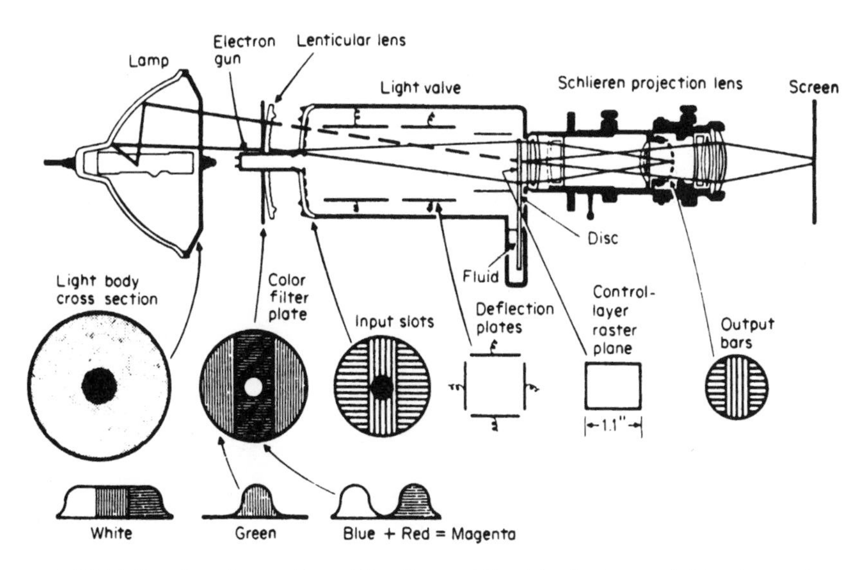

Figure 2.12.14 Functional operation of the General Electric single-gun light-valve system.

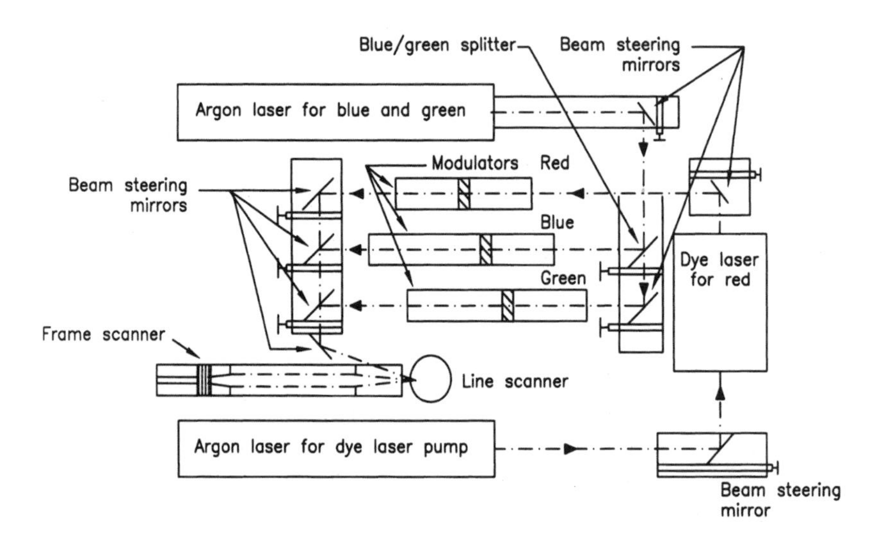

Figure 2.12.15 Mechanical configuration of a color laser projector. (*After: Pease, Richard W.: "An Overview of Technology for Large Wall Screen Projection Using Lasers as a Light Source,"* Large Screen Projection Displays II, *William P. Bleha, Jr. (ed.), Proc. SPIE 1255, SPIE, Bellingham, Wash., pp. 93-103, 1990.*)

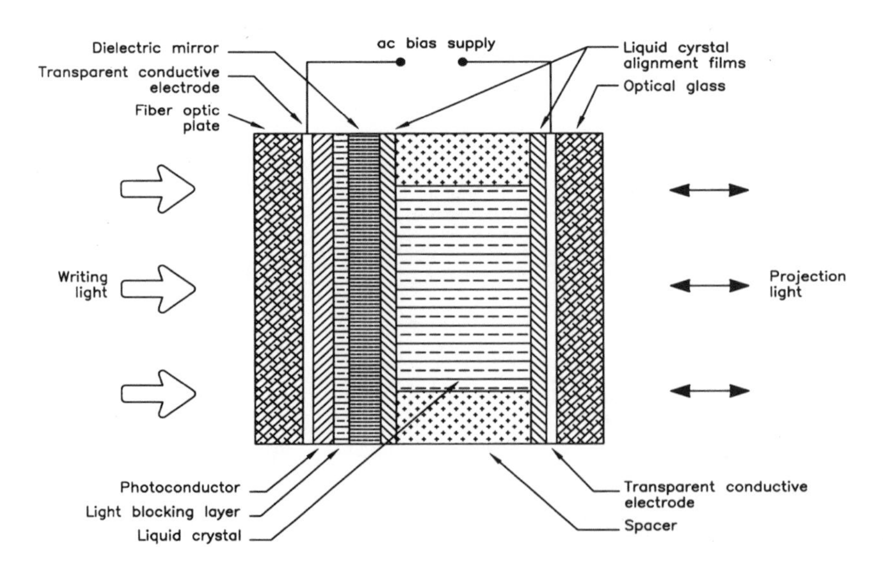

Figure 2.12.16 Mechanical configuration of a homeotropic liquid crystal light valve. (*After: Fritz, Victor J.: "Full-Color Liquid Crystal Light Valve Projector for Shipboard Use,"* Large Screen Projection Displays II, *William P. Bleha, Jr. (ed.), Proc. SPIE 1255, SPIE, Bellingham, Wash., pp. 59–68, 1990.*)

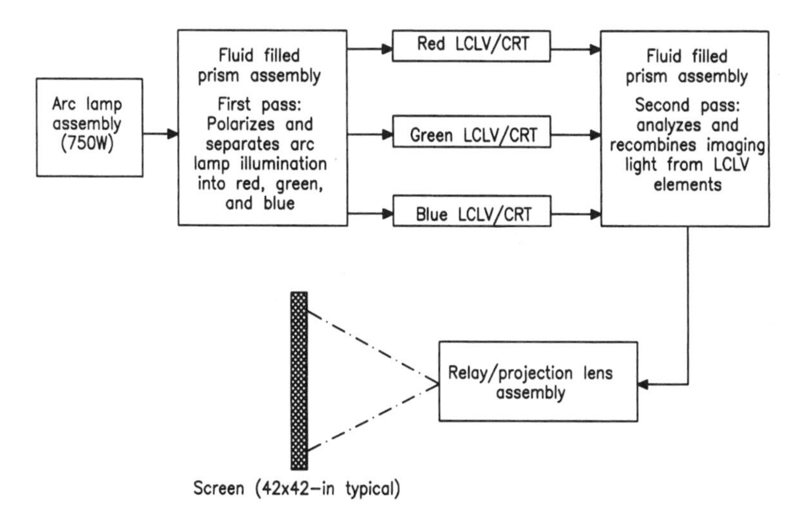

Figure 2.12.17 Mechanical configuration of a color homeotropic liquid crystal light valve projector. (*After: Fritz, Victor J.: "Full-Color Liquid Crystal Light Valve Projector for Shipboard Use,"* Large Screen Projection Displays II, *William P. Bleha, Jr. (ed.), Proc. SPIE 1255, SPIE, Bellingham, Wash., pp. 59–68, 1990.*)

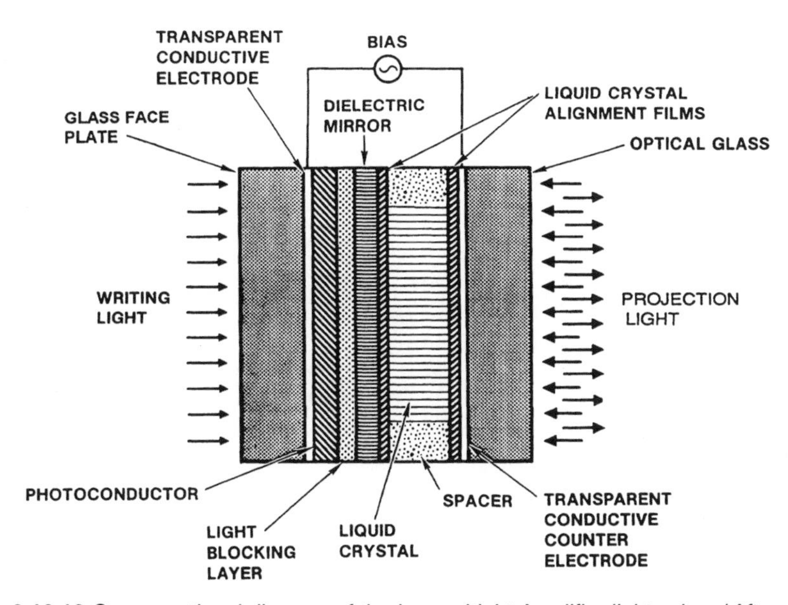

Figure 2.12.18 Cross-sectional diagram of the Image Light Amplifier light valve. (*After: Bleha, W. P.: "Image Light Amplifier (ILA) Technology for Large-Screen Projection,"* SMPTE Journal, *SMPTE, White Plains, N.Y., pp. 710–717, October 1997. Courtesy of Hughes-JVC Technology.*)

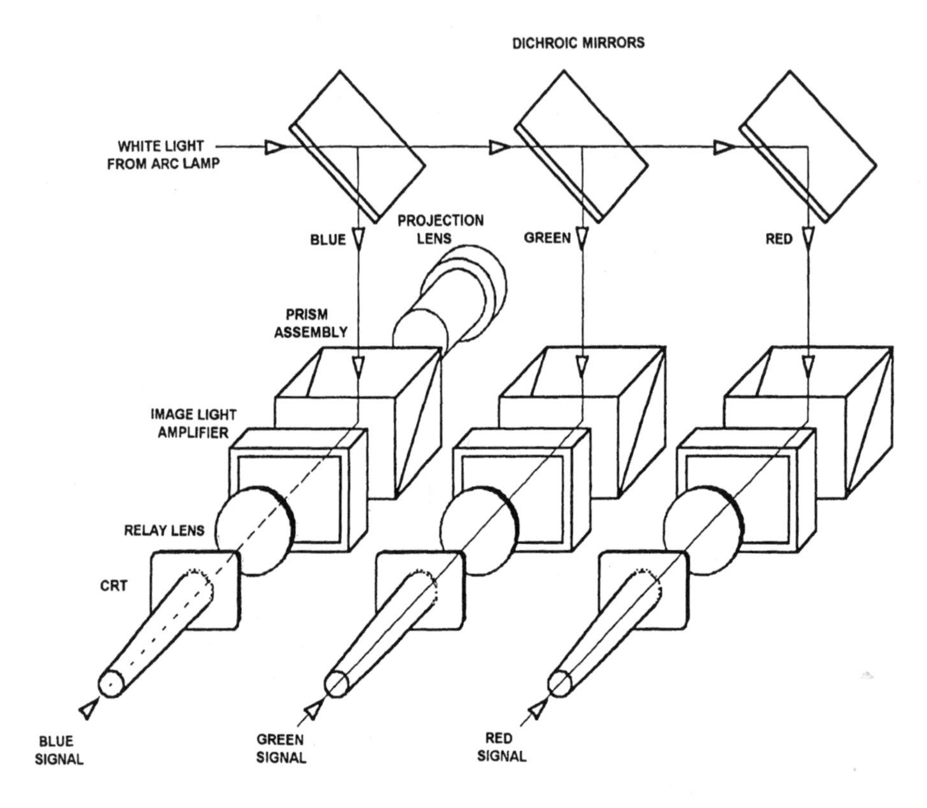

Figure 2.12.19 Full-color optical system for the ILA projector. (*After: Bleha, W. P.: "Image Light Amplifier (ILA) Technology for Large-Screen Projection,"* SMPTE Journal, *SMPTE, White Plains, N.Y., pp. 710–717, October 1997. Courtesy of Hughes-JVC Technology.*)

Table 2.12.9 Minimum Projector Specifications for Consumer and Theater Displays (*After: Glenn, W. E., C. E. Holton, G. J. Dixon, and P. J. Bos: "High-Efficiency Light Valve Projectors and High-Efficiency Laser Light Sources,"* SMPTE Journal, *SMPTE, White Plains, N.Y., pp. 210–216, April 1997.*)

Parameter	Consumer Display	Theater Display
Resolution	Greater than 750 TV lines	Greater than 750 TV lines
Light output	1000 lm	10,000 lm
Cost	Less than $2000	Less than $50,000
Response time	Less than 10 ms	Less than 10 ms
Power consumption	Less than 300 W	Less than 3000 W
Small area uniformity	±0.25 percent	±0.25 percent
Contrast ratio	Greater than 90:1	Greater than 90:1
Flicker	Undetectable	Undetectable

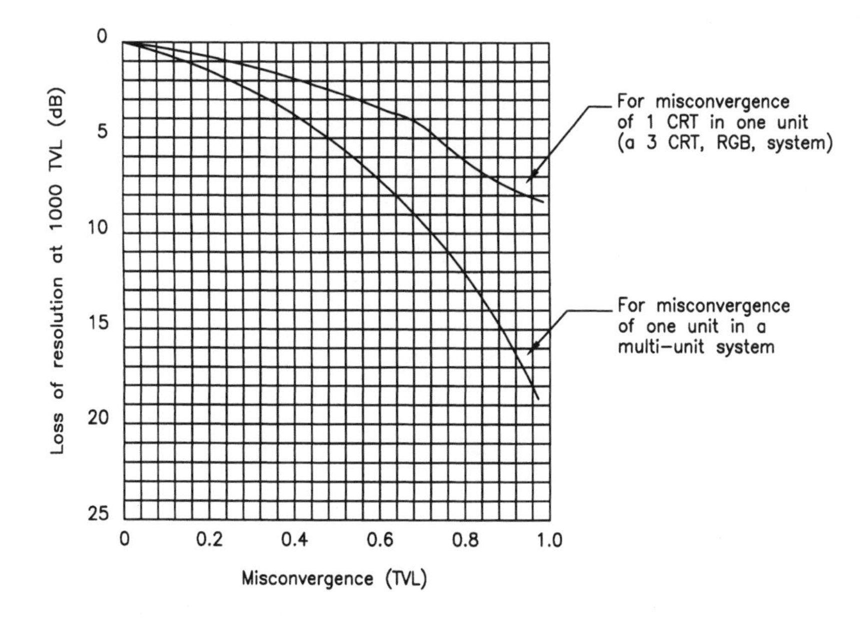

Figure 2.12.20 The relationship between misconvergence and display resolution. (*After: Mitsuhashi, Tetsuo: "HDTV and Large Screen Display,"* Large-Screen Projection Displays II*, William P. Bleha, Jr. (ed.), Proc. SPIE 1255, SPIE, Bellingham, Wash., pp. 2–12, 1990.*)

Table 2.12.10 Appearance of Scan Line Irregularities in a Projection Display (*After: Mitsuhashi, Tetsuo: "HDTV and Large Screen Display,"* Large-Screen Projection Displays II*, William P. Bleha, Jr. (ed.), Proc. SPIE 1255, SPIE, Bellingham, Wash., pp. 2–12, 1990.*)

Condition of Adjacent Lines	S/N (p-p)
Clearly overlapped	Less than 69 dB
Just before overlapped	75 dB
No irregularities	Greater than 86 dB

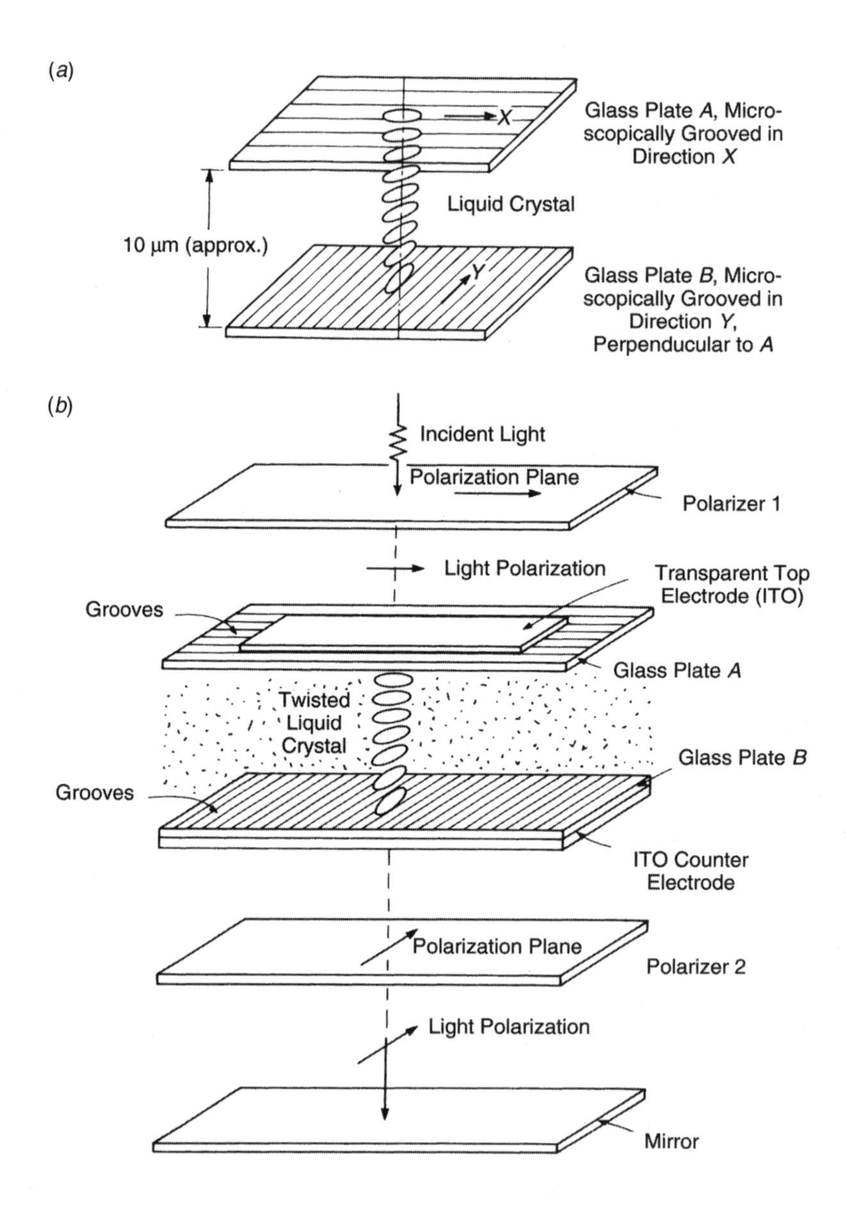

Figure 2.12.21 LCD architecture: (*a*) twisted nematic cell, (*b*) LCD element. (*After: Allison, J.:* Electronic Engineering Semiconductors and Devices, *2nd ed., McGraw-Hill, London, pg. 308–309, 1990.*)

2.13 Audio/Video Recording Systems

Table 2.13.1 Conversion Factors from CGS to MKSA or SI Units

Parameter	CGS units†	Multiply by	To obtain MKSA or SI units†
Flux Φ	Maxwell	10^{-8}	Webers (Wb)
Flux density B	Gauss	10^{-4}	Webers/meter² = 1 tesla (T)
Magnetization M	Gauss (1 gauss = 1 emu/cm^3)	10^3	Ampere turns/meter (At/m)
Permeability μ_0 of free space	1	$4\pi \times 10^{-7}$	Henry/meter (H/m)
Magnetomotive force F	Gilbert	$\frac{1}{0.4\pi}$	Ampere turns (At)
Field H (magnetomotive force per unit length)	Oersted	$\frac{10^3}{4\pi}$	Ampere turns/meter (At/m)

†Unit system abbreviations: CGS (centimeter-gram-second), MKSA (meter-kilogram-second-ampere), and SI (International System).

Table 2.13.2 Properties of Typical Recording Head Core Materials

Material	Permeability μ	Coercivity H_c, Oe	Saturation induction, G	Resistivity ρ	Curie temp., °C	Coef. of expan., 10^{-6} cm/(cm·°C)
Alfesil	See Fig. 15-35	0.06	10,000	90 μΩ·cm	500	18
Hot-pressed ferrite Mn Zn	See Fig. 15-34	0.02–0.15	4000–6000	Approx. 10^5 μΩ·cm	90–300	9.5–11.5
Single-crystal ferrite	Approx. 300–500, at 5 MHz	0.05	4000–5000	Approx. 10^5 μΩ·cm	140–250	9–11

Table 2.13.3 Properties of Soft and Hard Magnetic Materials

Material	M_s, G	$B_s = 4\pi M_s$, G	H_c, O	B_r, G	μ (dc) initial	Resistivity, $\Omega \cdot$cm	Thermal expn.	Curie temp.	Vickers hardness
Soft magnetic materials									
Iron Fe	1700	21,362	1	......	20,000				
Hi-Mu 80 80% Ni, 20% Fe	661	8,300	0.02	......	50,000	65	12.9×10^{-6} cm/(cm°C)	733 °C	127
Alfesil (Sendust) 85% Fe, 6% Al, 9% Si	796	10,000	0.06	......	10,000	90	11.3×10^{-6} cm/(cm°C)	773 °C	496
Mn Zn, Hot-pressed ferrite	358	4,500	0.02–0.2	≈900	2000–5000	10^4	$10\text{–}15 \times 10^{-6}$ cm/(cm°C)	100–300 °C	650–750
Ni Zn, Hot-pressed ferrite	238	3,000	0.15–3	≈1800	100–2000	10^{10}	$7\text{–}9 \times 10^{-6}$ cm/(cm°C)	150–200 °C	700–750
Hard magnetic materials									
					Squareness ratio				
γ-Ferric oxide	400	5,026	300–350	1300†	0.75†				
Chromium dioxide	470	6,000	300–700	1600†	0.9†				
Metal particles	800	10,000	1000	3500†	0.8†				

†Value typical for finished tape.

Table 2.13.4 Physical Properties of Poly(ethyleneterephthalate) Base

Property	Balanced	Tensilized
Tensile strength, lb/in^2 N/m^2	25,000 172.38×10^6	40,000 275.8×10^6
Force to elongate 5%, lb/in^2 N/m^2	14,000 96.53×10^6	22,000 151.69×10^6
Elastic modulus, lb/in^2 N/m^2	550,000 3.79×10^9	1,100,000 7.58×10^9
Elongation,%	130	40
Thermal coefficient of linear expansion, per deg C	1.7×10^{-5}	1.7×10^{-5}
Shrinkage at 100°C, % (per 30 min interval)	0.4	2.5
Note: Measurements in machine direction		

Table 2.13.5 Physical Properties of a Poly(esterurethane)

Parameter	Value
Tensile strength, lb/in^2 N/m^2	8000 55.16×10^6
Stress at 100% elongation, lb/in^2 N/m^2	300 2.07×10^6
Ultimate elongation,%	450
Glass transition temperature, deg C	12
Hardness, shore A	76
Density, g/cm^3	1.17
Viscosity at 15% solids/tetrahydrofuron, cP	800
Pa· s	0.8

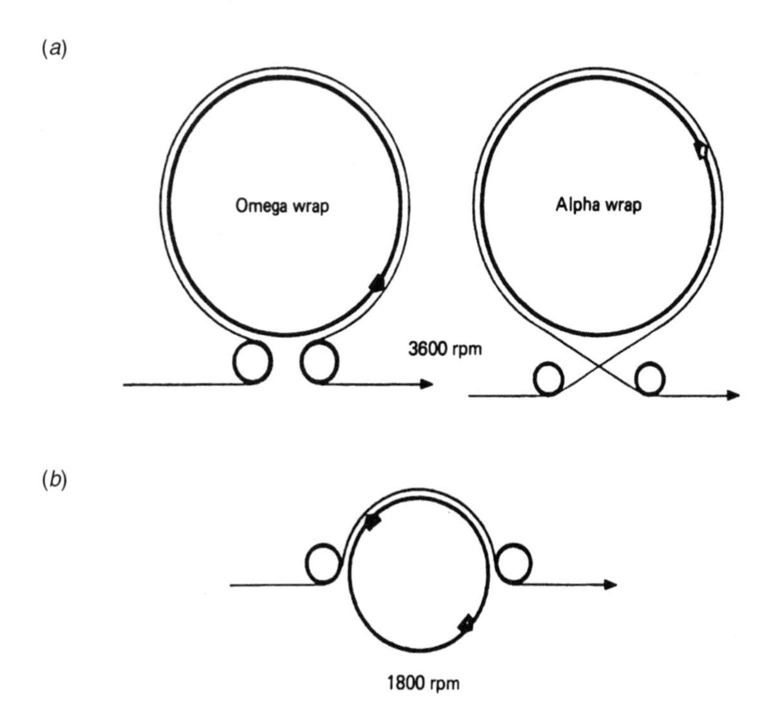

Figure 2.13.1 Typical helical geometries: (*a*) full wrap field per scan, (*b*) half wrap field per scan or segmented.

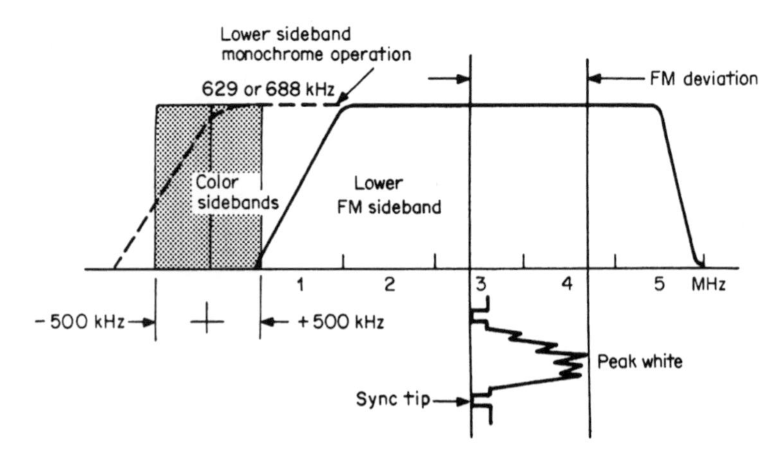

Figure 2.13.2 The recording spectrum used by video tape machines that use the color-under recording process. The AM chrominance subcarrier is added to the FM luminance carrier and recorded on tape. Part of the lower sideband of the FM luminance signal is removed to allow for the chrominance signal.

Table 2.13.6 Professional Analog VTR Specifications (*From: Epstein, Steve: "Videotape Storage Systems,"* The Electronics Handbook, *Jerry C. Whitaker (ed.), CRC Press, Boca Raton, Fla., 1996. Used with permission.*)

	2-in Quad	1-in type C	3/4-in U-matic	BetacamSP	M-II	S-VHS	Hi-8
Signal system	FM, composite	FM, composite	FM, composite luminance, color under chrominance	FM, component	FM, component	FM, composite luminance, color under chrominance	FM, composite luminance, color under chrominance
Scanning method	Transverse, 4 heads	Helical, 3 heads +3 sync heads	Helical, 2 heads, 2 tracks/revolution	Helical, 1 track pair per field	Helical, 1 track pair per field	Helical, minimum of 2 heads	Helical, minimum of 2 heads
Peak-white freq., MHz	10	10	5.4	7.7	7.7	7.0	7.7
Black level freq. MHz	7.9	7.9	3.8	5.7	6.2	5.4	5.7
Chroma carrier, kHz	N/A	N/A	688	CTDM recording	CTDM recording	629	743.4
Media	2-in gamma ferric oxide, oriented trans versely, open reel	1-in cobalt modified gamma ferric oxide, open reel	3/4-in wide, 1.1-mil cobalt modified gamma ferric oxide, cassette 2 shell sizes	12.7 mm, 14-μm-thick metal particle cassettes, 2 shell sizes	12.7 mm, 13-μm-thick metal particle cassettes, 2 shell sizes	1/2-in, 0.8-mil-thick, cobalt modified gamma ferric oxide, cassette	1/4-in, 10-μm-thick, enhanced metal particle, cassette
Tape speed	75 ft/min	48 ft/min	18.75 ft/min	7.1 m/min	4.06 m/min	0.667 m/min	0.86 m/min
Tip to tape speed	1550 i/s	1009 i/s	410 i/s	7.07 m/s	7/09 m/s	5.8 m/s	3.76 m/s
Track width	10 mil	5.1 mil	3.35 mil	42 μm	42 μm	19.3 μm	20.5 μm
Track pitch	15 mil	7.2 mil	5.39 mil	84.5 μm	84.5 μm	19.3 μm	20.5 μm
Track length	1.818 in	16.1718 in	6.68 in (calculated)	115 mm	118.25 mm	97 mm	62.6 mm
Track angle	89.43°	2° 24'	4° 57'	4.67°	4.25°	6.0°	4° 53'
Head azimuth angle	Perpendicular to plane of scanner	Perpendicular to track	±7° from perpendicular	±15° from perpendicular	±15° from perpendicular	±6° from perpendicular	±10° from perpendicular
Scanner diameter	2.64 in	5.35 in	4.34 in	74.49 mm	76 mm	62 mm	40 mm
Scanner rotation, rev/s	240	59.94	29.97	29.97	29.97	29.97	29.97
Wrap angle	115°	330°	180°	180°	180°	180°	180°
Audio channels	1 plus a cue track	3 longitudinal, time code on A3	2 longitudinal plus an address track for time code	2 longitudinal, 2 FM centered on 310 and 540 kHz recorded by chrominance video head, plus timecode	2 longitudinal, 2 FM centered on 400 and 700 kHz recorded by chrominance video head, plus timecode	2 longitudinal	2 longitudinal

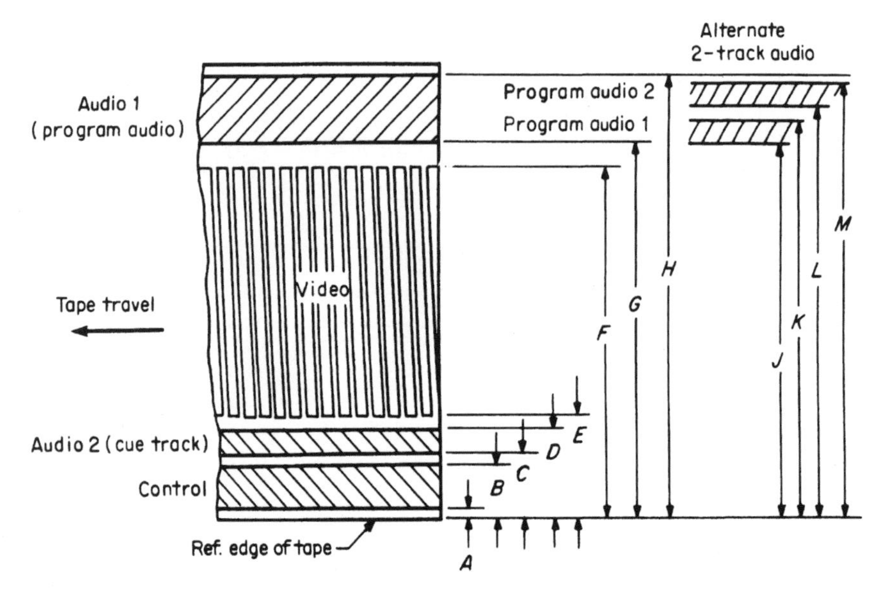

Dimensions	in Min.	in Max.	mm Min.	mm Max.	Dimensions	in Min.	in Max.	mm Min.	mm Max.
A	0.000	0.004	0.00	0.10	*G*	1.921	1.930	48.79	49.02
B	0.040	0.049	1.02	1.24	*H*	1.988	1.996	50.50	50.70
C	0.058	0.062	1.47	1.57	*J*	1.920	1.928	48.77	48.97
D	0.078	0.085	1.98	2.16	*K*	1.945	1.951	49.40	49.56
E	0.087	0.094	2.21	2.39	*L*	1.965	1.971	49.91	50.06
F	1.902	1.914	48.31	48.62	*M*	1.988	1.996	50.50	50.70

Figure 2.13.3 The recorded format used in 2-in quadruplex machines. Audio 2, as shown, is also used as the cue track. (*From: Fink, D. G., and D. Christiansen (eds.):* Electronic Engineers' Handbook, *2nd ed., McGraw-Hill, New York, N.Y., 1982. Used with permission.*)

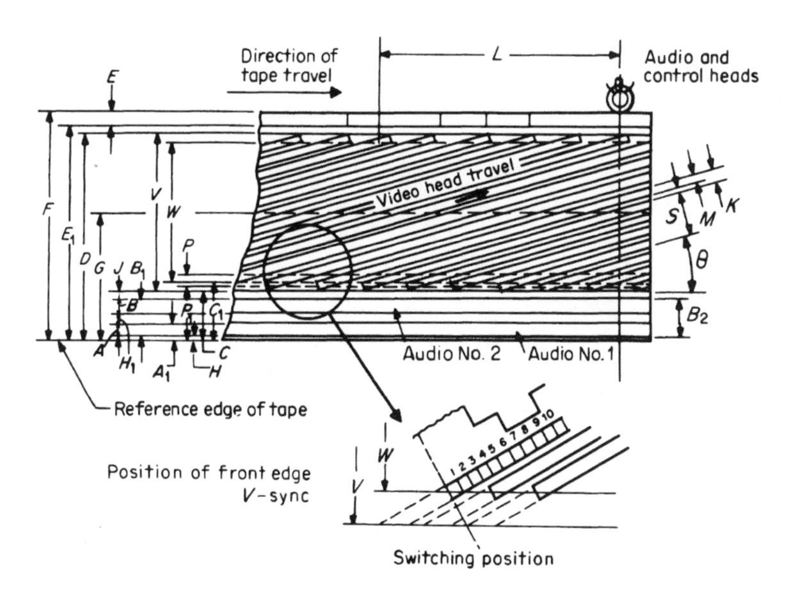

Dimensions	mm	in
A Audio 1 width	0.80 ± 0.05	0.0315 ± 0.0020
A_1 Audio 1 reference	1.00 nom.	0.0394 nom.
B Audio 2 width	0.80 ± 0.05	0.0315 ± 0.0020
B_1 Audio 2 reference	2.50 nom.	0.0984 nom.
B_2 Audio track total width	2.30 ± 0.08	0.0906 ± 0.0031
C Video area lower limit	2.70 min.	0.1063 min.
C_1 Video effective area lower limit	3.05 min.	0.1201 min.
D Video area upper limit	18.20 max.	0.7165 max.
E Control track width	0.60 nom.	0.0236 nom.
E_1 Control track reference	18.40 + 0.28 − 0.18	0.7244 + 0.7244 − 0.0071
F Tape width	19.00 ± 0.03	0.7480 ± 0.0012
G Video trace center from reference edge	10.45 ± 0.05	0.4114 ± 0.0020
H Audio guard band to tape edge	0.2 ± 0.1	0.008 ± 0.004
H_1 Audio-to-audio guard band	0.7 nom.	0.028 nom.
J Audio-to-video guard band	0.2 nom.	0.008 nom.
K Video track pitch (calculated)	0.137 nom.	0.00539 nom.
L Audio and control head position from end of 180° scan	74.0 nom.	2.913 nom.
M Video track width	0.085 ± 0.007	0.00335 ± 0.00028
P† Address track width	0.50 ± 0.05	0.0197 ± 0.0020
P_1 Address track lower limit	2.90 ± 0.15	0.1142 ± 0.0059
S Video guard band width	0.052 nom.	0.00205 nom.
Y Video width	15.5 nom.	0.610 nom.
W Video effective width	14.8 nom.	0.583 nom.
θ Video track angle, moving tape	4°57′ 33.2″	
stationary tape	4°54′ 49.1″	

†For reference value only.

Figure 2.13.4 Track locations and dimensions specified for the 3/4-in U-matic format. (*From: Fink, D. G., and D. Christiansen (eds.):* Electronic Engineers' Handbook, *2nd ed., McGraw-Hill, New York, N.Y., 1982. Used with permission.*)

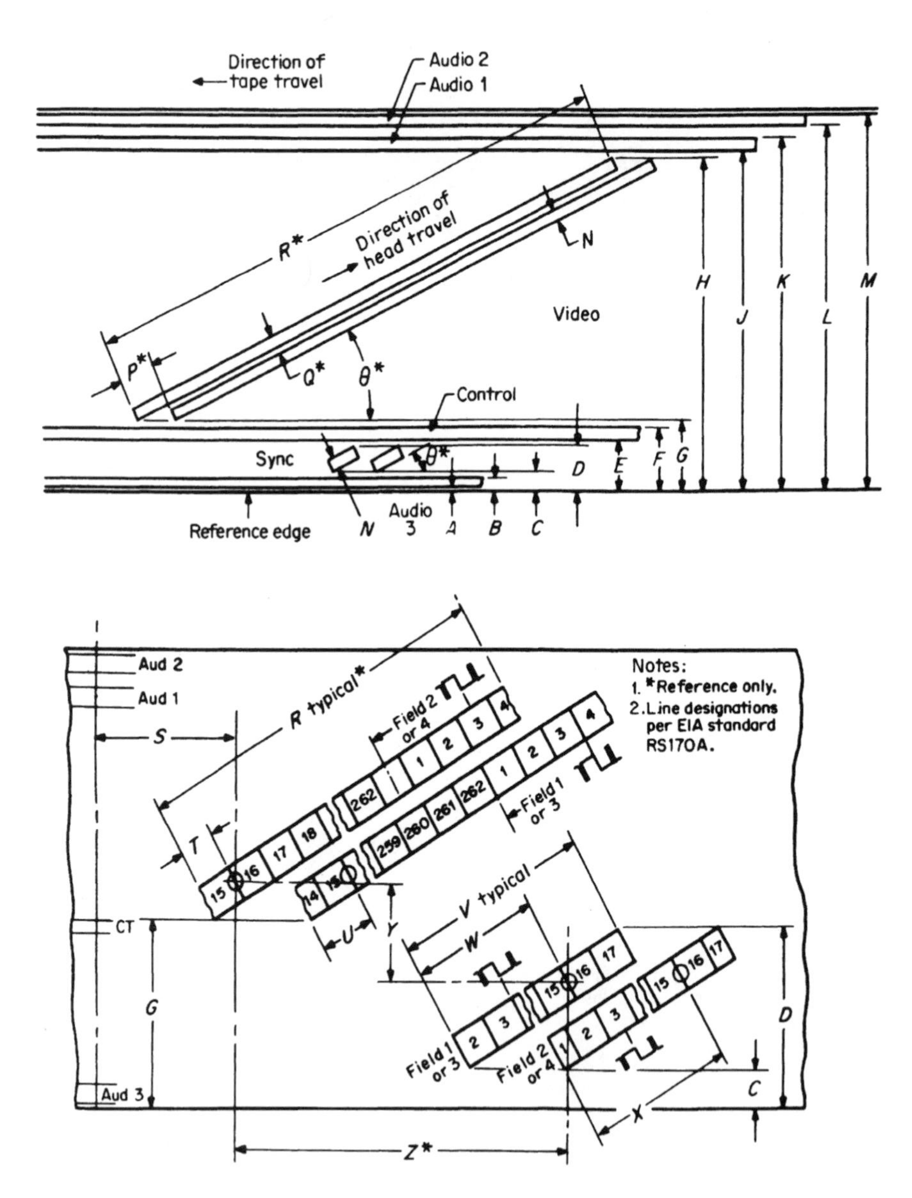

Figure 2.13.5 Type C track layout showing record locations and principle dimensions. (*From: Fink, D. G., and D. Christiansen (eds.):* Electronic Engineers' Handbook, *2nd ed., McGraw-Hill, New York, N.Y., 1982. Used with permission.*)

Table 2.13.7 Type C Video and Sync Specifications (*From: Fink, D. G., and D. Christiansen (eds.):* Electronic Engineers' Handbook, *2nd ed., McGraw-Hill, New York, N.Y., 1982. Used with permission.*)

	Dimensions	mm		in	
		Minimum	Maximum	Minimum	Maximum
A	Audio 3 lower edge	0.000	0.200	0.00000	0.00787
B	Audio 3 upper edge	0.775	1.025	0.03051	0.04035
C	Sync track lower edge	1.385	1.445	0.05453	0.05689
D	Sync track upper edge	2.680	2.740	0.10551	0.10787
E	Control tract lower edge	2.870	3.130	0.11299	0.12323
F	Control track upper edge	3.430	3.770	0.13504	0.14843
G	Video track lower edge	3.860	3.920	0.15197	0.15433
H	Video track upper edge	22.355	22.475	0.88012	0.88484
J	Audio 1 lower edge	22.700	22.900	0.89370	0.90157
K	Audio 1 upper edge	23.475	23.725	0.92421	0.93406
L	Audio 2 lower edge	24.275	24.525	0.95571	0.96555
M	Audio 2 upper edge	25.100	25.300	0.98819	0.99606
N	Video and sync track width	0.125	0.135	0.00492	0.00531
P	Video offset	4.067 (2.5*H*) ref.		0.16012 nom.	
Q	Video track pitch	0.1823 ref.		0.007177 nom.	
R	Video track length	410.764 (252.5*H*) ref.		16.17181 nom.	
S	Control track head distance	101.60	102.40	4.0000	4.0315
T	Vertical phase odd field	1.220 (0.75*H*)	2.030 (1.25*H*)	0.04803	0.07992
U	Vertical phase even field	2.030 (1.25*H*)	2.850 (1.75*H*)	0.07992	0.11220
V	Sync track length	25.620 (15.75*H*)	26.420 (16.25*H*)	1.00866	1.04016
W	Vertical phase odd sync field	22.360 (13.75*H*)	23.170 (14.25*H*)	0.88031	0.91220
X	Vertical phase even sync field	23.170 (14.25*H*)	28.980 (14.75*H*)	0.91220	0.94409
Y†	Vertical head offset	1.529 nom.		0.06020 nom.	
Z†	Horizontal head offset	35.350 nom.		1.39173 nom.	
θ	Track angle	2°34′ ref.			

†Reference value only.

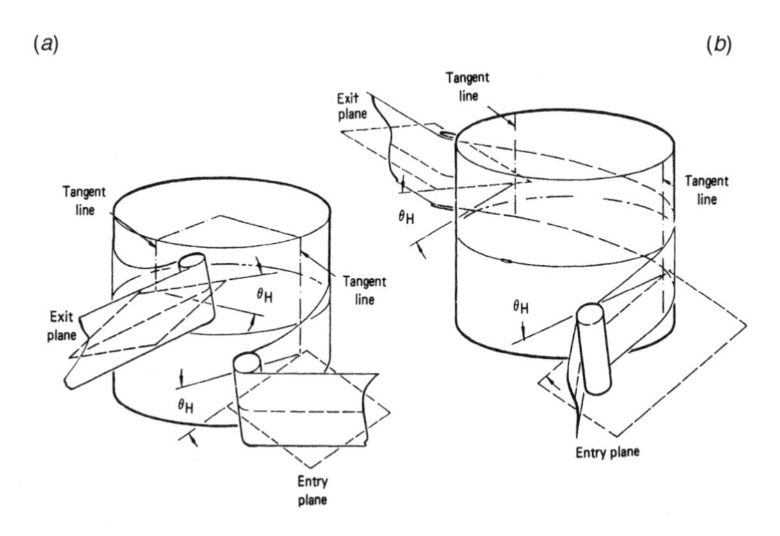

Figure 2.13.6 Helical scan tape paths: (*a*) full wrap (omega), (*b*) half wrap.

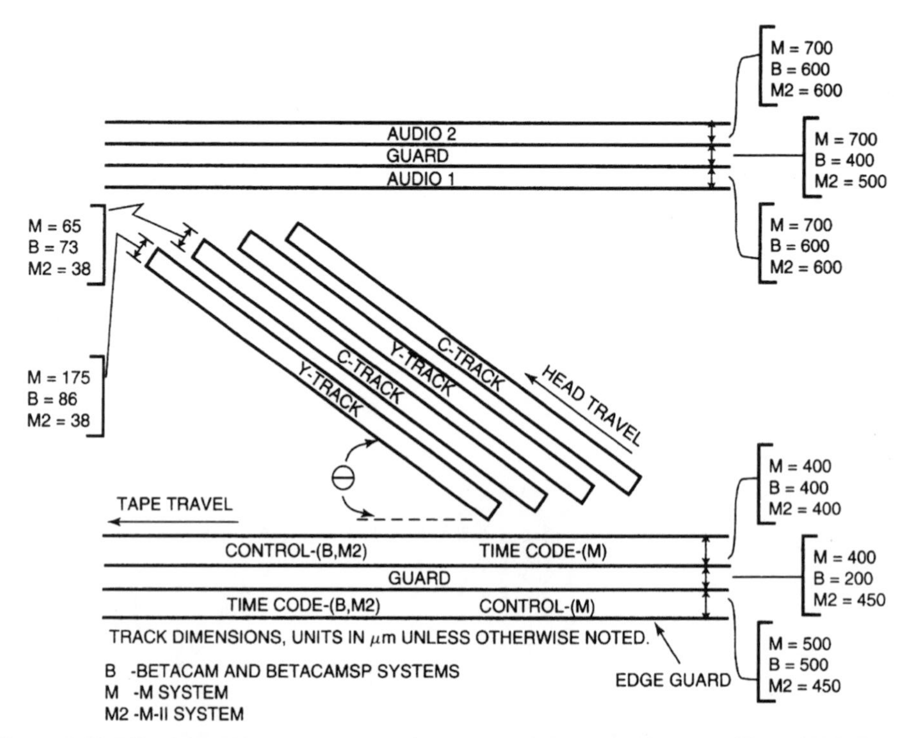

Figure 2.13.7 Track locations used in the 1/2-in component analog formats. (*From:* NAB Engineering Handbook*, 8th ed., National Association of Broadcasters, Washington, D.C., pg. 890, 891, 1992. Used with permission.*)

Table 2.13.8 Video Storage Capabilities of Digital VCRs

Recorder Type	Video Type	Recorded Lines	Audio Type	Audio Tracks	Additional Data Space
Digital Betacam	Digital 10-bit compressed	505/525 596/625	Digital 20-bit	4 Track	2,880 Byte/frame
D2	Digital 8-bit component	500/525 600/625	Digital 20-bit	4 Track	None
D2	Digital 8-bit composite	512/525 608/625	Digital 20-bit	4 Track	None
D3	Digital 8-bit composite	505/525 596/625	Digital 20-bit	4 Track	None
D5	Digital 10-bit component	505/525 596/625	Digital 20-bit	4 Track	None
DV25 (DVCAM DVCPRO)	Digital 8 bit compressed	480/525 576/625	Digital 16-bit	2 Track	738 kbit/s
DVCPRO 50 (DVCPRO 50, Digi-tal-S)	Digital 8-bit compressed	480/525 576/625	Digital 16-bit	4 Track	1.47 Mbits/s
Betacam SX	Digital 8-bit compressed	512/525 608/625	Digital 16-bit	4 Track	2,880 Byte/frame
HDCAM	Digital 10-bit compressed	1440 × 1080I	Digital 20-bit	4 Track	None
HDD-1000	Digital 8-bit	1920 × 1035I	Digital 20-bit	8 Track	None
HDD-2700 HDD-2000	Digital 10-bit compressed	1280 × 720P 1920 × 1080I	Digital 20-bit	4 Track	2,880 Byte/frame

Table 2.13.9 Mechanical Characteristics of 4:2:2 Component Digital VTRs (*From: Robin, Michael, and Michel Poulin:* Digital Television Fundamentals, *McGraw-Hill, New York, N.Y.,*

Parameter	Sony D1	Sony Digital Betacam	Panasonic D5	AMPEX DCT 700d
Track width, μm	32–40	20	20	39.1
Guard band, μm	5–10	1.7	0	0
Total wrap angle, deg	258	196	197	188
Eff. wrap angle, deg	238	173.8	176.9	178.2
Track angle, deg	5.4005	4.62652	4.9384	6.12894
Tracks per field	10	6	12	6
Track length, mm	170	123.335	116.397	148.06
Azimuth, deg	0	±15° 15′	±20	±15
Tape tension, N	0.7	0.3	0.3	0.7
Tape speed, cm/s	28.66	9.67	16.7228	13.17
Writing speed, m/s	35.5	22.9	21.5	27.39
Drum rotation, rps	150/1.001	90/1.001	90/1.001	90/1.001
Drum diameter, mm	75	81.4	76.0	96.4
No. of heads	16	14+4	18	10
No. of record heads	4	4	8	4
No. of RF channels	4	2	4	2
Tape thickness, μm	16	13.5	11/14	13
Tape width, inch (mm)	¾ (19.01)	½ (12.65)	½ (12.65)	¾ (19.01)
Coercivity, Oe	850	1500	1500	1500
Max. play time, min	76	124	123	208
Picture in shuttle	±40×play	±50 × play	±50 × play	±20 × play
Variable speed range	−0.25 to 0.25	−1 to +3	−1 to +2	−1 to +3

Table 2.13.10 Electrical Characteristics of 4:2:2 Component Digital VTRs (*From: Robin, Michael, and Michel Poulin:* Digital Television Fundamentals, *McGraw-Hill, New York, N.Y., 1997.*)

Parameter	Sony D1	Sony Digital Betacam	Panasonic D5	AMPEX DCT 700d
		Video		
Y sampling rate, MHz	13.5	13.5	13.5	13.5
B-Y/R-Y sampling rate, MHz	6.75	6.75	6.75	6.75
Sample resolution, bits	8	10	10	8
Y bandwidth	5.75	5.75	5.75	5.75
B-Y/R-Y bandwidth	2.75	2.75	2.75	2.75
Y SNR, dB	56	61	62	55*
Recorded lines per frame	500	507/512†	510	504/512†
Compression ratio	1:1	2.3:1	1:1	2:1
Tape data rate, Mbps	225	127.76	~300	124.7
Video bit rate, Mbps	172.0	95‡	220	88‡
Min. recorded wavelength, μm	0.9	0.692	0.63	0.854
		Audio		
Sampling rate, kHz	48	48	48	48
Sample resolution, bits§	16–20	16–20	16–20	16–20
Bandwidth	20 Hz–20 kHz	20 Hz–20 kHz	20 Hz–20 kHz	20 Hz–20 kHz
Dynamic range, dB	90	90	90	90
AES channels	2	2	2	2
RF channel coding	Random NRZ	Partial resp.	8–14 code	Miller square

*NTSC in/out measurement.
†The first figure indicates the number of lines with video information.
‡The active picture area is compressed by a ratio of about 2:1.
§The 20-bit recording mode discards V, U, C, and P bits.

Table 2.13.11 SMPTE Documents Relating to Tape Recording Formats (*Courtesy of SMPTE.*)

	B	C	D-1	D-2	D-3	D-5	D-6	D-7 (1)	D-9 (2)	E (3)	G (4)	H (5)	L (6)	M-2
Basic system parameters														
525/60	15M	18M	EG10	EG20	264M	279M	277M	306M	316M	21M			RP144	RP158
625/50					265M	279M	277M	306M	316M					
Record dimensions	16M	19M	224M	245M	264/5M	279M	277M			21M		32M	229M	249M
Characteristics														
Video signals	RP84	RP86								RP87		32M	230M	251M
Audio and control signals	17M	20M	RP155	RP155	264/5M	279M	278M			RP87		32M	230M	251M
Data and control record			227M	247M	264/5M	279M	278M							
Tracking control record	RP83	RP85				279M	277M							
Pulse code modulation audio														252M
Time and control recording	RP93		228M	248M	264/5M	279M	278M						230M	251M
Audio sector time code, equipment type information			RP181											
Nomenclature		18M	EG21	EG21						21M		32M		
Index of documents				EG22										
Stereo channels	RP142	RP142								RP142	RP142	RP142	RP142	
Relative polarity	RP148	RP148	RP148	RP148						RP148	RP148	RP148	RP148	
Tape	25M	25M	225M	246M	264/5M	279M	277M				35M	32M	238M	250M
Reels	24M	24M												
Cassettes			226M	226M	263M	263M	226M	307M	317M	22M	35M	32M	238M	250M
Small										31M				
Bar code labeling			RP156	RP156										
Dropout specifications	RP121	RP121												
Reference tape and recorder														
System parameters	29M													
Tape	26M	26M												

Notes:
1DVCPRO, 2 Digital S, 3 U-matic, 4 Beta, 5 VHS, 6 Betacam

Table 2.13.12 Storage Space Requirements for Audio and Video Data Signals (*After: Robin, Michael, and Michel Poulin: "Multimedia and Television," in* Digital Television Fundamentals*, McGraw-Hill, New York, N.Y., pp. 455–488, 1997.*)

Media Signals	Specifications	Data Rate
Voice-grade audio	1 ch; 8-bit @ 8 kHz	64 kbits/s
MPEG audio Layer II	1 ch; 16-bit @ 48 kHz	128 kbits/s
MPEG audio Layer III	1 ch; 16-bit @ 48 kHz	64 kbits/s
AC-3	5.1 ch; 16-bit @ 48 kHz	384 kbits/s
CD	2 ch; 16-bit @ 44.1 kHz	1.4 Mbits/s
AES/EBU	2 ch; 24-bit @ 48 kHz	3.07 Mbits/s
MPEG-1 (video)	352 × 288, 30 f/s, 8-bit	1.5 Mbits/s
MPEG-2 (MP@ML)	720 × 576, 30 f/s, 8-bit	15 Mbits/s, max.
MPEG-2 (4:2:2 P@ML)	720 × 608, 30 f/s, 8-bit	50 Mbits/s, max.
ITU-R Rec. 601	720 × 480, 30 f/s, 8-bit	216 Mbits/s
HDTV	1920 × 1080, 30 f/s, 8-bit	995 Mbits/s

Table 1.13.13 Production and Broadcast Server Bandwidth Requirements for 4:2:2 Video Signals (*After: Robin, Michael, and Michel Poulin: "Multimedia and Television," in* Digital Television Fundamentals*, McGraw-Hill, New York, N.Y., pp. 455–488, 1997.*)

Production Applications	Bandwidth, Mbits/s	Sample Resolution	Compression Ratio
High-end post-production	270	10	1
Typical post-production	90	10	2.3:1
Low-end post-production	25–50	8	6.6:1–3.3:1
News (compressed data)	18–25	8	9:1–6.6:1
HDTV broadcast	20	8	10:1–50:1
Good-quality SDTV broadcast	8	8	20:1
Medium-quality SDTV broadcast	3	8	55:1
Low-quality SDTV broadcast	1.5	8	110:1

Table 2.13.14 Summary of RAID Level Properties (*After: Whitaker, Jerry C.: "Data Storage Systems," in* The Electronics Handbook, *Jerry C. Whitaker (ed.), CRC Press, Boca Raton, Fla., pp. 1445–1459, 1996.*)

RAID Level	Capacity	Data Availability	Data Throughput	Data Integrity
0	High	Read/Write High	High I/O Transfer Rate	
1		Read/Write High		Mirrored
2	High		High I/O Transfer Rate	ECC
3	High		High I/O Transfer Rate	Parity
4	High	Read High		Parity
5	High	Read/Write High		Parity
6		Read/Write High		Double Parity
10		Read/Write High	High I/O Transfer Rate	Mirrored
53			High I/O Transfer Rate	Parity

Table 2.13.15 Fibre Channel Features and Capabilities

Parameter	Range of Capabilities
Line rate	266, 531, or 1062.5 Mbits/s
Data transfer rate (maximum)	640–720 Mbits/s @ 1062.5 line rate
Frame size	2112 byte payload
Protocol	SCSI, IP, ATM, SDI, HIPPI, 802.3, 802.5
Topology	Loop, switch
Data integrity	10E–12 BER
Distance	Local and campus; up to 10 km

Table 2.13.16 Basic Parameters of FC-AL (*After: Whitaker, Jerry C.: "Data Storage Systems," in* The Electronics Handbook, *Jerry C. Whitaker (ed.), CRC Press, Boca Raton, Fla., pp. 1445–1459, 1996.*)

Parameter	Range of Capabilities
Number of devices	126
Data rate	100 MB/s (1.062 GHz using an 8B/10B code)
Cable distance	30 m between each device using copper (longer, with other cabling options)
Cable types	Backplane, twinaxial, coaxial, optical
Fault tolerance	Dual porting, hot plugging

2.14 Audio/Video Production Standards, Equipment, and Design

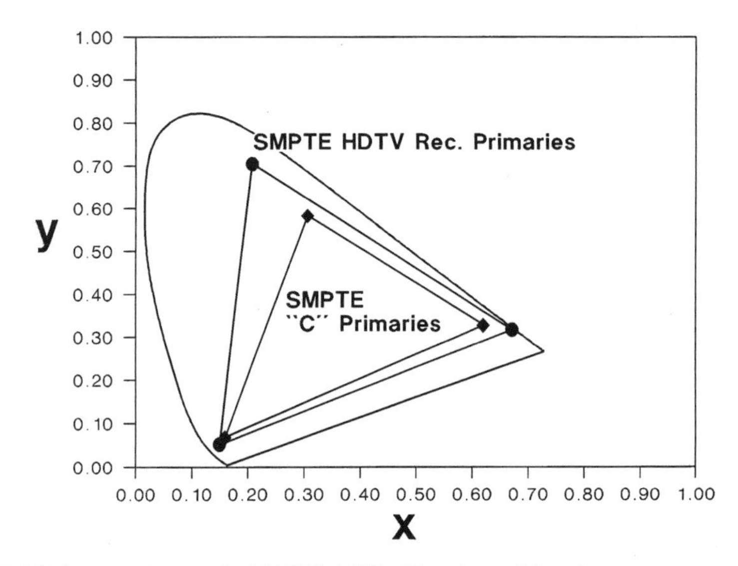

Figure 2.14.1 Color gamut curve for SMPTE 240M. (*Courtesy of Sony.*)

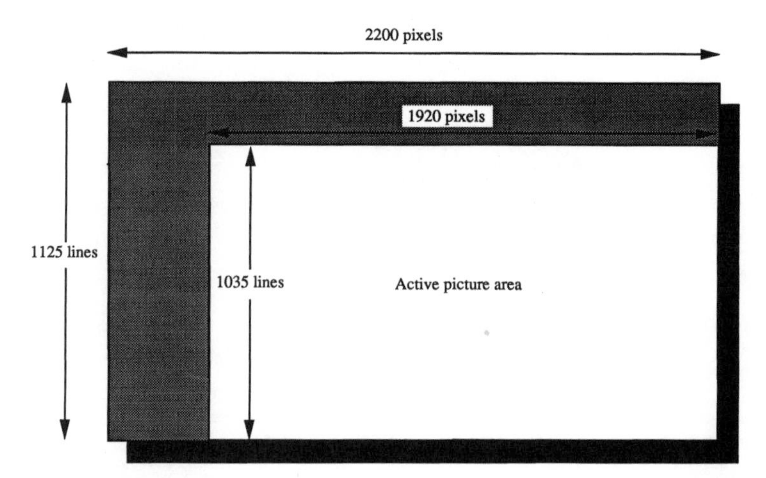

Figure 2.14.2 Overall pixel count for the digital representation of the 1125/60 HDTV production standard.

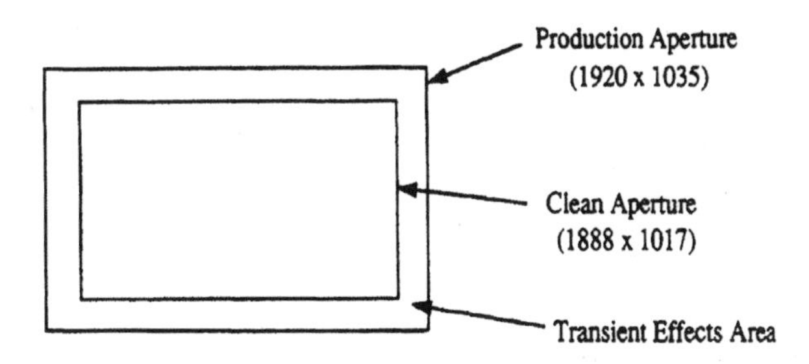

Figure 2.14.3 Production and clean aperture recommendations in SMPTE-260M. (*Courtesy of SMPTE.*)

Table 2.14.1 Quantizing S/N Associated with Various Quantization Levels

Number of Bits	Quantization S/N Levels
8 bits	S/N = 10.8 + (6 × 8) = 58.8 dB
9 bits	S/N = 10.8 + (6 × 9) = 64.8 dB
10 bits	S/N = 10.8 + (6 × 10) = 70.8 dB
11 bits	S/N = 10.8 + (6 × 11) = 76.8 dB
12 bits	S/N = 10.8 + (6 × 12) = 82.8 dB

Table 2.14.2 Encoding Parameter Values for SMPTE-260M

Parameter		Value	
Matrix formulas	E_Y', E_{CB}', E_{CR}' E_G', E_B', E_R'	E_Y', E_{CB}', E_{CR}' are derived from gamma-corrected values of E_G', E_B', E_R' as defined by the linear matrix specified in SMPTE-240M	
Number of samples/line	Video components	E_Y' = 2200 E_{CB}' = 1100 E_{CR}' = 1100	E_G' = 2200 E_B' = 2200 E_R' = 2200
	Auxiliary channel	2200	
Sampling structure	E_G', E_B', E_R', Luminance E_Y', Auxiliary channel	Identical sampling structures: orthogonal sampling, line, field, and frame repetitive.	
	Color difference signals (E_{CB}', E_{CR}')	Samples are co-sited with odd (1st, 3rd, 5th, ...) $E_{Y'}$ samples in each line.	
Sampling frequency (tolerance ±10 ppm)	Video components	E_Y' = 74.25 MHz E_{CB}' = 37.125 MHz E_{CR}' = 37.125 MHz	E_G' = 74.25 MHz E_B' = 74.25 MHz E_R' = 74.25 MHz
	Auxiliary channel	74.25 MHz	
Form of encoding	Uniformly quantized, PCM 8- or 10-bits/sample for each of the video component signals and the auxiliary channel.		
Active number of samples/line	Video components	E_Y' = 1920 E_{CB}' = 960 E_{CR}' = 960	E_G' = 1920 E_B' = 1920 E_R' = 1920
	Auxiliary channel	1920	
Timing relationship between video data and the analog synchronizing waveform	The time duration between the *end of active video* (EAV) timing reference code and the reference point 0H of the horizontal sync waveform = 88 clock intervals.		
Correspondence between video signal levels and quantization levels[1]	8-bit system: E_G', E_B', E_R', Luminance E_Y', Auxiliary channel	220 quantization levels with the black level corresponding to level 16 and the peak white level corresponding to level 235.	
	Each color difference signal (E_{CB}', E_{CR}')	225 quantization levels symmetrically distributed about level 128, which corresponds to the zero signal.	
	10-bit system: E_G', E_B', E_R', Luminance E_Y', Auxiliary channel	877 quantization levels with the black level corresponding to level 64 and the peak white level corresponding to level 940.	
	Each color difference signal (E_{CB}', E_{CR}')	877 quantization levels symmetrically distributed about level 512, which corresponds to the zero signal.	
Quantization level assignment[1]	8-bit system	254 of the 256 levels (digital levels 1 through 254) of the 8-bit word used to express quantized values. Data levels 0 and 255 are reserved to indicate timing references.	
	10-bit system	1016 of the 1024 levels (digital levels 4 through 1019) of the 10-bit word used to express quantized values. Data levels 0–3 and 1020–1023 are reserved to indicate timing references.	

[1] These values refer to precise nominal video signal levels. Signal processing may occasionally cause the signal level to deviate outside this range.

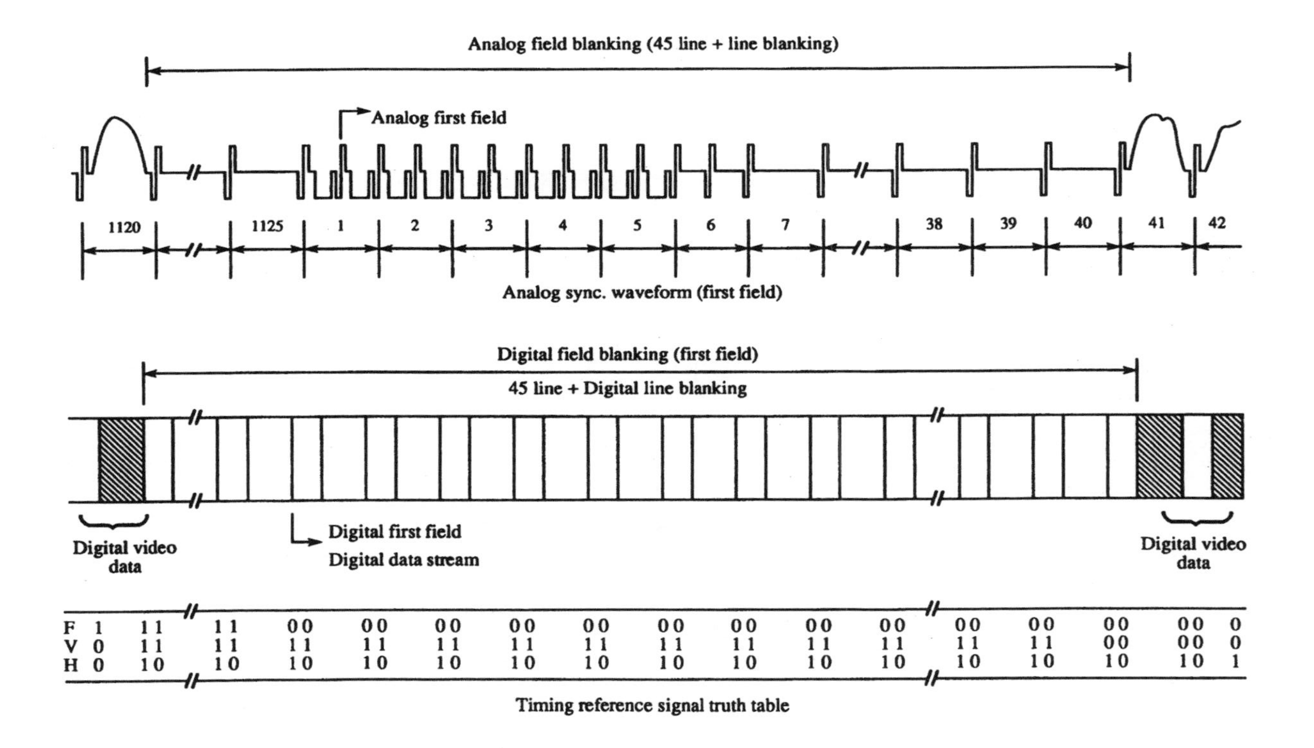

Figure 2.14.4 SMPTE 260M line numbering scheme, first digital field ($F = 0$). (*Courtesy of SMPTE.*)

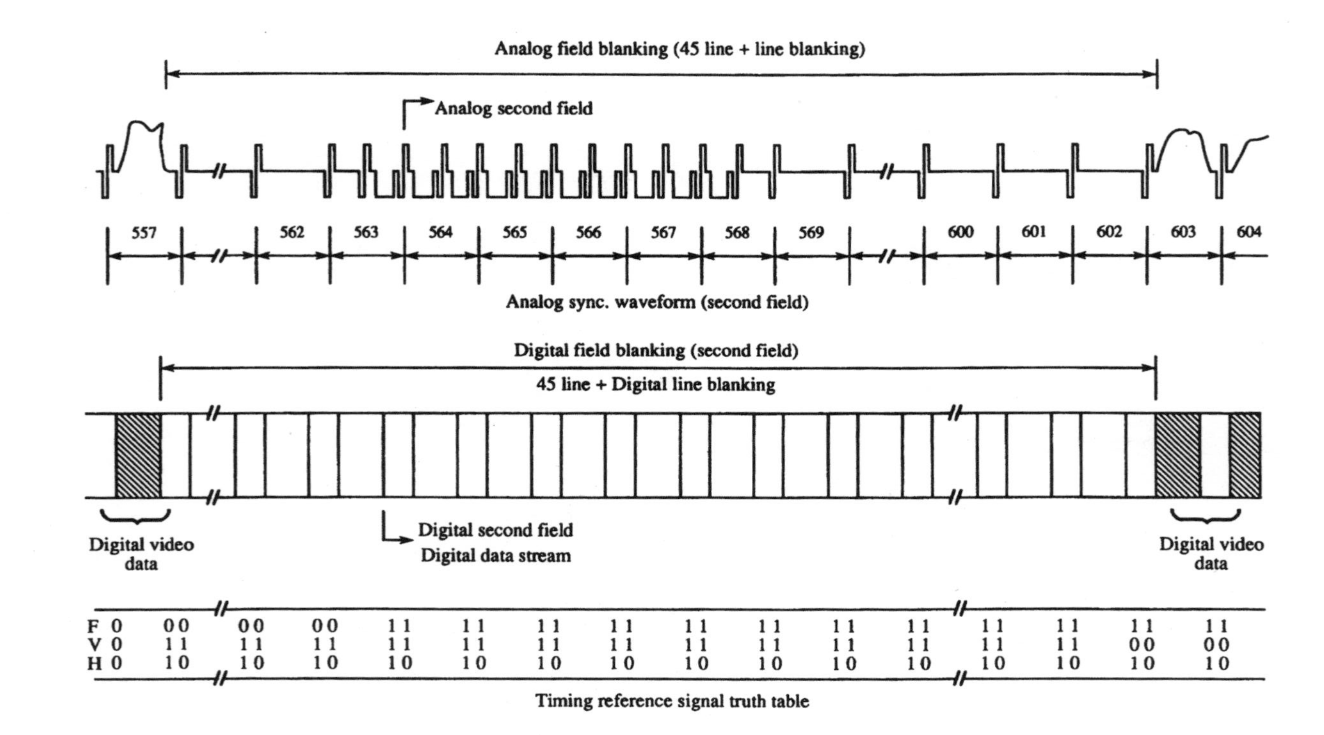

Figure 2.14.5 SMPTE 260M line numbering scheme, second digital field ($F = 1$). (*Courtesy of SMPTE.*)

Table 2.14.3 Scanned Raster Characteristics of SMPTE-240M (*After: Thorpe, Laurence,: "A New Global HDTV Program Origination Standard: Implications for Production and Technology,"* Proceedings of DTV99, *Intertec Publishing, Overland Park, Kan., 1999.*)

Parameter	1125/60 System	1125/59.94 System
Total scan lines/frame	1125	1125
Active lines/frame	1035	1035
Scanning format	Interlaced 2:1	Interlaced 2:1
Aspect ratio	16:9	16:9
Field repetition rate	60.00 Hz ± 10 ppm	59.94 Hz[1] ± 10 ppm
Line repetition rate (derived)	33750.00 Hz	33716.28 Hz[2]
[1] The 59.94 Hz notation denotes an approximate value. The exact value is 60/1.001. [2] The 33716.28 Hz notation denotes an approximate value. The exact value is (60 × 1125)/(2 × 1.001).		

Table 2.14.4 SMPTE 240M Analog Video Signal Levels (*After: "SMPTE Standard for Television—Signal Parameters—1125-Line High-Definition Production Systems," SMPTE 240M-1995, SMPTE, White Plains, N.Y., 1995.*)

E_Y', E_G', E_B', E_R' Signals	
Reference black level	0 mV
Reference white level	700 mV
Synchronizing level	± 300 mV
E_{PB}', E_{PR}' Signals	
Reference zero signal level	0 mV
Reference peak levels	± 350 mV
Synchronizing level	± 300 mV
All Signals	
Sync pulse amplitude	300 ± 6 mV
Amplitude difference between positive- and negative-going sync pulses	< 6 mV

Table 2.14.5 Scanning System Parameters for SMPTE-274M (incorporating SMPTE RP 211)

System Description	Samples per Active Line (S/AL)	Active Lines per Frame	Frame Rate (Hz)	Scanning Format	Interlace Sampling Frequency f_s (MHz)	Samples per Total Line (S/TL)	Total Lines per Frame
1: 1920 × 1080/ 60/1:1	1920	1080	60	Progressive	148.5	2200	1125
2: 1920 × 1080/ 59.94/1:1	1920	1080	60/1.001	Progressive	148.5/1.001	2200	1125
3. 1920 × 1080/ 50/1:1	1920	1080	50	Progressive	148.5	2640	1125
4: 1920 × 1080/ 60/2:1	1920	1080	30	2:1 Interlace	74.25	2200	1125
5: 1920 × 1080/ 59.94/2:1	1920	1080	30/1.001	2:1 Interlace	74.25/10.0	2200	1125
6. 1920 × 1080/ 50/2:1	1920	1080	25	2:1 Interlace	74.25	2540	1125
7: 1920 × 1080/ 30/1:1	1920	1080	30	Progressive	74.25	2200	1125
8: 1920 × 1080/ 29.97/1:1	1920	1080	30/1.001	Progressive	74.25/1.001	2200	1125
9: 1920 × 1080/ 25/1:1	1920	1080	25	Progressive	74.25	2640	1125
10: 1920 × 1080/24/1:1	1920	1080	24	Progressive	74.25	2750	1125
11: 1920 × 1080/23.98/1:1	1920	1080	24/1.001	Progressive	74.25/1.001	2750	1125
12: 1920 × 1080/30/sF	1920	1080	30	P (sF)	74.25	2200	1125
13: 1920 × 1080/29.97/sF	1920	1080	30/1.001	P (sF)	74.25/1.001	2200	1125
14: 1920 × 1080/25/sF	1920	1080	25	P (sF)	74.25	2640	1125
15: 1920 × 1080/24/sF	1920	1080	24	p (sF)	74.25	2750	1125
16: 1920 × 1080/23.98/sF	1920	1080	24/1.001	P (sF)	74.25/1.001	2750	1125

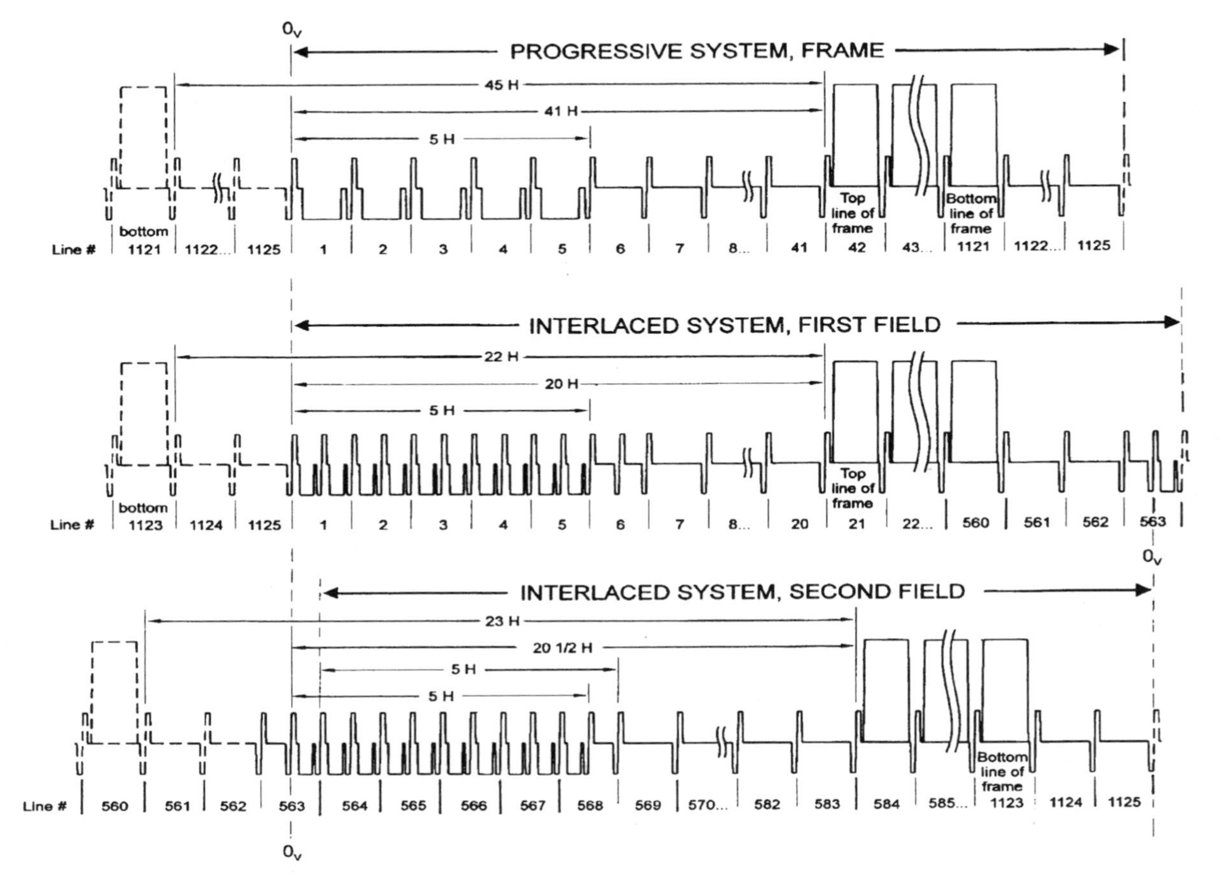

Figure 2.14.6 SMPTE 274M analog interface vertical timing. (*Courtesy of SMPTE.*)

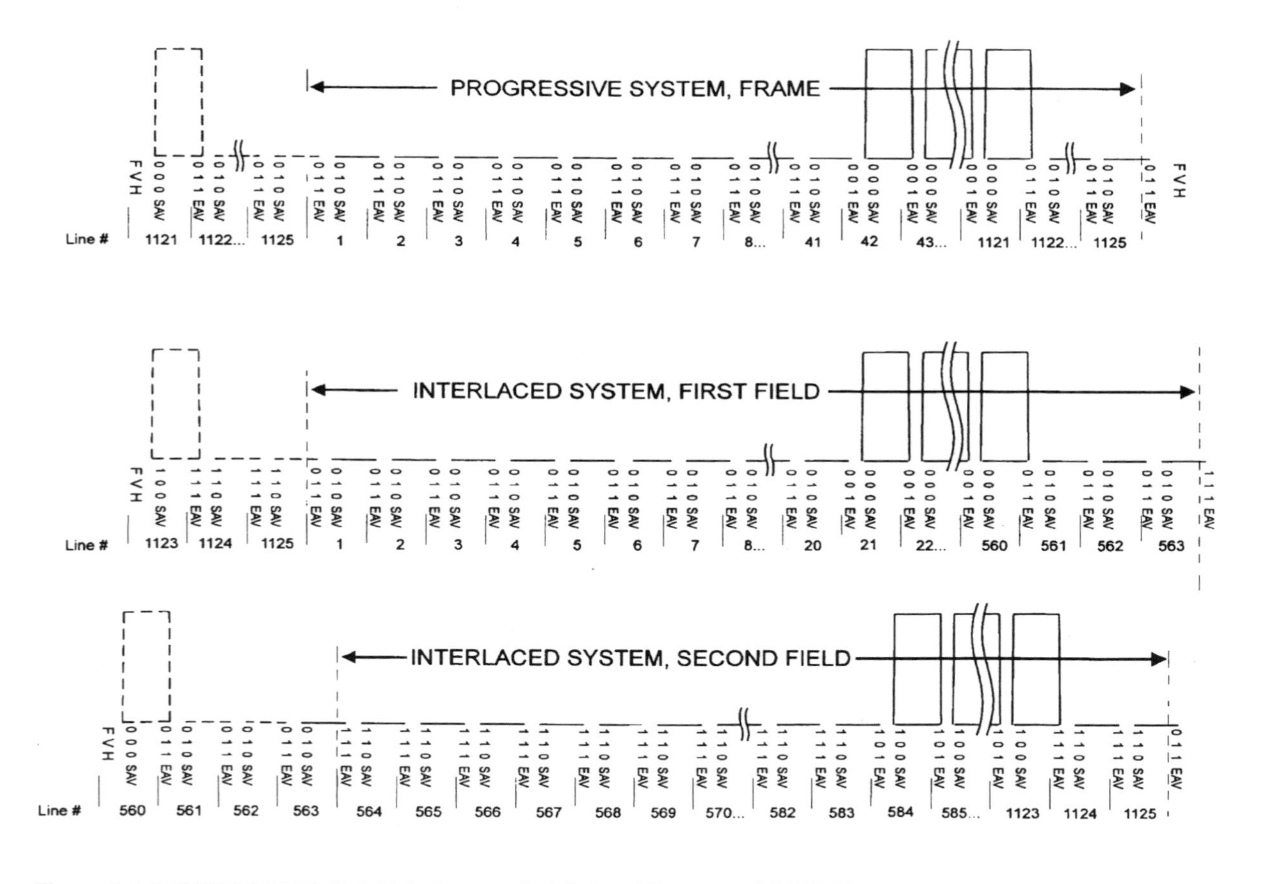

Figure 2.14.7 SMPTE 274M digital interface vertical timing. (*Courtesy of SMPTE.*)

Table 2.14.6 Scanning Systems Parameters for SMPTE-295M-1997 (*After: "SMPTE Standard for Television—1920 × 1080 50 Hz Scanning and Interfaces," SMPTE 295M-1997, SMPTE, White Plains, N.Y., 1997.*)

System Designation	Samples per Active Line (S/AL)	Active Lines per Frame	Frame Rate (Hz)	Scanning Format	Interlace Sampling Frequency f_s (MHz)	Samplesper Total Line (S/TL)	Total Lines per Frame
1: 1920 × 1080/50/ 1:1	1920	1080	50	Progressive	148.5	2376	1250
2: 1920 × 1080/50/ 2:1	1920	1080	25	2:1 interlace	74.25	2376	1250

Table 2.14.7 Scanning System Parameters for SMPTE 293M-1996 (*After: "SMPTE Standard for Television—720 × 483 Active Line at 59.94 Hz Progressive Scan Production—Digital Representation," SMPTE 293M-1996, SMPTE, White Plains, N.Y., 1996.*)

System Type	Samples per Digital Active Line (S/AL)	Lines per Active Image	Frame Rate	Sampling Frequency f_s (MHz)	Samples per Total Line	Total Lines per Frame	Colorimetry
720 × 483/59.94	720	483	60/1.001	27.0	858	525	ANSI/SMPTE 170M

Table 2.14.8 SMPTE 296M Imaging Sampling Systems

	System Nomenclature	Luma or R′ G′ B′ Samples per Active Line	Active Lines per Frame	Frame Rate (Hz)	Luma or R′ G′ B′ Sampling Frequency (MHz)	Luma Sample Periods per Total Line	Total Lines per Frame
1	1280 × 720/60	1280	720	60	74.25	1650	750
2	1280 × 720/59.94	1280	720	60/1.001	74.25/1.001	1650	750
3	1280 × 720/50	1280	720	50	74.25	1980	750
4	1280 × 720/30	1280	720	30	74.25	3300	750
5	1280 × 720/29.97	1280	720	30/1.001	74.25/1.001	3300	750
6	1280 × 720/25	1280	720	25	74.25	3960	750
7	1280 × 720/24	1280	720	24	74.25	4125	750
8	1280 × 720/23.98	1280	720	24/1.001	74.25/1.001	4125	750

Note: Systems 3 through 8 are specified in a proposed revision of SMPTE 296 (pending at this writing); for systems 4 through 8, the analog video interface is not preferred.

Table 2.14.9 SMPTE 294M-1997 Interface Parameters (*After: "SMPTE Standard for Television—720 × 483 Active Line at 59.94-Hz Progressive Scan Production Bit-Serial Interfaces," SMPTE 294M-1997, SMPTE, White Plains, N.Y., 1997.*)

System (Total Serial Data Rate)	**4:2:2 P (2 x 270 Mbits/s) Dual Link**	**4:2:0 P (360 Mbits/s) Single Link**
Frame rate	60/1.001 Hz	60/1.001 Hz
Word length	10 bits	10 bits
Parallel and multiplexed word rate: channels Y', Y'' and C_B'/C_R', C_B''/C_R'' + SAV, EAV and aux data	2 × 27 Mwords/s	36 Mword/s
Active lines per frame	483	483
Words per active line (channels Y' and Y'')	720 and 720	720 and 720
Words per active line (channels C_B' and C_R')	2 × (360 and 360)	360 and 360
Words per horizontal blanking area (SAV/EAV and auxiliary data)	2 × 276 (Total: 2 × (483 × 276 = 2 × 133,308/frame)	128 (Total: 483 × 128 = 61,824/frame)
Words in the active picture area	2 × (1440 × 483) = 2 x 695,520	2160 × 483 = 1,043,280
Words in the vertical blanking interval (SAV/EAV and auxiliary data)	2 × (1716 × 42) = 2 x 72,072	2288 × 42 = 96,096

Table 2.14.10 Active and Total Production Format Bit Rates Used at the Input of the ATSC System (*From: Robin, Michael, and Michel Poulin:* Digital Television Fundamentals, *McGraw-Hill, New York, N.Y., 1997. Used with permission.*)

Active/total video formats, pixels×lines*	Frame rates	Bit rates, Mbps†			
		8-bit 4:2:2	10-bit 4:2:2	8-bit 4:2:0	10-bit 4:2:0
640×480/	30I	147/(202)	184/(252)	110/(151)	138/(189)
(800×525)	30P	147/(202)	184/(252)	110/(151)	138/(189)
	60P	294/(404)	368/(504)	220/(302)	276/(378)
720×480/	30I	166/(216)	207/(270)	124/(162)	156/(202)
(858×525)	30P	166/(216)	207/(270)	124/(162)	156/(202)
	60P	332/(432)	414/(540)	248/(324)	312/(404)
1280×720/	30P	442/(594)	553/(742)	331/(445)	415/(556)
(1650×750)	60P	884/(1188)	1106/(1485)	662/(890)	830/(1113)
1920×1080/	30I	995/(1188)	1244/(1485)	746/(891)	933/(1113)
(2200×1125)	30P	995/(1188)	1244/(1485)	746/(891)	933/(1113)

*Values in parentheses are total scanning formats corresponding to the active video scanning formats defined in the ATSC standard documents.

†The 10-bit 6:3:3 sampling structure, as produced by an 18-MHz sampling frequency, for a 720×480 active video format and not shown in this table, will generate a bit rate of 360 Mbps. This production format is not included in the ATSC video input format standards.

Table 2.14.11 Summary of NTSC Signal Characteristics (*From: Robin, Michael, and Michel Poulin:* Digital Television Fundamentals, *McGraw-Hill, New York, N.Y., 1997. Used with permission.*)

Assumed chromaticity coordinates for primary colors of receiver	x / y: Green 0.310 0.596; Blue 0.155 0.070; Red 0.630 0.340
Chromaticity coordinates for equal primary signals	Illuminant D_{65}: $x = 0.3127; y = 0.3290$
Assumed receiver gamma value	2.2
Luminance signal	$E'_Y = 0.587\,E'_G + 0.114\,E'_B + 0.299\,E'_R$
Chrominance signals	$E'_{B\text{-}Y} = 0.877\,(E'_B - E'_Y)$ and $E'_{R\text{-}Y} = 0.493\,(E'_R - E'_Y)$ or $E'_Q = E'_{B\text{-}Y} \cos 33° + E'_{R\text{-}Y} \sin 33°$ and $E'_I = -E'_{B-Y} \sin 33° + E'_{R\text{-}Y} \cos 33°$
Equation of complete color signal	$E_M = 0.925\,E'_Y + 7.5 + 0.925\,E'_{B\text{-}Y} \sin (2\pi f_{SC} t) + 0.925\,E'_{R-Y} \cos (2\pi f_{SC} t)$ or $E_M = 0.925\,E'_Y + 7.5 + 0.925\,E'_Q \sin (2\pi f_{SC} t + 33°) + 0.925\,E'_I \cos (2\pi f_{SC} t + 33°)$
Type of chrominance subcarrier modulation	Suppressed-carrier amplitude modulation of two subcarriers in quadrature
Chrominance subcarrier frequency, Hz	Nominal value and tolerance: $f_{SC} = 3{,}579{,}545 \pm 10$ Relationship to line frequency f_H: $f_{SC} = (455/2)f_H$
Bandwidth of transmitted chrominance sidebands, kHz	$f_{SC} \pm 620$ or f_{SC} +620/−1300
Amplitude of chrominance subcarrier	$G = \sqrt{(E'^{\,2}_{B-Y} + E'^{\,2}_{R-Y})}$ or $G = \sqrt{(E'^{\,2}_I + E'^{\,2}_Q)}$
Synchronization of subcarrier	Subcarrier burst on blanking backporch

Table 2.14.12 Coding Parameters for $4f_{SC}$ NTSC Composite Digital Signals (*From: Robin, Michael, and Michel Poulin:* Digital Television Fundamentals, *McGraw-Hill, New York, N.Y.,*

Input signal	NTSC
Number of samples per total line	910
Number of samples per active digital line	768
Sampling frequency	$4f_{SC} = 14.32818$ MHz
Sampling structure	Orthogonal
Sampling instant	+33°, +123°, +213°, +303°
Coding	Uniformly quantized
Quantizing resolution	8 or 10 bits per sample

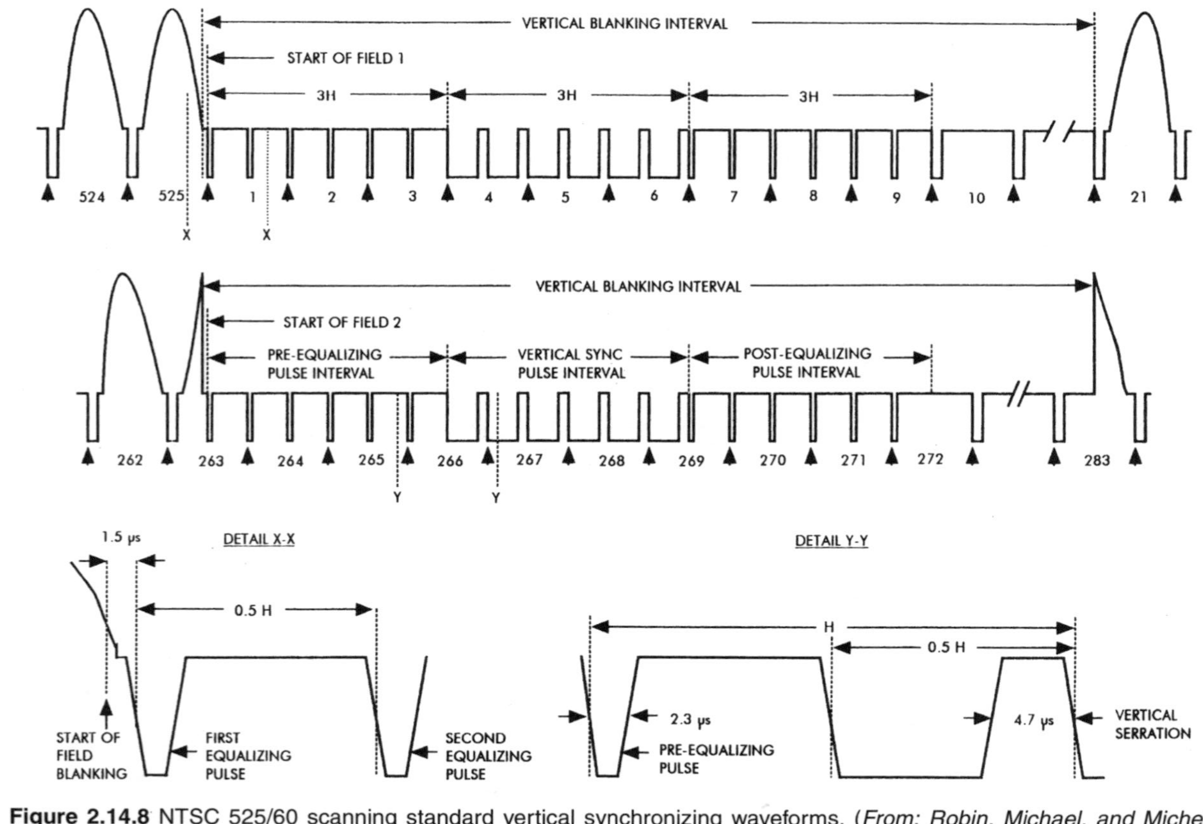

Figure 2.14.8 NTSC 525/60 scanning standard vertical synchronizing waveforms. (*From: Robin, Michael, and Michel Poulin:* Digital Television Fundamentals, *McGraw-Hill, New York, N.Y., 1997. Used with permission.*)

Table 2.14.13 PAL Signal Characteristics (*From: Robin, Michael, and Michel Poulin:* Digital Television Fundamentals, *McGraw-Hill, New York, N.Y., 1997. Used with permission.*)

Assumed chromaticity coordinates for primary colors of receiver	x y Green 0.29 0.60 Blue 0.15 0.06 Red 0.64 0.33
Chromaticity coordinates for equal primary signals	Illuminant D_{65}: $x = 0.3127$; $y = 0.3290$
Assumed receiver gamma value	2.8
Luminance signal	$E'_Y = 0.587E'_G + 0.144E'_B + 0.299E'_R$
Chrominance signals	$E'_U = 0.877\,(E'_B - E'_Y)$ and $E'_V = 0.493\,(E'_R - E'_Y)$
Equation of complete color signal	$E_M = E'_Y + E'_U \sin(2\pi f_{SC} t) \pm E'_V \cos(2\pi f_{SC} t)$
Type of chrominance subcarrier modulation	Suppressed-carrier amplitude modulation of two subcarriers in quadrature
Chrominance subcarrier frequency, Hz	Nominal value and tolerance: f_{SC} = 4,433,618.75 Hz ± 5 (CCIR B, D, G, H) ± 1 (CCIR I) Relationship to line frequency f_H: $f_{SC} = (1135/4 + 1/625)\, f_H$
Bandwidth of transmitted chrominance sidebands, kHz	f_{SC} +570/−1300 (CCIR B, D, G, H) f_{SC} +1066/−1300 (CCIR I)
Amplitude of chrominance subcarrier	$G = \sqrt{(E'^2_U + E'^2_V)}$
Synchronization of subcarrier	Subcarrier burst on blanking backporch

Table 2.14.14 Coding Parameters for $4f_{SC}$ PAL Composite Digital Signals (*From: Robin, Michael, and Michel Poulin:* Digital Television Fundamentals, *McGraw-Hill, New York, N.Y., 1997. Used with permission.*)

Input signal	PAL
Number of samples per total line	1135*
Number of samples per active line	948
Sampling frequency	$4f_{SC}$ = 17.734475 MHz
Sampling instant	+45°,+135°,+225°,+315°
Coding	Uniformly quantized
Quantizing resolution	8 or 10 bits per sample

*Except lines 313 and 625, which have 1137 samples each.

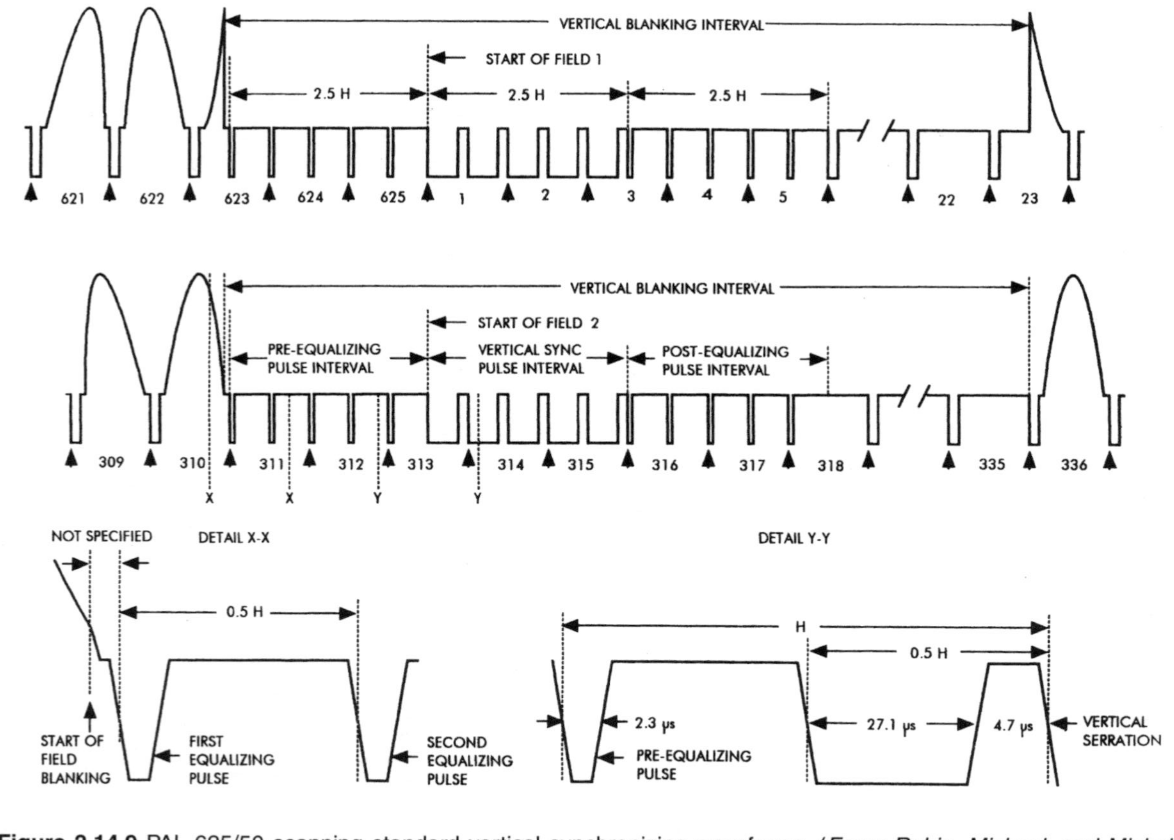

Figure 2.14.9 PAL 625/50 scanning standard vertical synchronizing waveforms. (*From: Robin, Michael, and Michel Poulin:* Digital Television Fundamentals, *McGraw-Hill, New York, N.Y., 1997. Used with permission.*)

Table 2.14.15 SECAM Signal Characteristics (*From: Robin, Michael, and Michel Poulin:* Digital Television Fundamentals, *McGraw-Hill, New York, N.Y., 1997. Used with permission.*)

Assumed chromaticity coordinates for primary colors of receiver	x y Green 0.29 0.60 Blue 0.15 0.06 Red 0.64 0.33
Chromaticity coordinates for equal primary signals	Illuminant D_{65}: $x = 0.3127$; $y = 0.3290$
Assumed receiver gamma value	2.8
Luminance signal	$E'_Y = 0.587E'_G + 0.114E'_B + 0.299E'_R$
Chrominance signals	$D'_B = 1.505\,(E'_B - E'_Y)$ and $D'_R = -1.902\,(E'_R - E'_Y)$
Low-frequency precorrection of color-difference signals	$D'_{B*} = A_{BF}(f)\,D'_B$ and $D'_{R*} = A_{BF}(f)\,D'_R$ where $A_{BF}(f) = [1+j(f/f_1)]/[1+j(f/3f_1)]$ f = signal frequency, kHz f_1 = 85 kHz
Equation of complete color signal	$E_M = E'_Y + G\cos 2\pi\,(f_{0B} + \Delta f_{0B} f_0 D_{B*} dt)$ or $E_M = E'_Y + G\cos 2\pi\,(f_{0R} + \Delta f_{0R} f_0 D_{R*} dt)$ alternately from line to line
Chrominance subcarrier modulation	Frequency modulation
Chrominance subcarrier frequency, Hz	Nominal value: $f_{0B} = 4{,}250{,}000 \pm 2000$ and $f_{0R} = 4{,}406{,}250 \pm 2000$ Relationship to line frequency f_H : $f_{0B} = 272 f_H$ and $f_{0R} = 282 f_H$
Maximum subcarrier deviation, kHz	$\Delta f_{0R} = +350/-506$ and $\Delta f_{0B} = +506/-350$
Amplitude of chrominance subcarrier	$G = M_0\,[(1+j\,16F)/(1+j\,1.26F)]$ where $F = (f/f_0) - (f_0/f)$, f_0 = 4286 kHz and the peak-to-peak amplitude, $2M_0$, is 23% of the peak luminance amplitude
Subcarrier switching synchronization	By undeviated chrominance subcarrier reference on the line-blanking backporch

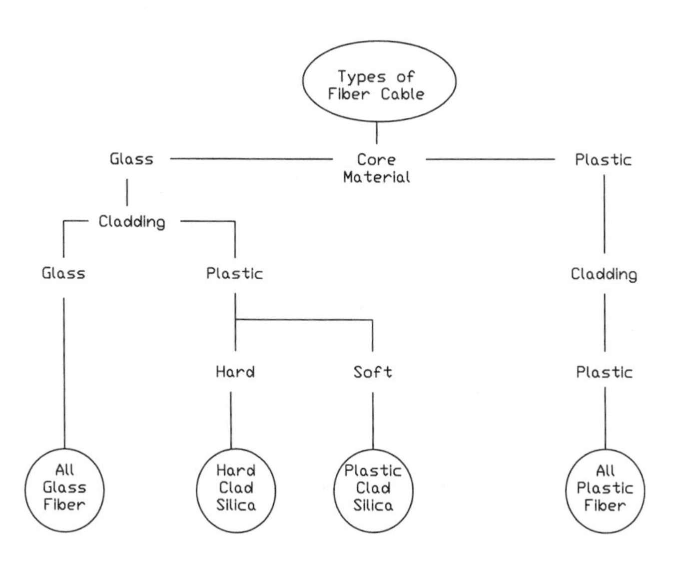

Figure 2.14.10 Primary types of fiber-optic cables, based on construction elements.

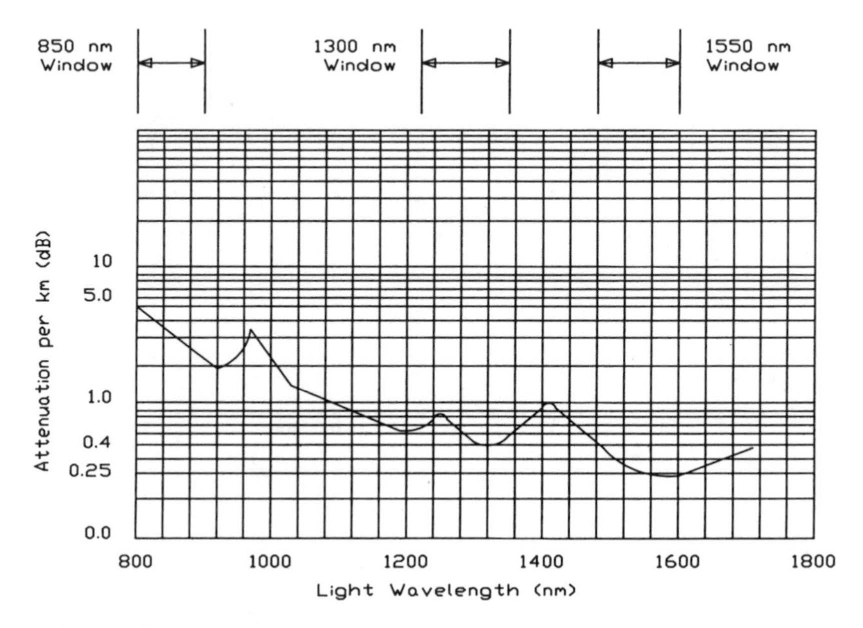

Figure 2.14.11 Fiber attenuation vs. light wavelength characteristics. Attenuation has been reduced steadily in the last two decades through improved fiber drawing techniques and a reduction in impurities. It has now approached the theoretical limits of silica-based glass at the 1,300 and 1,550 nm wavelengths.

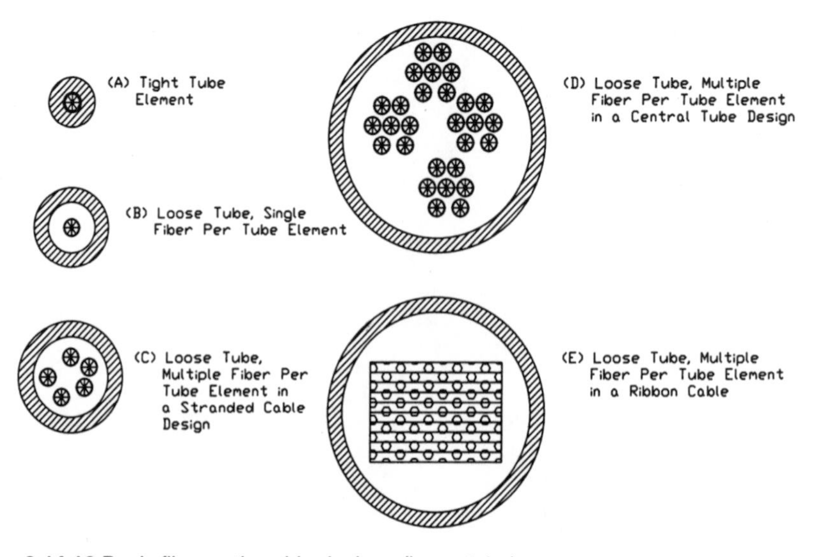

Figure 2.14.12 Basic fiber optic cable designs (loose tube).

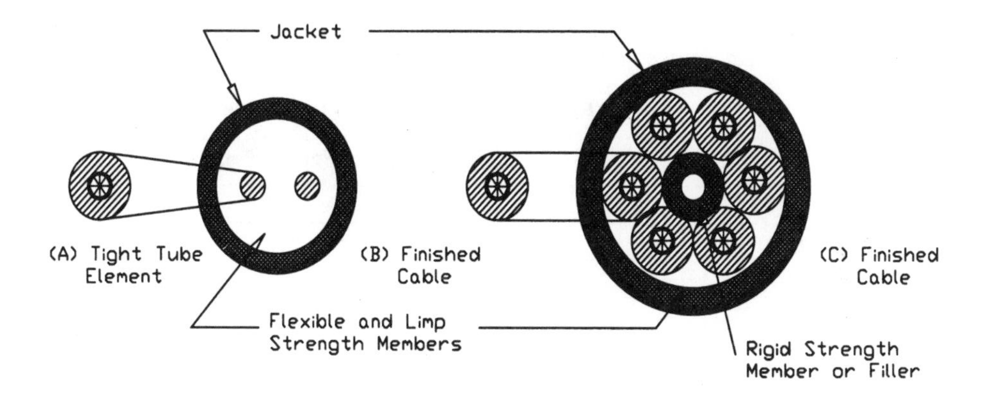

Figure 2.14.13 Basic fiber optic cable designs (tight tube).

Table 2.14.16 Maximum Installation Loads that Fiber Cable May be Exposed to in Various Applications

Typical Maximum Recommended Installation Loads	
Application	**Pounds Force**
1 fiber in raceway or tray	67 lb
1 fiber in duct or conduit	125 lb
2 fibers in duct or conduit	250–500 lb
Multi-fiber (6–12) cables	500 lb
Direct burial cables	600–800 lb
Lashed aerial cables	300 lb
Self-supported aerial cables	600 lb

Table 2.14.17 Typical Maximum Vertical Lengths for Common Types of Fiber-Optic Cable

Typical Maximum Vertical Rise Distances	
Application	**Distance**
1 fiber in raceway or tray	90 ft
2 fibers in duct or conduit	50–90 ft
Multi-fiber (6–12) cables	50–375 ft
Heavy-duty cables	1,000–1,640 ft

Table 2.14.18 Typical Maximum Long-Term Use Loads that a Fiber-Optic Cable May be Exposed to in Various Applications

Typical Maximum Recommended Long-Term Use Loads	
Application	**Pounds Force**
1 fiber in raceway or tray	23–35 lb
1 fiber in duct or conduit	67 lb
2 fibers in duct or conduit	67 lb
Multi-fiber (6–12) cable	33–330 lb
Direct burial cable	132–180 lb

Table 2.14.19 Crush Strength Ratings for Various FO Cable Types

Typical Cable Crush Strength Ratings		
Characteristic	**Type of Cable**	**Force lb/in**
Long-term crush load	6 fibers/cable	57–400 lb/in
	1–2 fiber cables	314–400 lb/in
	Armored cables	450 lb/in
Short-term crush load	6 fibers/cable	343–900 lb/in
	1–2 fiber cables	300–800 lb/in
	Armored cables	600 lb/in

2.15 Film for Video Applications

Table 2.15.1 Conversions of Filter Factors to Exposure Increase in Stops

Filter Factor	+ Stops	Filter Factor	+ Stops	Filter Factor	+ Stops
1.25	+ 1/3	4	+ 2	12	+ 3-2/3
1.5	+ 2/3	5	+ 2-1/3	40	+ 5-1/3
2	+ 1	6	+ 2-2/3	100	+ 6-2/3
2.5	+ 1-1/3	8	+ 3	1000	+ 10
3	+ 1-2/3	10	+ 3-1/3		

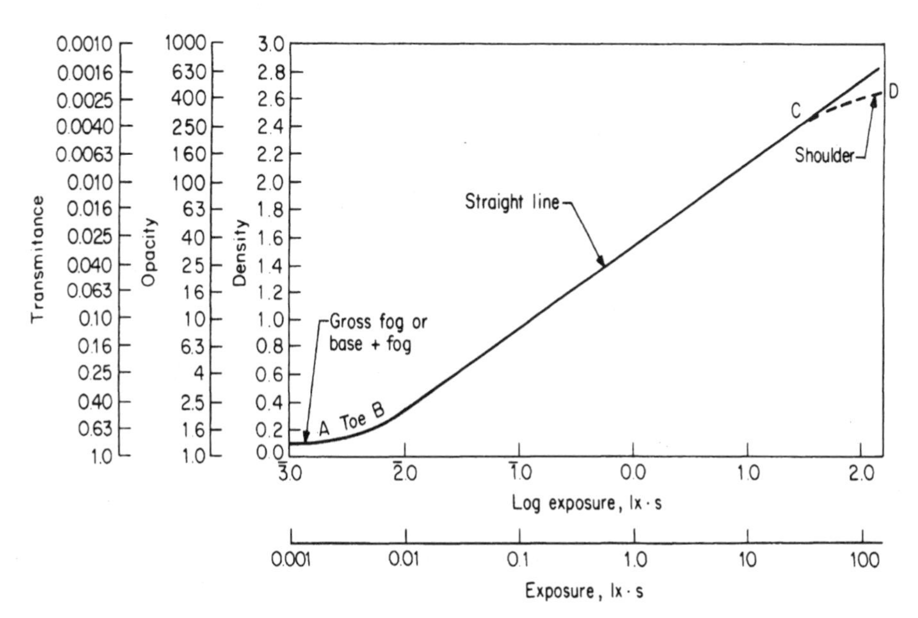

Figure 2.15.1 Typical sensitometric, or Hurter and Driffield (HD), characteristic curve. (*Courtesy of Eastman Kodak Company.*)

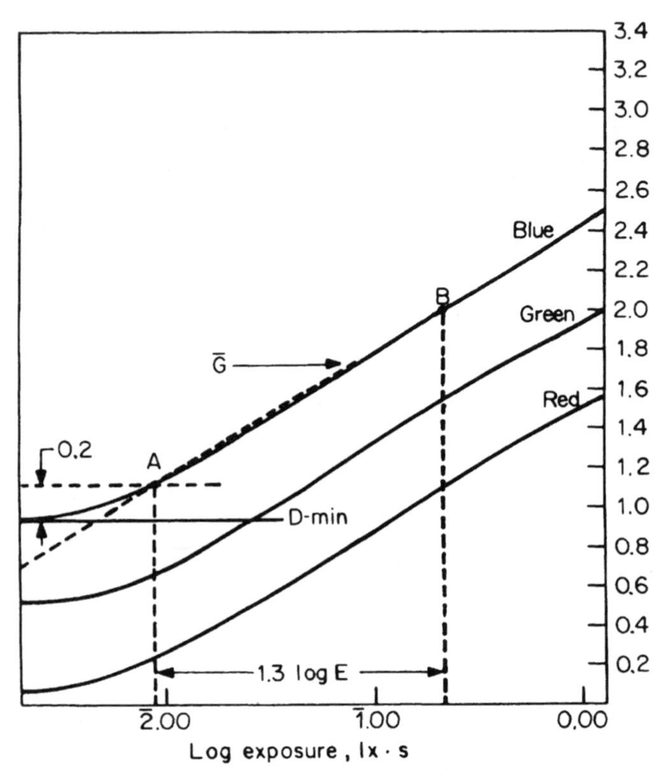

Figure 2.15.2 Typical characteristic curve of negative color film. (*Courtesy of Eastman Kodak Company.*)

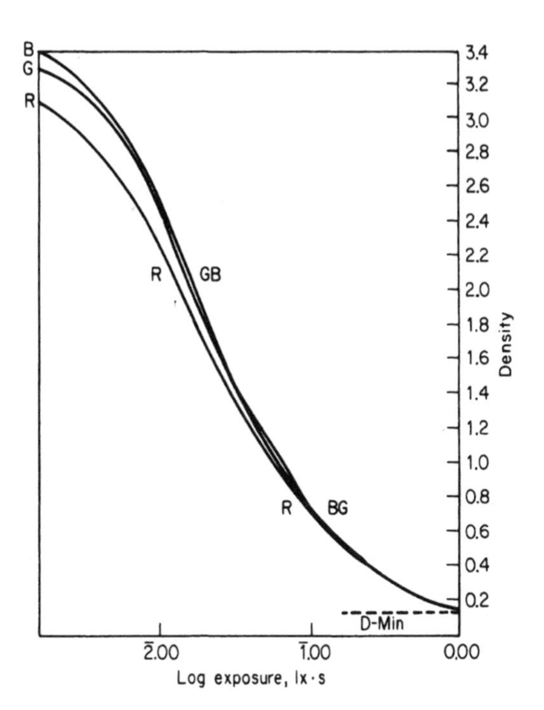

Figure 2.15.3 Typical characteristic curve of positive color film. (*Courtesy of Eastman Kodak Company.*)

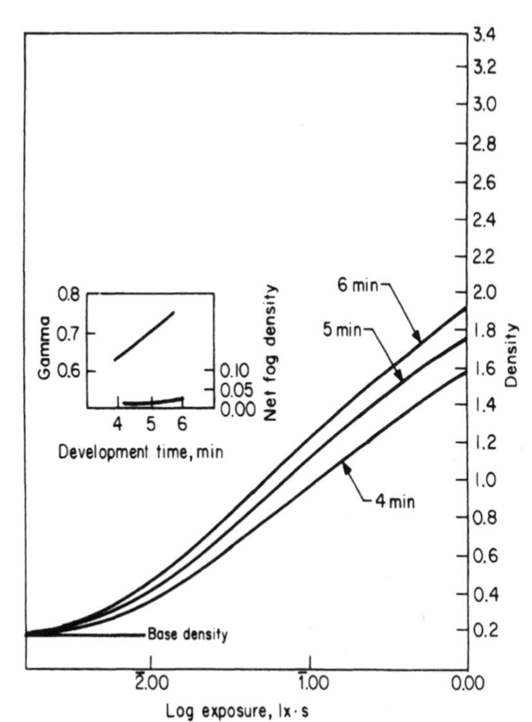

Figure 2.15.4 Curves for a development-time series on a typical black-and-white negative film. (*Courtesy of Eastman Kodak Company.*)

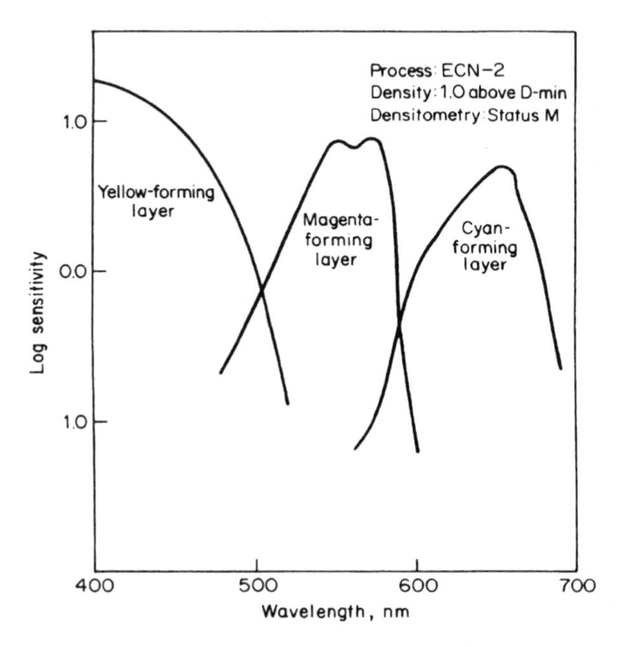

Figure 2.15.5 Typical color film spectral sensitivity curves. (*Courtesy of Eastman Kodak Company.*)

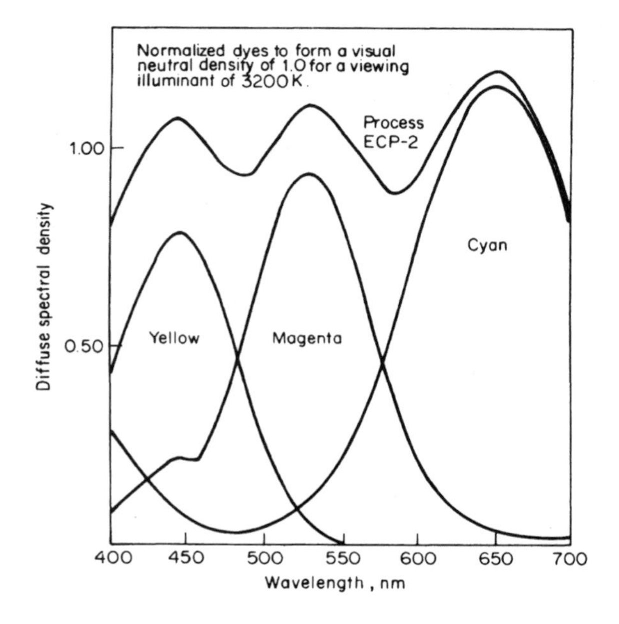

Figure 2.15.6 Typical spectral dye-density curves.

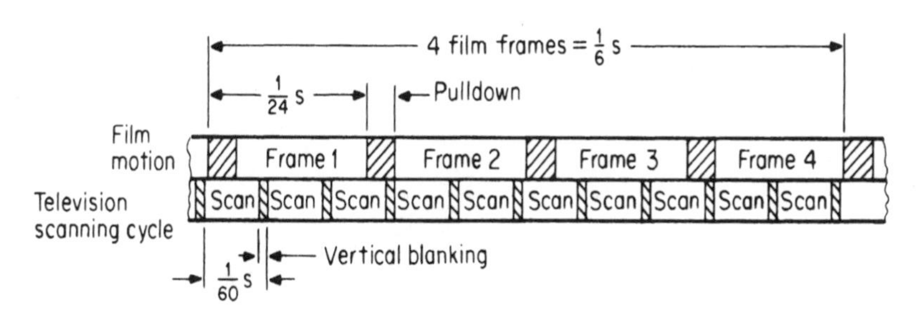

Figure 2.15.7 Standard projector time cycle.

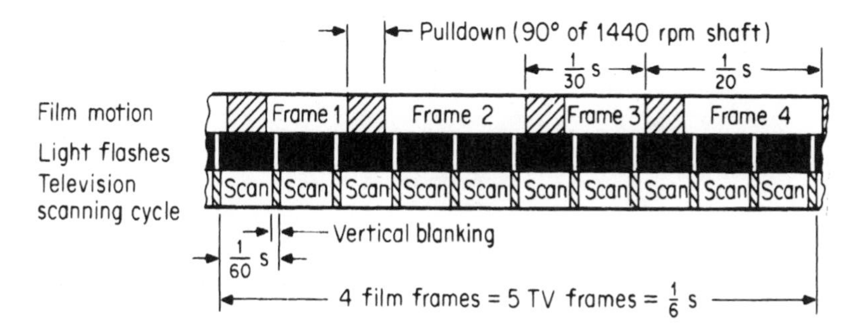

Figure 2.15.8 Short-application projector time-cycle.

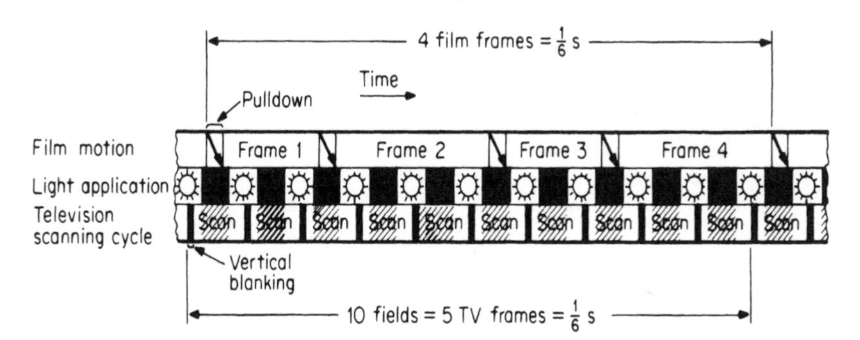

Figure 2.15.9 Long-application projector time-cycle.

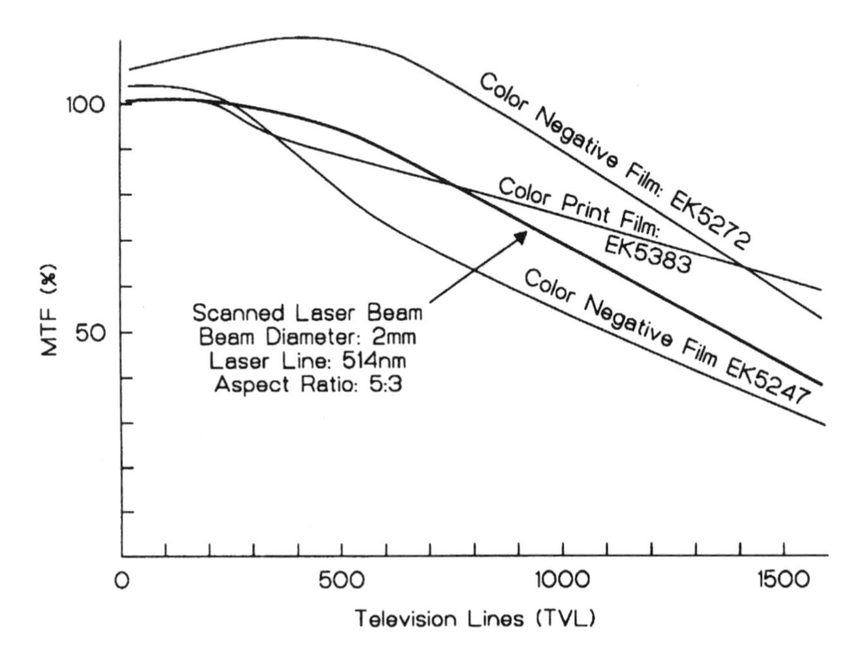

Figure 2.15.10 Modulation transfer function (MTF) of typical films and scanned laser beams. (*After: Benson, K. B., and D. G. Fink:* HDTV: Advanced Television for the 1990s, *McGraw-Hill, New York, 1990.*)

2.16 Audio/Video Compression Systems

Table 2.16.1 Storage Space Requirements for Video and Audio Data Signals (*From: Robin, Michael, and Michel Poulin:* Digital Television Fundamentals, *McGraw-Hill, New York, N.Y., 1997. Used with permission.*)

Media signals	Specifications	Data rate
Quality-audio voice	1 ch; 8-bit@8 kHz	64 kbps
MPEG audio Layer II	1 ch; 16-bit@48 kHz	128 kbps
MPEG audio Layer III	1 ch; 16-bit@48 kHz	64 kbps
AC-3	6 ch; 16-bit@48 kHz	384 kbps
CD	2 ch; 16-bit@44.1 kHz	1.4 Mbps
AES/EBU	2 ch; 24-bit@48 kHz	3.07 Mbps
MPEG-1 (Video)	352×288, 30 fps, 8-bit	1.5 Mbps
MPEG-2 (MP@ML)	720×576, 30 fps, 8-bit	Max. 15 Mbps
MPEG-2 (4:2:2 P@ML)	720×608, 30 fps, 8-bit	Max. 50 Mbps
CCIR-601	720×480, 30 fps, 8-bit	216 Mbps
HDTV	1920×1080, 30 fps, 8-bit	995 Mbps

Table 2.16.2 Common Bit-Rate-Reduction Applications *(From: Robin, Michael, and Michel Poulin:* Digital Television Fundamentals, *McGraw-Hill, New York, N.Y., 1997. Used with permission.)*

Applications	Coding standard	Maximum spatial resolution	Temporal resolution, bps	Maximum bit rate
Videophone	H.261	176 × 144 QCIF	7.5	64–128 kbps
Videoconferencing	H.261	352 × 288 CIF	10–30	0.384–1.554 Mbps
Entertainment TV	MPEG-1	352 × 288	30	<1.554 Mbps
Cable	MPEG-2	720 × 576	30	4–9 Mbps
Contribution	DPCM MPEG-2	720 × 576	30	<50 Mbps
Studio/production	Motion-JPEG MPEG-2 (@4:2:2)	720 × 576	30	<50 Mbps
HDTV production	MPEG-2	1920 × 1280	30	100 Mbps
HDTV transmission	MPEG-2	1920 × 1280	30	20 Mbps

Table 2.16.3 MPEG-1 Bit-Stream Encoding Constraints *(From: Robin, Michael, and Michel Poulin:* Digital Television Fundamentals, *McGraw-Hill, New York, N.Y., 1997. Used with permission.)*

Parameters	Maximum value
Picture width	768 pixels
Picture height	576 lines
Picture rate	30 picture/s
Number of macroblocks	396
Range-of-motion vectors	±64 pixels
Input buffer size	327,680 bits
Bit rate	1.8 Mbps

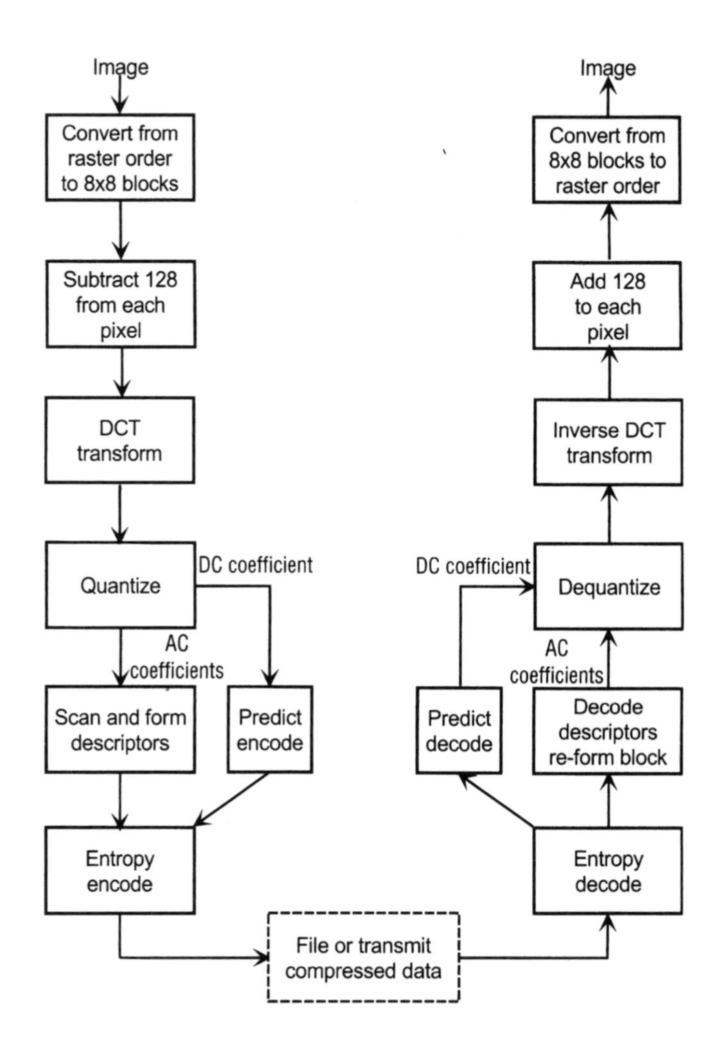

Figure 2.16.1 The basic JPEG encode/decode process.

Table 2.16.4 Layers of the MPEG-2 Video Bit-Stream Syntax (*After: Solari, Steve. J.:* Digital Video and Audio Compression, *McGraw-Hill, New York, 1997.*)

Syntax layer	Functionality
Video sequence layer	Context unit
Group of pictures (GOP) layer	Random access unit: video coding
Picture layer	Primary coding unit
Slice layer	Resynchronization unit
Macroblock layer	Motion-compensation unit
Block layer	DCT unit

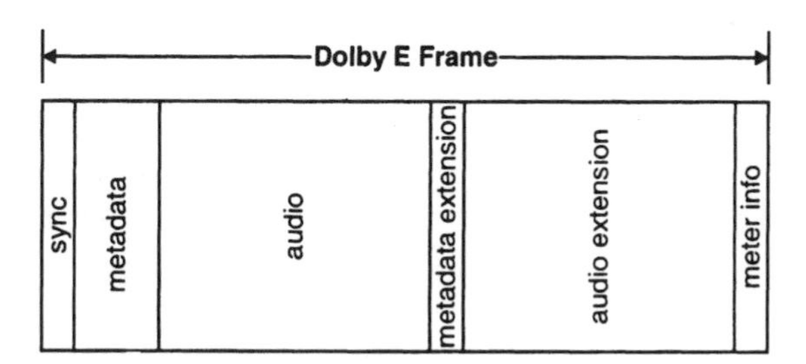

Figure 2.16.2 Basic frame structure of the Dolby E coding system. (*After. Terry, K. B., and S. B. Lyman: "Dolby E—A New Audio Distribution Format for Digital Broadcast Applications,"* International Broadcasting Convention Proceedings, *IBC, London, England, pp. 204–209, September 1999.*)

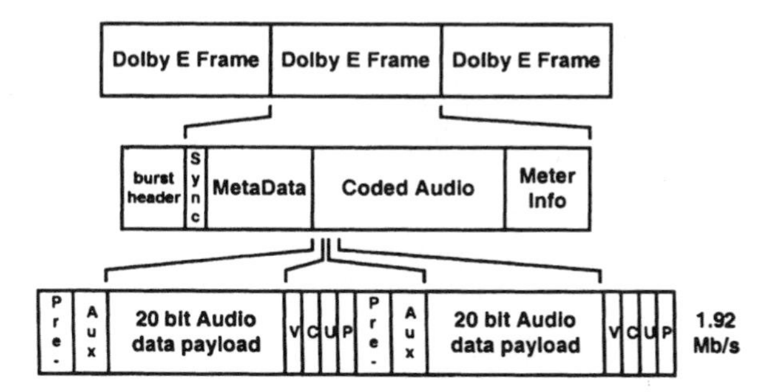

Figure 2.16.3 Overall coding scheme of Dolby E. (*After. Terry, K. B., and S. B. Lyman: "Dolby E—A New Audio Distribution Format for Digital Broadcast Applications,"* International Broadcasting Convention Proceedings, *IBC, London, England, pp. 204–209, September 1999.*)

Table 2.16.5 Common MPEG Profiles and Levels in Simplified Form

Profile	General Specifications	Parameter	Level			
			Low	Main (ITU 601)	High 1440 (HD, 4:3)	High (HD, 16:9)
Simple	Pictures: *I*, *P* Chroma: 4:2:0	Image size[1]		720× 576		
		Image frequency[2]		30		
		Bit rate[3]		15		
Main	Pictures: *I*, *P*, *B* Chroma: 4:2:0	Image size	325× 288	720× 576	1440×1152	1920× 1152
		Image frequency	30	30	60	60
		Bit rate	4	15	100	80
SNR-Scalable	Pictures: *I*, *P*, *B* Chroma: 4:2:0	Image size	325× 288	720× 576		
		Image frequency	30	30		
		Bit rate	3, 4[4]	15		
Spatially-Scalable	Pictures: *I*, *P*, *B* Chroma: 4:2:0	Image size			720×576	
		Image frequency			30	
		Bit rate			15	
	Enhancement Layer[5]	Image size			1440×1152	
		Image frequency			60	
		Bit rate			40, 60[6]	
High[7]	Pictures: *I*, *P*, *B* Chroma: 4:2:2	Image size		720× 576	1440×1152	1920× 1152
		Image frequency		30	60	60
		Bit rate		20	80	100
Studio	Pictures: *I*, *P*, *B* Chroma: 4:2:2	Image size		720× 608		
		Image frequency		30		
		Bit rate		50		

Notes:

[1] Image size specified as samples/line × lines/frame

[2] Image frequency in frames/s

[3] Bit rate in Mbits/s

[4] For *Enhancement Layer 1*

[5] For *Enhancement Layer 1*, except as noted by [6] for *Enhancement Layer 2*

[7] For simplicity, *Enhancement Layers* not specified individually

Table 2.16.6 Recommended MPEG-2 Coding Ranges for Various Video Formats (*After: SMPTE Recommended Practice: RP 202 (Proposed), "Video Alignment for MPEG-2 Coding," SMPTE, White Plains, N.Y., 1999.*)

Format	Resolution Pels x Lines	Coded Pels	Coded Lines			MPEG-2 Profile and Level
			Field 1	Field 2	Frame	
480I	720 × 480	0–719	23–262	286–525		MP@ML
480P	720 × 480	0–719			46–525	MP@HL
512I	720 × 512	0–719	7–262	270–525		422P@ML
512P	720 × 512	0–719			14–525	422P@HL
576I	720 × 576	0–719	23–310	336–623		MP@ML
608I	720 × 608	0–719	7–310	320–623		422P@ML
720P	1280 × 720	0–1279			26–745	MP@HL
720P	1280 × 720	0–1279			26–745	422P@HL
1080I	1920 × 1088[1]	0–1919	21–560	584–1123		MP@HL
1080I	1920 × 1088[1]	0–1919	21–560	584–1123		422P@HL
1080P	1920 × 1088[1]	0–1919			42–1121	MP@HL
1080P	1920 × 1088[1]	0–1919			42–1121	422P@HL

1 The active image only occupies the first 1080 lines.

Table 2.16.7 Comparison on Common Audio Compression Coding Schemes (*From: Robin, Michael, and Michel Poulin:* Digital Television Fundamentals, *McGraw-Hill, New York, N.Y., 1997. Used with permission.*)

Audio schemes	Total bit rates, kbps	Filter bank	Frequency resolution @48 kHz	Temporal resolution @48 kHz, ms	Frame length @48 kHz, msec	Bit rate target, kbps per channel
MPEG Layer I	32–448	PQMF	750 Hz	0.66	8	128
MPEG Layer II	32–384	PQMF	750 Hz	0.66	24	128
MPEG Layer III	32–320	PQMF/ MDCT	41.66 Hz	4	24	64
apt-X	Fixed 4:1 compression*	PQMF	12 kHz	<0.1	2.54	192
apt-Q	Fixed 12:1 & 18:1 compression†	MDCT	23.44 Hz	5.34	42.66	64
AC-3	32–640	MDCT	93.75 Hz	2.66	32	64

*Bit rates are from 56 to 384 kbps when the sampling frequency varies from 16 to 48 kHz.

†Bit rates can vary from 56 kbps for a mono signal sampled at 32 kHz to 128 kbps for a stereo signal sampled at 48 kHz.

Table 2.16.8 Operational Parameters of Subband APCM Algorithm (*After: Wylie, Fred: "Audio Compression Techniques,"* The Electronics Handbook, *Jerry C. Whitaker (ed.), CRC Press, Boca Raton, Fla., pp. 1260–1272, 1996.*)

Coding System	Compres-sion Ratio	Subbands	Bit Rate, kbits/s	A to A Delay, ms[1]	Audio Bandwidth, kHz
Dolby AC-2	6:1	256	256	45	20
ISO Layer 1	4:1	32	384	19	20
ISO Layer 2	Variable	32	192–256	>40	20
IOS Layer 3	12:1	576	128	>80	20
MUSICAM	Variable	32	128–384	>35	20

[1] The total system delay (encoder-to-decoder) of the coding system.

Table 2.16.9 Target Applications for ITU-R Rec. BS.1116 PEAQ

Category	Application	Version
Diagnostic	Assessment of implementations	Both
	Equipment or connection status	Advanced
	Codec identification	Both
Operational	Perceptual quality line-up	Basic
	On-line monitoring	Basic
Development	Codec development	Both
	Network planning	Both
	Aid to subjective assessment	Advanced

Table 2.16.10 Theoretical S/N as a Function of the Number of Sampling Bits

Number of Sampling Bits	Resolution (number of quantizing steps)	Maximum Theoretical S/N
18	262,144	110 dB
20	1,048,576	122 dB
24	16,777,216	146 dB

2.17 Audio/Video Networking

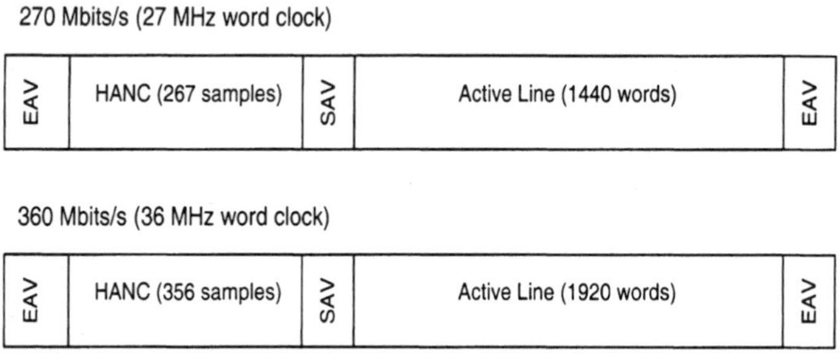

270 Mbits/s (27 MHz word clock)

EAV	HANC (267 samples)	SAV	Active Line (1440 words)	EAV

360 Mbits/s (36 MHz word clock)

EAV	HANC (356 samples)	SAV	Active Line (1920 words)	EAV

Figure 2.17.1 The basic SDI bit stream.

Table 2.17.1 SMPTE 272M Mode Definitions

A (default)	Synchronous 48 kHz, 20 bit audio, 48 sample buffer
B	Synchronous 48 kHz, composite video only, 64 sample buffer to receive 20 bits from 24 bit audio data
C	Synchronous 48 kHz, 24-bit audio and extended data packets
D	Asynchronous (48 kHz implied, other rates if so indicated)
E	44.1 kHz audio
F	32 kHz audio
G	32-48 kHz variable sampling rate audio
H	Audio frame sequence (inherent in 29.97 frame/s video systems, except 48 kHz synchronous audio—default A mode)
I	Time delay tracking
J	Non-coincident channel status Z bits in a pair.

Table 2.17.2 SDI Reference Source Format Parameters (*After: Gaggioni, H., M. Ueda, F. Saga, K. Tomita, and N. Kobayashi, "The Development of a High-Definition Serial Digital Interface," Sony Technical Paper, Sony Broadcast Group, San Jose, Calif., 1998.*)

Reference Document	SMPTE-260M	SMPTE-274M	SMPTE-274M
Parallel word rate (each channel Y, C_R/C_B)	74.25 Mword/s	74.25 Mword/s	74.25/1.001 Mword/s
Lines per frame	1125	1125	1125
Words per active line (each channel Y, C_R/C_B)	1920	1920	1920
Total active line	1035	1080	1080
Words per total line (each channel Y, C_R/C_B)	2200	2200	2200
Frame rate	30 Hz	30 Hz	30/1.001 Hz
Total Fields per frame	2	2	2
Total data rate	1.485Gbits/s	1.485Gbits/s	1.485/1.001 Gbits/s
Field 1 EAV V = 1	Line 1121	Line 1124	Line 1124
Field 1 EAV V = 0	Line 41	Line 21	Line 21
Field 2 EAV V = 0	Line 558	Line 561	Line 561
Field 2 EAV V = 0	Line 603	Line 584	Line 584
EAV F = 0	Line 1	Line 1	Line 1
EAV F = 1	Line 564	Line 563	Line 563

Table 2.17.3 Source Format Parameters for SMPTE 292M (*After: "SMPTE Standard for Television—Bit-Serial Digital Interface for High-Definition Television Systems," SMPTE 292M-1998, SMPTE, White Plains, N.Y., 1998.*)

Reference SMPTE Standard	260M		295M	274M								296M	
Format	A	B	C	D	E	F	G	H	I	J	K	L	M
Lines per frame	1125	1125	1250	1125	1125	1125	1125	1125	1125	1125	1125	750	750
Words per active line (each channel Y, C_B/C_R)	1920	1920	1920	1920	1920	1920	1920	1920	1920	1920	1920	1280	1280
Total active lines	1035	1035	1080	1080	1080	1080	1080	1080	1080	1080	1080	720	720
Words per total line (each channel Y, C_B/C_R)	2200	2200	2376	2200	2200	2640	2200	2200	2640	2750	2750	1650	1650
Frame rate (Hz)	30	30/M	25	30	30/M	25	30	30/M	25	24	24/M	60	60/M
Fields per frame	2	2	2	2	2	2	1	1	1	1	1	1	1
Data rate divisor	1	M	1	1	M	1	1	M	1	1	M	1	M

Table 2.17.4 Summary of SD Video Formats Referenced in SMPTE 346M (*After: "SMPTE Standard for Television—Signals and Generic Data over High-Definition Interfaces," SMPTE 346M-2000, SMPTE, White Plains, N.Y., 2000.*)

SD System/Sampling Structure	525 × 60 or 625 × 50, 4 × 3 13.5 MHz	525 × 59.94 or 625 × 50, 16 × 9 18 MHz
4:0:0I	135 Mbits/s	180 Mbits/s
4:2:2I	270 Mbits/s ANSI/SMPTE 267M	360 Mbits/s ANSI/SMPTE 267M
4:2:2:4I	360 Mbits/s	540 Mbits/s
4:4:4I	360 Mbits/s ITU-R BT.601-5	540 Mb/s ITU-R BT.601-5
4:2:0P	360 Mbits/s	540 Mbits/s
4:4:4:4I	540 Mbits/s SMPTE RP 174	720 Mbits/s
8:4:4I	540 Mbits/s	720 Mbits/s
4:2:2P	540 Mbits/s ANSI/SMPTE 293M	720 Mbits/s

Table 2.17.5 Parameters of the SMPTE 349M Source Format (*After "SMPTE Standard for Television (Proposed)—Transport of Alternate Source Image Formats through SMPTE 292M, SMPTE 349M, SMPTE, White Plains, N.Y., 2001.*)

No.	Source Format						
	Total Lines per Frame	Total Words per Line	Frame Rate (Hz)	Fields per Frame	Parallel and Multiplexed Word Rate	Signal Type	Reference Standards
1	525	858	30/M	2	27 Mb/s	4:2:2i	SMPTE 125M SMPTE 259M
2		1144	30/M	2	36 Mb/s	4:2:2i	SMPTE 267M SMPTE 259M
3		858	30/M	2	54 Mb/s	4:4:4:4i	SMPTE RP 174 SMPTE 344M
4		858	60/M	1	27 Mb/s × 2	4:2:2p	SMPTE 293M SMPTE 294M
5		858	60/M	1	54 Mb/s	4:2:2p	SMPTE 293M SMPTE 344M
6		858	60/M	1	36 Mb/s	4:2:0p	SMPTE 293M SMPTE 294M
7	625	864	25	2	27 Mb/s	4:2:2i	ITU-R BT.601 ITU-R BT.656
8		1152	25	2	36 Mb/s	4:2:2i	ITU-R BT.601 ITU-R BT.656
9		864	25	2	54 Mb/s	4:4:4:4i	ITU-R BT.799 SMPTE 344M
10		864	50	1	27 Mb/s × 2	4:2:2p	ITU-R BT.1358 ITU-R BT.1362
11		864	50	1	54 Mb/s	4:2:2p	ITU-R BT.1358 SMPTE 344M
12		864	50	1	36 Mb/s	4:2:0p	ITU-R BT.1358 ITU-R BT.1362

Note: M = 1.001

Table 2.17.6 Source Image Format Parameters Specified in SMPTE 347M (*After "SMPTE Standard for Television (proposed)—540 Mb/s Serial Digital Interface: Source Image Format Mapping," SMPTE 347M, SMPTE, White Plains, N.Y., 2001.*)

	Nomenclature	Total Lines per Frame	Frame Rate	Fields per Frame	Active Lines per Frame	Samples per Total Line (each component)	Samples per Active Line (each component)	Reference Standard
1	Single link 4:4:4:4 525i/59.94	525	30/1.001	2	483	858	720	SMPTE RP 174
2	Single link 4:4:4:4 625i/50	625	25	2	576	864	720	ITU-R BT.799
3	Single link 4:2:2P 525p/59.94	525	60/1.001	1	483	858	720	SMPTE 293M
4	Single link 4:2:2P 625p/50	625	50	1	576	864	720	ITU-R BT.1358
5	Dual link 4:2:2P 525p/59.94	525	60/1.001	1	483	858	720	SMPTE 294M
6	Dual link 4:2:2P 625p/50	625	50	1	576	864	720	ITU-R BT.1362

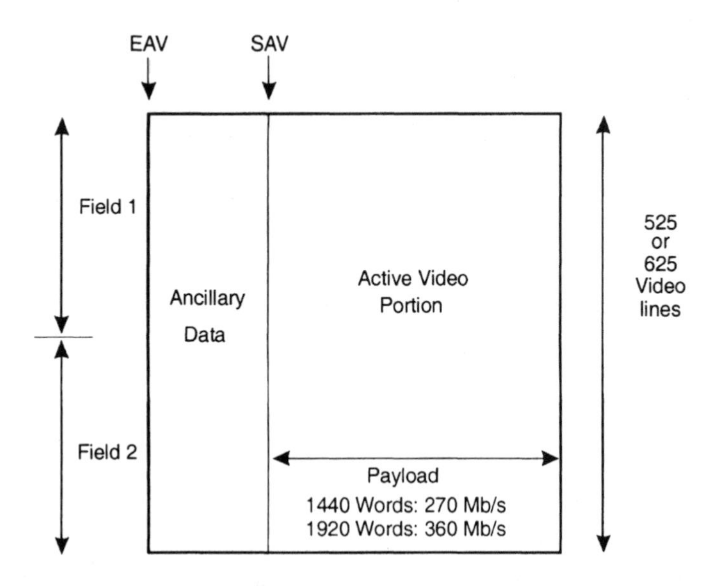

Figure 2.17.2 SMPTE 259M principal video timing parameters. (*From: Legault, Alain, and Janet Matey: "Interconnectivity in the DTV Era—The Emergence of SDTI,"* Proceedings of Digital Television '98, *Intertec Publishing, Overland Park, Kan., 1998. Used with permission.*)

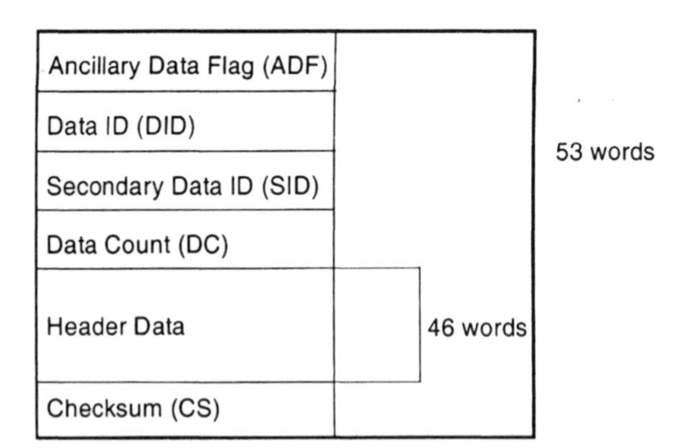

Figure 2.17.3 SDTI header data packet structure. (*From: Legault, Alain, and Janet Matey: "Interconnectivity in the DTV Era—The Emergence of SDTI,"* Proceedings of Digital Television '98, *Intertec Publishing, Overland Park, Kan., 1998. Used with permission.*)

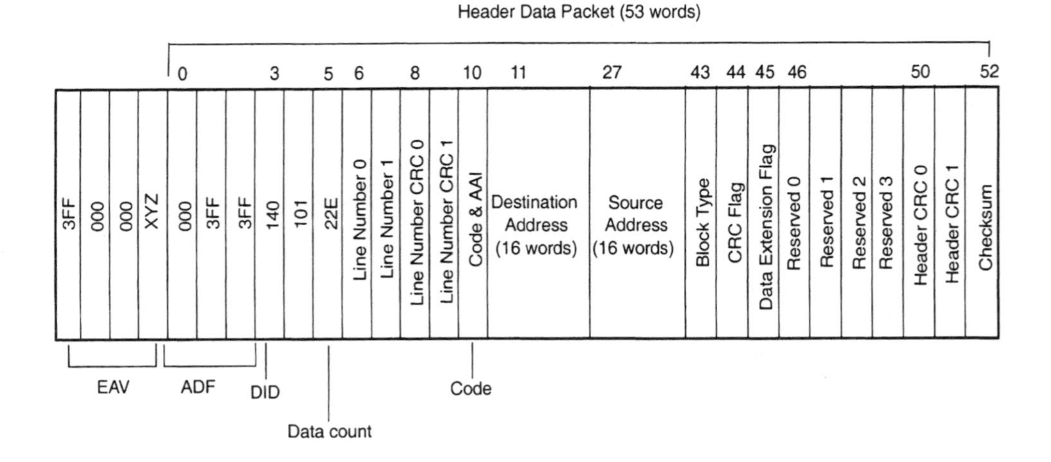

Figure 2.17.4 SDTI header data structure. (*From: Legault, Alain, and Janet Matey: "Interconnectivity in the DTV Era—The Emergence of SDTI,"* Proceedings of Digital Television '98, *Intertec Publishing, Overland Park, Kan., 1998. Used with permission.*)

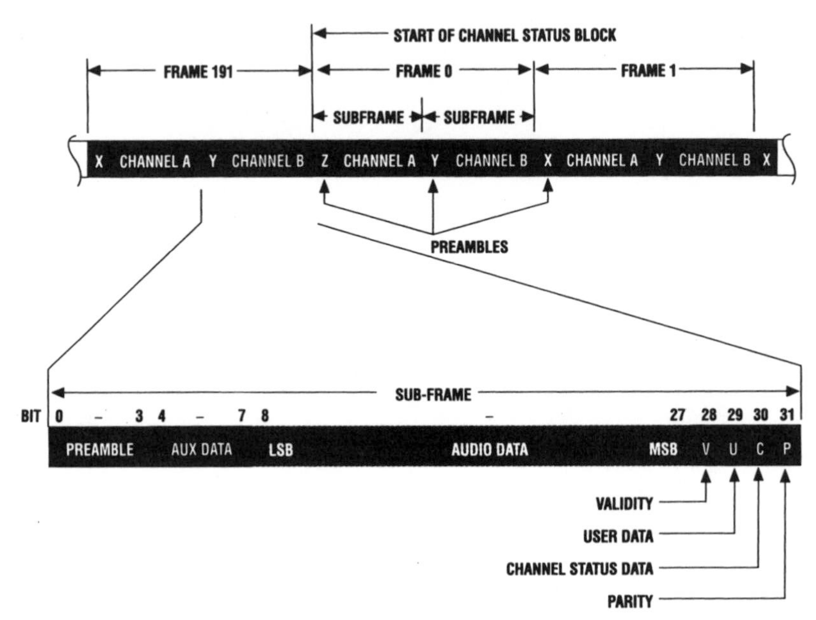

Figure 2.17.5 The typical packing of an internodal ATM trunk. (*After: Wu, Tsong-Ho: "Network Switching Concepts,"* The Electronics Handbook, *Jerry C. Whitaker (ed.), CRC Press, Boca Raton, Fla., p. 1513, 1996.*)

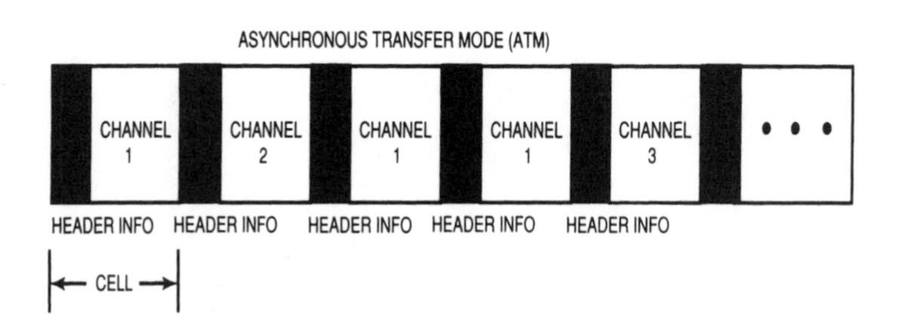

Figure 2.17.6 The ATM cell format. (*After: Wu, Tsong-Ho: "Network Switching Concepts,"* The Electronics Handbook, *Jerry C. Whitaker (ed.), CRC Press, Boca Raton, Fla., p. 1513, 1996.*)

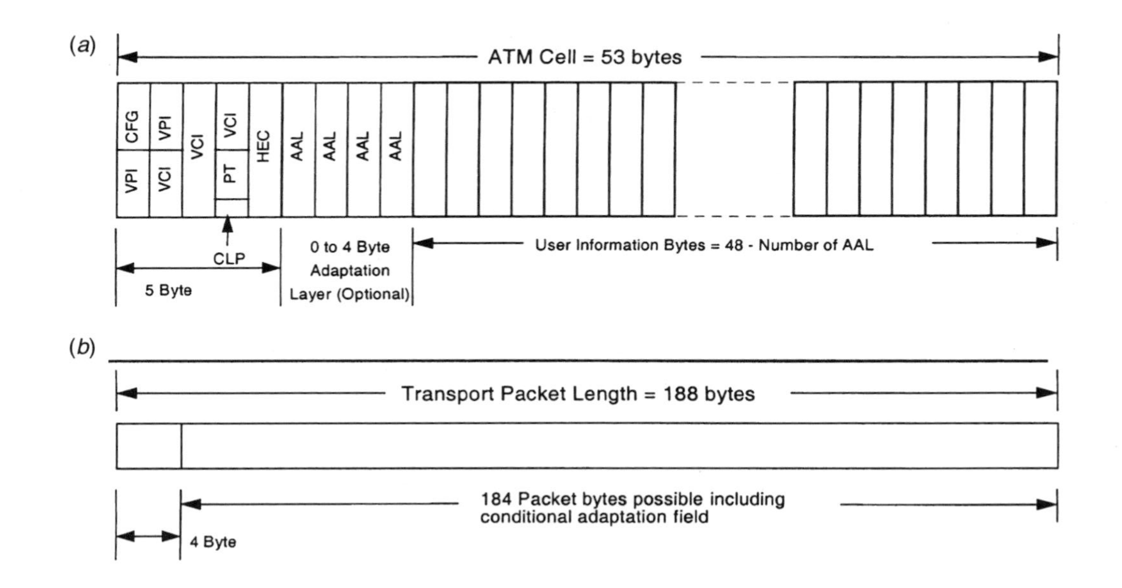

Figure 2.17.7 Comparison of the ATM cell structure and the MPEG-2 transport packet structure: (*a*) structure of the ATM cell, (*b*) structure of the transport packet. (*After: ATSC, "Guide to the Use of the Digital Television Standard," Advanced Television Systems Committee, Washington, D.C., Doc. A/54, Oct. 4, 1995.*)

Table 2.17.7 ATM Cell Header Fields (*After: Piercy, John: "ATM Networked Video: Moving From Leased-Lines to Packetized Transmission,"* Proceedings of the Transition to Digital Conference, *Intertec Publishing, Overland Park, Kan., 1996.*)

GFC	A 4-bit *generic flow control* field: used to manage the movement of traffic across the user network interface (UNI).
VPI	An 8-bit network *virtual path identifier.*
VCI	A 16-bit network *virtual circuit identifier.*
PT	A 3-bit *payload type* (i.e., user information type ID).
CLP	A 1-bit *cell loss priority* flag (eligibility of the cell for discard by the network under congested conditions).
HEC	An 8-bit *header error control* field for ATM header error correction.
AAL	ATM *adaptation-layer* bytes (user-specific header).

Table 2.17.8 Basic Specifications of Ethernet Performance (*After: Gallo and Hancock:* Networking Explained, *Digital Press, pp. 191–235, 1999.*)

Parameter	10 Mbits/s		100 Mbits/s		1000 Mbits/s	
Frame size	Minimum	Maximum	Minimum	Maximum	Minimum	Maximum
Frames/s	14.8 k	812	148 k	8.1 k	1.48 M	81 k
Data rate	5.5 Mbits/s	9.8 Mbits/s	55 Mbits/s	98 Mbits/s	550 Mbits/s	980 Mbits/s
Frame interval	67 μs	1.2 ms	6.7 μs	120 μs	0.7 μs	12 μs

Table 2.17.9 Fibre Channel General Performance Specifications (*After: "Technology Brief—Networking and Storage Strategies," Omneon Video Networks, Campbell, Calif., 1999.*)

Media	Speed		Distance
Electrical Characteristics			
Coax/twinax	ECL	1.0625 Gigabits/s	24 Meters
	ECL	266 Megabits/s	47 Meters
Optical Characteristics			
9 micrometer single mode fiber	Longwave laser	1.0625 Gigabits/s	10 Kilometers
50 micrometer multi-mode fiber	Shortwave laser	1.0625 Gigabits/s	300 Meters
	Shortwave Laser	266 Megabits/s	2 Kilometer
62.5 micrometer multi-mode fiber	Longwave LED	266 Megabits/s	1 Kilometer
	Longwave LED	132 Megabits/s	500 Meters

Note: In FC-AL configurations, the distance numbers represents the distance between nodes, not the total distance around the loop.
Note: In fabric configurations, the distance numbers represent the distance from the fabric to a node, not the distance between nodes.

2.18 Digital Broadcast Transmission Systems

Table 2.18.1 Standardized Video Input Formats

Video Standard	Active Lines	Active Samples/Line
SMPTE 274M-1995 SMPTE 295M-1997 (50 Hz)	1080	1920
SMPTE 296M-1997	720	1280
ITU-R Rec. 601-4 SMPTE 293M-1996 (59.94, P) SMPTE 294M-1997 (59.94, P)	483	720

Table 2.18.2 ATSC DTV Compression Format Constraints (*After: ATSC, "ATSC Digital Television Standard," Advanced Television Systems Committee, Washington, D.C., Doc. A/53, Sep.16, 1995.*)

Vertical Size Value	Horizontal Size Value	Aspect Ratio Information	Frame-Rate Code	Progressive Sequence
1080[1]	1920	16:9, square pixels	1,2,4,5	Progressive
			4,5	Interlaced
720	1280	16:9, square pixels	1,2,4,5,7,8	Progressive
480	704	4:3, 16:9	1,2,4,5,7,8	Progressive
			4,5	Interlaced
	640	4:3, square pixels	1,2,4,5,7,8	Progressive
			4,5	Interlaced

Frame-rate code: 1 = 23.976 Hz, 2 = 24 Hz, 4 = 29.97 Hz, 5 = 30 Hz, 7 = 59.94 Hz, 8 = 60 Hz

[1] Note that 1088 lines actually are coded in order to satisfy the MPEG-2 requirement that the coded vertical size be a multiple of 16 (progressive scan) or 32 (interlaced scan).

Table 2.18.3 Picture Formats Under the DTV Standard (*After: ATSC, "Guide to the Use of the Digital Television Standard," Advanced Television Systems Committee, Washington, D.C., Doc. A/54, Oct. 4, 1995.*)

Vertical Lines	Pixels	Aspect Ratio	Picture Rate
1080	1920	16:9	60I, 30P, 24P
720	1280	16:9	60P, 30P, 24P
480	704	16:9 and 4:3	60P, 60I, 30P, 24P
480	640	4:3	60P, 60I, 30P, 24P

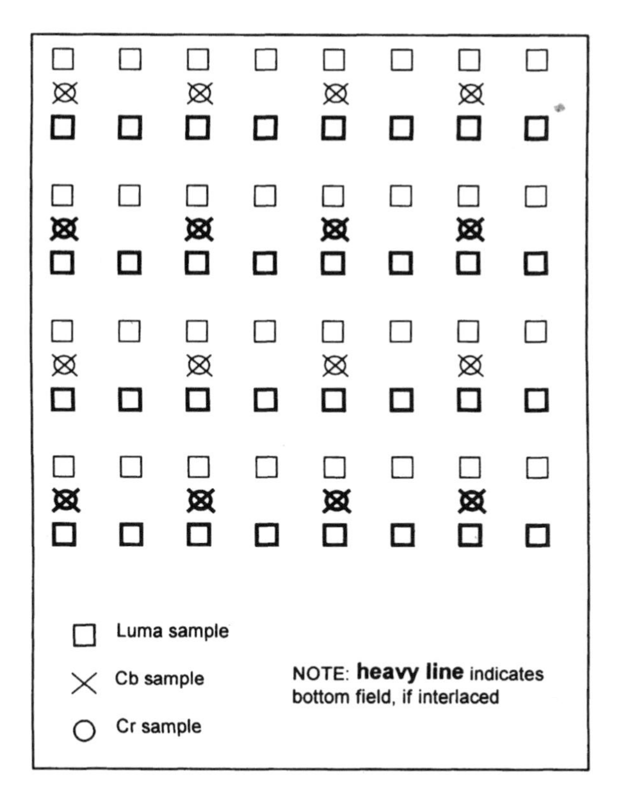

Figure 2.18.1 Placement of luma/chroma samples for 4:2:0 sampling. (*From: ATSC, "Guide to the Use of the Digital Television Standard," Advanced Television Systems Committee, Washington, D.C., Doc. A/54, Oct. 4, 1995. Used with permission.*)

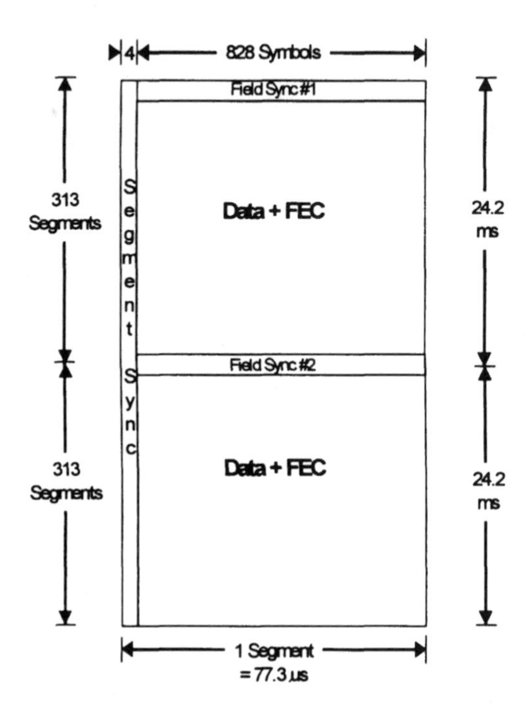

Figure 2.18.2 The basic VSB data frame. (*From: ATSC, "ATSC Digital Television Standard," Advanced Television Systems Committee, Washington, D.C., Doc. A/53, Sep.16, 1995.*)

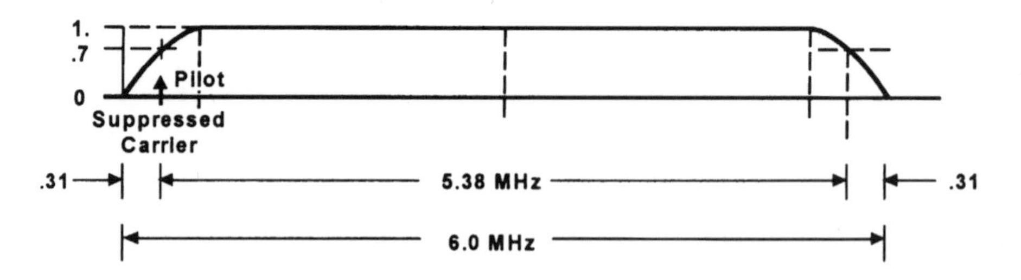

Figure 2.18.3 Nominal VSB channel occupancy. (*From: ATSC, "ATSC Digital Television Standard," Advanced Television Systems Committee, Washington, D.C., Doc. A/53, Sep.16, 1995.*)

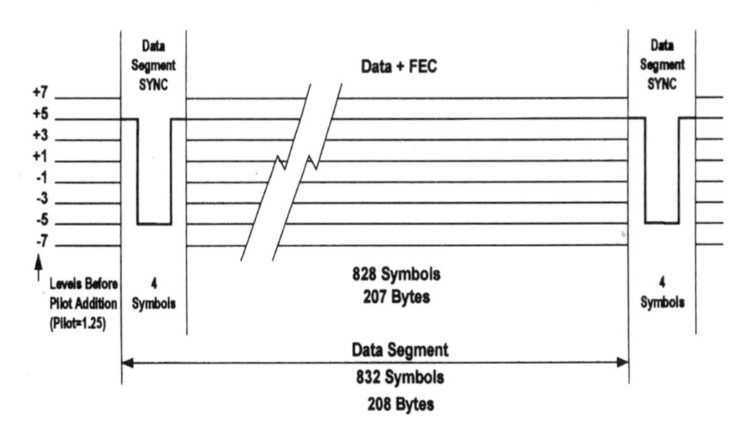

Figure 2.18.4 The 8-VSB data segment. (*From: ATSC, "ATSC Digital Television Standard," Advanced Television Systems Committee, Washington, D.C., Doc. A/53, Sep.16, 1995. Used with permission.*)

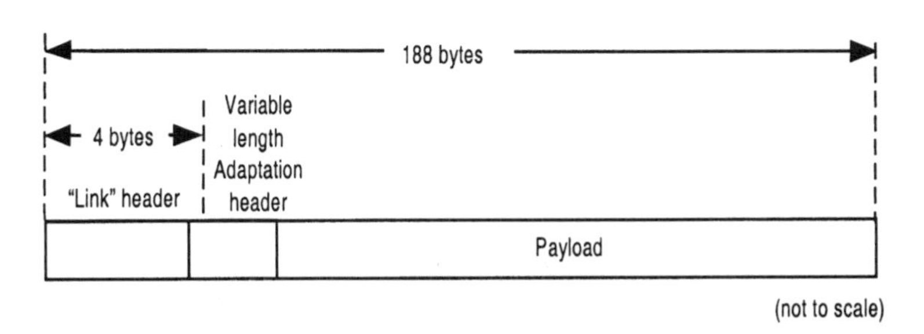

Figure 2.18.5 Basic transport packet format for the DTV standard. (*From: ATSC, "Guide to the Use of the Digital Television Standard," Advanced Television Systems Committee, Washington, D.C., Doc. A/54, Oct. 4, 1995. Used with permission.*)

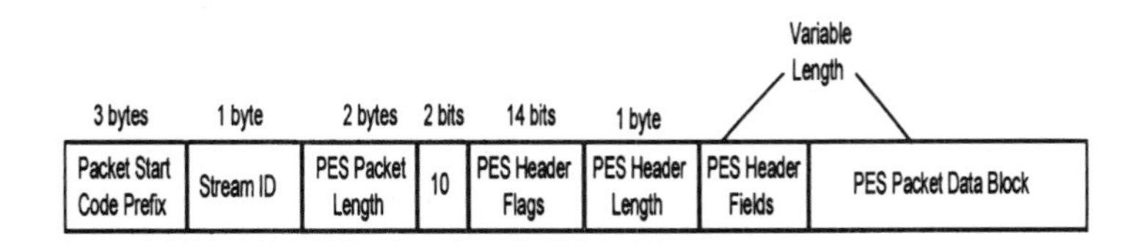

Figure 2.18.6 Structural overview of a PES packet. (*From: ATSC, "ATSC Digital Television Standard," Advanced Television Systems Committee, Washington, D.C., Doc. A/53, Sep.16, 1995. Used with permission.*)

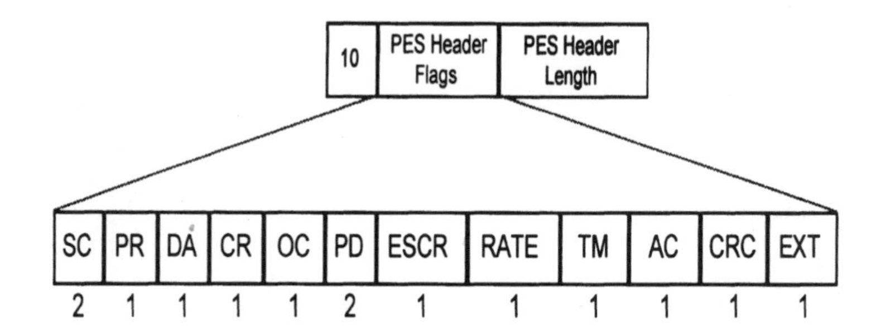

Figure 2.18.7 PES header flags in their relative positions (all sizes in bits). (*From: ATSC, "ATSC Digital Television Standard," Advanced Television Systems Committee, Washington, D.C., Doc. A/53, Sep.16, 1995. Used with permission.*)

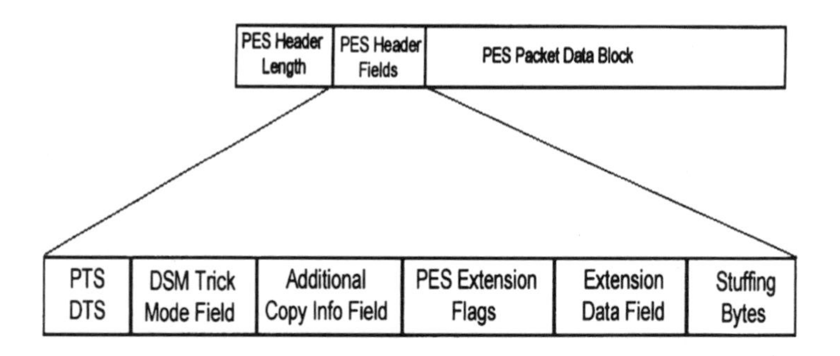

Figure 2.18.8 Organization of the PES header. (*From: ATSC, "ATSC Digital Television Standard," Advanced Television Systems Committee, Washington, D.C., Doc. A/53, Sep.16, 1995. Used with permission.*)

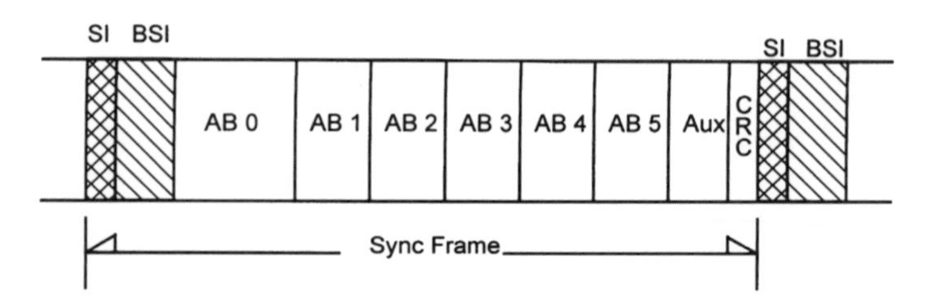

Figure 2.18.9 The AC-3 synchronization frame. (*From: ATSC, "Guide to the Use of the Digital Television Standard," Advanced Television Systems Committee, Washington, D.C., Doc. A/54, Oct. 4, 1995. Used with permission.*)

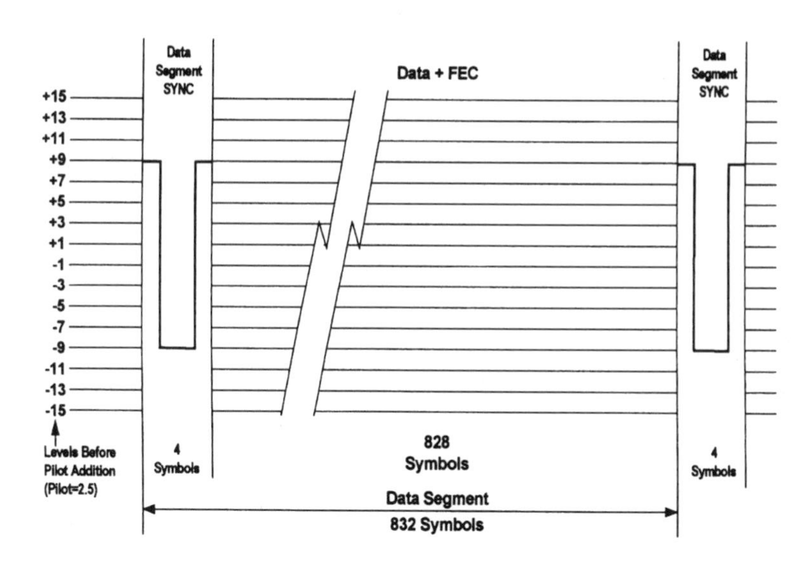

Figure 2.18.10 Typical data segment for the 16-VSB mode. (*From: ATSC, "Guide to the Use of the Digital Television Standard," Advanced Television Systems Committee, Washington, D.C., Doc. A/54, Oct. 4, 1995. Used with permission.*)

Table 2.18.4 Typical Bit Rates for Various Services (*After: ATSC, "Digital Television Standard," Advanced Television Systems Committee, Washington, D.C., Doc. A/53, Sept.16, 1995.*)

Type of Service	Number of Channels	Typical Bit Rates
CM, ME, or associated audio service containing all necessary program elements	5	320–384 kbits/s
CM, ME, or associated audio service containing all necessary program elements	4	256–384 kbits/s
CM, ME, or associated audio service containing all necessary program elements	3	192–320 kbits/s
CM, ME, or associated audio service containing all necessary program elements	2	128–256 kbits/s
VI, narrative only	1	48–128 kbits/s
HI, narrative only	1	48–96 kbits/s
D	1	64–128 kbits/s
D	2	96–192 kbits/s
C, commentary only	1	32–128 kbits/s
E	1	32–128 kbits/s
VO	1	64–128 kbits/s

Table 2.18.5 Language Code Table for AC-3 (*After: ATSC, "Digital Television Standard," Advanced Television Systems Committee, Washington, D.C., Doc. A/53, Sept.16, 1995.*)

Code	Language	Code	Language	Code	Language	Code	Language
0x00	unknown/not applicable	0x20	Polish	0x40	background sound/ clean feed	0x60	Moldavian
0x01	Albanian	0x21	Portuguese	0x41		0x61	Malaysian
0x02	Breton	0x22	Romanian	0x42		0x62	Malagasay
0x03	Catalan	0x23	Romansh	0x43		0x63	Macedonian
0x04	Croatian	0x24	Serbian	0x44		0x64	Laotian
0x05	Welsh	0x25	Slovak	0x45	Zulu	0x65	Korean
0x06	Czech	0x26	Slovene	0x46	Vietnamese	0x66	Khmer
0x07	Danish	0x27	Finnish	0x47	Uzbek	0x67	Kazakh
0x08	German	0x28	Swedish	0x48	Urdu	0x68	Kannada
0x09	English	0x29	Turkish	0x49	Ukrainian	0x69	Japanese
0x0A	Spanish	0x2A	Flemish	0x4A	Thai	0x6A	Indonesian
0x0B	Esperanto	0x2B	Walloon	0x4B	Telugu	0x6B	Hindi
0x0C	Estonian	0x2C		0x4C	Tatar	0x6C	Hebrew
0x0D	Basque	0x2D		0x4D	Tamil	0x6D	Hausa
0x0E	Faroese	0x2E		0x4E	Tadzhik	0x6E	Gurani
0x0F	French	0x2F		0x4F	Swahili	0x6F	Gujurati
0x10	Frisian	0x30	reserved	0x50	Sranan Tongo	0x70	Greek
0x11	Irish	0x31	"	0x51	Somali	0x71	Georgian
0x12	Gaelic	0x32	"	0x52	Sinhalese	0x72	Fulani
0x13	Galician	0x33	"	0x53	Shona	0x73	Dari
0x14	Icelandic	0x34	"	0x54	Serbo-Croat	0x74	Churash
0x15	Italian	0x35	"	0x55	Ruthenian	0x75	Chinese
0x16	Lappish	0x36	"	0x56	Russian	0x76	Burmese
0x17	Latin	0x37	"	0x57	Quechua	0x77	Bulgarian
0x18	Latvian	0x38	"	0x58	Pustu	0x78	Bengali
0x19	Luxembourgian	0x39	"	0x59	Punjabi	0x79	Belorussian
0x1A	Lithuanian	0x3A	"	0x5A	Persian	0x7A	Bambora
0x1B	Hungarian	0x3B	"	0x5B	Papamiento	0x7B	Azerbijani
0x1C	Maltese	0x3C	"	0x5C	Oriya	0x7C	Assamese
0x1D	Dutch	0x3D	"	0x5D	Nepali	0x7D	Armenian
0x1E	Norwegian	0x3E	"	0x5E	Ndebele	0x7E	Arabic
0x1F	Occitan	0x3F	"	0x5F	Marathi	0x7F	Amharic

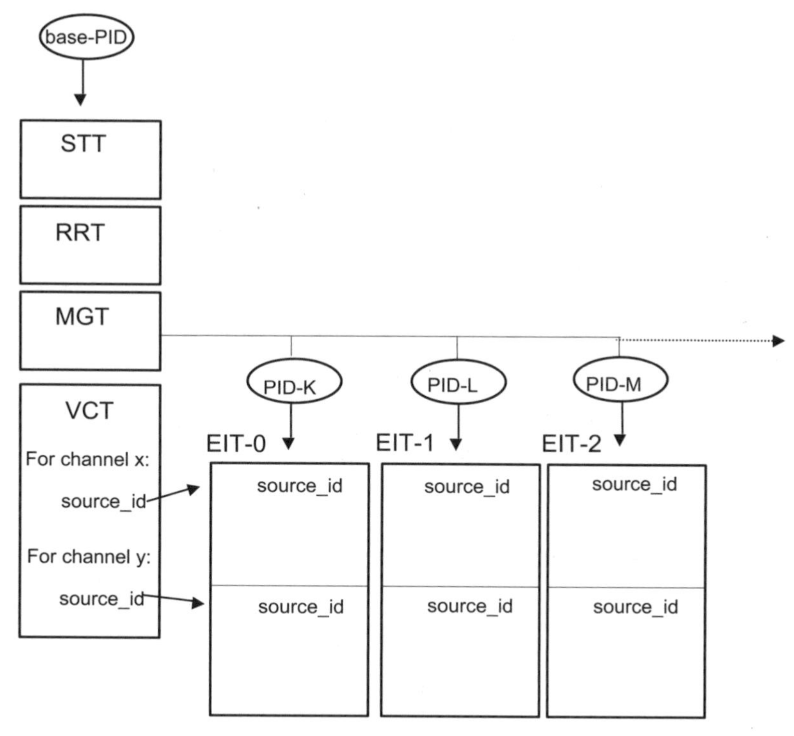

Figure 2.18.11 Overall structure for the PSIP tables. (*From: ATSC: "Program and System Information Protocol for Terrestrial Broadcast and Cable," Advanced Television Systems Committee, Washington, D.C., Doc. A/65, February 1998. Used with permission.*)

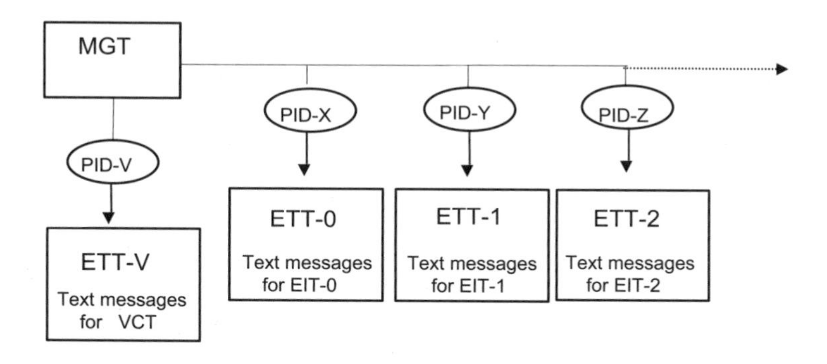

Figure 2.18.12 Extended text tables in the PSIP hierarchy. (*From: ATSC: "Program and System Information Protocol for Terrestrial Broadcast and Cable," Advanced Television Systems Committee, Washington, D.C., Doc. A/65, February 1998. Used with permission.*)

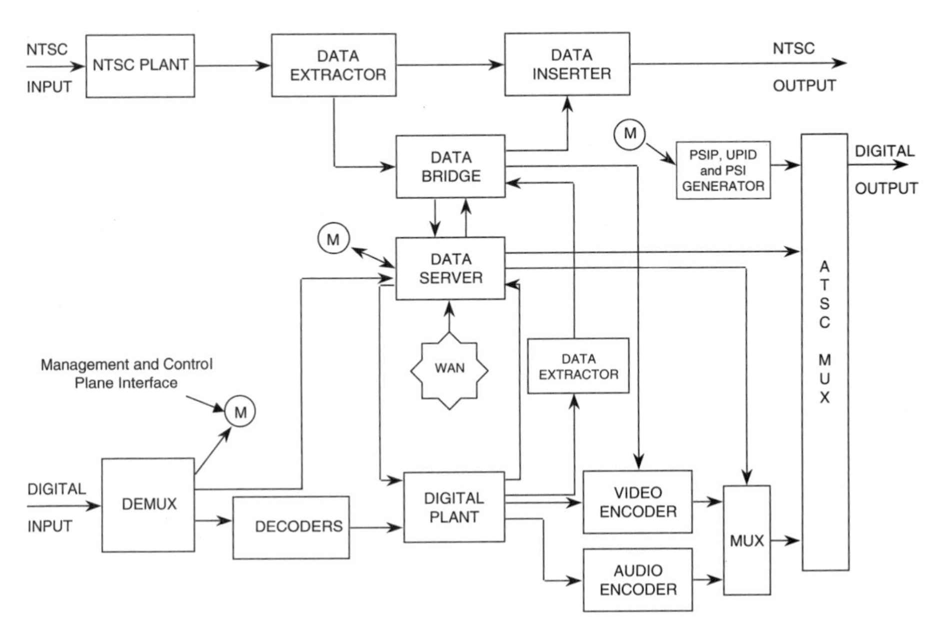

Figure 2.18.13 Block diagram of the PSIP generation and insertion process.

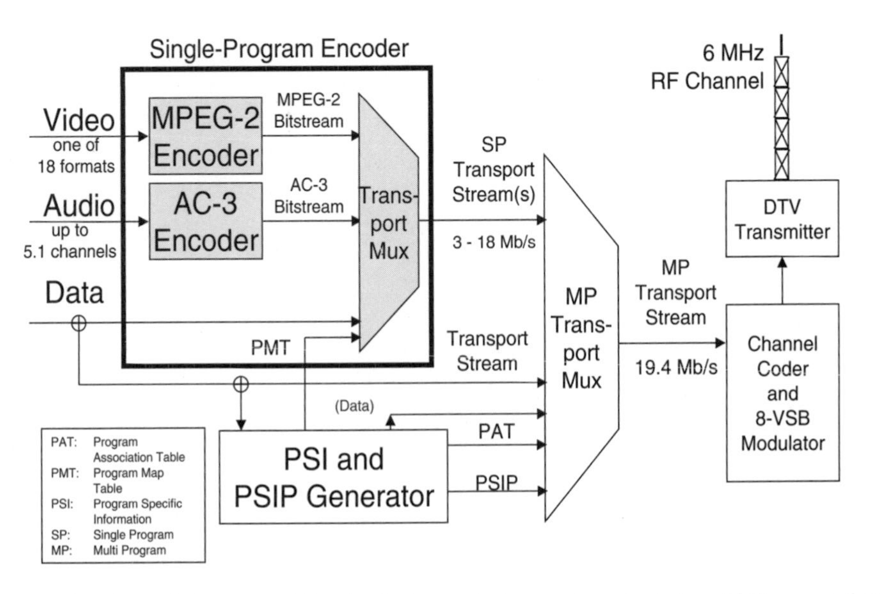

Figure 2.18.14 Block diagram of the ATSC transmission system, including PSIP generation and insertion.

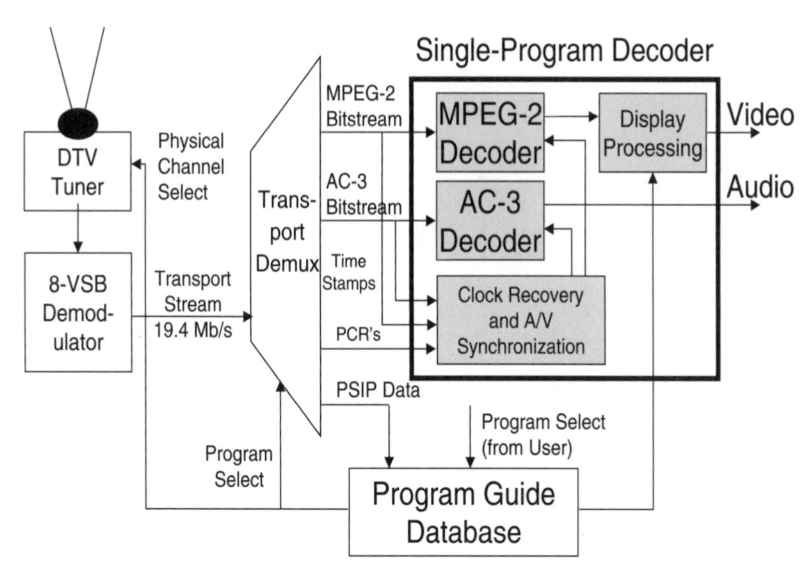

Figure 2.18.15 Bock diagram of an ATSC decoder, including PSIP extraction and program guide generation.

Table 2.18.6 PSIP Tables Required for Transmission in the Broadcast and Cable Modes

Table	Required for Broadcast?	Required for Cable?
STT	Yes	Yes
MGT	Yes	Yes
VCT	Yes (TVCT)	Yes (CVCT)
RRT	Yes	Yes
EIT	Yes (EIT-0, -1, -2, -3). All others optional	Optional
ETT	Optional	Optional

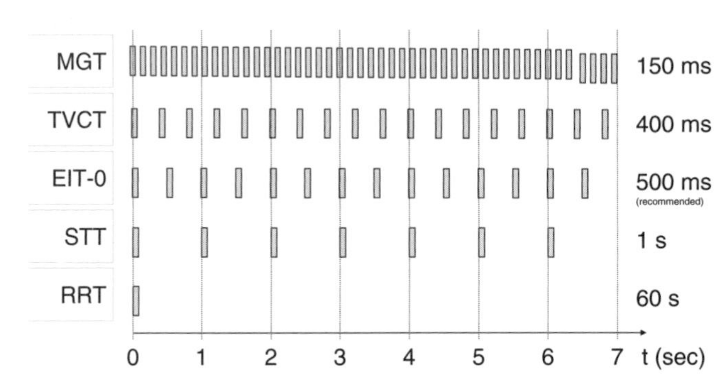

Figure 2.18.16 Maximum cycle times for PSIP tables.

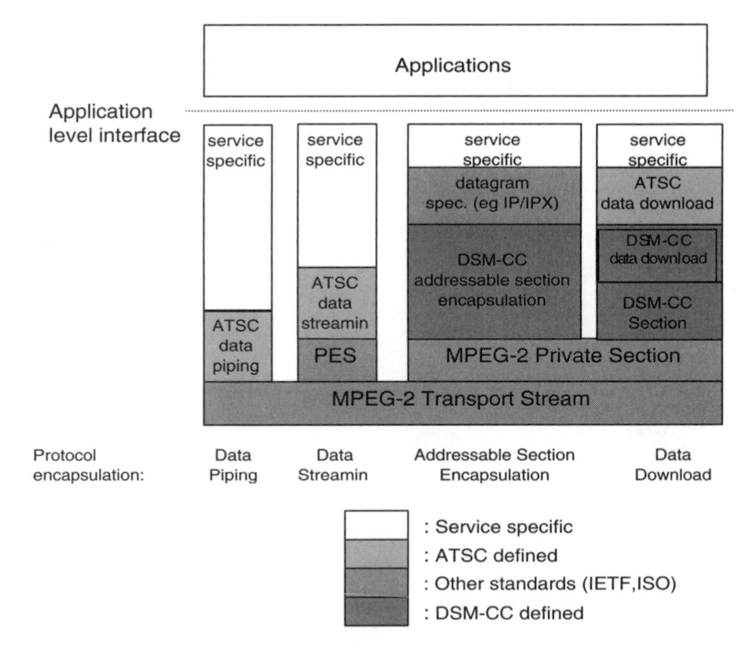

Figure 2.18.17 ATSC data broadcast standard graphical encapsulation overview and relation to other standards. (*From: ATSC: "Recommended Practice—Implementation Guidelines for the ATSC Data Broadcast Standard," Advanced Television Systems Committee, Washington, D.C., Doc. A/91, 2001. Used with permission.*)

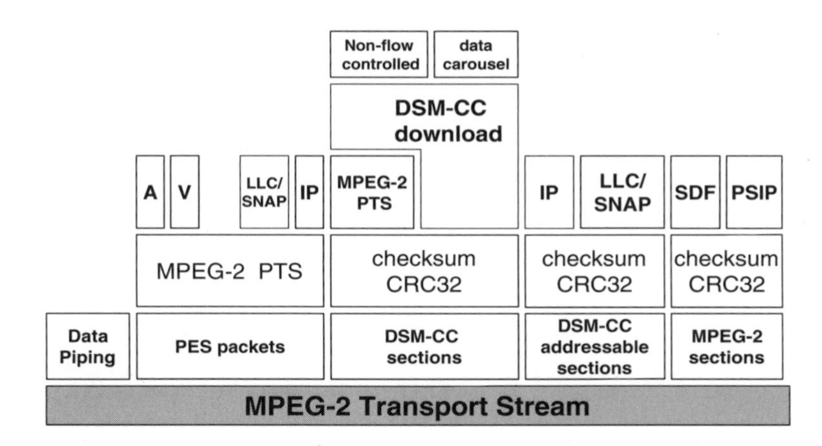

Figure 2.18.18 ATSC data broadcast protocol packetization, synchronization, and protection layers. (*From: ATSC: "Recommended Practice—Implementation Guidelines for the ATSC Data Broadcast Standard," Advanced Television Systems Committee, Washington, D.C., Doc. A/91, 2001. Used with permission.*)

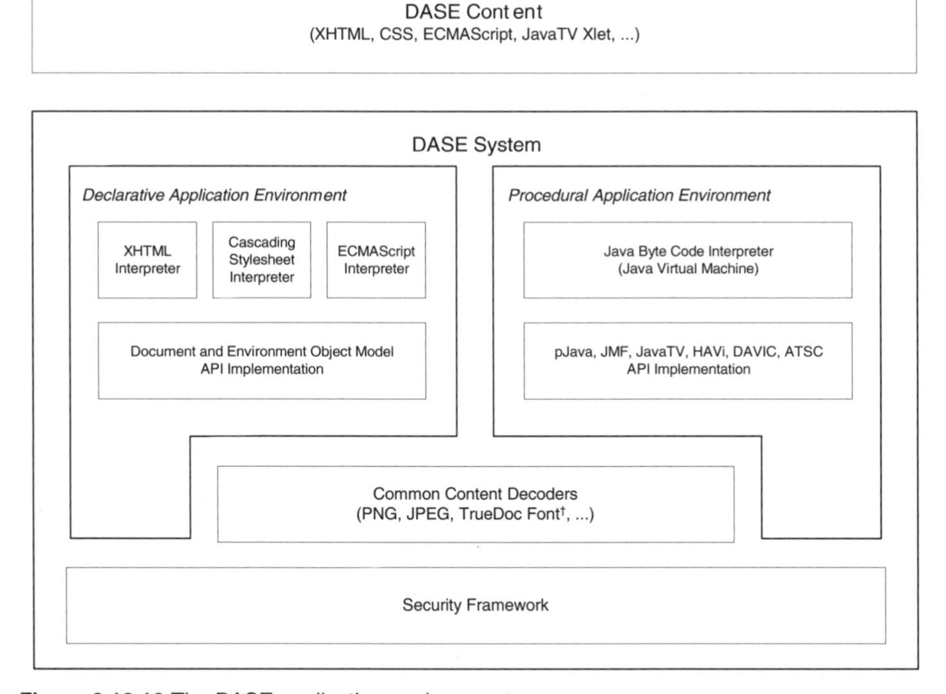

Figure 2.18.19 The DASE application environment.

Table 2.18.7 General System Comparison of ATSC 8-VSB and DVB-T COFDM (*After: ITU Radiocommunication Study Groups, Special Rapporteur's Group: "Guide for the Use of Digital Television Terrestrial Broadcasting Systems Based on Performance Comparison of ATSC 8-VSB and DVB-T COFDM Transmission Systems," International Telecommunications Union, Geneva, Document 11A/65-E, May 11, 1999.*)

System Parameter	ATSC 8-VSB	DVB-T COFDM (ITU mode M3)	Comments
Signal peak-to-average power ratio	7 dB	9.5 dB	99.99% of time
E_b/N_o AWGN channel			
Theoretical	10.6 dB	11.9 dB	A 0.8 dB correction factor is used to compensate the measurement threshold difference
RF back-to-back test	11.0 dB	14.6 dB	
Static multipath distortion			
> 4 dB	Better	Worse	
< 4 dB	Worse	Better	
Dynamic multipath	Worse	Much better	
Mobile reception	No	2k-mode	
Spectrum efficiency	Better	Worse	
HDTV capability	Yes	Yes*	*6 MHz DVB-T might have difficulty, because of low data rate
Interference into analog TV system	Low	Medium	ATSC E_b/N_o is low, which require less transmission power
Single frequency networks			
Large scale SFN	No	Yes	DVB-T 8k mode.
On-channel repeater	Yes	Yes	ATSC and DVB-T 2k mode
Impulse noise	Better	Worse	
Tone interference	Worse	Better	
Co-channel analog TV interference into DTV	Same	Same	Assuming ATSC system has comb-filter
Co-channel DTV	Better	Worse	
Phase noise sensitivity	Better	Worse	
Noise figure	Same	Same	
Indoor reception	N/A	N/A	Needs more investigation
System for different channel bandwidth	Same	Same	ATSC might need different comb-filter, and DVB-T 6 MHz (8k mode) might be sensitive to phase noise

Table 2.18.8 DTV Protection Ratios for Frequency Planning (*After: ITU Radiocommunication Study Groups, Special Rapporteur's Group: "Guide for the Use of Digital Television Terrestrial Broadcasting Systems Based on Performance Comparison of ATSC 8-VSB and DVB-T COFDM Transmission Systems," International Telecommunications Union, Geneva, Document 11A/65-E, May 11, 1999.*)

System Parameters (protection ratios)	Canada [16]	USA [9]	EBU [12, 15] ITU-mode M3
C/N for AWGN Channel	+19.5 dB (16.5dB[1])	+15.19 dB	+19.3 dB
Co-Channel DTV into Analog TV	+33.8 dB	+34.44 dB	+34 ~ 37 dB
Co-Channel Analog TV into DTV	+7.2 dB	+1.81 dB	+4 dB
Co-Channel DTV into DTV	+19.5 dB (16.5 dB[1])	+15.27 dB	+19 dB
Lower Adjacent Channel DTV into Analog TV	–16 dB	–17.43 dB	–5 ~ –11 dB
Upper Adjacent Channel DTV into Analog TV	–12 dB	–11.95 dB	–1 ~ –10 dB
Lower Adjacent Channel Analog TV into DTV	–48 dB	–47.33 dB	–34 ~ –37 dB
Upper Adjacent Channel Analog TV into DTV	–49 dB	–48.71 dB	–38 ~ –36 dB
Lower Adjacent Channel DTV into DTV	–27 dB	–28 dB	N/A
Upper Adjacent Channel DTV into DTV	–27 dB	–26 dB	N/A

[1] The Canadian parameter, *C*/(*N*+I) of noise plus co-channel DTV interference should be 16.5 dB.

2.19 Frequency Bands and Propagation

Table 2.19.1 Numerical Designation of Television Channels in the U.S.

Channel no.	MHz	Channel no.	MHz	Channel no.	MHz	Channel no.	MHz
2	54–60	22	518–524	42	638–644	63	764–770
3	60–66	23	524–530	43	644–650	64	770–776
4	66–72	24	530–536	44	650–656	65	776–782
5	76–82	25	536–542	45	656–662	66	782–788
6	82–88	26	542–548	46	662–668	67	788–794
7	174–180	27	548–554	47	668–674	68	794–800
8	180–186	28	554–560	48	674–680	69	800–806
9	186–192	29	560–566	49	680–686	70	806–812
10	192–198	20	566–572	50	686–692	71	812–818
11	198–204	31	572–578	51	692–696	72	818–824
12	204–210	32	578–584	52	698–704	73	824–830
13	210–216	33	584–590	53	704–710	74	830–836
14	470–476	34	590–596	54	710–176	75	836–842
15	476–482	35	596–602	55	716–722	76	842–848
16	482–488	36	602–608	56	722–728	77	848–854
17	488–494	37	608–614	57	728–734	78	854–860
18	494–500	38	614–620	58	734–740	79	860–866
19	500–506	39	620–626	59	740–746	80	866–872
20	506–512	40	626–632	60	746–752	81	872–878
21	519–518	41	632–638	62	758–764	83	884–890

Table 2.19.2 Applications of Interest in the RF Band

Band	Frequency
Longwave broadcasting band	150–290 kHz
AM broadcasting band	550–1640 kHz (1.640 MHz), 107 channels, 10 kHz separation
International broadcasting band	3–30 MHz
Shortwave broadcasting band	5.95–26.1 MHz (8 bands)
VHF TV (Channels 2 - 4)	54–72 MHz
VHF TV (Channels 5 - 6)	76–88 MHz
FM broadcasting band	88–108 MHz
VHF TV (Channels 7 - 13)	174–216 MHz
UHF TV (Channels 14 - 69)	512–806 MHz

Table 2.19.3 Frequency Band Designations

Description	Band Designation	Frequency	Wavelength
Extremely Low Frequency	ELF (1) Band	3 Hz up to 30 Hz	100 Mm down to 10 Mm
Super Low Frequency	SLF (2) Band	30 Hz up to 300 Hz	10 Mm down to 1 Mm
Ultra Low Frequency	ULF (3) Band	300 Hz up to 3 kHz	1 Mm down to 100 Km
Very Low Frequency	VLF (4) Band	3 kHz up to 30 kHz	100 Km down to 10 Km
Low Frequency	LF (5) Band	30 kHz up to 300 kHz	10 Km down to 1 Km
Medium Frequency	MF (6) Band	300 kHz up to 3 MHz	1 Km down to 100 m
High Frequency	HF (7) Band	3 MHz up to 30 MHz	100 m down to 10 m
Very High Frequency	VHF (8) Band	30 MHz up to 300 MHz	10 m down to 1 m
Ultra High Frequency	UHF (9) Band	300 MHz up to 3 GHz	1 m down to 10 cm
Super High Frequency	SHF (10) Band	3 GHz up to 30 GHz	10 cm down to 1 cm
Extremely High Frequency	EHF (11) Band	30 GHz up to 300 GHz	1 cm down to 1 mm
—	(12) Band	300 GHz up to 3 THz	1 mm down to 100 μ

Table 2.19.4 Radar Band Designations

Band	Frequency	Wavelength
A Band	0 Hz up to 250 MHz	∞down to 1.2 m
B Band	250 MHz up to 500 MHz	1.2 m down to 60 cm
C Band	500 MHz up to 1 GHz	60 cm down to 30 cm
D Band	1 GHz up to 2 GHz	30 cm down to 15 cm
E Band	2 GHz up to 3 GHz	15 cm down to 10 cm
F Band	3 GHz up to 4 GHz	10 cm down to 7.5 cm
G Band	4 GHz up to 6 GHz	7.5 cm down to 5 cm
H Band	6 GHz up to 8 GHz	5 cm down to 3.75 cm
I Band	8 GHz up to 10 GHz	3.75 cm down to 3 cm
J Band	10 GHz up to 20 GHz	3 cm down to 1.5 cm
K Band	20 GHz up to 40 GHz	1.5 cm down to 7.5 mm
L Band	40 GHz up to 60 GHz	7.5 mm down to 5 mm)
M Band	60 GHz up to 100 GHz	5 mm down to 3 mm
N Band	100 GHz up to 200 GHz	3 mm down to 1.5 mm
O Band	200 GHz up to 300 GHz	1.5 mm down to 1 mm

Table 2.19.5 Detail of UHF, SHF, and EHF Band Designations

Band	Frequency	Wavelength
L Band	1.12 GHz up to 1.7 GHz	26.8 cm down to 17.6 cm
LS Band	1.7 GHz up to 2.6 GHz	17.6 cm down to 11.5 cm
S Band	2.6 GHz up to 3.95 GHz	11.5 cm down to 7.59 cm
C(G) Band	3.95 GHz up to 5.85 GHz	7.59 cm down to 5.13 cm
XN(J, XC) Band	5.85 GHz up to 8.2 GHz	5.13 cm down to 3.66 cm
XB(H, BL) Band	7.05 GHz up to 10 GHz	4.26 cm down to 3 cm
X Band	8.2 GHz up to 12.4 GHz	3.66 cm down to 2.42 cm
Ku(P) Band	12.4 GHz up to 18 GHz	2.42 cm down to 1.67 cm
K Band	18 GHz up to 26.5 GHz	1.67 cm down to 1.13 cm
V(R, Ka) Band	26.5 GHz up to 40 GHz	1.13 cm down to 7.5 mm
Q(V) Band	33 GHz up to 50 GHz	9.09 mm down to 6 mm
M(W) Band	50 GHz up to 75 GHz	6 mm down to 4 mm
E(Y) Band	60 GHz up to 90 GHz	5 mm down to 3.33 mm
F(N) Band	90 GHz up to 140 GHz	3.33 mm down to 2.14 mm
G(A)	140 GHz p to 220 GHz	2.14 mm down to 1.36 mm
R Band	220 GHz up to 325 GHz	1.36 mm down to 0.923 mm

Table 2.19.6 Applications of Interest in the Microwave Band

Application	Frequency
Aero Navigation	0.96–1.215 GHz
GPS Down Link	1.2276 GHz
Military COM/Radar	1.35–1.40 GHz
Miscellaneous COM/Radar	1.40–1.71 GHz
L-Band Telemetry	1.435–1.535 GHz
GPS Down Link	1.57 GHz
Military COM (Troposcatter/Telemetry)	1.71–1.85 GHz
Commercial COM & Private LOS	1.85–2.20 GHz
Microwave Ovens	2.45 GHz
Commercial COM/Radar	2.45–2.69 GHz
Instructional TV	2.50–2.69 GHz
Military Radar (Airport Surveillance)	2.70–2.90 GHz
Maritime Navigation Radar	2.90–3.10 GHz
Miscellaneous Radars	2.90–3.70 GHz
Commercial C-Band SAT COM Down Link	3.70–4.20 GHz
Radar Altimeter	4.20–4.40 GHz
Military COM (Troposcatter)	4.40–4.99 GHz
Commercial Microwave Landing System	5.00–5.25 GHz
Miscellaneous Radars	5.25–5.925 GHz
C-Band Weather Radar	5.35–5.47 GHz
Commercial C-Band SAT COM Up Link	5.925–6.425 GHz
Commercial COM	6.425–7.125 GHz
Mobile TV Links	6.875–7.125 GHz
Military LOS COM	7.125–7.25 GHz
Military SAT COM Down Link	7.25–7.75 GHz
Military LOS COM	7.75–7.9 GHz
Military SAT COM Up Link	7.90–8.40 GHz
Miscellaneous Radars	8.50–10.55 GHz
Precision Approach Radar	9.00–9.20 GHz
X-Band Weather Radar	9.30–9.50 GHz
Police Radar	10.525 GHz
Commercial Mobile COM (LOS & ENG)	10.55–10.68 GHz
Common Carrier LOS COM	10.70–11.70 GHz
Commercial COM	10.70–13.25 GHz
Commercial Ku-Band SAT COM Down Link	11.70–12.20 GHz
DBS Down Link & Private LOS COM	12.20–12.70 GHz
ENG & LOS COM	12.75–13.25 GHz
Miscellaneous Radars & SAT COM	13.25–14.00 GHz
Commercial Ku-Band SAT COM Up Link	14.00–14.50 GHz
Military COM (LOS, Mobile, &Tactical)	14.50–15.35 GHz
Aero Navigation	15.40–15.70 GHz
Miscellaneous Radars	15.70–17.70 GHz
DBS Up Link	17.30–17.80 GHz
Common Carrier LOS COM	17.70–19.70 GHz
Commercial COM (SAT COM & LOS)	17.70–20.20 GHz
Military SAT COM	20.20–21.20 GHz
Commercial COM	27.50–30.00 GHz
Commercial COM	31.00–31.20 GHz

Table 2.19.7 Frequencies Associated with Terrestrial Broadcast Television Channels in the U.S.

Channel number	Frequency limits	Picture carrier	Sound carrier	Chrominance subcarrier	Local oscillator	First-order sound image (+)	Second-order sound image (−)	First-order picture image (+)	Second-order picture image (−)
2	54–60	55.25	59.75	58.829545	101.00	142.25	160.75	142.75–147.25	155.75–160.25
3	60–66	61.25	65.75	64.829545	107.00	148.25	172.75	148.75–153.25	167.75–172.25
4	66–72	67.25	71.75	70.829545	113.00	154.25	184.75	154.75–159.25	179.75–184.25
5	76–82	77.25	81.75	80.829545	123.00	164.25	204.75	164.75–169.25	199.75–204.25
6	82–88	83.25	87.75	86.829545	129.00	170.25	216.75	170.75–175.25	211.75–216.25
7	174–180	175.25	179.75	178.829545	221.00	262.25	401.75	262.75–267.25	396.75–401.25
8	180–186	181.25	185.75	184.829545	227.00	268.25	413.75	268.75–273.25	408.75–413.25
9	186–192	187.25	191.75	190.829545	233.00	274.25	425.75	274.75–279.25	420.75–425.25
10	192–198	193.25	197.75	196.829545	239.00	280.25	437.75	280.75–285.25	432.75–437.25
11	198–204	199.25	203.75	202.829545	245.00	286.25	449.75	286.75–291.25	444.75–449.25
12	204–210	205.25	209.75	208.829545	251.00	292.52	461.75	292.75–297.25	456.75–461.25
13	210–216	211.25	215.75	214.829545	257.00	298.25	473.75	298.75–303.25	468.75–473.25
14	470–476	471.25	475.75	474.829545	517.00	558.25	992.75	558.75–563.25	987.75–992.25
15	476–482	477.25	481.75	480.829545	523.00	564.25	1004.75	564.75–569.25	999.75–1004.25
16	482–488	483.25	487.75	486.829545	529.00	570.25	1016.75	570.75–575.25	1011.75–1016.25
17	488–494	489.25	493.75	492.829545	535.00	576.25	1028.75	576.75–581.25	1023.75–1028.25
18	494–500	495.25	499.75	498.829545	541.00	582.25	1040.75	582.75–587.25	1035.75–1040.25
19	500–506	501.25	505.75	504.829545	547.00	588.25	1052.75	588.75–593.25	1047.75–1052.25
20	506–512	507.25	511.75	510.829545	553.00	594.25	1064.75	594.75–599.25	1059.75–1064.25
21	512–518	513.25	517.75	516.829545	559.00	600.25	1076.75	600.75–605.25	1071.75–1076.25
22	518–524	519.25	523.75	522.829545	565.00	606.25	1088.75	606.75–611.25	1083.75–1088.25
23	524–530	525.25	529.75	528.829545	571.00	612.25	1100.75	612.75–617.25	1095.75–1100.25
24	530–536	531.25	535.75	534.829545	577.00	618.25	1112.75	618.75–623.25	1107.75–1112.25
25	536–542	537.25	541.75	540.829545	583.00	624.25	1124.75	624.75–629.25	1119.75–1124.25
26	542–548	543.25	547.75	546.829545	589.00	630.25	1136.75	630.75–635.25	1131.75–1136.25
27	548–554	549.25	553.75	552.829545	595.00	636.25	1148.75	636.75–641.25	1143.75–1148.25
28	554–560	555.25	559.75	558.829545	601.00	642.25	1160.75	642.75–647.25	1115.75–1160.25

Table 2.19.7 Frequencies Associated with Terrestrial Broadcast Television Channels in the U.S. (Continued)

Channel number	Frequency limits	Picture carrier	Sound carrier	Chrominance subcarrier	Local oscillator	First-order sound image (+)	Second-order sound image (−)	First-order picture image (+)	Second-order picture image (−)
29	560–566	561.25	565.75	564.829545	607.00	648.25	1172.75	648.75–653.25	1167.75–1172.25
30	566–572	567.25	571.75	570.829545	613.00	654.25	1184.75	654.75–659.25	1179.75–1184.25
31	572–578	573.25	577.75	576.829545	619.00	660.25	1196.75	660.75–665.25	1191.75–1196.25
32	578–584	579.25	583.75	582.829545	625.00	666.25	1208.75	666.75–671.25	1203.75–1208.25
33	584–590	585.25	589.75	588.829545	631.00	672.25	1220.75	672.75–677.25	1215.75–1220.25
34	590–596	591.25	595.75	594.829545	637.00	678.25	1232.75	678.75–683.25	1227.75–1232.25
35	596–602	597.25	601.75	600.829545	643.00	684.25	1244.75	684.75–689.25	1239.75–1244.25
36	602–608	603.25	607.75	606.829545	649.00	690.25	1256.75	690.75–695.25	1251.75–1256.25
37	608–614	609.25	613.75	612.829545	655.00	696.25	1268.75	696.75–701.25	1263.75–1268.25
38	614–620	615.25	619.75	618.829545	661.00	702.25	1280.75	702.75–707.25	1275.75–1280.25
39	620–626	621.25	625.75	624.829545	667.00	708.25	1292.75	708.75–713.25	1287.75–1292.25
40	626–632	627.25	631.75	630.829545	673.00	714.25	1304.75	714.75–719.25	1299.75–1304.25
41	632–638	633.25	637.75	636.829545	679.00	720.25	1316.75	720.75–725.25	1311.75–1316.25
42	638–644	639.25	643.75	642.829545	685.00	726.25	1328.75	726.75–731.25	1323.75–1328.25
43	644–650	645.25	649.75	648.829545	691.00	732.25	1340.75	732.75–737.25	1335.75–1340.25
44	650–656	651.25	655.75	654.829545	697.00	738.25	1352.75	738.75–743.25	1347.75–1352.25
45	656–662	657.25	661.75	660.829545	703.00	744.25	1364.75	744.75–749.25	1359.75–1364.25
46	662–668	663.25	667.75	666.829545	709.00	750.25	1376.75	750.75–755.25	1371.75–1376.25
47	668–674	669.25	673.75	672.829545	715.00	756.25	1388.75	756.75–761.25	1383.75–1388.25
48	674–680	675.25	679.75	678.829545	721.00	762.25	1400.75	762.75–767.25	1395.75–1400.25
49	680–686	681.25	685.75	684.829545	727.00	768.25	1412.75	768.75–773.25	1407.75–1412.25
50	686–692	687.25	691.75	690.829545	733.00	774.25	1424.75	774.75–779.25	1419.75–1424.25
51	692–698	693.25	697.75	696.829545	739.00	780.25	1436.75	780.75–785.25	1431.75–1436.25
52	698–704	699.25	703.75	702.829545	745.00	786.25	1448.75	786.75–791.25	1443.75–1448.25
53	704–710	705.25	709.75	708.829545	751.00	792.25	1460.75	792.75–797.25	1455.75–1460.25
54	710–716	711.25	715.75	714.829545	757.00	798.25	1472.75	798.75–803.25	1467.75–1472.25
55	716–722	717.25	721.75	720.829545	763.00	704.25	1484.75	804.75–809.25	1479.75–1484.25
56	722–728	723.25	727.75	726.829545	769.00	810.25	1496.75	810.75–815.25	1491.75–1496.25
57	728–734	729.25	733.75	732.829545	775.00	816.25	1508.75	816.75–821.25	1503.75–1508.25

Table 2.19.7 Frequencies Associated with Terrestrial Broadcast Television Channels in the U.S. (Continued)

58	734–740	735.25	739.75	738.829545	781.00	822.25	1520.75	822.75–827.25	1515.75–1520.25
59	740–746	741.25	745.75	744.829545	787.00	828.25	1532.75	828.75–833.25	1527.75–1532.25
60	746–752	747.25	751.75	750.829545	793.00	834.25	1544.75	834.75–839.25	1539.75–1544.25
61	752–758	753.25	757.75	756.829545	799.00	840.25	1556.75	840.75–845.25	1551.75–1556.25
62	758–764	759.25	763.75	762.829545	805.00	846.25	1568.75	846.75–851.25	1563.75–1568.25
63	764–770	765.25	769.75	768.829545	811.00	852.25	1580.75	852.75–857.25	1575.75–1580.25
64	770–776	771.25	775.75	774.829545	817.00	858.25	1592.75	858.75–863.25	1587.75–1592.25
65	776–782	775.25	781.75	780.829545	823.00	864.25	1604.75	864.75–869.25	1599.75–1604.25
66	782–788	783.25	787.75	786.829545	829.00	870.25	1616.75	870.75–875.25	1611.75–1616.25
67	788–794	789.25	793.75	792.829545	835.00	876.25	1628.75	876.75–881.25	1623.75–1628.25
68	794–800	795.25	799.75	798.829545	841.00	882.25	1640.75	882.75–887.25	1635.75–1640.25
69	800–806	810.25	805.75	804.829545	847.00	888.25	1652.75	888.75–893.25	1647.75–1652.25
70	806–812	807.25	811.75	810.829545	853.00	894.25	1664.75	894.75–899.25	1659.75–1664.25
71	812–818	813.25	817.75	816.829545	859.00	900.25	1676.75	900.75–905.25	1671.75–1676.25
72	818–824	819.25	823.75	822.829545	865.00	906.25	1688.75	906.75–911.25	1683.75–1688.25
73	824–830	825.25	829.75	828.829545	871.00	912.25	1700.75	912.75–917.25	1695.75–1700.25
74	830–836	831.25	835.75	834.829545	877.00	918.25	1712.75	918.75–923.25	1707.75–1712.25
75	836–842	837.25	841.75	840.829545	883.00	924.25	1724.75	924.75–929.25	1719.75–1724.25
76	842–848	843.25	847.75	846.829545	889.00	930.25	1736.75	930.75–935.25	1731.75–1736.25
77	848–854	849.25	853.75	852.829545	895.00	936.25	1748.75	936.75–941.25	1743.75–1748.25
78	854–860	855.25	859.75	858.829545	901.00	942.25	1760.75	942.75–947.25	1755.75–1760.25
79	860–866	861.25	865.75	864.829545	907.00	948.25	1772.75	948.75–953.25	1767.75–1772.25
80	866–872	867.25	871.75	870.829545	913.00	954.25	1784.75	954.75–959.25	1779.75–1784.25
81	872–878	873.25	877.75	876.829545	919.00	960.25	1796.75	960.75–965.25	1791.75–1796.25
82	878–884	879.25	883.75	882.829545	925.00	966.25	1808.75	966.75–971.25	1803.75–1808.25
83	884–890	885.25	889.75	888.829545	931.00	972.25	1820.75	972.75–977.25	1815.75–1820.25

Table 2.19.8 Basic Characteristics of the Radiated Signals for Worldwide Television Systems

	CCIR system designation										
Characteristics	A	M	N	C	B, G	H	I	D, K	K1	L	E
Channel bandwidth, MHz	......	6	6	7	B; 7 G; 8	8	8	8	8	8	14
Vision to sound carrier spacing, MHz	−3.5	+4.5	+4.5	+5.5	+5.5 ±0.001	+5.5	+5.9996 ±0.0005	+6.5 ±0.001	+6.5	+6.5	+11.15
Vision carrier to near-channel edge, MHz	+1.25	−1.25	−1.25	−1.25	−1.25	−1.25	−1.25	−1.25	−1.25	−1.25	±2.83
Main sideband width, MHz	3	4.2	4.2	5	5	5	5.5	6	6	6	10
Vestigial sideband width, MHz	0.75	0.75	0.75	0.75	0.75	1.25	1.25	0.75	1.25	1.25	2.0
Vestigial sideband attenuation, dB at MHz	No spec.	20(1.25) 42(3.58)	20(1.25) 42(3.5)	20(1.25) 20(3.0)	20(1.25) 20(3.0) 30(4.43)	20(1.75) 20(3.0)	20(3.0) 30(4.43)	20(1.25) 30(4.43 ± 0.1)	30(4.3) 20(2.7) 0(0.8)	30(4.3) 15(2.7) 0(0.8)	15(2.7) 0(0.8)

Visual modulation type and polarity	A5C pos.	A5C neg.	A5C neg.	A5C pos.	A5C neg.	A5C neg.	A5C neg.	A5C neg.	A5C neg.	A5C pos.	A5C pos.
Sync pulse level (% of peak carrier)	<3	100	100	<3	100	100	100	100	100	<6	<3
Blanking level	30	72.5–77.5	72.5–77.5	22.5–27.5	75 ± 2.5	72.5–77.5	76 ± 2	75 ± 2.5	75 ± 2.5	30 ± 2	30 ± 2
Setup	0	2.88–6.75	2.88–6.75	0	0–2	0–7	0	0–4.5	0–4.5	0–4.5	0–4.5
Peak white	100	10–15	10–15	100	10–12.5	10–12.5	20 ± 2	12.5	10–12.5	100–110	100
Sound modulation	A3	F3	F3	A3	F3	F3	F3	F3	F3	A3	A3
Frequency deviation, kHz	.....	±25	±25	....	±50	±50	±50	±50	±50		
Preemphasis, μs	.....	75	75	50	50	50	50	50	50		
Visual/sound ratio	4/1	10/1 5/1	10/1 5/1	4/1	10/1	5/1 10/1	5/1	10/1 5/1	10/1	10/1	10/1
Precorrection group delay, ns	No spec.	−170	....	....	−170						

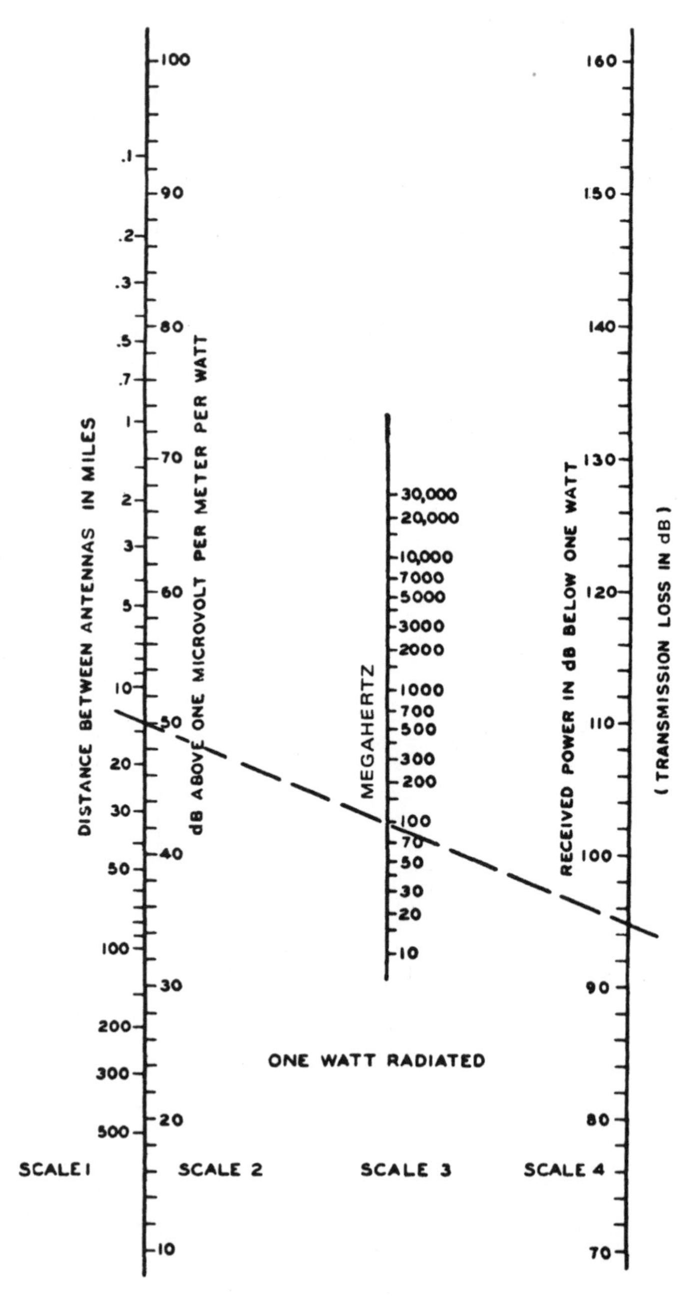

Figure 2.19.1 Free-space field intensity and received power between half-wave dipoles. (*From: Fink, D. G., (ed.):* Television Engineering Handbook, *McGraw-Hill, New York, N.Y., 1957. Used with permission.*)

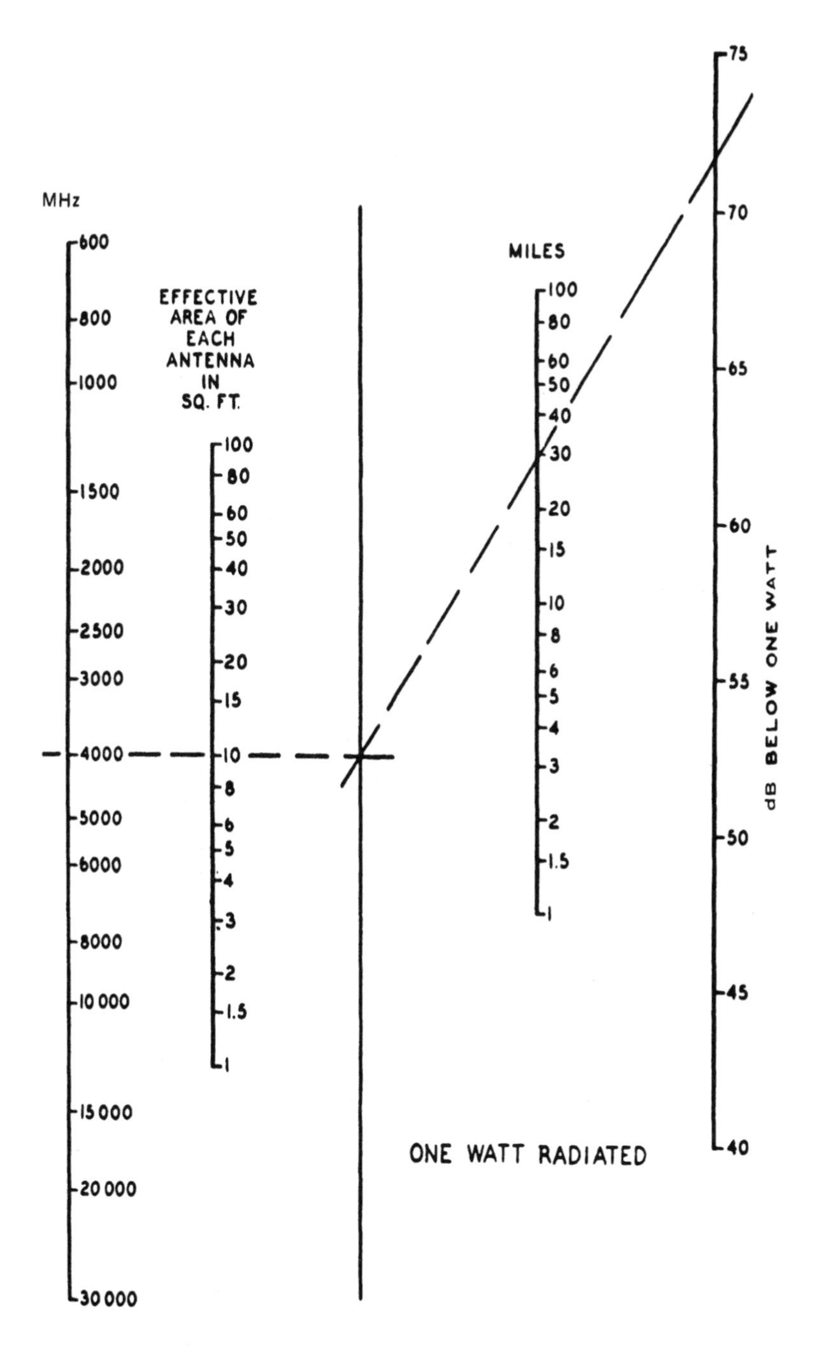

Figure 2.19.2 Received power in free space between two antennas of equal effective areas. (*From: Fink, D. G., (ed.):* Television Engineering Handbook, *McGraw-Hill, New York, N.Y., 1957. Used with permission.*)

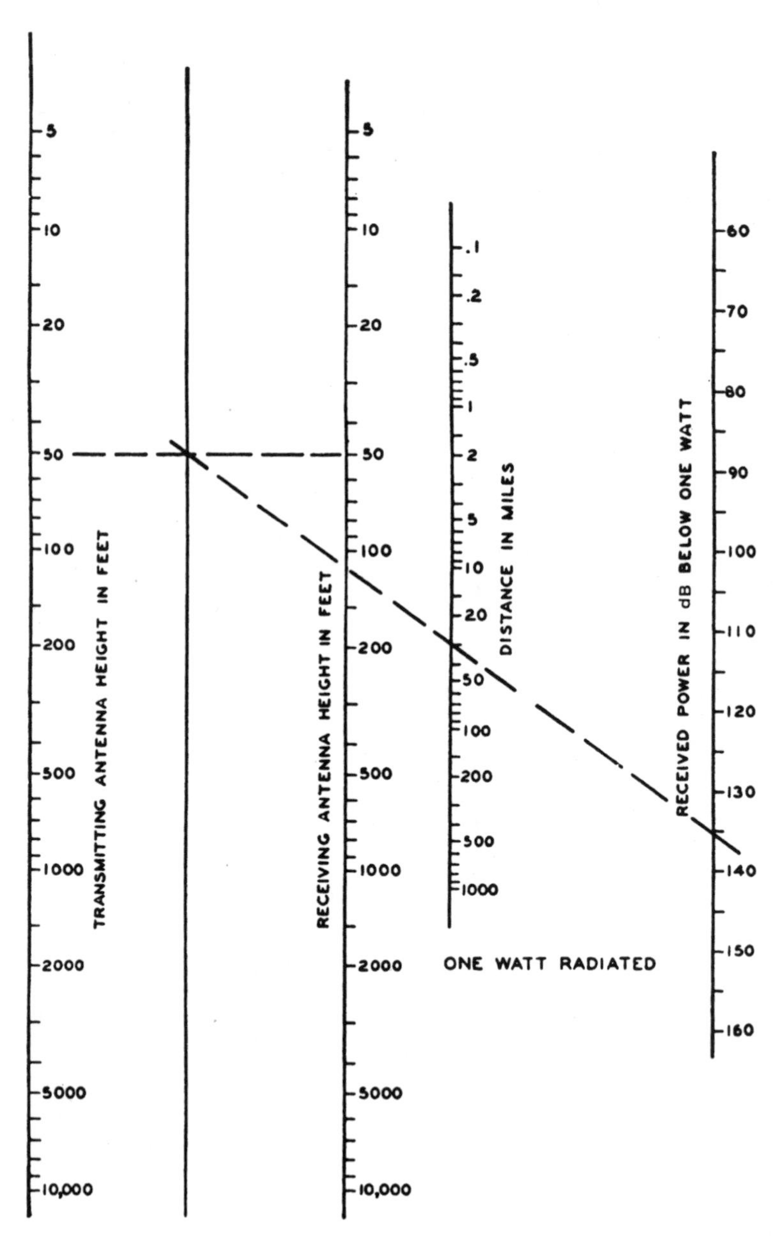

Figure 2.19.3 Received power over plane earth between half-wave dipoles. (*From: Fink, D. G., (ed.):* Television Engineering Handbook, *McGraw-Hill, New York, N.Y., 1957. Used with permission.*)

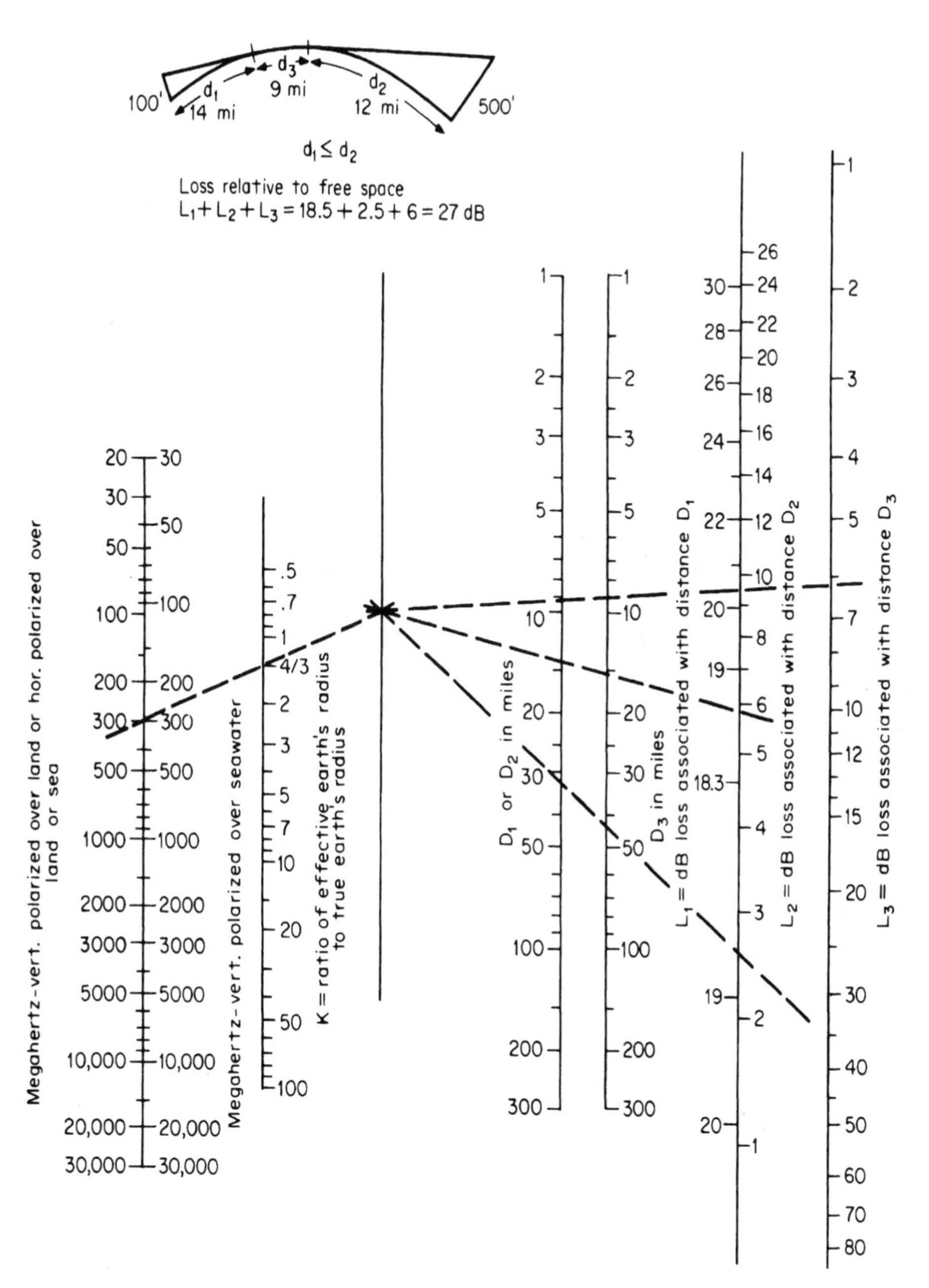

Figure 2.19.4 Loss beyond line of sight in decibels. (*From: Fink, D. G., (ed.):* Television Engineering Handbook, *McGraw-Hill, New York, N.Y., 1957. Used with permission.*)

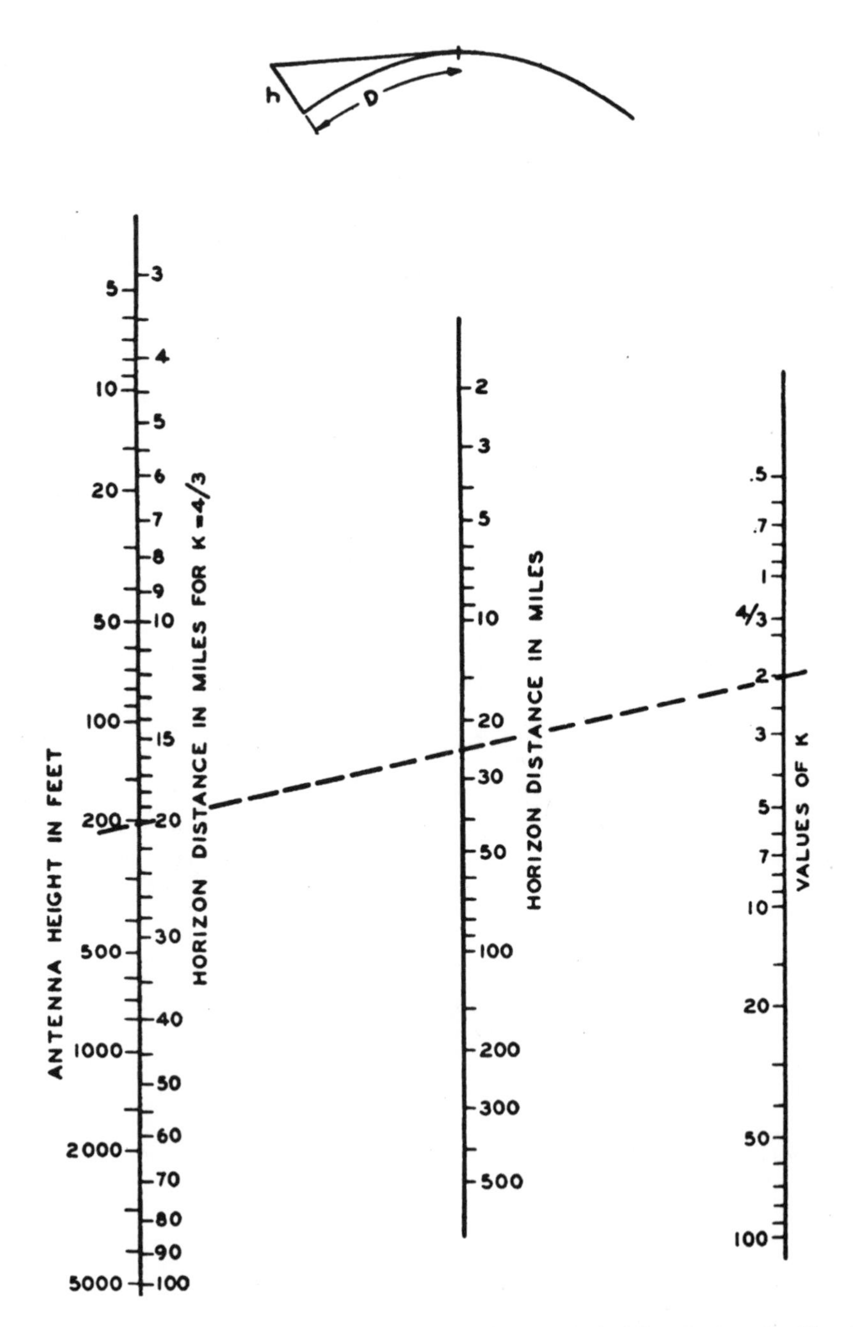

Figure 2.19.5 Distance to the horizon. (*From: Fink, D. G., (ed.):* Television Engineering Handbook, *McGraw-Hill, New York, N.Y., 1957. Used with permission.*)

Table 2.19.9 PLL Filter Characteristics

Circuit and Transfer Characteristics of Several PLL Filters

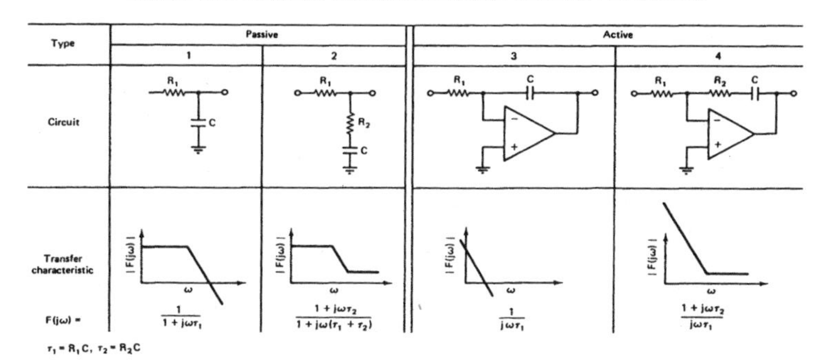

Implementation of Different Loop Filters

Passive Lead-Lag	Passive Lead Lag with Pole	Active Integrator	Active Integrator with Pole
R1, R2, C2	R1, R2, C3, C2	R1, R2, C2	R1, R2, C2, C3
$F(s) = \frac{s\tau_2 + 1}{[s(\tau_1 + \tau_2) + 1]}$ $\tau_1 = R_1C_2; \tau_2 = R_2C_2$	$F(s) = \frac{s\tau_2 + 1}{[s(\tau_1 + \tau_2) + 1](s\tau_3 + 1)}$ $\tau_1 = R_1C_2; \tau_2 = R_2C_2;$ $\tau_3 = (R_2 \| R_1)C_3$	$F(s) = \frac{s\tau_2 + 1}{s\tau_1}$ $\tau_1 = R_1C_2; \tau_2 = R_2C_2;$	$F(s) = \frac{s\tau_2 + 1}{s\tau_1(s\tau_3 + 1)}$ $\tau_1 = R_1(C_2 + C_3); \tau_2 = R_2C_2;$ $\tau_3 = R_2(C_3 \| C_2)$
Type 1.5, 2nd Order (Low Gain)	Type 1.5, 3rd Order (Low Gain)	Type 2, 2nd Order (High Gain)	Type 2, 3rd Order (High Gain)

Recommended Passive Filters for Charge Pumps

Integrator	Integrator With Poles	Integrator With 2 Poles
R1, C1	R1, C1, C2	R1, C1, C3, R2, C2
$F(s) = R_1\frac{s\tau_1 + 1}{s\tau_1}$ $\tau_1 = R_1C_1$	$F(s) = R_1\frac{s\tau_1 + 1}{s\tau_1(s\tau_2 + 1)}$ $\tau_1 = R_1C_1; \tau_2 = R_1(\frac{C_1C_2}{C_1 + C_2})$	$F(s) = R_1\frac{s\tau_1 + 1}{s\tau_1(s\tau_2 + 1)(s\tau_3 + 1)}$ $\tau_1 = R_1C_1; \tau_2 = R_1\frac{C_1C_3}{C_1 + C_3};$ $\tau_3 = R_2C_2$
Type 2, 2nd Order	Type 2, 3rd Order	Type 2, 4th Order

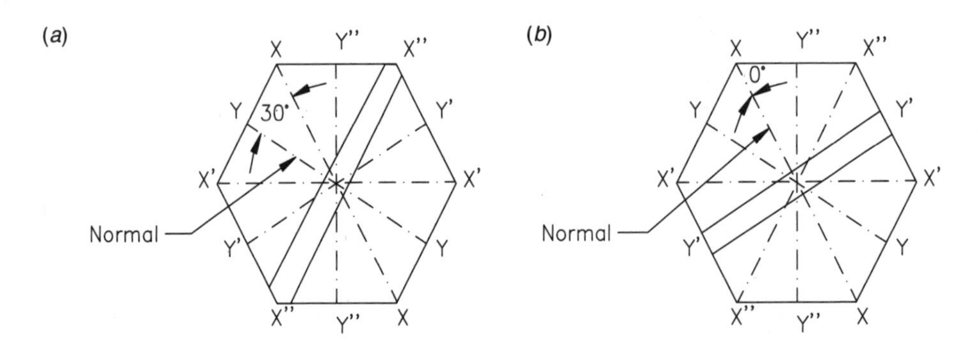

Figure 2.19.6 Cross section of a quartz crystal taken in the plane perpendicular to the optical axis: (*a*) *Y*-cut plate, (*b*) *X*-cut plate.

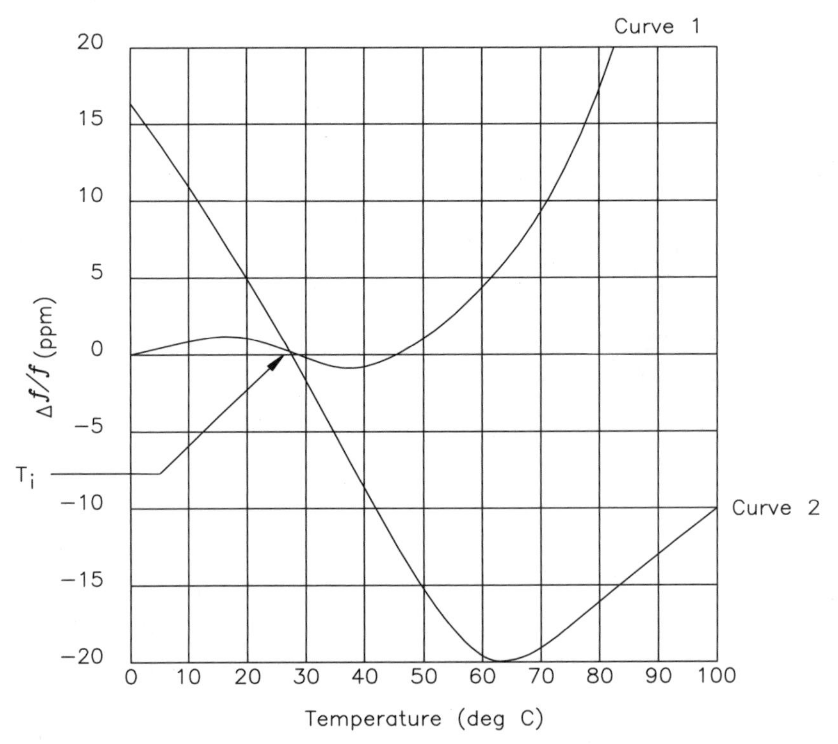

Figure 2.19.7 The effects of temperature on two types of AT-cut crystals.

2.20 Radio/Television Transmission Systems

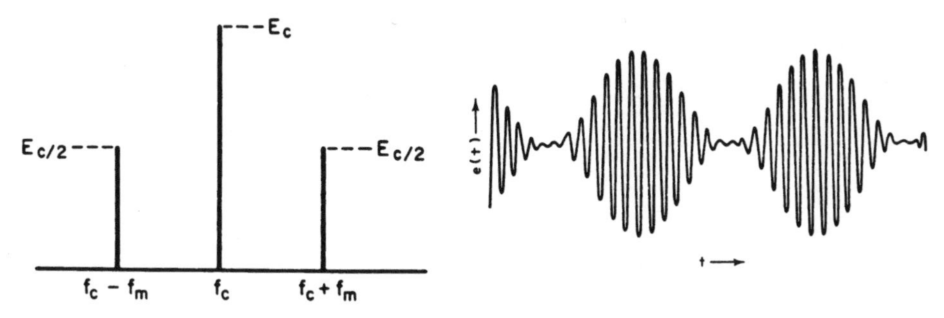

Figure 2.20.1 Double-sideband amplitude modulation.

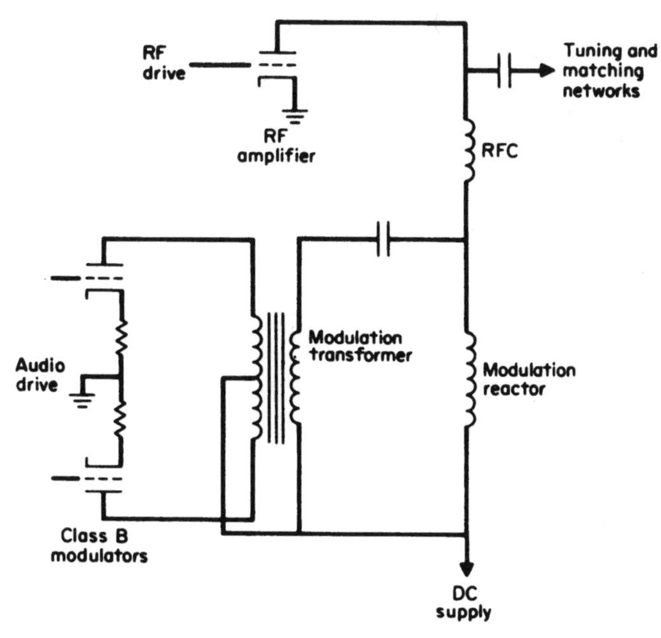

Figure 2.20.2 Basic configuration for high-level amplitude modulation in the standard broadcast band.

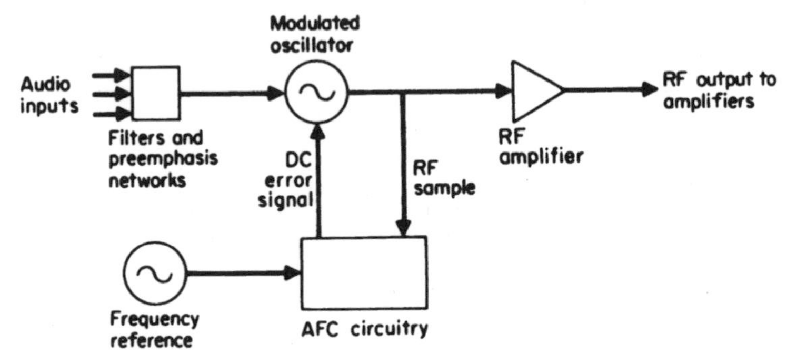

Figure 2.20.3 Block diagram of an FM exciter.

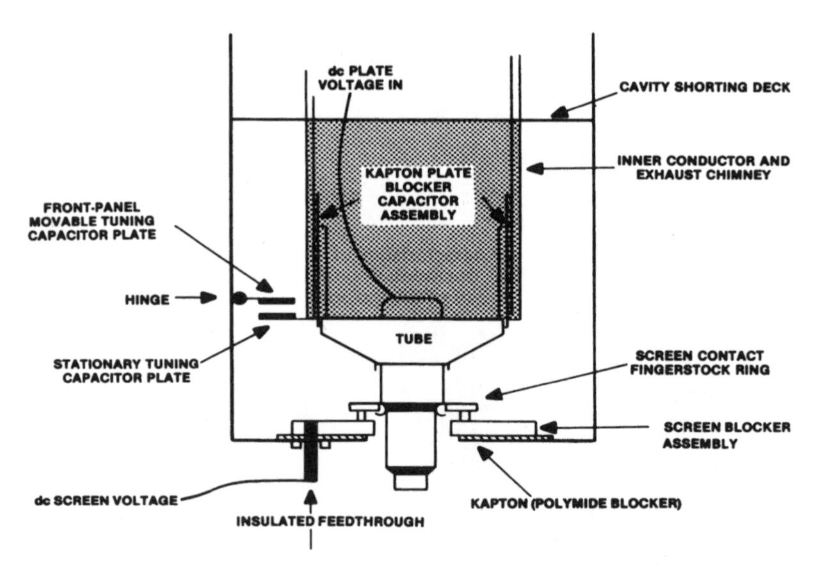

Figure 2.20.4 Physical layout of a common type of 1/4-wave PA cavity for FM broadcast service.

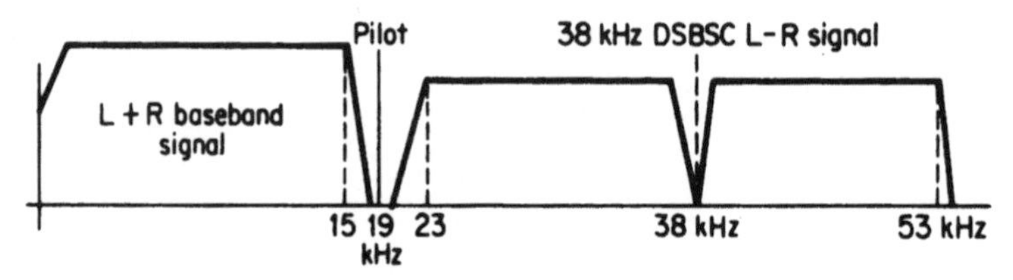

Figure 2.20.5 Composite stereo FM signal.

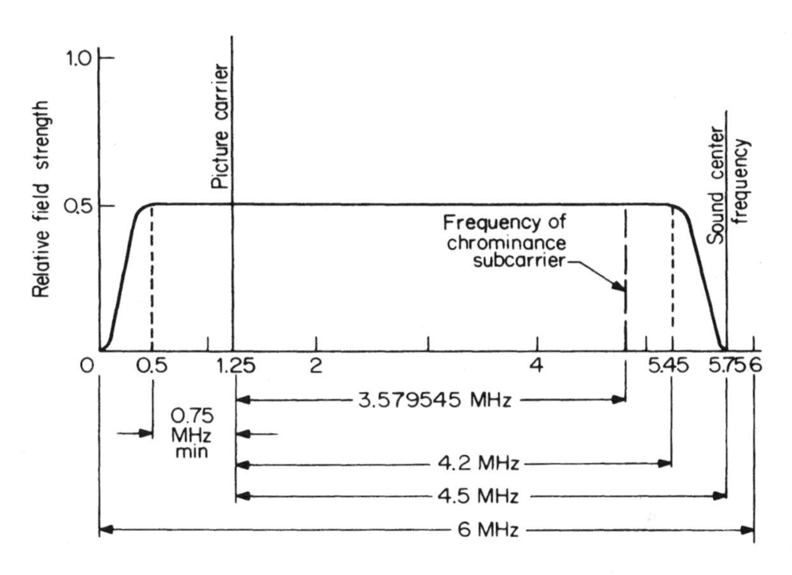

Figure 2.20.6 Idealized picture transmission amplitude characteristics for VHF and UHF broadcast transmitters in the U.S. (*From FCC.*)

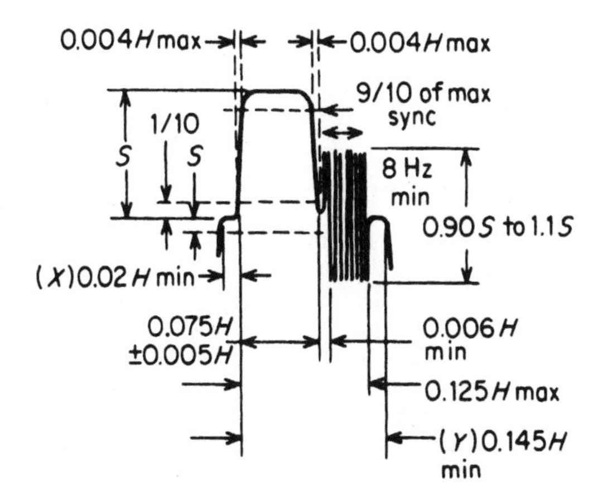

Figure 2.20.7 Detail of the NTSC color burst synchronizing signal. (*From FCC.*)

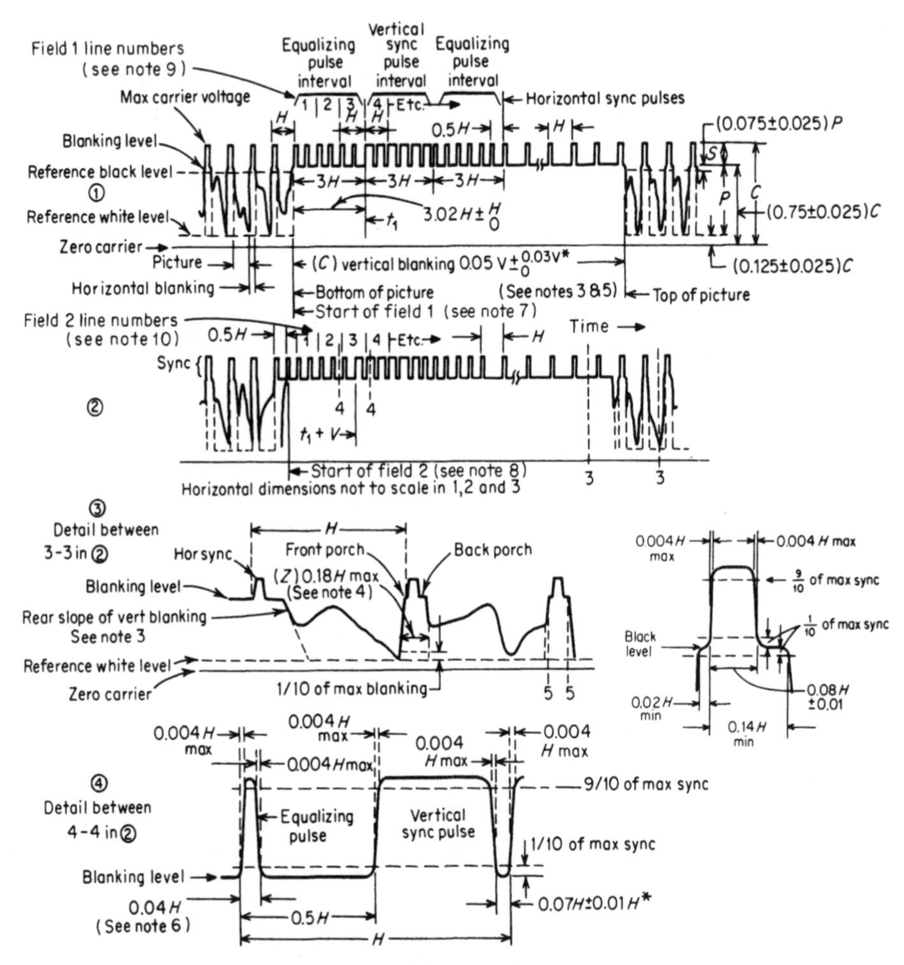

FCC television synchronizing waveforms for monochrome transmission. *Notes:* (1) H = time from start of one line to start of next line. (2) V = time from start of one field to start of next field. (3) Leading and trailing edges of vertical blanking should be complete in less than $0.1H$. (4) Leading and trailing slopes of horizontal blanking must be steep enough to preserve minimum and maximum values of $(x + y)$ and (z) under all conditions of picture content. *(5) Dimensions marked with an asterisk indicate that tolerances are permitted only for long time variations and not for successive cycles. (6) Equalizing pulse area shall be between 0.45 and 0.5 of the area of a horizontal synchronizing pulse. (7) Start of field 1 is defined by a whole line between the first equalizing pulse and preceding H sync pulses. (8) Start of field 2 is defined by a half line between first equalizing pulse and preceding H sync pulses. (9) Field 1 line numbers start with first equalizing pulse in field 1. (10) Field 2 line numbers start with second equalizing pulse in field 2.

Figure 2.20.8 Principle elements of the NTSC signal. (*From FCC.*)

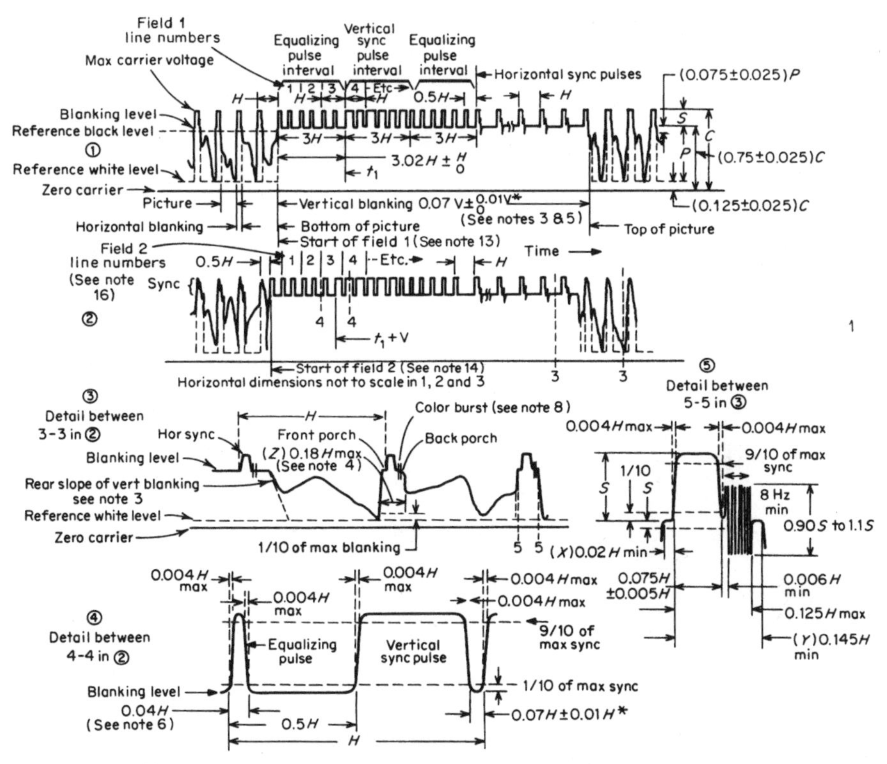

FCC television synchronizing waveforms for color transmission. *Notes:* (1) H = time from start of one line to start of next line. (2) V = time from start of one field to start of next field. (3) Leading and trailing edges of vertical blanking should be complete in less than $0.1H$. (4) Leading and trailing slopes of horizontal blanking must be steep enough to preserve minimum and maximum values of $(x + y)$ and (z) under all conditions of picture content. *(5) Dimensions marked with an asterisk indicate that tolerances are permitted only for long time variations and not for successive cycles. (6) Equalizing pulse duration shall be between 0.45 and 0.55 of the horizontal synchronizing pulse duration. (7) Color burst follows each horizontal pulse but is omitted following the equalizing pulses and during the broad vertical pulses. (8) Color bursts to be omitted during monochrome transmission. (9) The burst frequency shall be 3.579545 MHz. The tolerance on the frequency shall be ±10 Hz with a maximum rate of change not to exceed 0.1 Hz/s. (10) The horizontal scanning frequency shall be 2/455 times the burst frequency. (11) The dimensions specified for the burst determine the times of starting and stopping the burst but not its phase. The color burst consists of amplitude modulation of a continuous sine wave. (12) Dimension P represents the peak excursion of the luminance signal from blanking level but does not include the chrominance signal. Dimension S is the synchronizing pulse amplitude above blanking level. Dimension C is the peak carrier amplitude. (13) Start of field 1 is defined by a whole line between first equalizing pulse and preceding H sync pulses. (14) Start of field 2 is defined by a half line between the first equalizing pulse and the preceding H sync pulses. (15) Field 1 line numbers start with the first equalizing pulse in field 1. (16) Field 2 line numbers start with second equalizing pulse in field 2. (17) Refer to text for further explanations and tolerances. (18) During color transmissions, the chrominance component of the picture signal may penetrate the synchronizing region and the color burst penetrate the picture region. *(FCC.)*

Figure 2.20.9 Synchronizing elements of the NTSC signal. (*From FCC.*)

Table 2.20.1 IRE Standard Scale

Level	IRE Units	Modulation, %
Zero carrier	120	0
Reference white	100	12.5
Blanking	0	75
Sync peaks (max. carrier)	–40	100

Table 2.20.2 Comparison of NTSC, PAL, and SECAM Systems

All systems:
Use three primary additive colorimetric principles
Use similar camera pickup and display technologies
Employ wide-band luminance and narrow-band chrominance
All are compatible with coexisting monochrome systems First-order differences are therefore:
Line and field rates
Component bandwidths
Frequency allocations
Major differences lie in color-encoding techniques:
NTSC: Simultaneous amplitude and phase quadrature modulation of an interlaced, suppressed subcarrier
PAL: Similar to NTSC but with line alternation of one color-modulation component
SECAM: Frequency modulation of line-sequential color subcarrier(s)

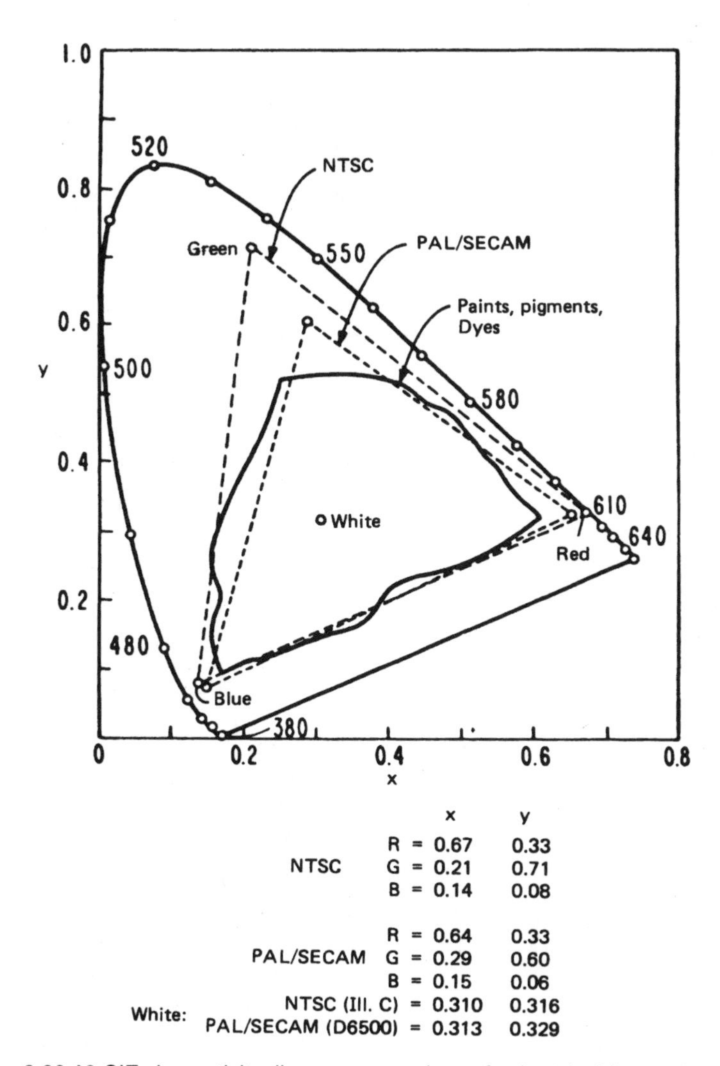

		x	y
NTSC	R =	0.67	0.33
	G =	0.21	0.71
	B =	0.14	0.08
PAL/SECAM	R =	0.64	0.33
	G =	0.29	0.60
	B =	0.15	0.06
White:	NTSC (Ill. C) =	0.310	0.316
	PAL/SECAM (D6500) =	0.313	0.329

Figure 2.20.10 CIE chromaticity diagram comparison of color television systems.

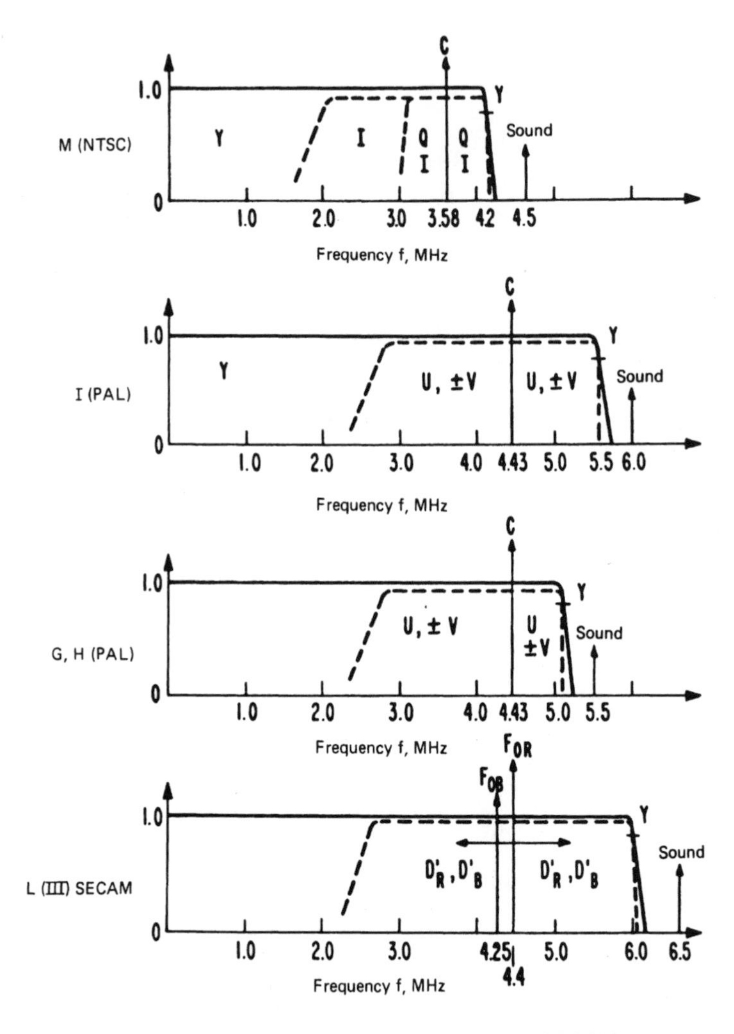

Figure 2.20.11 Bandwidth comparison among NTSC, PAL, and SECAM systems.

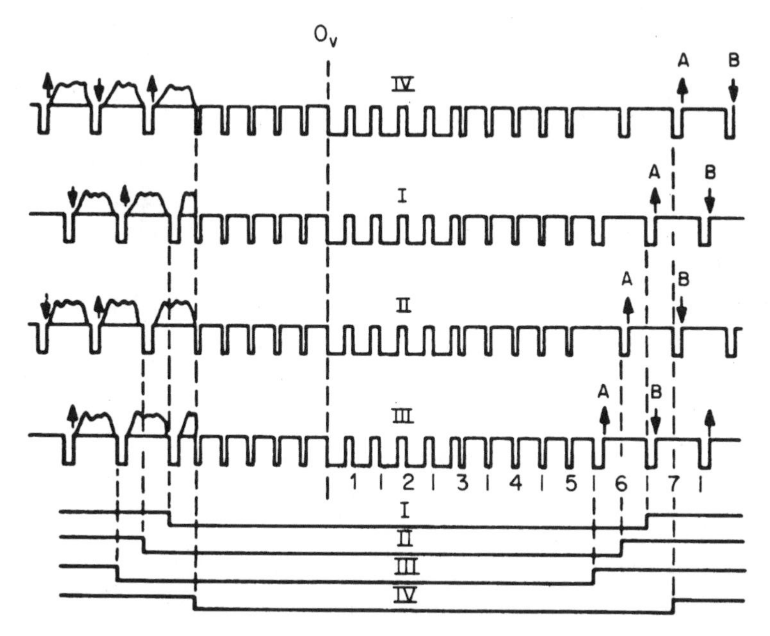

Figure 2.20.12 Meander burst-blanking gate timing diagram for systems B, G, H, and PAL/I. (*Source: CCIR.*)

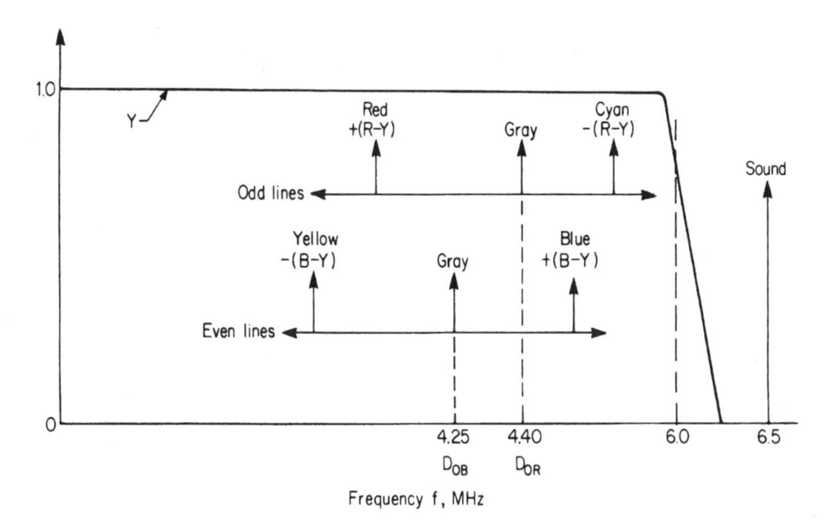

Figure 2.20.13 SECAM FM color modulation system.

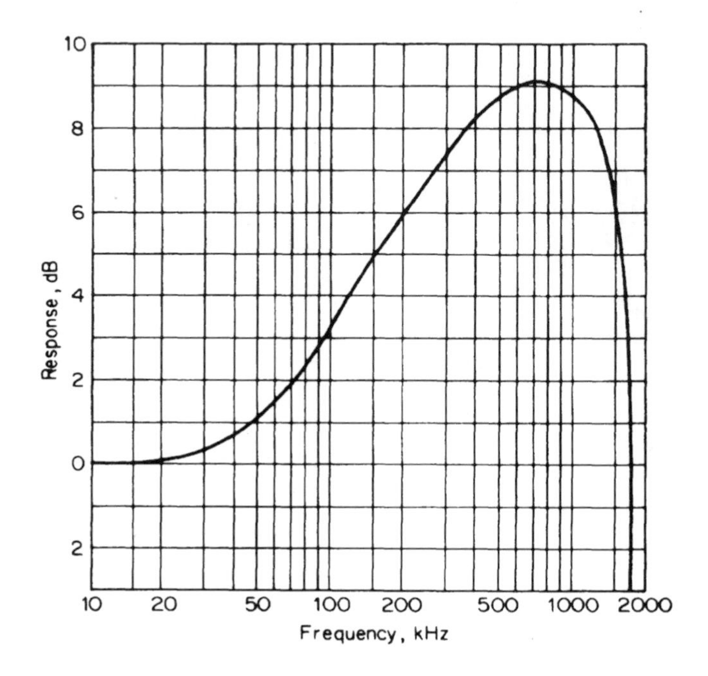

Figure 2.20.14 SECAM color signal low-frequency preemphasis. (*CCIR Rep. 624-2.*)

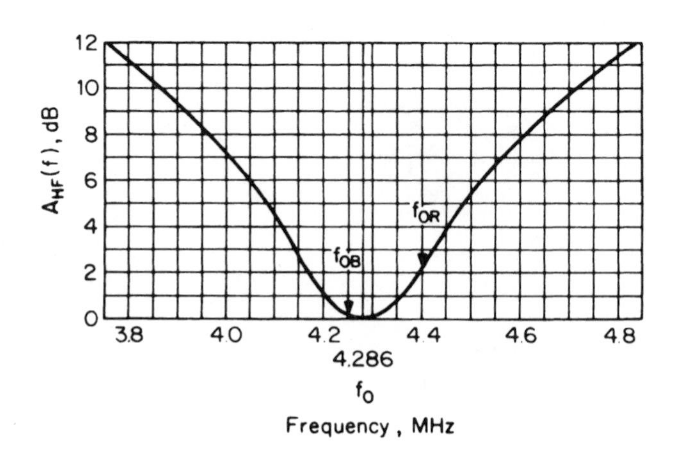

Figure 2.20.15 SECAM high-frequency subcarrier preemphasis. (*CCIR Rep. 624-2.*)

SECAM Line Sequential Color			
Field	Line #	Color	Subcarrier ϕ
Odd (1)	n	f_{OR}	0°
Even (2)	$n + 313$	f_{OB}	180°
Odd (3)	$n + 1$	f_{OB}	0°
Even (4)	$n + 314$	f_{OR}	0°
Odd (5)	$n + 2$	f_{OR}	180°
Even (6)	$n + 315$	f_{OB}	180°
Odd (7)	$n + 3$	f_{OB}	0°
Even (8)	$n + 316$	f_{OR}	180°
Odd (9)	$n + 4$	f_{OR}	0°
Even (10)	$n + 317$	f_{OB}	0°
Odd (11)	$n + 5$	f_{OB}	180°
Even (12)	$n + 318$	f_{OR}	180°

Figure 2.20.16 Color versus line-and-field timing relationship for SECAM. *Notes*: (1) Two frames (four fields) for picture completion; (2) subcarrier interlace is field-to-field and line-to-line of same color.

Table 2.20.3 Basic Characteristics of Video and Synchronizing Signals

	CCIR system identification										
Characteristic	A	M	N	C	B, G	H	I	D, K	K1	L	E
Number of lines per frame	405	525	625	625	625	625	625	625	625	625	819
Number of fields per second	50	60 (59.94)	50	50	50	50	50	50	50	50	50
Line frequency f_H, Hz, and tolerances	10,125	15,750 15,734 (±0.0003%)	15,625 ±0.15%	15,625 ±0.02%	15,625 ±0.02% (±0.0001%)	15,625 ±0.02% (±0.0001%)	15,625 (±0.0001%)	15,625 ±0.02% (±0.0001%)	15,625 ±0.02% (±0.0001)	15,625 ±0.02% (±0.0001)	20,475
Interlace ratio Aspect ratio	2/1 4/3	2/1 4/3	2/1 4/3	2/1 4/3	2/1 4/3	2/1 4/3	2/1 4/3	2/1 4/3	2/1 4/3	2/1 4/3	2/1 4/3
Blanking level, IRE units	0	0	0	0	0	0	0	0	0	0	
Peak-white level	100	100	100	100	100	100	100	100	100	100	100
Sync-pulse level	−43	−40	−40	−43	−43	−43	−43	−43	−43	−43	−43
Picture-black level to blanking level (setup)	0	7.5 ±2.5	7.5 ±2.5	0	0	0	0	0–7	0 color 0–7 mono	0 color 0–7 mono	0–5
Nominal video bandwidth, MHz	3	4.2	4.2	5	5	5	5.5	6	6	6	10
Assumed display gamma	2.8	2.2	2.2	2.8	2.8	2.8	2.8	2.8	2.8	2.8	2.8

Notes: (1) Systems A, C, and E are not recommended by CCIR for adoption by countries setting up a new television service. (2) Values of horizontal line rate tolerances in parentheses are for color television. (3) In the systems using an assumed display gamma of 2.8, an overall system gamma of 1.2 is assumed. All other systems assumed an overall transfer function of unity.

Table 2.20.4 Details of Synchronization Pulses for Worldwide Systems

Item	A†	M	N	C	E†	B, G, H, I, D, K, K1, L
Nominal line period, μs	98.8	63.555	64	64	48, 84	64
Line blanking, μs	17.5–19	10.9 ± 0.2	10.24–11.52	11.8–12.2	9.2–9.8	12 ± 0.3
Horizontal sync pulse, μs	8–10	4.7 ± 0.1	4.22–5.76	4.8–5.2	2.4–2.6	4.7 ± 0.2
Field period, ms	20	16.6833	20	20	20	20
Field blanking (lines)	(13–15.5) $H + a$	(19–21) $H + a$	(19–25) $H + a$	25 $H + a$	33 $H + a$	$25H + a$
Duration of first equalizing pulses	None	$3H$	$3H$	$2.5H$	None	$2.5H$
Duration of second equalizing pulses	None	$3H$	$3H$	$2.5H$	None	$2.5H$
Duration of single equalizing pulse, μs	....	2.3 ± 0.1	2.3–2.56	2.3–2.5	....	2.35 ± 0.1
Duration of field pulse, μs	38–42	27.1	26.52–28.16	26.8–272.	19–21	27.3
Interval between field pulses, μs	11.4–7.4	4.7 ± 0.1	3.84–5.63	4.8–5.2	....	4.7 ± 0.2

†Systems A and E do not use equalizing pulses.

Table 2.20.5 Scanning Constants of Television Systems

Scanning constant	NTSC 525/30 system		PAL 625/25 system		SECAM-III 625/25 system	NHK developmental 1125/30 system
	Monochrome	Color				
Nominal luminance bandwidth, MHz	4.2	4.2	5	5.5	6	20
Frame frequency, Hz	30	29.97	25	25	25	30
Frame period, ms	33.333	33.367	40	40	40	33.333
Active frame time, ms (maximum blanking tolerance)	30.644	30.674	36.775	36.775	36.775	32.000
Field frequency, Hz	60	59.94	50	50	50	60
Field period, ms	16.667	16.683	20	20	20	16.667
Active field time, ms (maximum blanking tolerance)	15.322	15.337	18.387	18.387	18.387	16.000
Line frequency, Hz	15,750	15,734.3	15,625	15,625	15,625	33,750
Line period, μs	63.492	63.555.4	64	64	64	29.630
Active line time, μs (maximum blanking tolerance)	52.092	52.456	51.7	51.7	51.7	24.692
Horizontal resolution factor (lines per megahertz)	78.1	78.7	77.6	77.6	77.6	29.65
Horizontal luminance resolution (lines per picture height)	328	330	388	427	466	593
Vertical luminance resolution (lines per picture height)	343	343	408	408	408	767
Picture elements per frame	150,000	151,000	211,000	232,000	254,000	758,000

†Data on NTSC, PAL, and SECAM systems based on CCIR Reports, Volume XI—Broadcasting Service (Television), International Telecommunications Union, Geneva, 1978. Data on the NHK high-definition system provided by Dr. Takashi Fujio, Deputy Director, Technical Research Laboratories, Japan Broadcasting Corporation. The latter data were those in use in the NHK-CBS HDTV demonstrations in the United States in February 1982. The vertical resolution figures are based on a number of lines equal to 71% of the active lines in the frame. The NHK system has an aspect ratio of 5/3, the other systems 4/3.

Table 2.20.6 Chrominance Frequencies and Resolutions

	NTSC 525/30 system		PAL 625/25 systems		SECAM-III 625/25 system	
Chrominance subcarrier (f_H = line-scanning frequency)	3.579545 MHz ± 10 Hz = 455 $f_H/2$		4.433618.75 MHz ± 5 Hz = 1135 $f_H/4 + f_H/625$		4.406250 MHz ± 2 kHz 4.250000 MHz ± 2 kHz	
	I signal	*Q* signal	E_U signal	E_V signal	D_R signal	D_B signal
Nominal bandwidth	0–1.3 MHz	0–0.62 MHz	0–0.57 MHz 0–1.07 MHz†	0–1.3 MHz 0–1.3 MHz†	0–1.3 MHz	0–1.3 MHz
Maximum video frequency	1.3 MHz	0.62 MHz	0.57 MHz 1.07 MHz†	1.3 MHz 1.3 MHz†	1.3 MHz	1.3 MHz
Horizontal resolution	102 lines	49 lines	44 lines 83 lines†	101 lines 101 lines†	101 lines	101 lines
Vertical resolution	343 lines	343 lines	204 lines	204 lines	204 lines	204 lines

†Values applicable to PAL system I, whose nominal luminance bandwidth (Table 22-6) is 5.5 MHz.

Source: CCIR Recommendations and Reports Volume XI—Broadcasting Service (Television), Geneva, 1982.

Table 2.20.7 Parameters for VSB Transmission Modes (*From: ATSC, "Guide to the Use of the Digital Television Standard," Advanced Television Systems Committee, Washington, D.C., Doc. A/54, Oct. 4, 1995. Used with permission.*)

Parameter	Terrestrial Mode	High-Data-Rate Mode
Channel bandwidth	6 MHz	6 MHz
Excess bandwidth	11.5 percent	11.5 percent
Symbol rate	10.76 Msymbols/s	10.76 Msymbols/s
Bits per symbol	3	4
Trellis FEC	2/3 rate	None
Reed-Solomon FEC	T = 10 (207,187)	T = 10 (207,187)
Segment length	832 symbols	832 symbols
Segment sync	4 symbols/segment	4 symbols/segment
Frame sync	1/313 segments	1/313 segments
Payload data rate	19.28 Mbits/s	38.57 Mbits/s
NTSC co-channel rejection	NTSC rejection filter in receiver	N/A
Pilot power contribution	0.3 dB	0.3 dB
C/N threshold	14.9 dB	28.3 dB

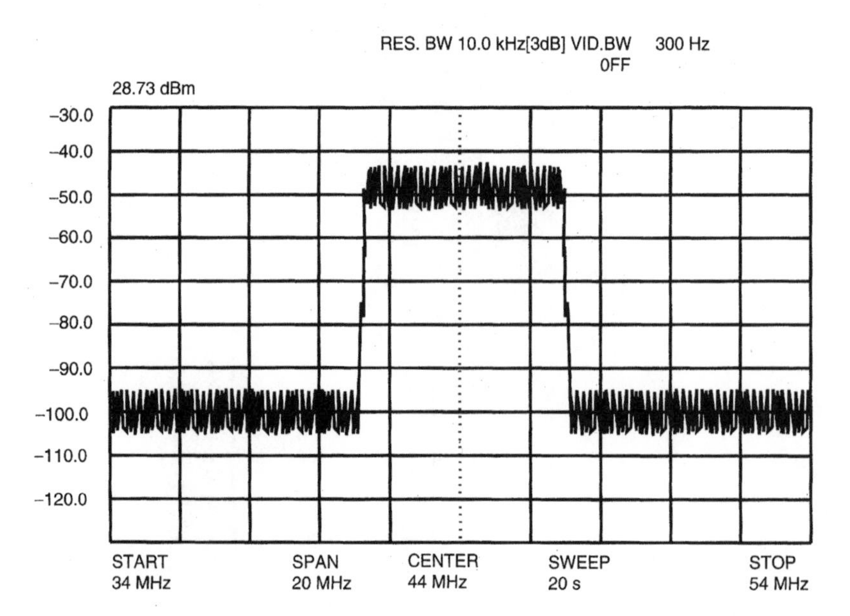

Figure 2.20.17 Typical spectrum of a digital television signal. (*From: Rhodes, Charles W.: "Terrestrial High-Definition Television,"* The Electronics Handbook, *Jerry C. Whitaker (ed.), CRC Press, Boca Raton, Fla., pp. 1599–1610, 1996. Used with permission.*)

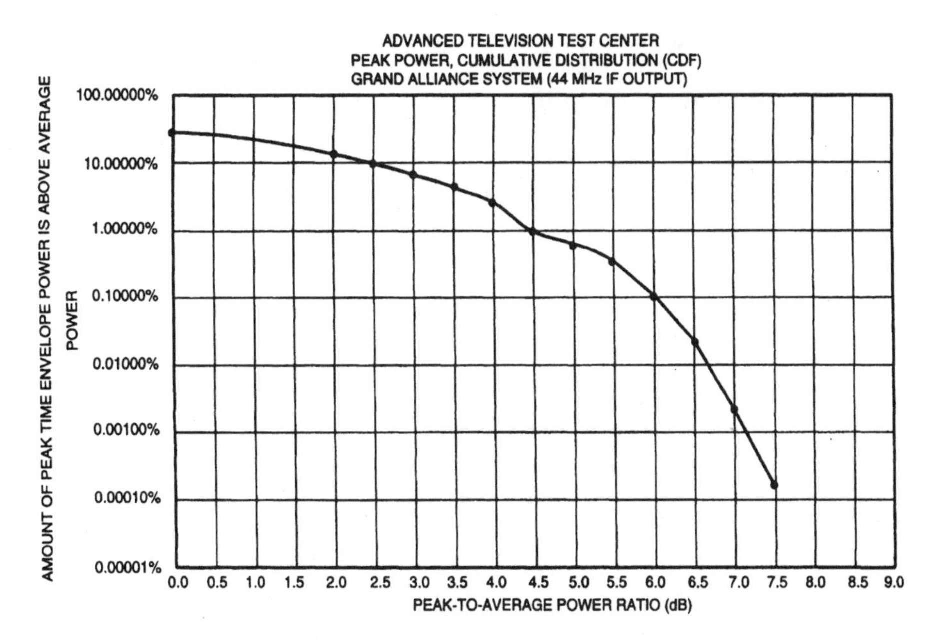

Figure 2.20.18 The cumulative distribution function transient peaks, relative to the average power of the 8-VSB DTV signal at the output of the exciter (IF). (*From: Rhodes, Charles W.: "Terrestrial High-Definition Television,"* The Electronics Handbook, *Jerry C. Whitaker (ed.), CRC Press, Boca Raton, Fla., pp. 1599–1610, 1996. Used with permission.*)

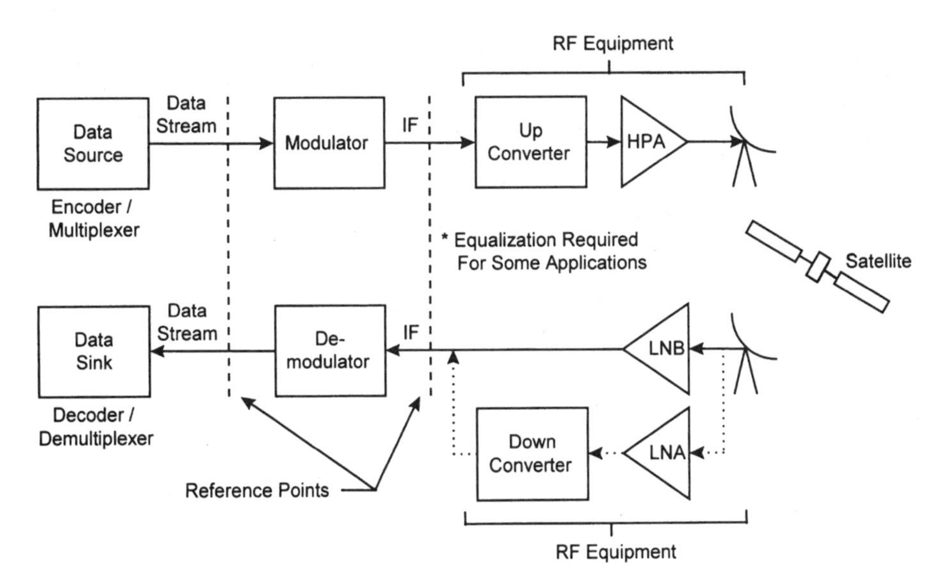

Figure 2.20.19 Overall system block diagram of a digital satellite system. The ATSC standard described in document A/80 covers the elements noted by the given reference points.

Table 2.20.8 System Interfaces Specified in A/80 (*After: ATSC Standard: "Modulation And Coding Requirements For Digital TV (DTV) Applications Over Satellite," Doc. A/80, ATSC, Washington, D.C., July, 17, 1999.*)

Location	System Inputs/Outputs	Type	Connection
Transmit station	Input	MPEG-2 transport (Note 1) or arbitrary	From MPEG-2 multiplexer or other device
	Output	70/140 MHz IF, L-band IF, RF (Note 2)	To RF devices
Receive installation	Input	70/140 MHz IF, L-band IF (Note 2)	From RF devices
	Output	MPEG-2 transport (Note 1) or arbitrary	To MPEG-2 de-multiplexer or other device

1 In accordance with ISO/IEC 13838-1
2 The IF bandwidth may impose a limitation on the maximum symbol rate.

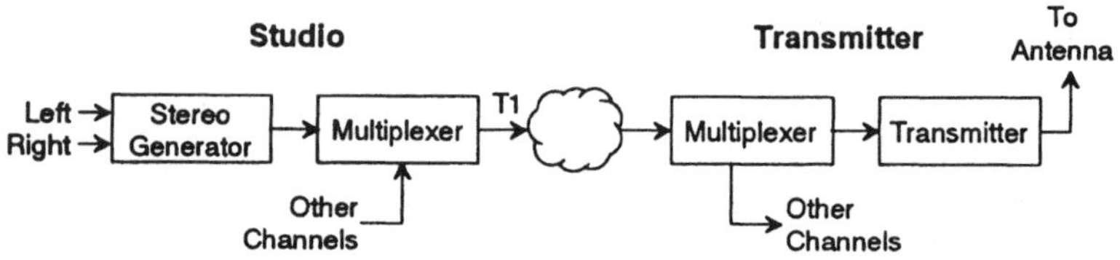

Figure 2.20.20 Composite T1 digital audio STL system. (*After: Rollins, William W., and Robert L. Band: "T1 Digital STL: Discrete vs. Composite Transmission,"* NAB 1996 Broadcast Engineering Conference Proceedings, *National Association of Broadcasters, Washington, D.C., pp. 356–359, 1996.*)

Table 2.20.9 Digital Audio Sampling Size vs. T1 Bandwidth Usage (*After: Rollins, William W., and Robert L. Band: "T1 Digital STL: Discrete vs. Composite Transmission,"* NAB 1996 Broadcast Engineering Conference Proceedings, *National Association of Broadcasters, Washington, D.C., pp. 356–359, 1996.*)

System Type	Samples/s (for 15 kHz stereo)	Sample Size	Digital Transmission Rate	Number of TI Time Slots Used for Broadcast Audio	Number of T1 Time Slots Available for Other Channels
Discrete	32,000 × 2	16-bit	1. 152 Mbits/s	18	6
Composite	112,000	16-bit	1.792 Mbits/s	Not usable	--
		15-bit	1.680 Mbits/s	Not usable	--
		14-bit	1.568 Mbits/s	Not usable	--
		13-bit	1.456 Mbits/s	23	1
		12-bit	1.344 Mbits/s	21	3

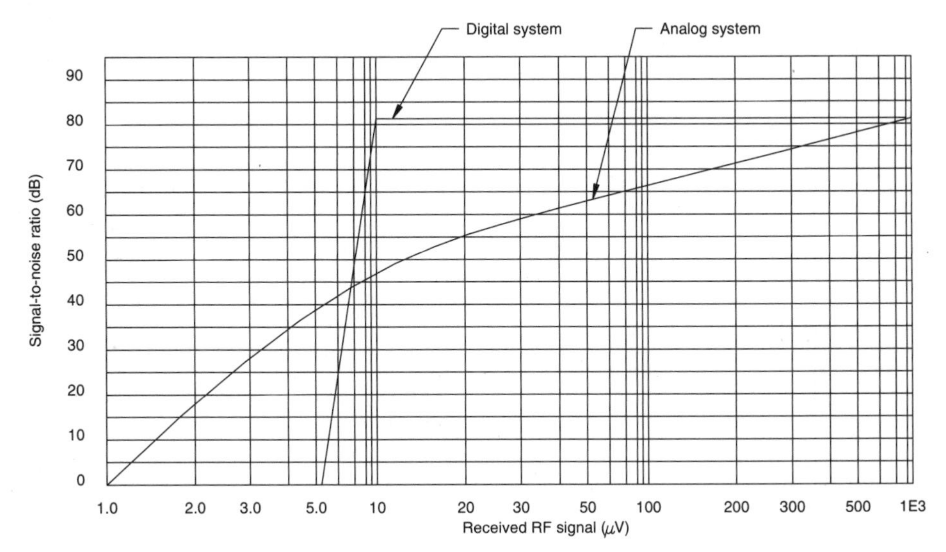

Figure 2.20.21 The benefits of digital vs. analog STL systems in terms of S/N and received RF level. (*After: Whitaker, Jerry C., (ed.):* A Primer: Digital Aural Studio to Transmitter Links, *TFT, Santa Clara, CA, 1994.*)

Table 2.20.10 Typical Coaxial Transmission Line Loss at 950 MHz

Cable Type	Loss (dB/100 Feet)
RG-8/U	8.5
RG-218/U	3.8
1/2-in foam dielectric	3.0
7/8-in foam dielectiic	2.0
1/2-in low-loss foam dielectric	2.4
7/8-in low-loss foam dielectric	1.4
1-5/8-in foam dielectric	1.4
1/2-in air dielectiic	2.7
7/8-in air dielectric	1.4
1-5/8-in air dielectric	0.7

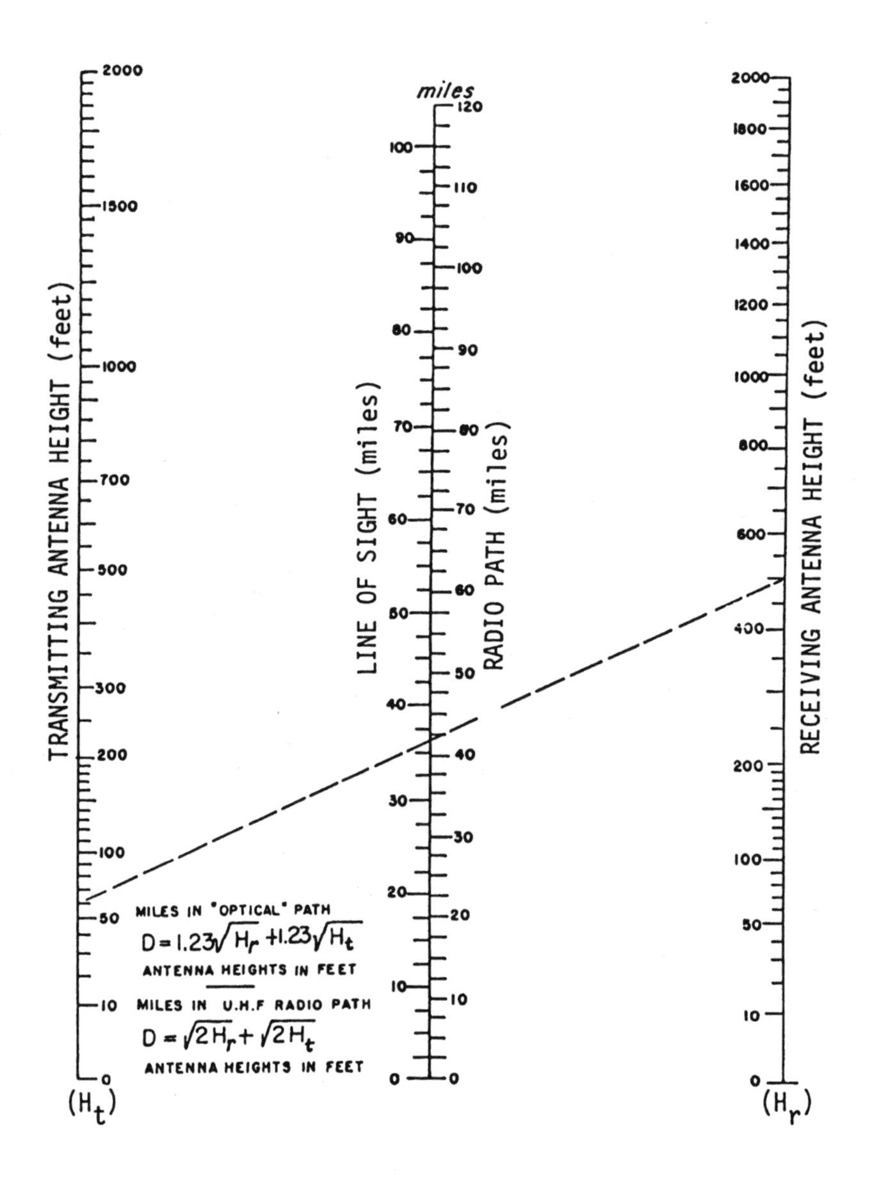

Figure 2.20.22 Nomogram to quickly determine if a radio STL path is viable. (*After: Whitaker, Jerry C., (ed.):* A Primer: Digital Aural Studio to Transmitter Links, *TFT, Santa Clara, CA, 1994.*)

Table 2.20.11 Conversion of Watts to dBm (50 Ω Impedance)

Power in Watts	dBm	Power in Watts	dBm
5.0	37.0	18.0	42.6
5.5	37.42	19.0	42.8
6.0	37.79	20.0	43.0
6.5	38.12	25.0	43.9
7.0	38.46	30.0	44.8
7.5	38.75	35.0	45.4
8.0	39.0	40.0	46.0
8.5	39.3	45.0	46.5
9.0	39.55	50.0	47.0
9.5	39.78	55.0	47.4
10.0	40.0	60.0	47.8
10.5	40.2	65.0	48.1
11.0	40.4	70.0	48.5
12.0	40.8	75.0	48.8
13.0	41.1	80.0	49.0
14.0	41.5	85.0	49.3
15.0	41.8	90.0	49.5
16.0	42.0	95.0	49.9
17.0	42.3	100.0	50

Table 2.20.12 Typical Isotropic Gain for Full Parabolic Antennas at 950 MHz

Antenna Diameter	Isotropic Gain
4 feet	18.9 dBi
5 feet	21.0 dBi
6 feet	22.0 dBi
8 f eet	25.0 dBi
10 feet	27.0 dBi

Table 2.20.13 Conversion of Microvolts to dBm (50 Ω Impedance)

Microvolts	dBm	Microvolts	dBm
0.10	–127	40	–74.9
0.12	–125.25	45	–73.9
0.14	–124	50	–72.9
0.16	–122.9	60	–71.25
0.18	–121.9	70	–70
0.20	–120.9	80	–68.9
0.25	–119	90	–67.9
0.30	–117.25	100	–66.9
0.35	–116	120	–65.25
0.40	–115	140	–64
0.60	–111.25	160	–62.9
0.70	–110	180	–61.9
0.80	–108.9	200	–60.9
0.90	–107.9	250	–59
1.0	–106.9	300	–57.5
1.2	–105.25	350	–56
1.4	–104	400	–54.9
1.6	–102.9	450	–53.9
1.8	–101.9	500	–52.9
2.0	–100.9	600	–51.25
2.5	–99	700	–50
3.0	–97.5	800	–49
3.5	–96	900	–48
4.0	–95	1000	–46.9
4.5	–93.9	1200	–45.25
5.0	–92.9	1400	–44
6.0	–91.25	1600	–42.9
7.0	–90	1800	–41.9
8.0	–88.9	2000	–40.9
9.0	–87.9	2500	–39
10	–86.9	3000	–37.25
1.1	–86	3500	–36
12	–85.25	4000	–34.9
14	–84	4500	–33.9
16	–82.9	5000	–33
18	–81.9	6000	–31.25
20	–80.9	7000	–30
25	–79	8000	–28.9
30	–77.25	9000	–27.9
35	–76	10,000	–26.9

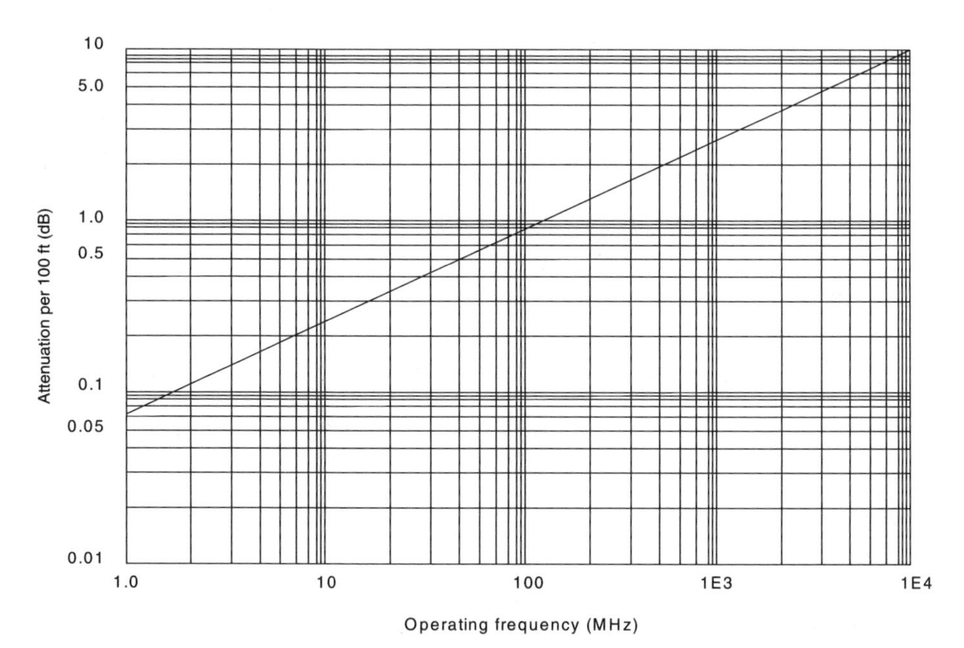

Figure 2.20.23 Loss vs. frequency for 1/2-inch foam dielectric transmission line. (*After: Whitaker, Jerry C., (ed.):* A Primer: Digital Aural Studio to Transmitter Links, *TFT, Santa Clara, CA, 1994.*)

Table 2.20.14 Path Attenuation at 950 MHz for an Aural STL System

Distance in Miles	Loss in dB	Distance in Miles	Loss in dB
1	–96.2	16	–120.2
2	–102.2	17	–120.7
3	–105.7	18	–121.3
4	–108.2	19	–121.7
5	–110.1	20	–122.2
6	–111.7	21	–122.6
7	–113.1	22	–123.0
8	–114.2	23	–123.4
9	–115.3	24	–123.8
10	–116.2	25	–124.1
11	–116.9	26	–124.5
12	–117.7	27	–124.8
13	–118.4	28	–125.1
14	–119.1	29	–125.4
15	–119.7	30	–125.7

Table 2.20.15 Relationship Between Fade Margin, Reliability, and Outage Time for Rayleigh Distributed Paths

Fade Margin (dB)	Path Reliability/ Availability	Outage Hours per Year	Outage Minutes per Month	Outage Seconds per Day
10	90.4837	834.20	4170.98	8222.05
20	99.0050	87.22	436.12	859.69
21	99.2088	69-35	346.77	683.58
22	99.3710	55.14	275.68	543.43
23	99.5001	43.82	219.12	431.94
24	99.6027	34.83	174.14	343.28
25	99-6843	27.68	138.38	272.79
26	99.7491	21.99	109.96	216.75
27	99.9007	17.47	87.37	172.22
28	99.8416	13.88	69.41	136.83
29	99.8742	11.03	55.14	108.70
30	99.9000	8.76	43.81	86.36
31	99.9206	6.96	34.80	68.60
32	99.9369	5.53	27.65	54.50
33	99.9499	4.39	21.96	43.29
34	99.9602	3.49	17.45	34.39
35	99.9684	2.77	13.86	27.32
36	99.9749	2.20	11.01	21.70
37	99-9800	1.75	8.74	17.24
38	99.9842	1.39	6.95	13.69
39	99.9874	1.10	5.52	10.88
40	99.9900	0.88	4.38	8.64
41	99.9921	0.70	3.48	6.86
42	99-9937	0.55	2.77	5.45
43	99.9950	0.44	2.20	4.33
44	99.9960	0.35	1.74	3.44
45	99.9968	0.28	1.39	2.73
50	99.9990	0.09	0.44	0.86
55	99.9997	0.03	0.14	0.27
60	99.9999	0.01	0.04	0.09

Table 2.20.16 Important Relations in Low-Loss Transmission Lines (*From: Feinstein, J.: "Passive Microwave Components," in* Electronic Engineers' Handbook, *D. Fink and D. Chris-*

Equation	Explanation
$r = \dfrac{1 + \lvert\Gamma\rvert}{1 - \lvert\Gamma\rvert}$	r = VSWR
$\lvert\Gamma\rvert = \dfrac{r - 1}{r + 1}$	$\lvert\Gamma\rvert$ = magnitude of reflection coefficient
$\Gamma = \dfrac{R - Z_0}{R + Z_0}$	Γ = reflection coefficient (real) at a point in a line where impedance is real (R)
$r = \dfrac{R}{Z_0}$	$R > Z_0$ (at voltage maximum)
$r = \dfrac{Z_0}{R}$	$R < Z_0$ (at voltage minimum)
$\dfrac{P_t}{P_i} = \lvert\Gamma\rvert^2 = \left(\dfrac{r - 1}{r + 1}\right)^2$	P_r = reflected power P_i = incident power
$\dfrac{P_t}{P_i} = 1 - \lvert\Gamma\rvert^2 = \dfrac{4r}{(r + 1)^2}$	P_t = transmitted power
$\dfrac{P_b}{P_m} = \dfrac{1}{r}$	P_b = net power transmitted to load at onset of breakdown in a line where VSWR = r exists P_m = same when line is matched, $r = 1$
$\dfrac{\alpha_t}{\alpha_m} = \dfrac{1 + \Gamma^2}{1 - \Gamma^2} = \dfrac{r^2 + 1}{2r}$	α_m = attenuation constant when $r = 1$, matched line α_t = attenuation constant allowing for increased ohmic loss caused by standing waves
$r_{\max} = r_1 r_2$	$r_{\max}$ = maximum VSWR when r_1 and r_2 combine in worst phase
$r_{\min} = \dfrac{r_2}{r_1}\ r_2 > r_1$	$r_{\min}$ = minimum VSWR when r_1 and r_2 are in best phase
$\lvert\Gamma\rvert = \dfrac{\lvert X\rvert}{\sqrt{X^2 + 4}}$	Relations for a normalized reactance X in series with resistance Z_0
$\lvert X\rvert = \dfrac{r - 1}{\sqrt{r}}$	
$\lvert\Gamma\rvert = \dfrac{\lvert B\rvert}{\sqrt{B^2 + 4}}$	Relations for a normalized susceptance B in shunt with admittance Y_0
$\lvert B\rvert = \dfrac{r - 1}{\sqrt{r}}$	

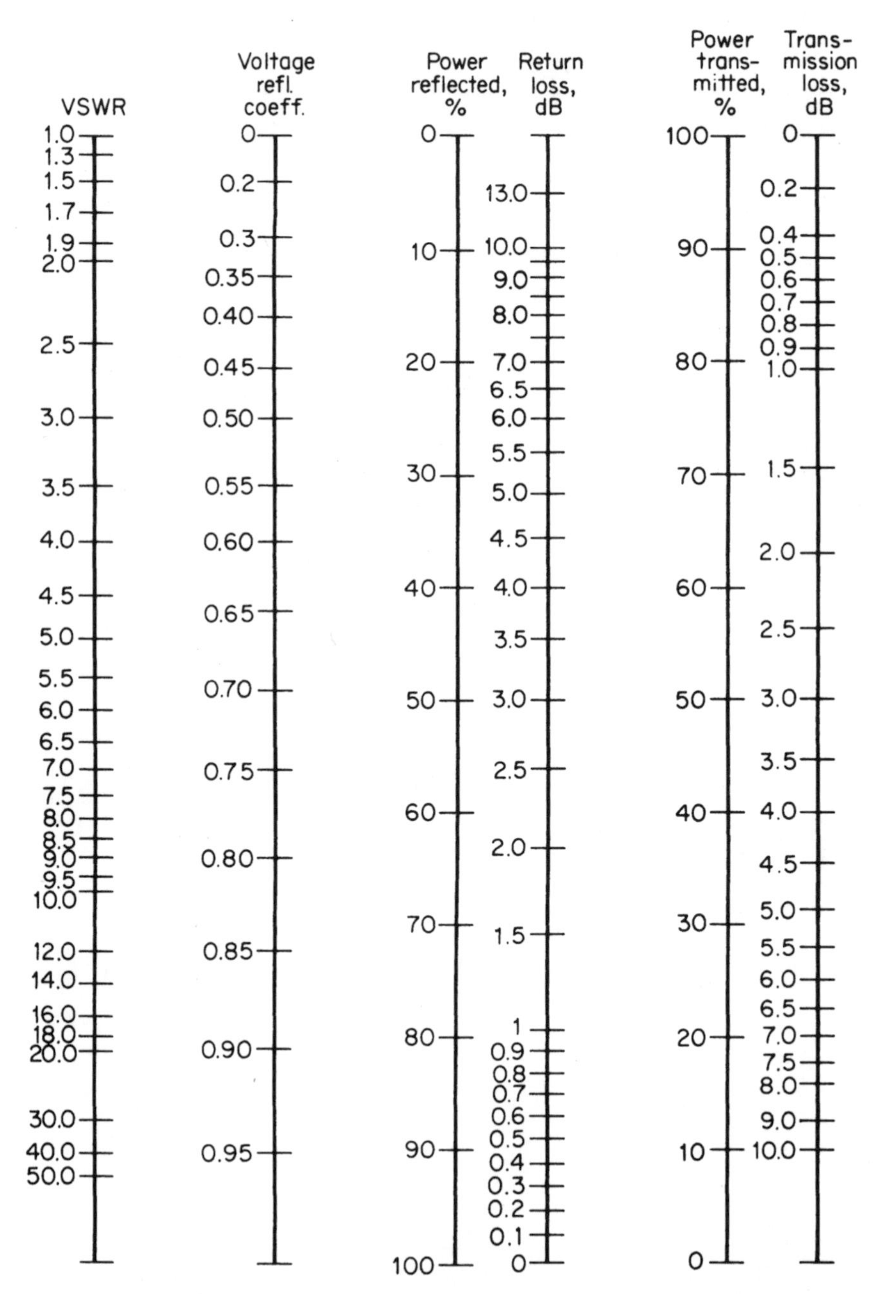

Figure 2.20.24 Nomograph for transmission and reflection of power at high-voltage standing-wave ratios (VSWR). (*From: Feinstein, J.: "Passive Microwave Components," in* Electronic Engineers' Handbook, *D. Fink and D. Christiansen (eds.), McGraw-Hill, New York, N.Y., 1982. Used with permission.*)

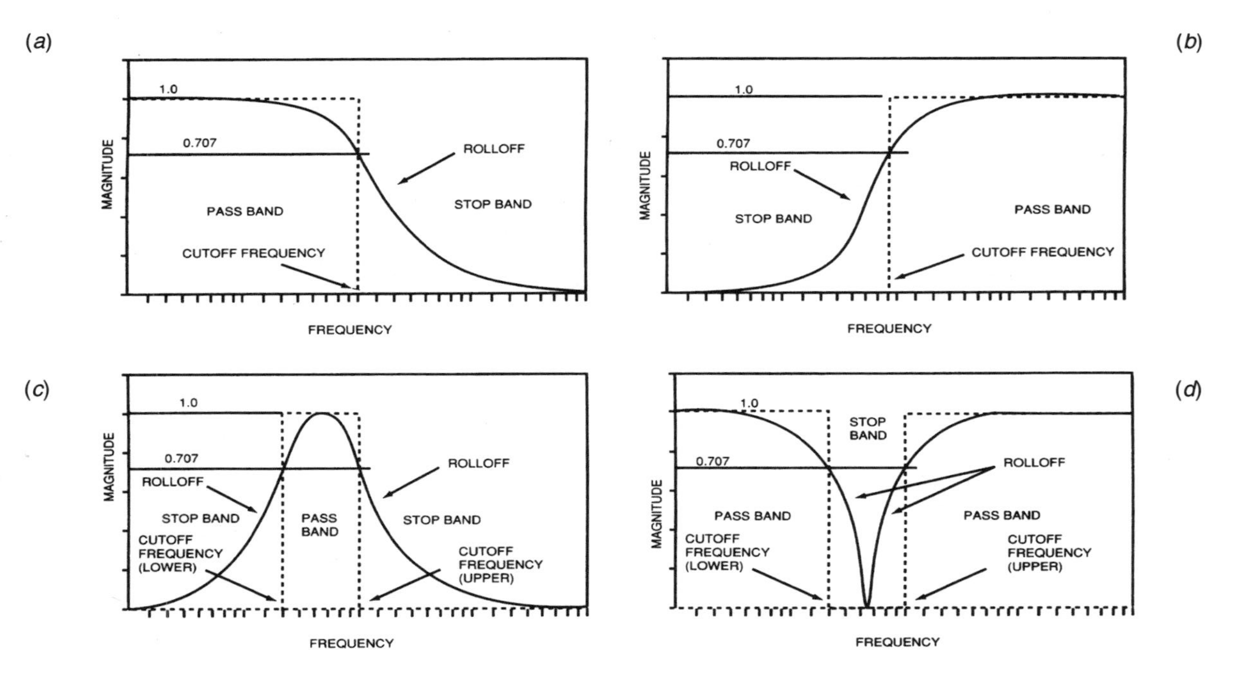

Figure 2.20.25 Filter characteristics by type: (*a*) low-pass, (*b*) high-pass, (*c*) bandpass, (*d*) bandstop. (*From: Harrison, Cecil: "Passive Filters," in* The Electronics Handbook, *Jerry C. Whitaker (ed.), CRC Press, Boca Raton, Fla., pp. 279–290, 1996. Used with permission.*)

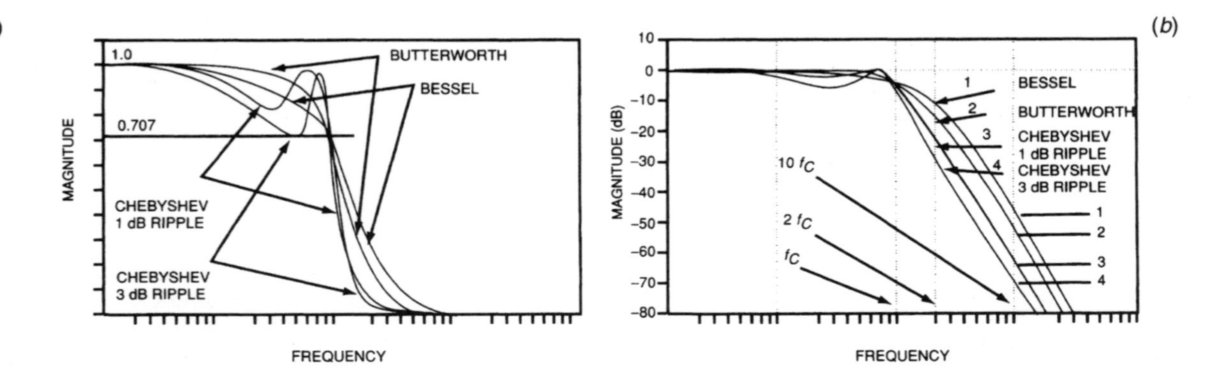

Figure 2.20.26 Filter characteristics by alignment, third-order, all-pole filters: (*a*) magnitude, (*b*) magnitude in decibels. (*From: Harrison, Cecil: "Passive Filters," in* The Electronics Handbook, *Jerry C. Whitaker (ed.), CRC Press, Boca Raton, Fla., pp. 279–290, 1996. Used with permission.*)

Table 2.20.17 Summary of Standard Filter Alignments (*After: Harrison, Cecil: "Passive Filters," in* The Electronics Handbook, *Jerry C. Whitaker (ed.), CRC Press, Boca Raton, Fla., pp. 279–290, 1996. Used with permission.*)

Alignment	Pass Band Description	Stop Band Description	Comments
Butterworth	Monotonic	Monotonic	All-pole; maximally flat
Chebyshev	Rippled	Monotonic	All-pole
Bessel	Monotonic	Monotonic	All-pole; constant phase shift
Inverse Chebyshev	Monotonic	Rippled	
Elliptic (Cauer)	Rippled	Rippled	

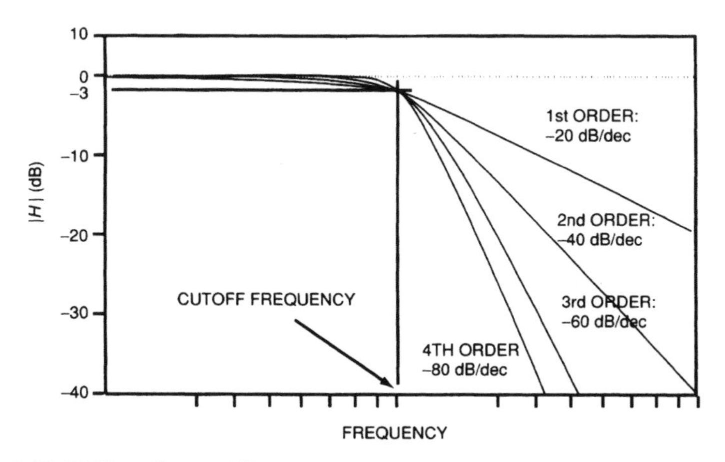

Figure 2.20.27 The effects of filter order on rolloff (Butterworth alignment). (*From: Harrison, Cecil: "Passive Filters," in* The Electronics Handbook, *Jerry C. Whitaker (ed.), CRC Press, Boca Raton, Fla., pp. 279–290, 1996. Used with permission.*)

2.21 Radio/Television Transmitting Antennas

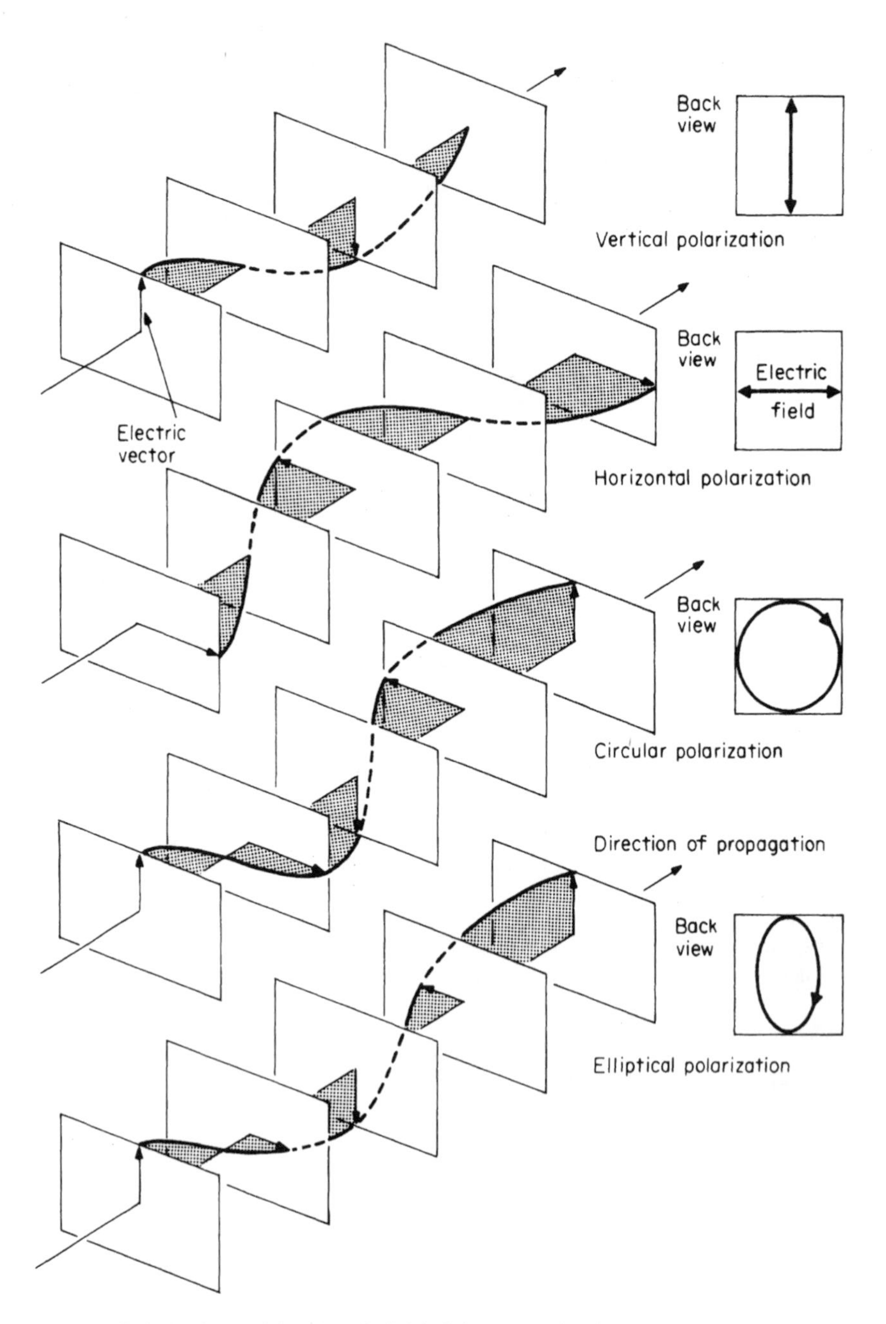

Figure 2.21.1 Polarizations of the electric field of the transmitted wave.

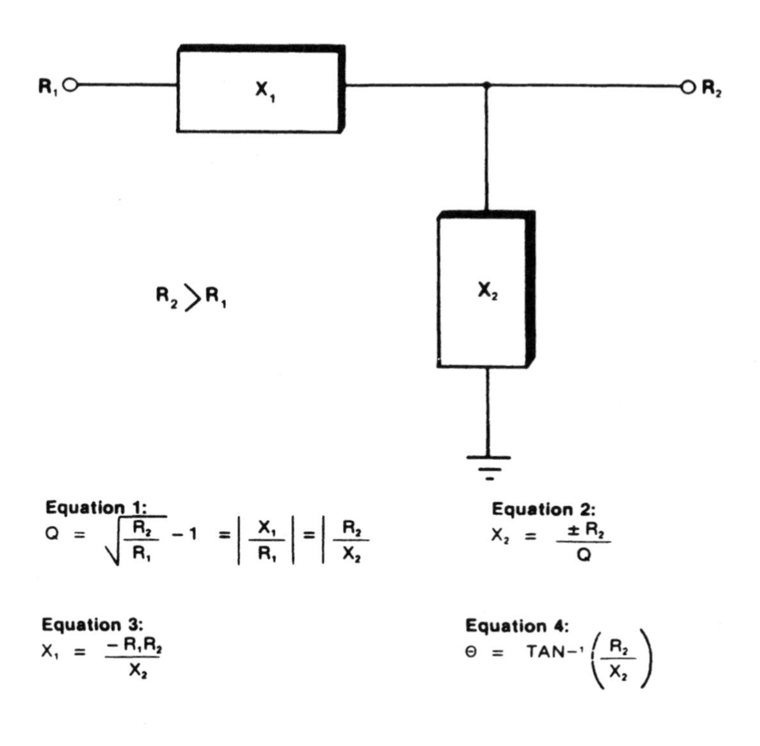

Where: R_1 = L network input resistance (ohms)

R_2 = L network output resistance (ohms)

X_1 = Series leg reactance (ohms)

X_2 = Shunt leg reactance (ohms)

Q = Loaded Q of the L network

Figure 2.21.2 L network parameters.

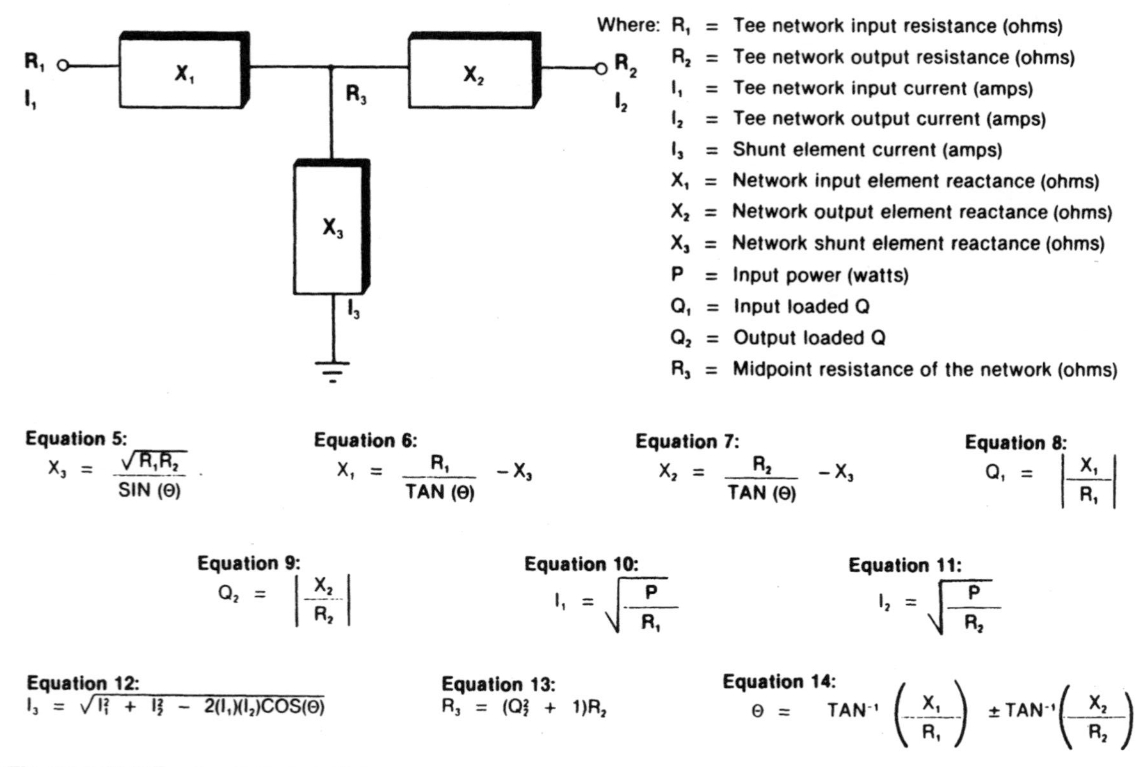

Figure 2.21.3 T network parameters.

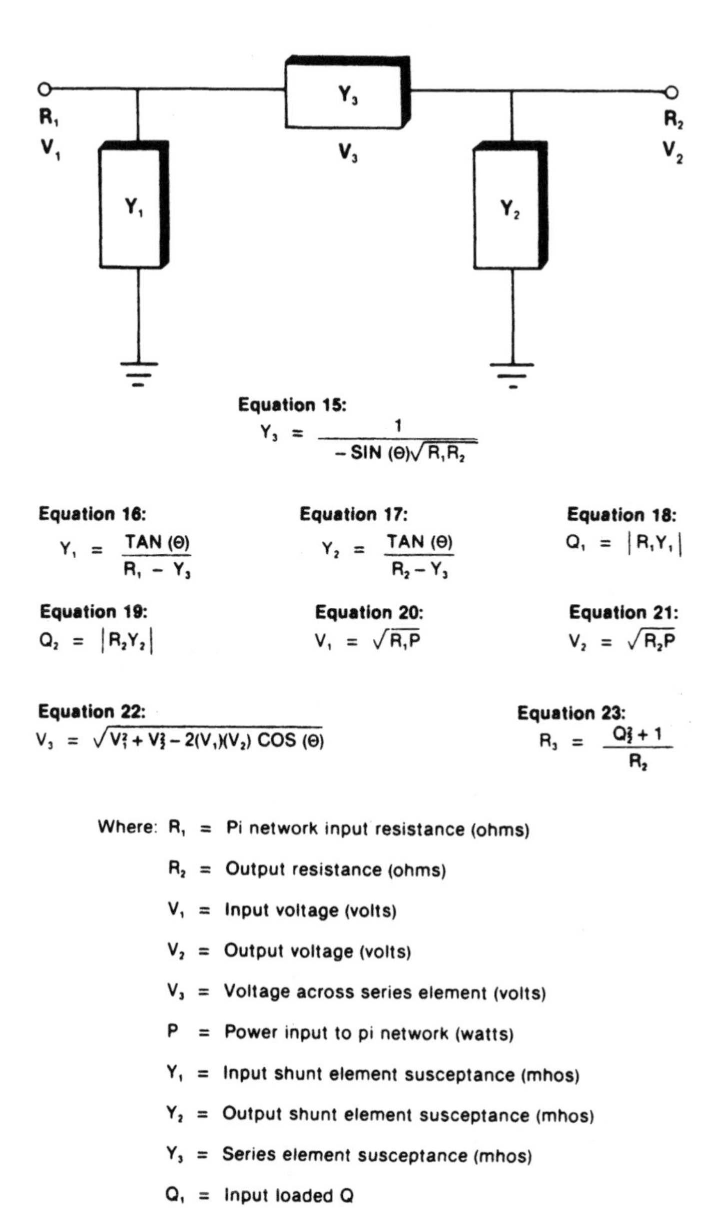

Figure 2.21.4 Pi network parameters.

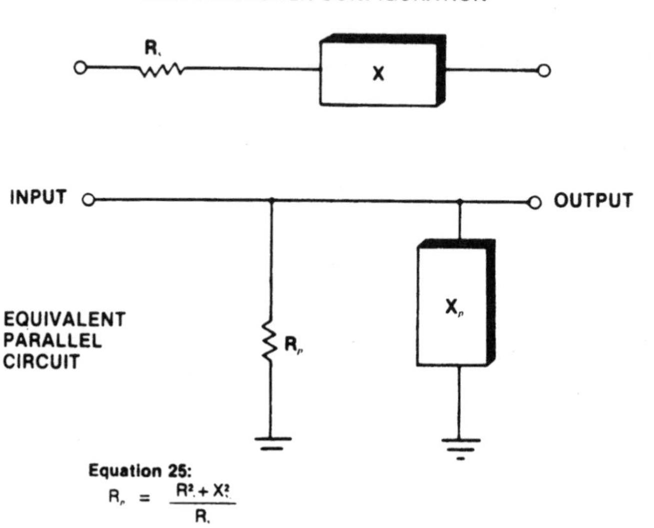

Equation 25:

$$R_p = \frac{R_s^2 + X_s^2}{R_s}$$

Equation 26:

$$X_p = \frac{R_s^2 + X_s^2}{X_s}$$

Where: R_s = Series configuration resistance (ohms)

R_p = Parallel configuration resistance (ohms)

X_s = Series reactance (ohms)

X_p = Parallel reactance (ohms)

Figure 2.21.5 Line stretcher configuration.

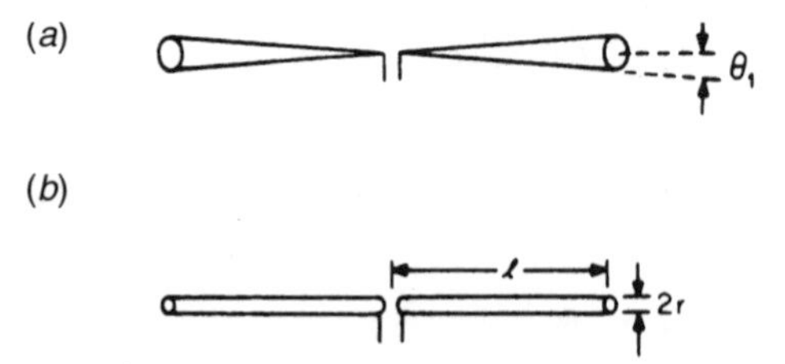

Figure 2.21.6 Half-wave dipole antenna: (*a*) conical dipole; (*b*) conventional dipole.

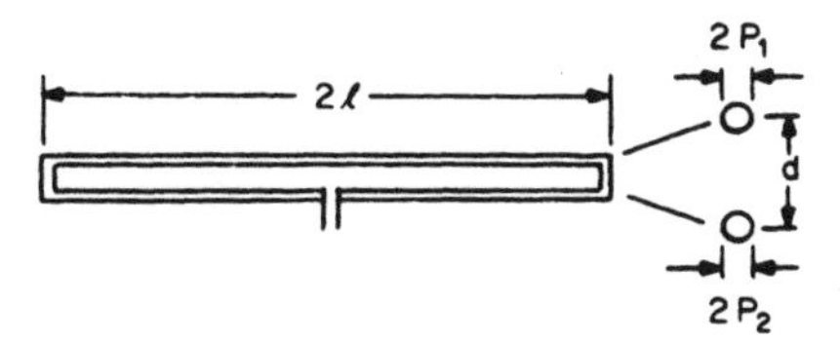

Figure 2.21.7 Folded dipole antenna.

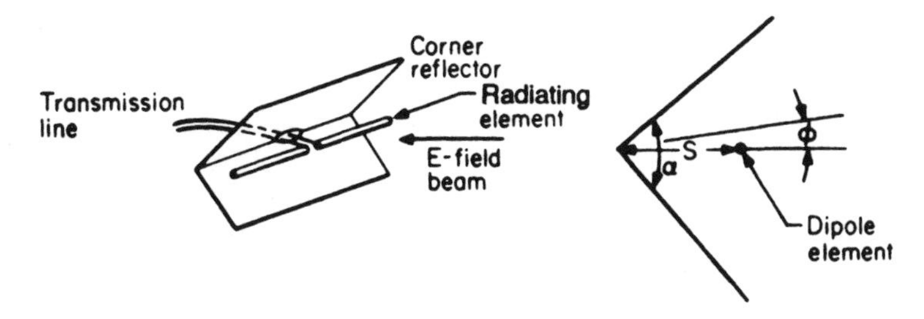

Figure 2.21.8 Corner-reflector antenna.

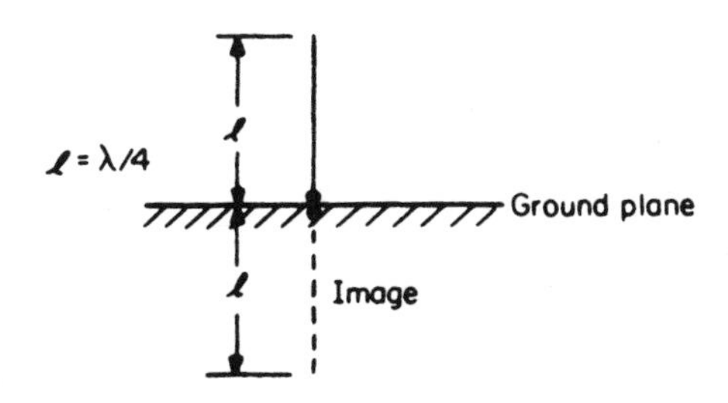

Figure 2.21.9 Vertical monopole mounted above a ground plane.

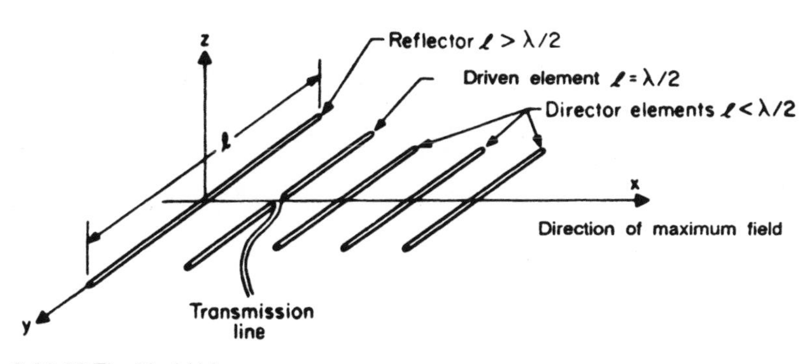

Figure 2.21.10 The Yagi-Uda array.

Table 2.21.1 Typical Characteristics of Single-Channel Yagi-Uda Arrays

Number of Elements	Gain, dB	Beamwidth, Degrees
2	3 to 4	65
3	6 to 8	55
4	7 to 10	50
5	9 to 11	45
9	12 to 14	37
15	14 to 16	30

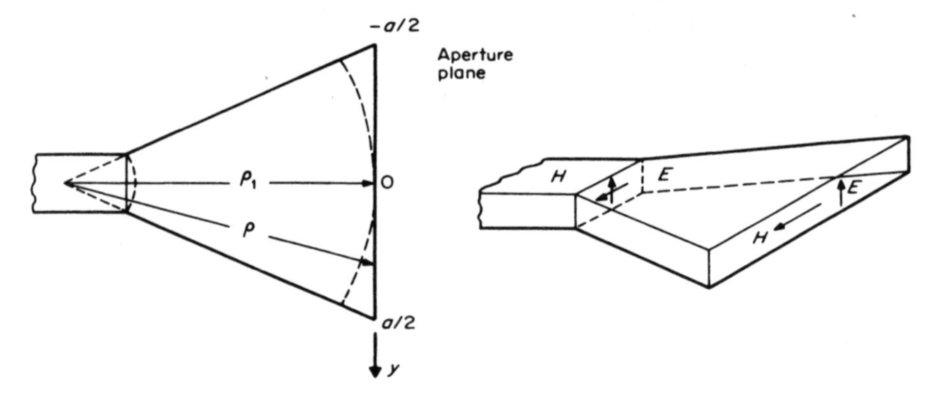

Figure 2.21.11 Geometry of an *H*-plane sectoral horn.

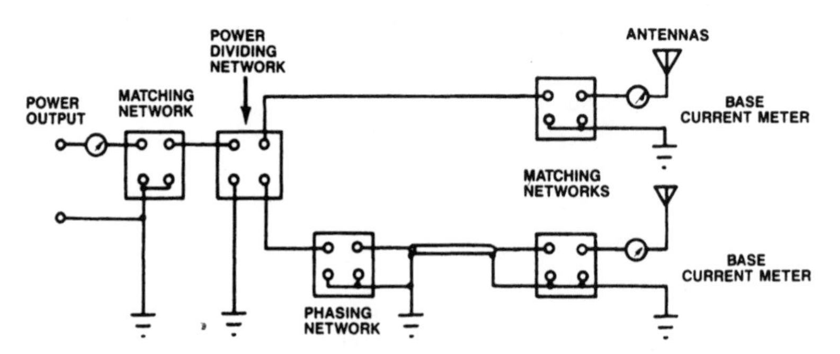

Figure 2.21.12 Block diagram of an AM directional antenna feeder system for a two tower array.

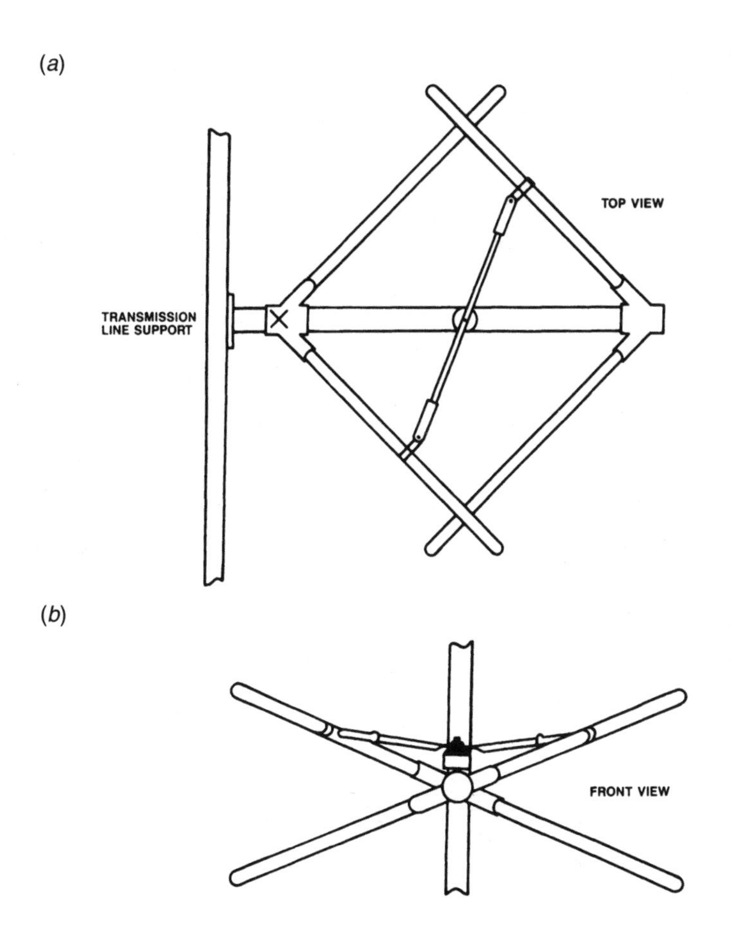

Figure 2.21.13 Mechanical configuration of one bay of a circularly-polarized FM transmitting antenna: (*a*) top view of the antenna, (*b*) front view.

Table 2.21.2 Various Combinations of Transmitter Power and Antenna Gain that will Produce 100 KW Effective Radiated Power (ERP) for an FM Station

Number of Bays	Antenna Gain	Required Transmitter Power (kW)*
3	1.5888	66.3
4	2.1332	49.3
5	2.7154	38.8
6	3.3028	31.8
7	3.8935	27.0
8	4.4872	23.5
10	5.6800	18.5
12	6.8781	15.3

* Effects of transmission line and related losses not included.

Table 2.21.3 Television Antenna Equivalent Heating Power

Parameter	Carrier Levels (percent)		Fraction of Time (percent)	Average Power (percent)
	Voltage	Power		
Sync	100	100	8	8
Blanking	75	56	92	52
Visual black-signal power				60
Aural power (percent of sync power)				20 (or 10)
Total transmitted power (percent peak-of-sync)				80 (or 70)

Table 2.21.4 Typical Circularities of Panel Antennas

Shape	Tower-face Size, ft (m)	Circularity, ± dB	
		Channels 2–6	Channels 7–13
Triangular	5 (1.5)	0.9	1.8
	6 (1.8)	1.0	2.0
	7 (2.1)	1.1	2.3
	10 (3.0)	1.3	3.0
	4 (1.2)	0.5	1.6
Square	5 (1.5)	0.6	1.9
	6 (1.8)	0.7	2.4
	7 (2.1)	0.8	2.7
	10 (3.0)	1.2	3.2

Table 2.21.5 VHF Transmitter Output in Kilowatts as a Function of Operating Characteristics

Line		Channels 2–6, ERP = 100 kW		Channels 7–13, ERP = 316 kW	
Size, in (cm)	Length, ft (m)	Antenna Gain = 3	Antenna Gain = 6	Antenna Gain = 6	Antenna Gain = 12
3-1/8 (8), 50 Ω	500 (152)	36.9	18.5	62.1	31.0
	1000 (305)	40.9	20.5	73.2	36.6
	1500 (457)	45.3	22.7	86.3	43.2
	2000 (610)	50.2	25.1	101.7	50.9
6-1/8 (15.6), 75 Ω	500	35.0	17.5	56.7	28.5
	1000	36.7	18.9	61.6	30.8
	1500	38.5	19.3	66.5	33.3
	2000	40.5	20.3	71.9	36.0

Table 2.21.6 UHF Transmitter Output in Kilowatts as a Function of Operating Characteristics

Line (75 Ω)		Antenna Gain = 30			
Size, in (cm)	Length, ft (m)	Channels 14–26	Channels 27–40	Channels 41–54	Channels 55–70
6-1/8 (15.6)	500 (152)	38.0	38.4	38.8	39.2
	1000 (305)	43.2	44.7	45.4	49.2
	1500 (457)	49.4	51.0	52.7	54.4
	2000 (610)	56.2	58.6	61.3	64.0
8-3/16 (21)	500	36.8	37.0	37.3	Waveguide
	1000	40.5	41.1	41.7	
	1500	44.7	45.7	46.6	
	2000	49.3	50.7	52.1	

Table 2.21.7 Circularities of Panel Antennas for VHF Operation

Shape	Tower-face Size, ft (m)	Circularity, ±dB[1]	
		Channels 2–6	Channels 7–13
Triangular	5 (1.5)	0.9	1.8
	6 (1.8)	1.0	2.0
	7 (2.1)	1.1	2.3
	10 (3.0)	1.3	3.0
	4 (1.2)	0.5	1.6
Square	5 (1.5)	0.6	1.9
	6 (1.8)	0.7	2.4
	7 (2.1)	0.8	2.7
	10 (3.0)	1.2	3.2

1 Add up to ±0.3 dB for horizontally polarized panels and ± 0.6 dB for circularly polarized panels. These values are required to account for tolerances and realizable phase patterns of practical hardware assemblies.

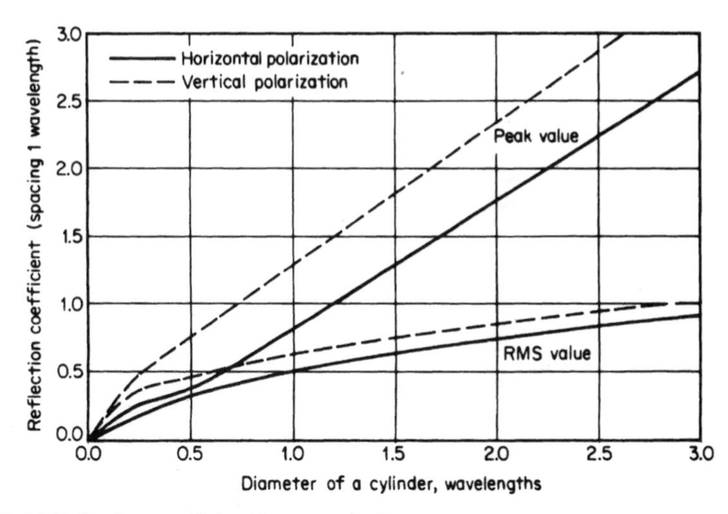

Figure 2.21.14 Reflection coefficient from a cylinder.

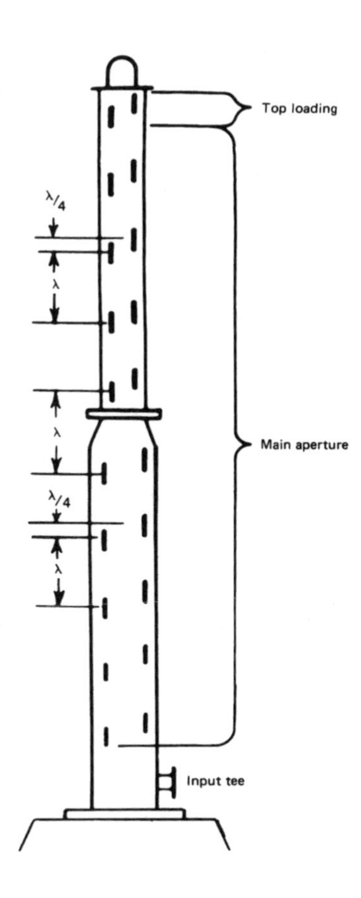

Figure 2.21.15 Schematic of a multislot traveling-wave antenna.

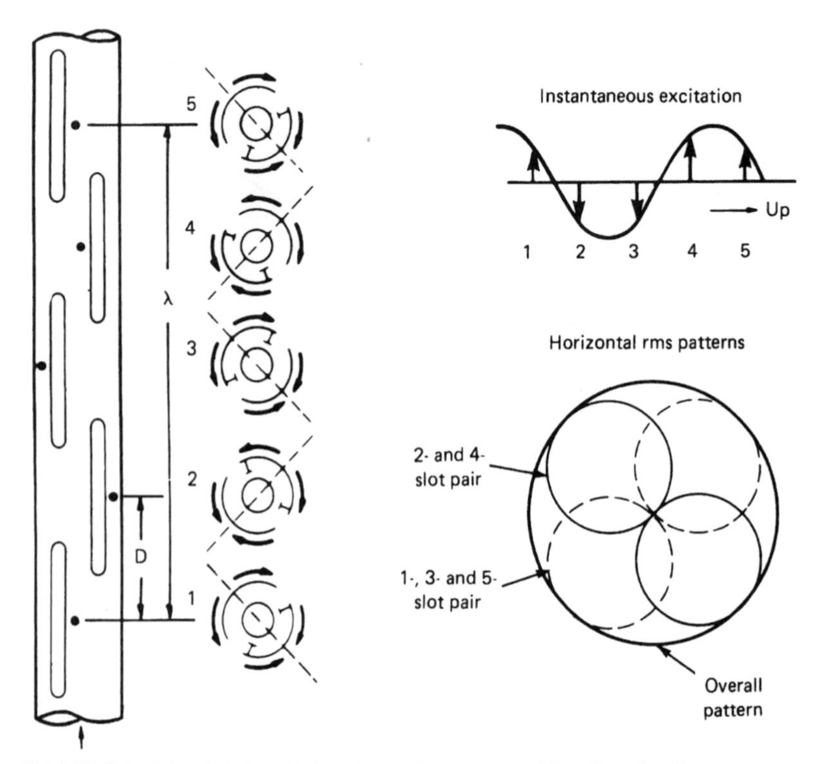

Figure 2.21.16 Principle of slot excitation to produce an omnidirectional pattern.

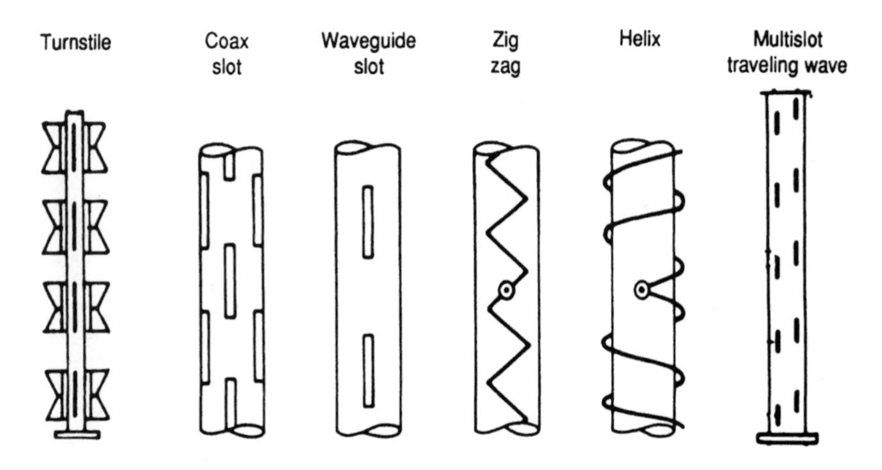

Figure 2.21.17 Various antenna designs for VHF and UHF broadcasting. Note that not all these designs have found commercial use.

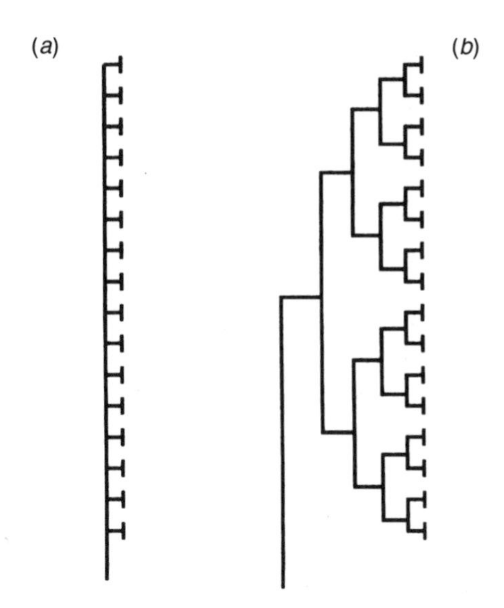

Figure 2.21.18 Antenna feed configurations: (*a*) series feed, (*b*) branch feed.

2.22 Radio/Television Receivers and Cable/Satellite Systems

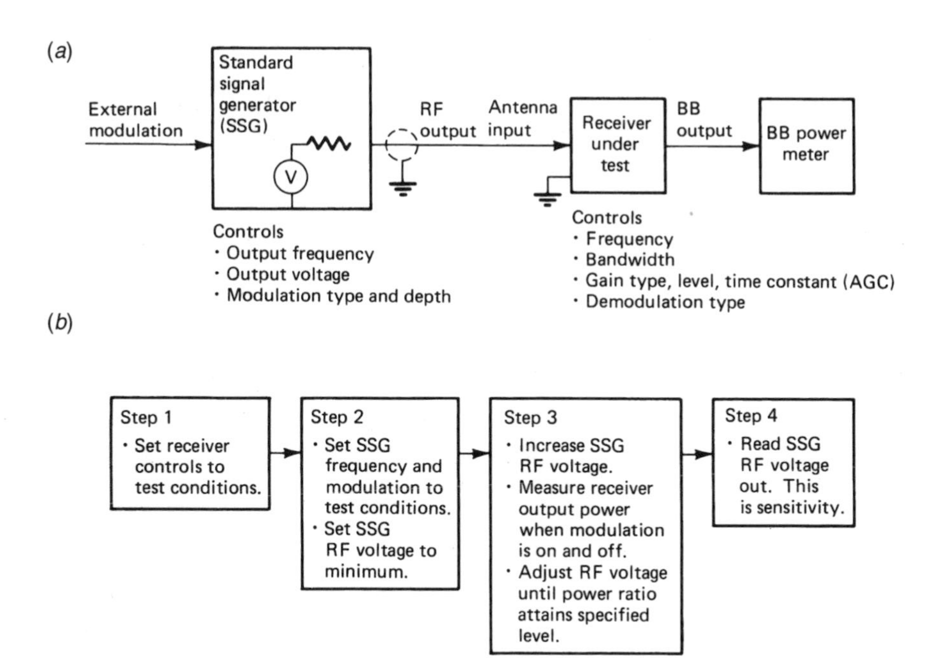

Figure 2.22.1 Receiver sensitivity measurement: (*a*) test setup, (*b*) procedure.

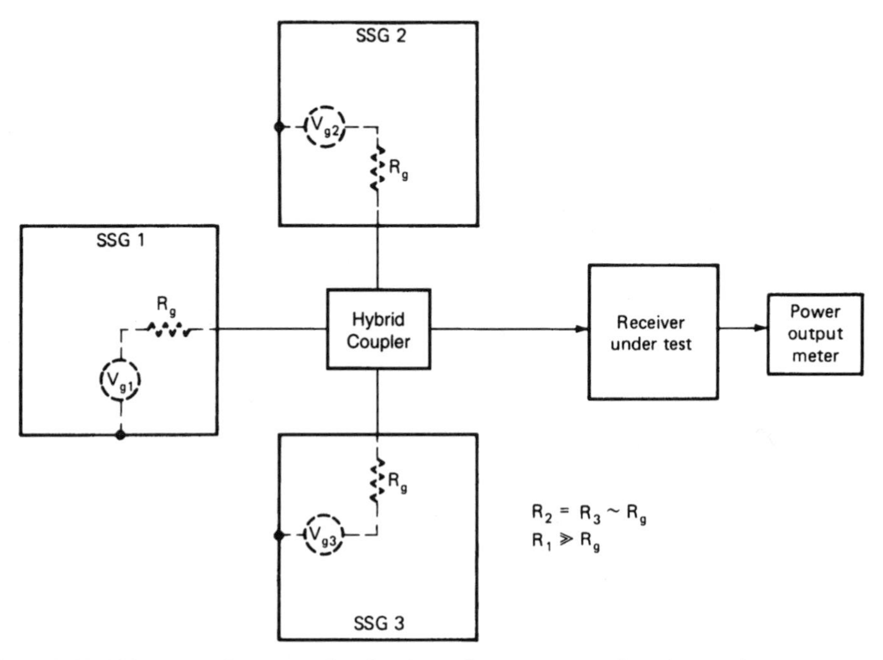

Figure 2.22.2 Test setup for measuring the dynamic range properties of a receiver.

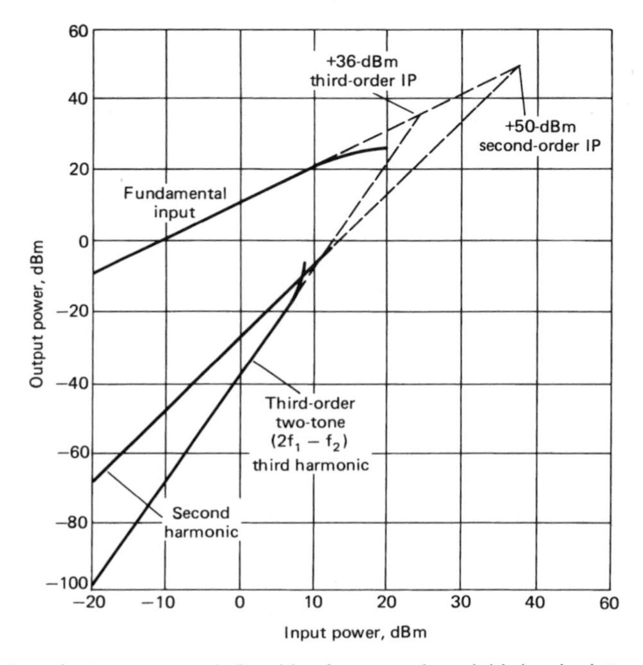

Figure 2.22.3 Input/output power relationships for second- and third-order intercept points.

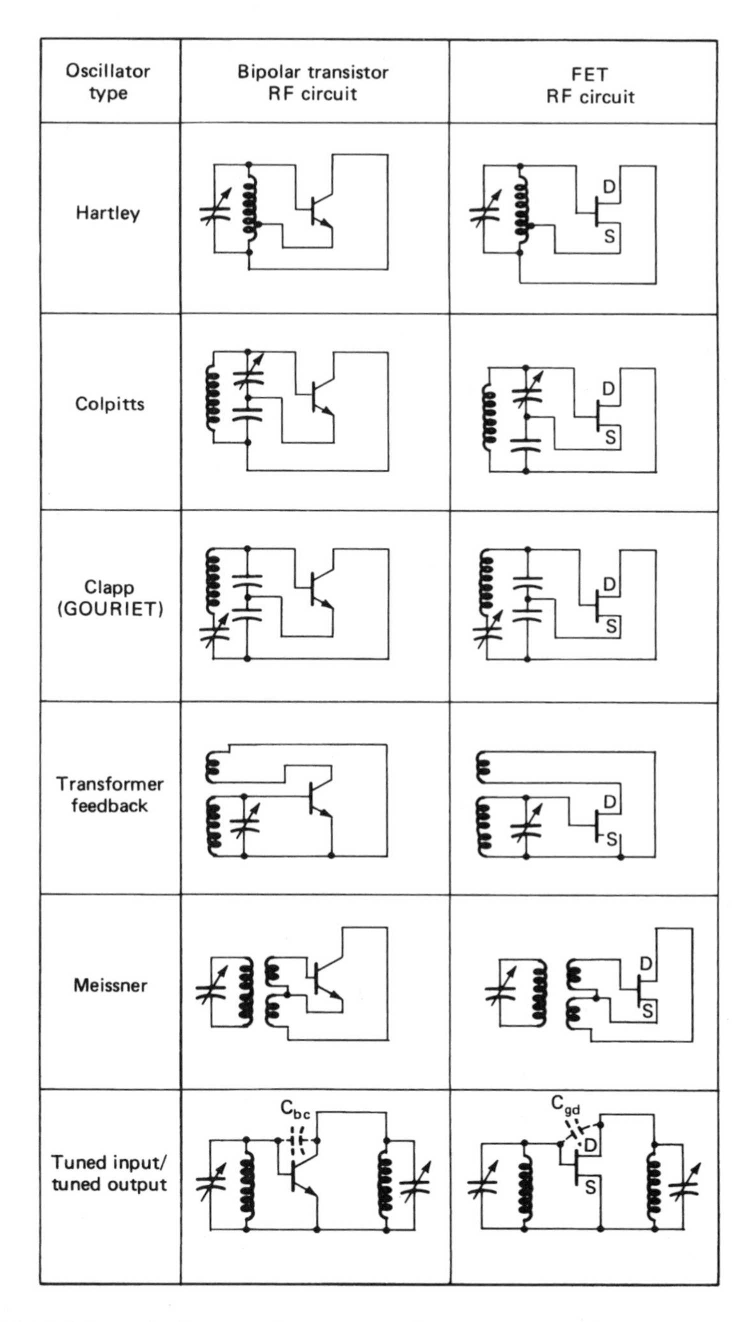

Figure 2.22.4 Schematic diagrams of common oscillator circuits using vacuum-tube, transistor, and FET active circuits. (*From: Rohde, Ulrich L, and Jerry C. Whitaker:* Communications Receivers, *3rd ed., McGraw-Hill, New York, N.Y., 2000. Used with permission.*)

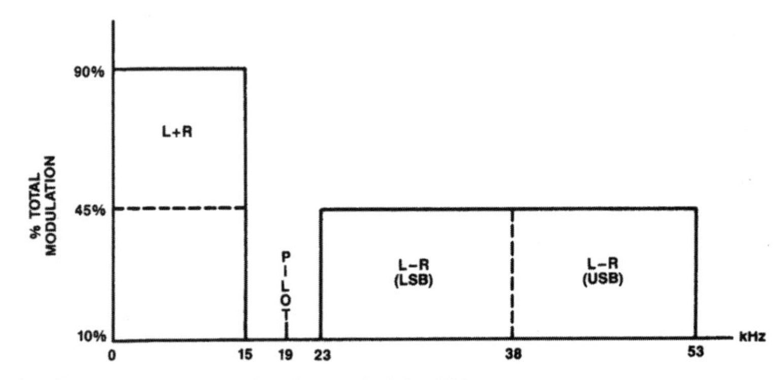

Figure 2.22.5 The composite baseband signal of the FM stereo system.

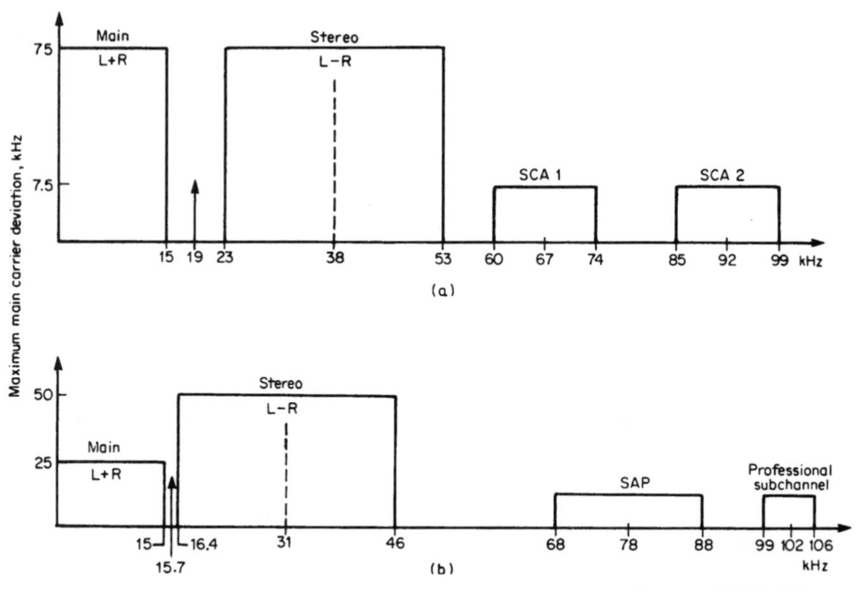

Figure 2.22.6 Baseband frequency allocations: (*a*) stereophonic FM, (*b*) BTSC TV stereophonic system.

Table 2.22.1 Modulation Standards for the TV MTS System

Service or Signal	Modulating Signal	Modulating Frequency Range, kHz	Audio Processing or Preemphasis	Subcarrier Frequency (f_H = 15.734 kHz)	Subcarrier Modulation Type	Subcarrier Deviation, kHz	Aural Carrier Peak Deviation, kHz
Monophonic	L + R	0.05–15	75 μs				25*
Pilot				f_H			5
Stereophonic	L – R	0.05–15	BTSC compression	$2f_H$	AM-DSB SC		50*
Second Audio Program		0.05–10	BTSC compression	$5f_H$	FM	10	15
Professional Channel	Voice	0.3–3.4	150 μs	$6.5f_H$	FM	3	3
	Data	0–1.5	0		FSK		

* Sum does not exceed 50 kHz

Table 2.22.2 Television Service Operating Parameters

Band and Channels	Frequency	City grade		Grade A		Grade B	
		μV/m	μV	μV/m	μV	μV/m	μV
VHF 2–6	54–88 MHz	5,010	7030	2510	3520	224	314
VHF 7–13	174–216 MHz	7,080	3550	3550	1770	631	315
UHF 14–69	470–806 MHz	10,000	1570	5010	787	1580	248
UHF 70–83[1]	806–890 MHz	10,000	1570	5010	571	1580	180

1. Receiver coverage of Channels 70 to 83 has been on a voluntary basis since July 1982. This frequency band was reallocated by the FCC to land mobile use in 1975 with the provision that existing transmitters could continue indefinitely.

Table 2.22.3 Potential VHF Interference Problems

Desired Channel	Interfering Signals	Mechanism
5	Channel. 11 picture	2 × ch. 5 osc. – ch. 11 pix = IF
6	Channel 13 picture	2 × ch. 6 osc. – ch. 13 pix = IF
7 and 8	Channel 5, FM (98–108 MHz)	Ch. 5 pix + FM = ch. 7 and 8
2–6	Channel 5, FM (97–99 MHz)	2 × (FM – ch. 5) = IF
7–13	FM (88–108 MHz)	2 × FM = ch. 7–13
6	FM (89–92 MHz)	Ch. 6 pix + FM – ch. 6 osc. = IF
2	6 m amateur (52–54 MHz)	2 × ch. 2 pix – 6 m = ch. 2
2	CB (27 MHz)	2 × CB = ch. 2
5 and 6	CB (27 MHz)	3 × CB = ch. 5 and 6

Table 2.22.4 Potential UHF Interference Problems

Interference Type	Interfering Channels	Channel 30 Example
IF beat	*N* ± 7, ± 8	22, 23, 37, 38
Intermodulation	*N* ± 2, ± 3, ± 4, ± 5	25–28, 32–35
Adjacent channel	*N* + 1, – 1	29, 31
Local oscillator	*N* ± 7 ×	23, 37
Sound image	*N* + 1/6 (2 × 41.25)	44
Picture image	*N* + 1/6 (2 × 45.75)	45

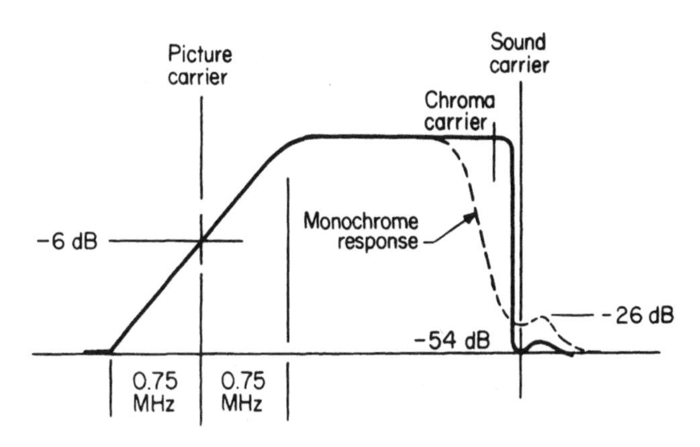

Figure 2.22.7 Ideal IF amplitude response for color and monochrome reception.

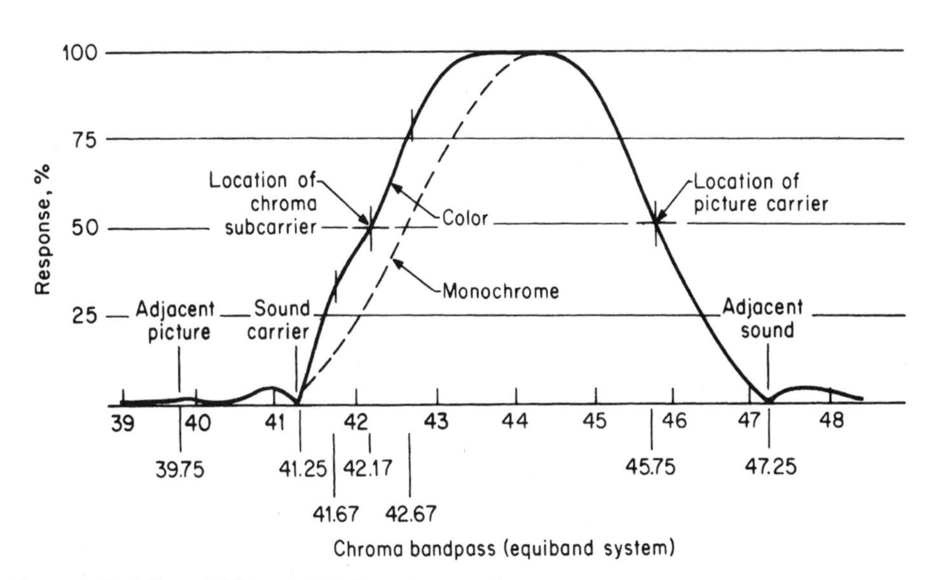

Figure 2.22.8 Overall IF bandwidth for color reception.

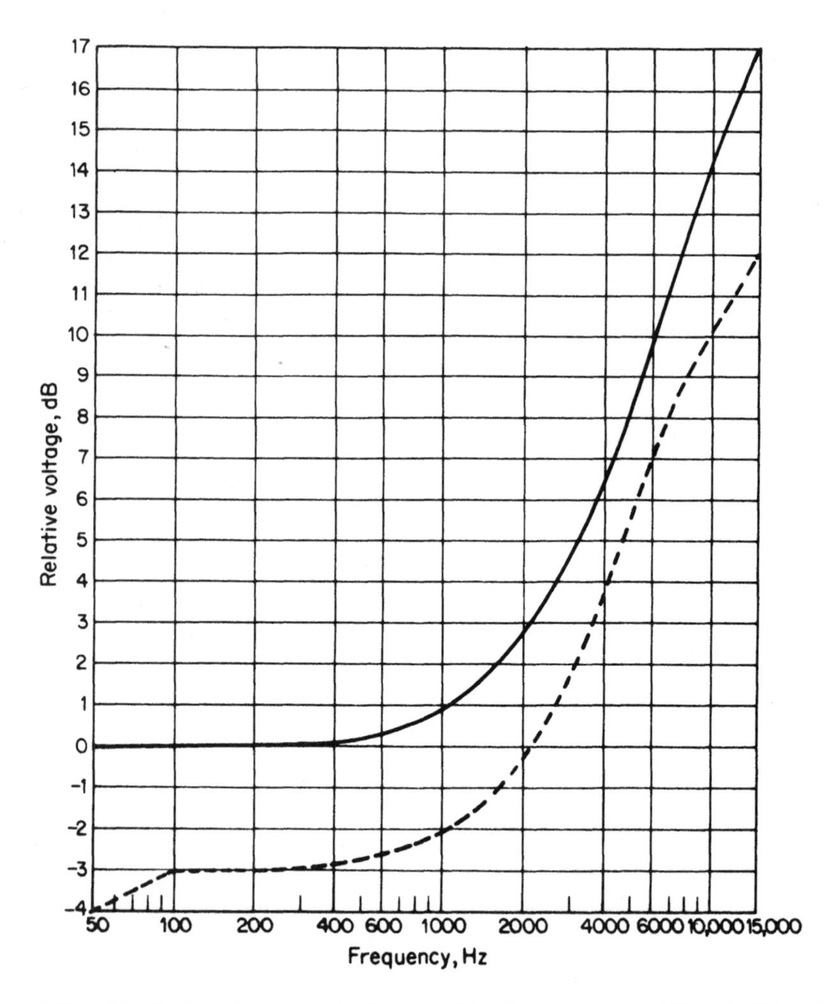

Figure 2.22.9 Standard audio preemphasis curve. The time constant is 75 µs (solid line). The lower frequency response limits are shown by the dashed line. (*Source: FCC.*)

Table 2.22.5 ATSC DTV Receiver Planning Factors Used by the FCC (*After: "Receiver Planning Factors Applicable to All ATV Systems," Final Report of PS/WP3, Advanced Television Systems Committee, Washington, D.C., Dec. 1, 1994.*)

Planning Factors	Low VHF	High VHF	UHF
Antenna impedance (ohms)	75	75	75
Bandwidth (MHz)	6	6	6
Thermal noise (dBm)	–106.2	–106.2	–106.2
Noise figure (dB)	10	10	7
Frequency (MHz)	69	194	615
Antenna factor (dBm/dBμ)	–111.7	–120.7	–130.7[1]
Line loss (dB)	1	2	4
Antenna gain (dB)	4	6	10
Antenna F/B ratio (dB)	10	12	14

1. See Appendix B of the Sixth Report and Order (MM 87-268), adopted April 3, 1997, for a discussion of the dipole factor.

Table 2.22.6 ATSC DTV Interference Criteria (*After: ATSC, "Guide to the Use of the ATSC Digital Television Standard," Advanced Television Systems Committee, Washington, D.C., Doc. A/54, Oct. 4, 1995.)*

Co-channel DTV-into-NTSC	33.8 dB
Co-channel NTSC-into-DTV	2.07 dB
Co-channel DTV-into-DTV	15.91 dB
Upper-adjacent DTV-into-NTSC	–16.17 dB
Upper-adjacent NTSC-into-DTV	–47.05 dB
Upper-adjacent DTV-into-DTV	–42.86 dB
Lower-adjacent DTV-into-NTSC	–17.95 dB
Lower-adjacent NTSC-into-DTV	–48.09 dB
Lower-adjacent DTV-into-DTV	–42.16 dB

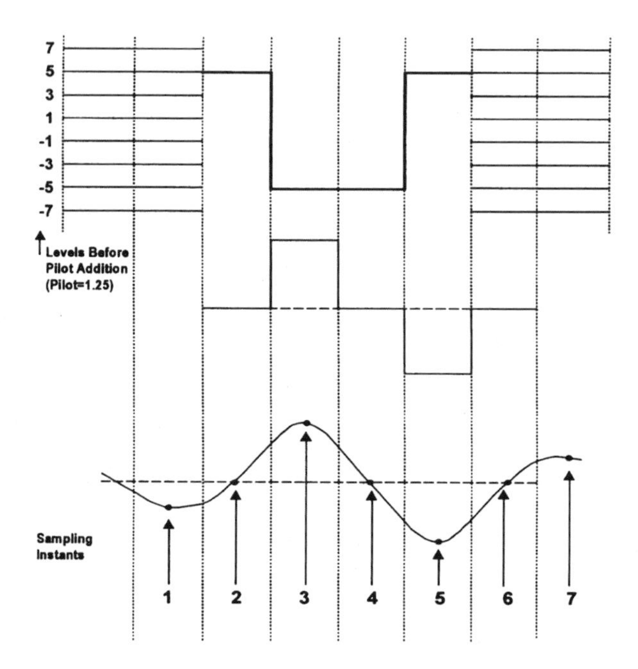

Figure 2.22.10 ATSC DTV data segment sync waveforms. (*From: ATSC, "Guide to the Use of the ATSC Digital Television Standard," Advanced Television Systems Committee, Washington, D.C., Doc. A/54, Oct. 4, 1995. Used with permission.)*

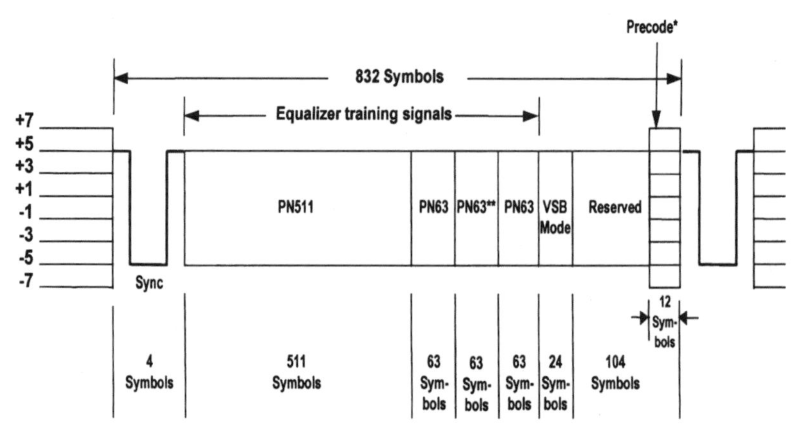

Figure 2.22.11 ATSC DTV data field sync waveform. (*From: ATSC, "Guide to the Use of the ATSC Digital Television Standard," Advanced Television Systems Committee, Washington, D.C., Doc. A/54, Oct. 4, 1995. Used with permission.)*

Table 2.22.7 Target DTV Receiver Parameters (*After: Sgrignoli, Gary: "Preliminary DTV Field Test Results and Their Effects on VSB Receiver Design," ICEE '99.*)

DTV Receiver Parameter	Target Value
Dynamic range: minimum level	< –80 dBm
Dynamic range: maximum level	> 0 dBm
Noise figure	< 10 dB
Synchronization white noise (lock) limit	< 3 dB
AGC speed (10 dB peak/valley fade)	> 75 Hz
White noise threshold of errors	< 15.5 dB
Phase noise threshold of errors	> –76 dBc/Hz @ 20 kHz offset
Gated white noise burst duration	> 185 µs
Co-channel N/D interference @ –45 dBm	< 3 dB, D/U
Co-channel D/D interference @ –45 dBm	< 15.5 dB, D/U
First adjacent N/D interference @ –60 dBm	< –40 dB, D/U
First Adjacent D/D Interference @ –60 dBm	< –28 dB, D/U
Inband CW Interference @ –45 dBm	< 10 dB, D/U
Out-of-band CW interference @ –45 dBm	< –40 dB, D/U
Equalizer length (pre-ghost)	< –3 µs
Equalizer length (post-ghost)	> +22 µs
Quasi-static multipath (1 µs @ < 0.2 Hz)	< +4 dB, D/U
Dynamic multipath: (1 µs @ < 5 Hz)	< +7 dB, D/U
Dynamic multipath: (1 µs @ <10 Hz)	< +10 dB, D/U

Table 2.22.8 Cable Television Channel Frequencies and Designations

Frequency, MHz	Channel	Designation
5–35	Reverse transmission	Sub-band HF
54–88	2–6	Low-band VHF
120–170	14–22	Mid-band VHF
174–216	7–13	High-band VHF
216–300	23–36	Super-band VHF
300–464	37–64	Hyper-band VHF
468–644	65–94	Ultra-band VHF

Table 2.22.9 Distortion-vs.-Number of Amplifiers Cascaded in a Typical Cable System

Number of Amplifiers in Cascade	Cross Modulation, dB	Second Order, dB	S/N, dB
1	–96	–86	60
2	–90	–81.5	57
4	–84	–77	54
8	–78	–725	51
16	–72	–68	48
32	–66	–63.5	45
64	–60	–59	42*
128	–54*	–54.5*	39

* Lower limit of acceptable system performance.

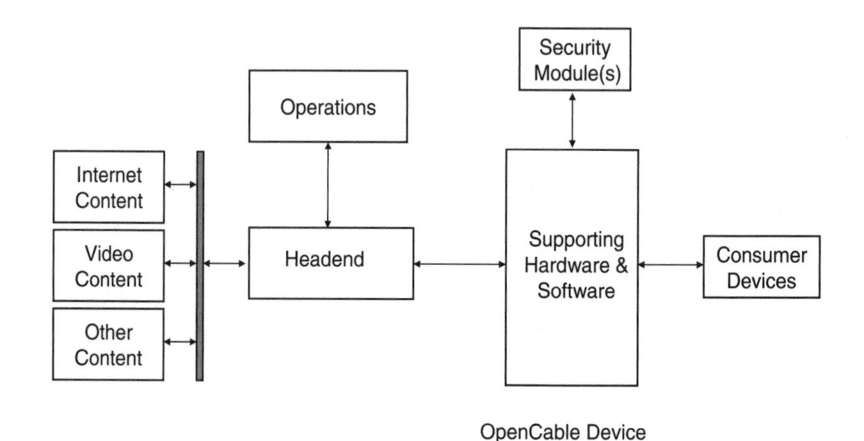

Figure 2.22.12 Basic OpenCable architecture.

Table 2.22.10 Typical Characteristics of Single-Channel Yagi-Uda Arrays

Number of Elements	Gain, dB	Beam Width, degrees
2	3–4	65
3	6–8	55
4	7–10	50
5	9–11	45
9	12–14	37
15	14–16	30

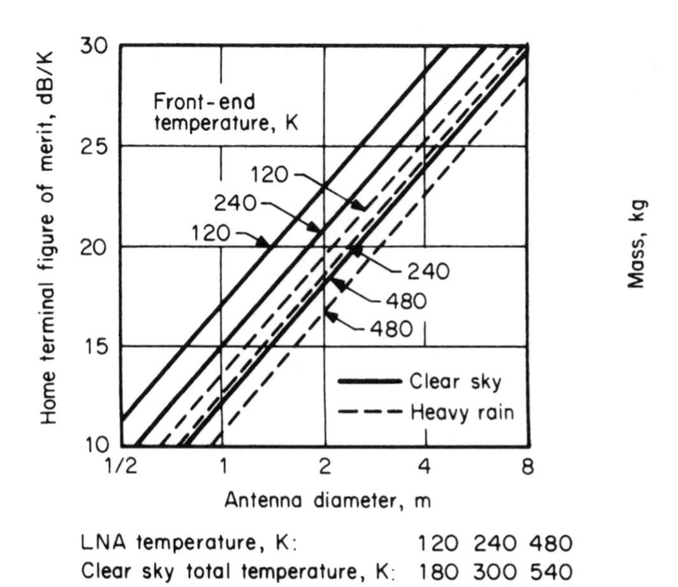

Figure 2.22.13 Receiving station G/T is a function of antenna diameter, receiver temperature, and antenna temperature. In heavy rain, the antenna temperature increases from a clear-weather value of perhaps 50K to a maximum value of 290K (ambient). (*After: Kase, C. A., and W. L. Pritchard: "Getting Set for Direct-Broadcast Satellites,"* IEEE Spectrum*, vol. 18, no. 8, pp. 22–28, 1981.*)

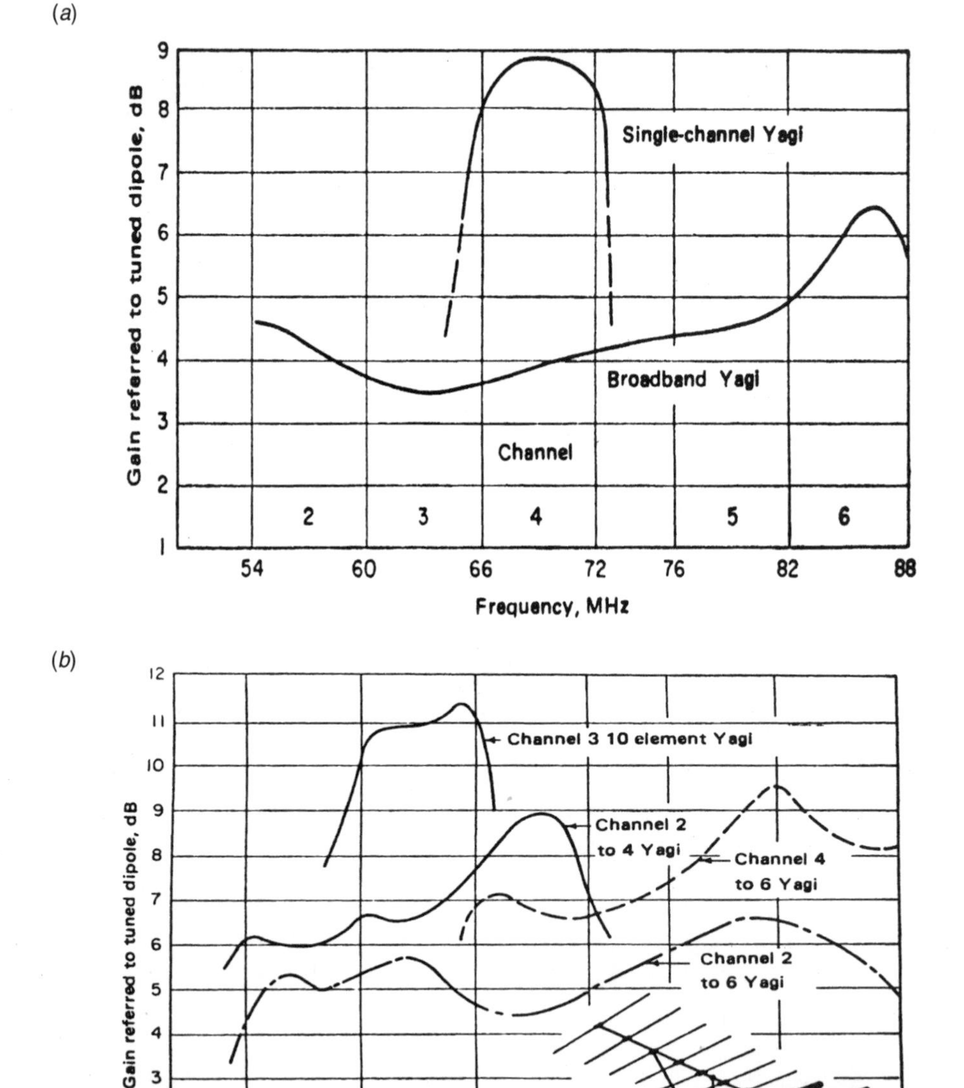

Figure 2.22.13 Gain-vs.-bandwidth for a Yagi antenna: (*a*) measured gain of a five-element single-channel Yagi and a broadband Yagi, (*b*) measured gain of three twin-driven 10-element yagi antennas and a single-channel 10-element Yagi. (*From: Jasik, H.,* Antenna Engineering Handbook, *McGraw-Hill, New York, Chapter 24, 1961. Used with permission.*)

Table 2.22.11 Summary of Transmission Line Equations

Quantity	General line expression†	Ideal line expression†
Propagation constant	$\gamma = \alpha + j\beta = \sqrt{(R + j\omega L)(G + j\omega C)}$	$\gamma = j\omega\sqrt{LC}$
Phase constant β	$\operatorname{Im} \gamma$	$\beta = \omega\sqrt{LC} = 2\pi/\lambda$
Attenuation constant α	$\operatorname{Re} \gamma$	0
Impedance characteristic	$Z_0 = \sqrt{\dfrac{R + j\omega L}{G + j\omega C}}$	$Z_0 = \sqrt{\dfrac{L}{C}}$
Input	$Z_{-l} = Z_0 \dfrac{Z_r + Z_0 \tanh \gamma l}{Z_0 + Z_r \tanh \gamma l}$	$Z_{-1} = Z_0 \dfrac{Z_r + jZ_0 \tan \beta l}{Z_0 + jZ_r \tan \beta l}$
Of short-circuited line, $Z_r = 0$	$Z_{\propto} = Z_0 \tanh \gamma l$	$Z_{\propto} = jZ_0 \tan \beta l$
Of open-circuited line, $Z_r = \infty$	$Z_{\propto} = Z_0 \coth \gamma l$	$Z_{\propto} = -jZ_0 \cot \beta l$
Of line an odd number of quarter wavelengths long	$Z = Z_0 \dfrac{Z_r + Z_0 \coth \alpha l}{Z_0 + Z_r \coth \alpha l}$	$Z = \dfrac{Z_0^2}{Z_r}$
Of line an integral number of half wavelengths long	$Z = Z_0 \dfrac{Z_r + Z_0 \tanh \alpha l}{Z_0 + Z_r \tanh \alpha l}$	$Z = Z_r$
Voltage along line	$V_{-l} = V_i(1 + \Gamma_0 e^{-2\lambda l})$	$V_{-l} = V_i(1 + \Gamma_0 e^{-2+\beta l})$
Current along line	$I_{-l} = I_i(1 - \Gamma_0 e^{-2\gamma l})$	$I_{-l} = I_i(1 - \Gamma_0 e^{-2j\beta l})$
Voltage reflection coefficient	$\Gamma = \dfrac{Z_r - Z_0}{Z_r + Z_0}$	$\Gamma = \dfrac{Z_r - Z_0}{Z_r + Z_0}$

†l = length of transmission line.

Source: From J. Feinstein, "Passive Microwave Components," in D. Fink and D. Christiansen (eds.), *Electronic Engineers Handbook,* McGraw-Hill, New York, 1982, chap. 9.

Table 2.22.12 Important Relations in Low-Loss Transmission Line

Equation	Explanation
$r = \dfrac{1 + \|\Gamma\|}{1 - \|\Gamma\|}$	r = VSWR
$\|\Gamma\| = \dfrac{r - 1}{r + 1}$	$\|\Gamma\|$ = magnitude of reflection coefficient
$\Gamma = \dfrac{R - Z_0}{R + Z_0}$	Γ = reflection coefficient (real) at a point in a line where impedance is real (R)
$r = \dfrac{R}{Z_0}$	$R > Z_0$ (at voltage maximum)
$r = \dfrac{Z_0}{R}$	$R < Z_0$ (at voltage minimum)
$\dfrac{P_t}{P_i} = \|\Gamma\|^2 = \left(\dfrac{r - 1}{r + 1}\right)^2$	P_r = reflected power P_i = incident power
$\dfrac{P_t}{P_i} = 1 - \|\Gamma\|^2 = \dfrac{4r}{(r + 1)^2}$	P_t = transmitted power
$\dfrac{P_b}{P_m} = \dfrac{1}{r}$	P_b = net power transmitted to load at onset of breakdown in a line where VSWR = r exists P_m = same when line is matched, $r = 1$
$\dfrac{\alpha_t}{\alpha_m} = \dfrac{1 + \Gamma^2}{1 - \Gamma^2} = \dfrac{r^2 + 1}{2r}$	α_m = attenuation constant when $r = 1$, matched line α_t = attenuation constant allowing for increased ohmic loss caused by standing waves
$r_{\max} = r_1 r_2$	$r_{\max}$ = maximum VSWR when r_1 and r_2 combine in worst phase
$r_{\min} = \dfrac{r_2}{r_1} r_2 > r_1$	$r_{\min}$ = minimum VSWR when r_1 and r_2 are in best phase
$\|\Gamma\| = \dfrac{\|X\|}{\sqrt{X^2 + 4}}$	Relations for a normalized reactance X in series with resistance Z_0
$\|X\| = \dfrac{r - 1}{\sqrt{r}}$	
$\|\Gamma\| = \dfrac{\|B\|}{\sqrt{B^2 + 4}}$	Relations for a normalized susceptance B in shunt with admittance Y_0
$\|B\| = \dfrac{r - 1}{\sqrt{r}}$	

Source: From J. Feinstein, "Passive Microwave Components," in D. Fink and D. Christiansen (eds.), *Electronic Engineers' Handbook,* McGraw-Hill, New York, 1982, chap. 9.

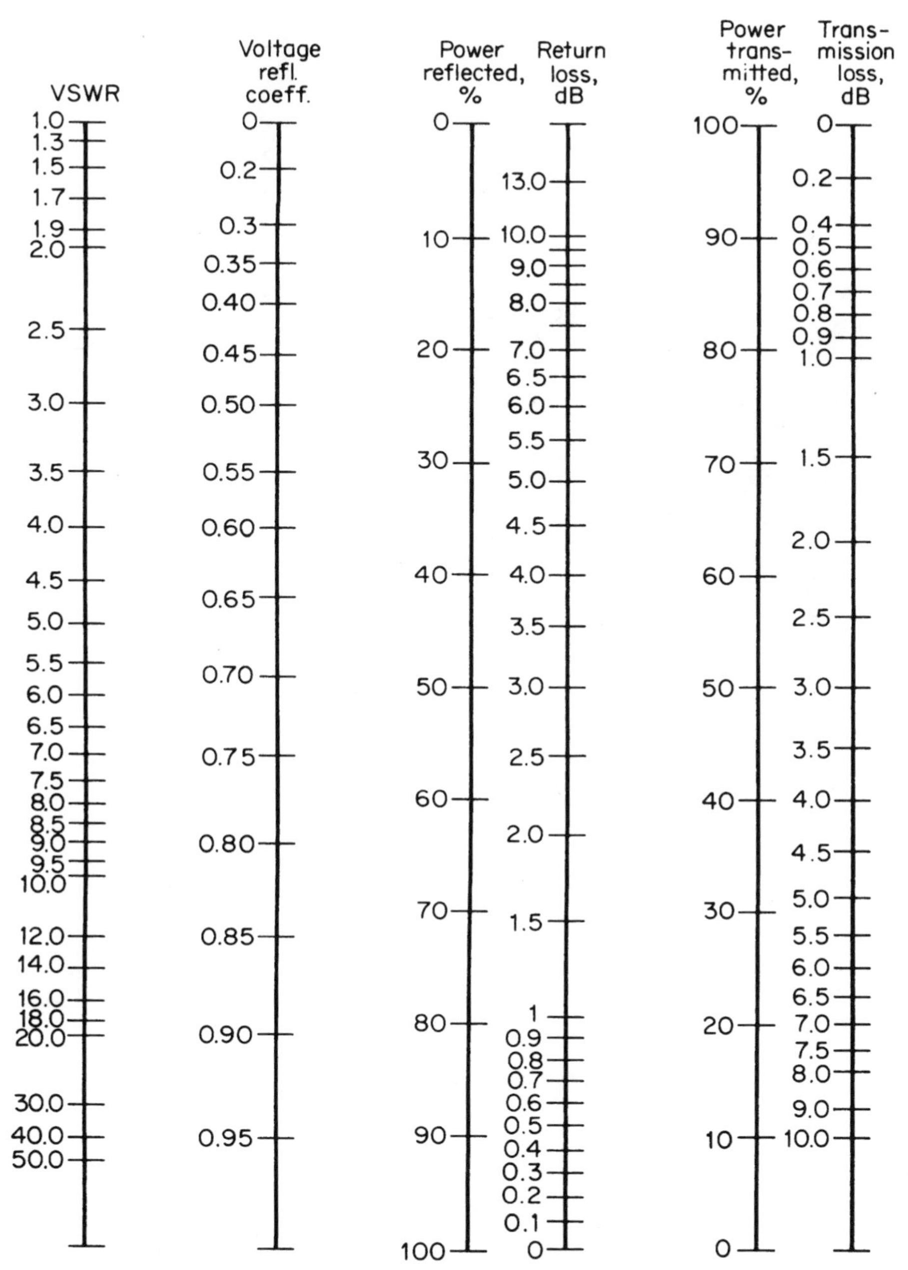

Figure 2.22.14 Nomograph for transmission and deflection of power at high-voltage standing-wave ratios (VSWR). (*From: Feinstein, J.: "Passive Microwave Components,"* Electronic Engineers' Handbook, *D. Fink and D. Christiansen (eds.), Handbook, McGraw-Hill, New York, N.Y., 1982.*)

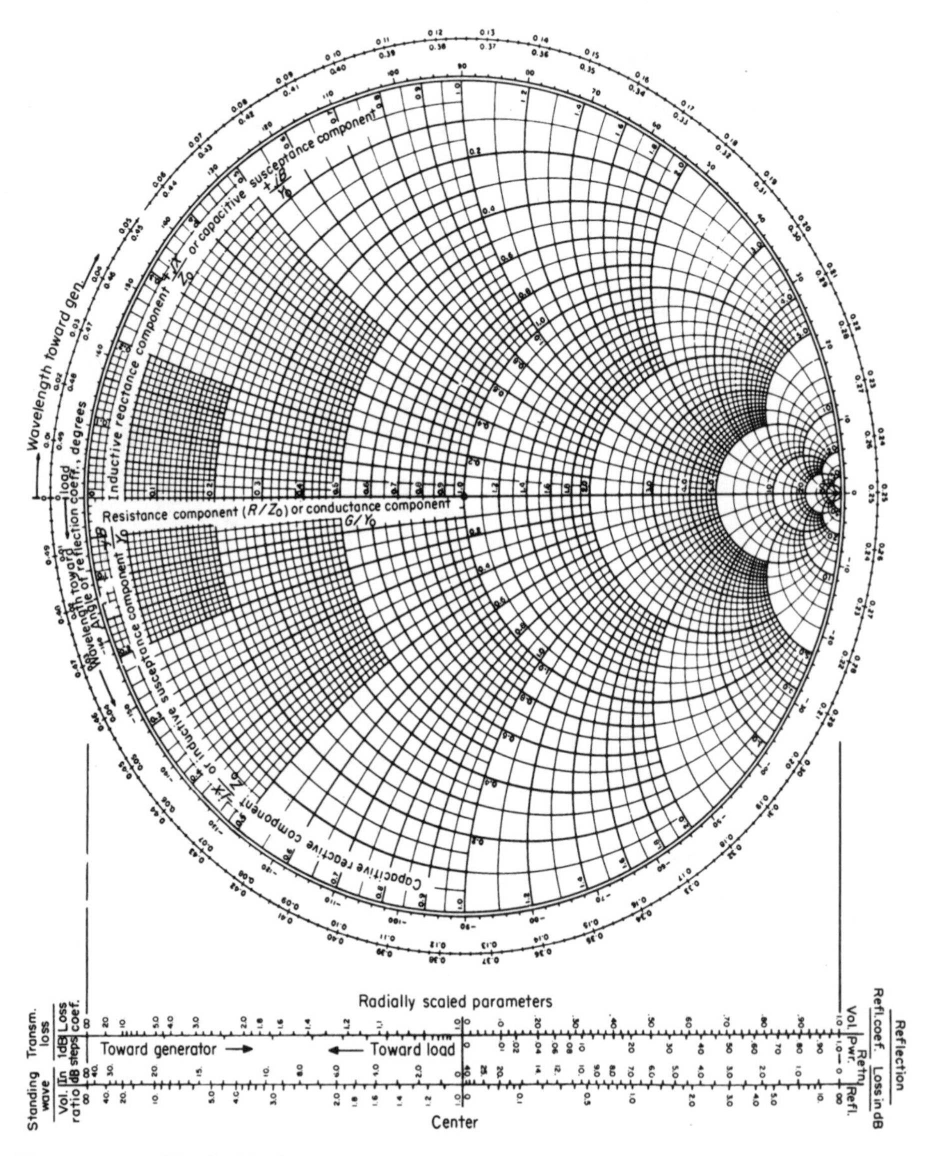

Figure 2.22.15 The Smith chart.

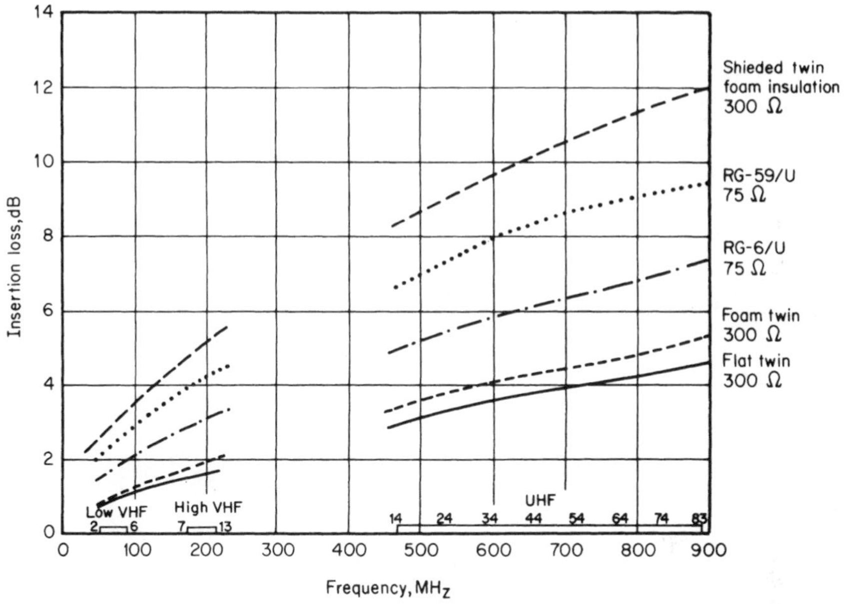

Figure 2.22.16 Insertion loss for common transmission line.

2.23 Audio/Video Signal Measurement and Analysis

Waveform		rms	Avg.	rms avg.	Crest factor
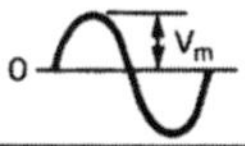	Sine wave	$\frac{V_m}{\sqrt{2}}$ 0.707 V_m	$\frac{2}{\pi}V_m$ 0.637V_m	$\frac{\pi}{2\sqrt{2}} = 1.111$	$\sqrt{2} = 1.414$
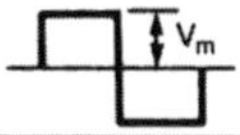	Symmetrical square wave or DC	V_m	V_m	1	1
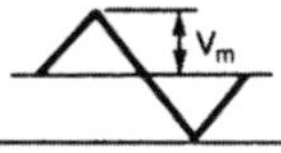	Triangular wave or sawtooth	$\frac{V_m}{\sqrt{3}}$	$\frac{V_m}{2}$	$\frac{2}{\sqrt{3}} = 1.155$	$\sqrt{3} = 1.732$
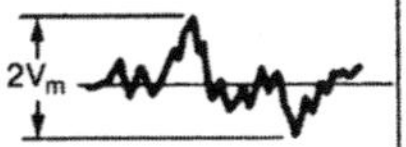	Gaussian noise 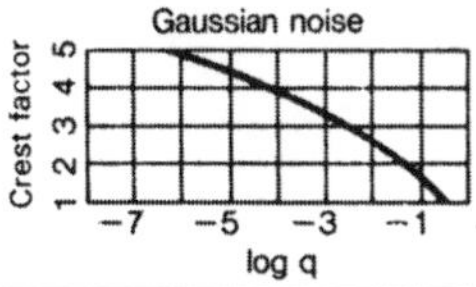	rms	$\frac{\sqrt{2}}{\pi}$ rms = 0.798 rms	$\sqrt{\frac{\pi}{2}}$ 1.253	C.F. / q: 1 / 32%; 2 / 4.6%; 3 / 0.37%
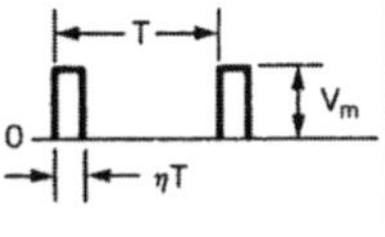	Pulse train (η / Mark/space): 1 / ∞; 0.25 / 0.3333; 0.0625 / 0.0667; 0.01 / 0.0101	$V_m\sqrt{\eta}$ V_m 0.5V_m 0.25V_m 0.1V_m	$V_m\eta$ V_m 0.25V_m 0.625V_m 0.01V_m	$\frac{1}{\sqrt{\eta}}$ 1 2 4 8 10	$\frac{1}{\sqrt{\eta}}$ 1 2 4 8 10

Figure 2.23.1 Comparison of rms and average measurement characteristics. (*After: EDN, January 20, 1982.*)

Table 2.23.1 Video Quality Definitions (*After: Fibush, David K.: "Picture Quality Measurements for Digital Television,"* Proceedings of the Digital Television '97 Summit, *Intertec Publishing, Overland Park, Kan., December 1997.*)

Parameter	In-Service	Out-of-Service
Indirect measurement		
Objective signal quality	Vertical interval test signals	Full-field test signals
Direct measurement		
Subjective picture quality	Program material	Test scenes
Objective picture quality	Program material	Test scenes

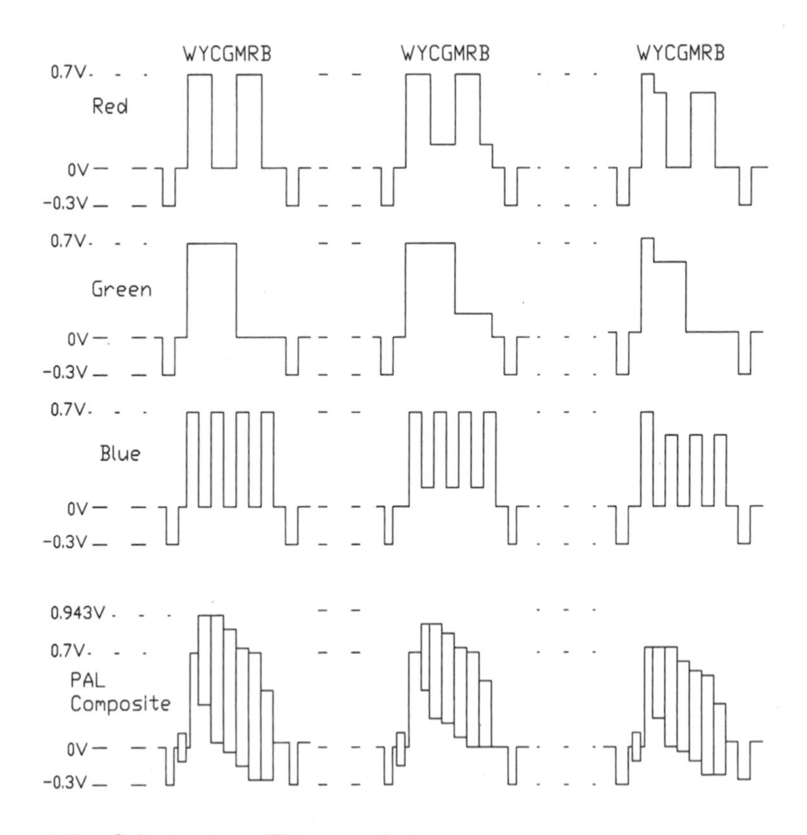

Figure 2.23.2 Color test bars, ITU nomenclature.

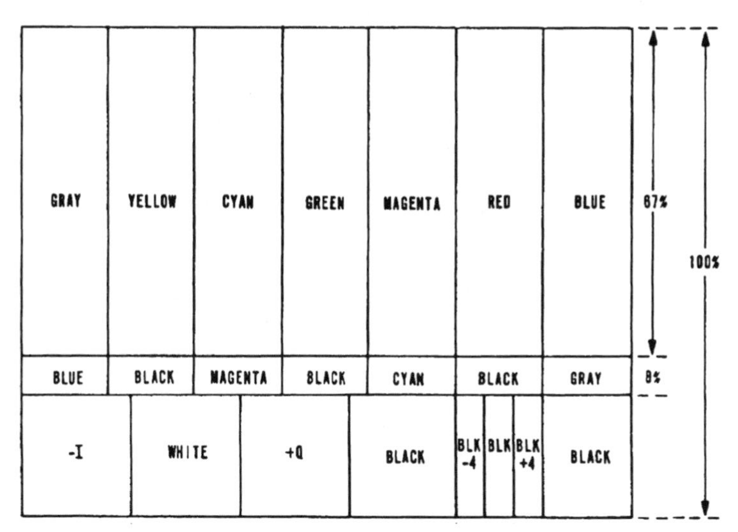

Figure 2.23.3 SMPTE EG 1-1990 color bar test signal.

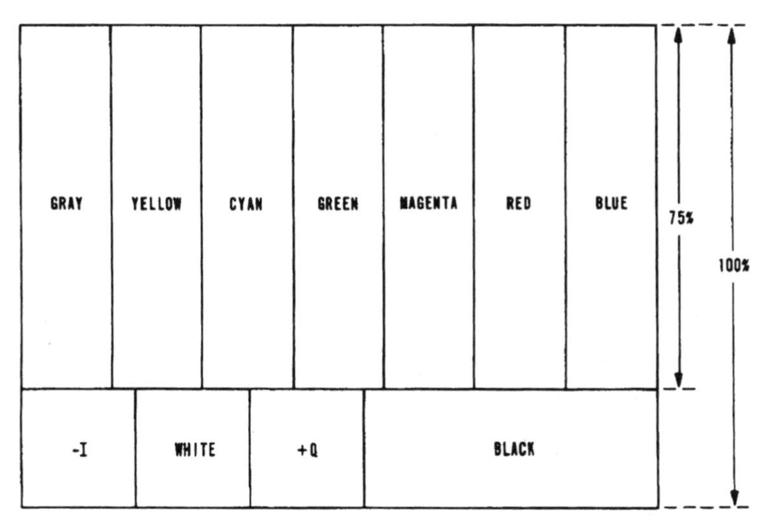

Figure 2.23.4 EIA RS-189A color bar test signal.

(*a*)

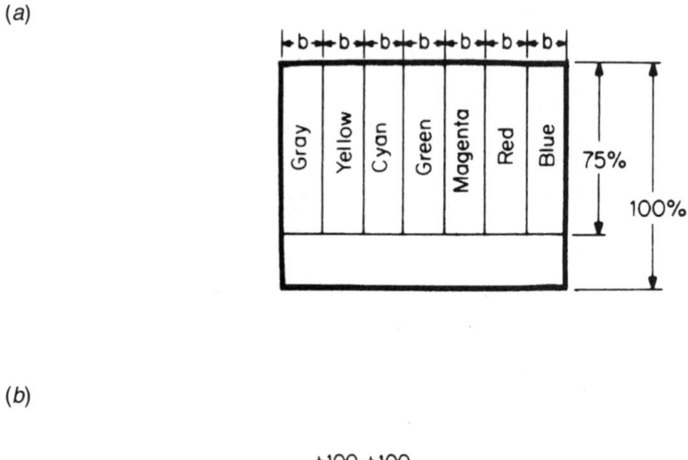

(*b*)

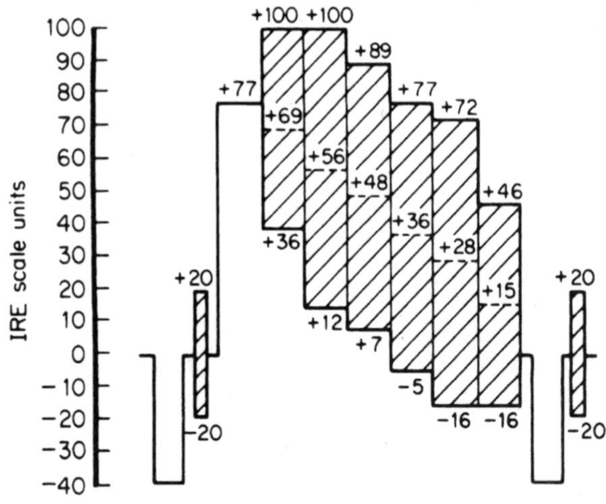

Figure 2.23.5 EIA RS-189-A color-bar displays: (*a*) color displays of gray and color bars, (*b*) waveform display of reference gray and primary/complementary colors, plus sync and burst.

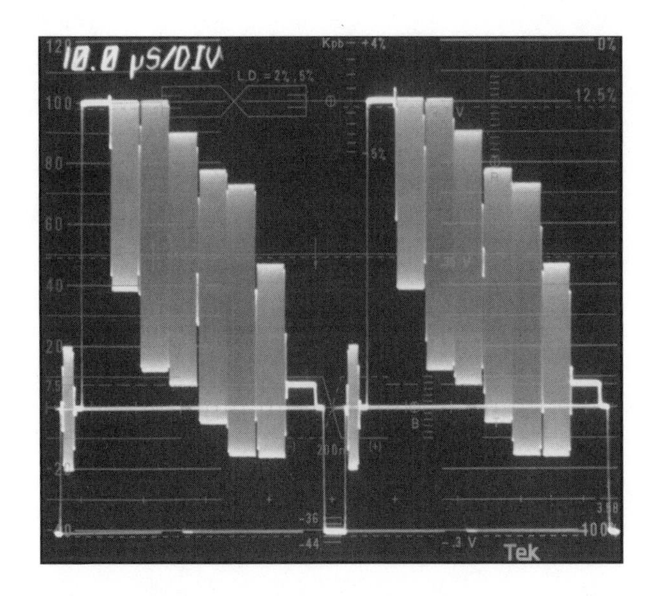

Figure 2.23.6 Waveform monitor display of a color-bar signal at the two-line rate. (*Courtesy of Tektronix.*)

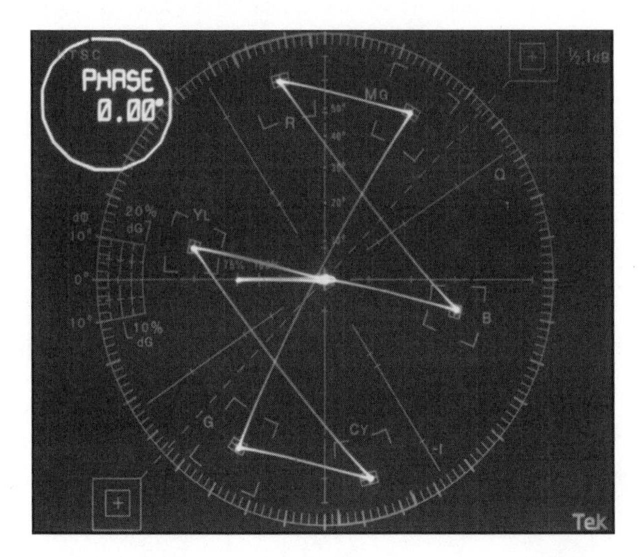

Figure 2.23.7 Vectorscope display of a color-bar signal. (*Courtesy of Tektronix.*)

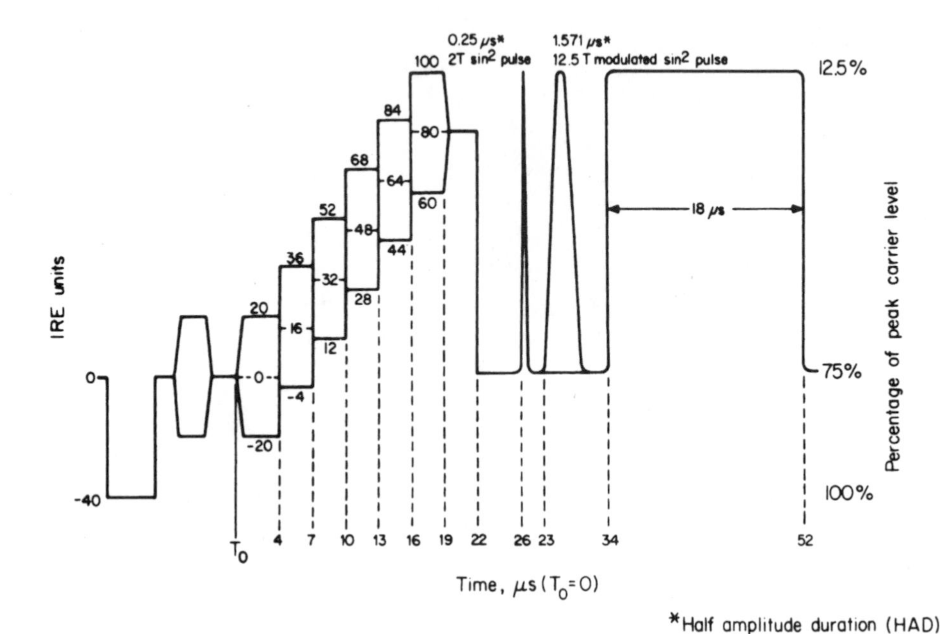

Figure 2.23.8 Composite vertical-interval test signal (VITS) inserted in field 1, line 18. The video level in IRE units is shown on the left; the radiated carrier signal is shown on the right.

Table 2.23.2 Error Frequency and Bit Error Rates (*After: Fibush, David K.: "Error Detection in Serial Digital Systems,"* NAB Broadcast Engineering Conference Proceedings, *National Association of Broadcasters, Washington, D.C., pp. 346-354, 1993.*)

Time Between	NTSC 143 Mbits/s	PAL 177 Mbits/s	Component 270 Mbits/s
1 television frame	2×10^{-7}	2×10^{-7}	1×10^{-7}
1 second	7×10^{-9}	6×10^{-9}	4×10^{-9}
1 minute	1×10^{-10}	9×10^{-11}	6×10^{-11}
1 hour	2×10^{-12}	2×10^{-12}	1×10^{-12}
1 day	8×10^{-14}	7×10^{-14}	4×10^{-14}
1 week	1×10^{-14}	9×10^{-15}	6×10^{-15}
1 month	3×10^{-15}	2×10^{-15}	1×10^{-15}
1 year	2×10^{-16}	2×10^{-16}	1×10^{-16}
1 decade	2×10^{-17}	2×10^{-17}	1×10^{-17}
1 century	2×10^{-18}	2×10^{-18}	1×10^{-18}

Table 2.23.3 Error Rate as a Function of S/N for Composite Serial Digital (*After: Fibush, David K.: "Error Detection in Serial Digital Systems,"* NAB Broadcast Engineering Conference Proceedings*, National Association of Broadcasters, Washington, D.C., pp. 346-354, 1993.*)

Time Between Errors	BER	SNR (dB)	S/N (volts ratio)
1 microsecond	7×10^{-3}	10.8	12
1 millisecond	7×10^{-6}	15.8	38
1 television frame	2×10^{-7}	17.1	51
1 second	7×10^{-9}	18.1	64
1 minute	1×10^{-10}	19.0	80
1 day	8×10^{-14}	20.4	109
1 month	3×10^{-15}	20.9	122
1 century	2×10^{-18}	21.8	150

Table 2.23.4 Error Rate as a Function of Cable Length Using 8281 Coax for Composite Serial Digital (*After: Fibush, David K.: "Error Detection in Serial Digital Systems,"* NAB Broadcast Engineering Conference Proceedings, *National Association of Broadcasters, Washington, D.C., pp. 346-354, 1993.*)

Time Between Errors	BER	Cable Length (meters)	Attenuation (dB) at 1/2 Clock Frequency
1 microsecond	7×10^{-3}	484	36.3
1 millisecond	7×10^{-6}	418	31.3
1 television frame	2×10^{-7}	400	30.0
1 second	7×10^{-9}	387	29.0
1 minute	1×10^{-10}	374	28.1
1 day	8×10^{-14}	356	26.7
1 month	3×10^{-15}	350	26.2
1 century	2×10^{-18}	338	25.3

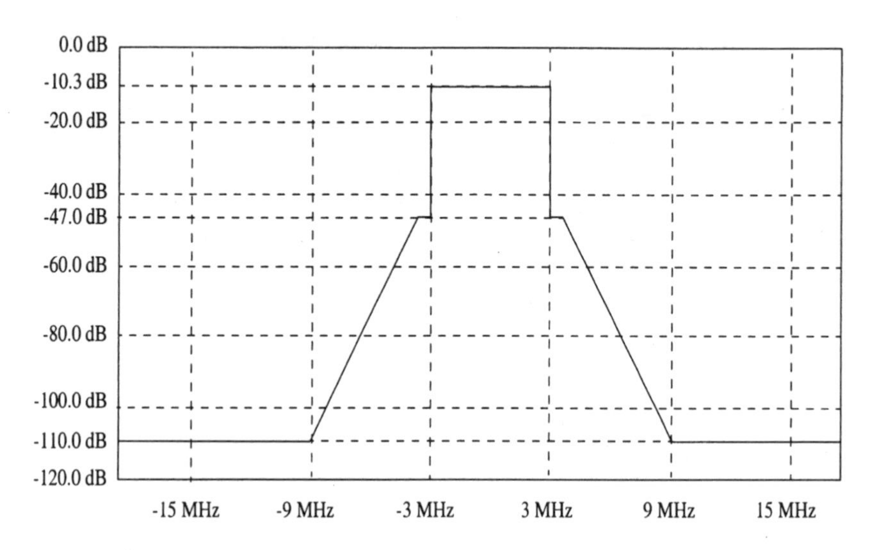

Figure 2.23.9 RF spectrum mask limits for DTV transmission. The mask is a contour that illustrates the maximum levels of out-of-band radiation from a transmitted signal permitted by the FCC. (*After:* DTV Express Training Manual on Terrestrial DTV Broadcasting*, Harris Corporation, Quincy, Ill., September 1998. Courtesy of Harris.*)

2.24 Standards and Practices

Table 2.24.1 The Effects of Current on the Human Body

Current	Effect
1 mA or less	No sensation, not felt
More than 3 mA	Painful shock
More than 10 mA	Local muscle contractions, sufficient to cause "freezing" to the circuit for 2.5 percent of the population
More than 15 mA	Local muscle contractions, sufficient to cause "freezing" to the circuit for 50 percent of the population
More than 30 mA	Breathing is difficult, can cause unconsciousness
50 mA to 100 mA	Possible ventricular fibrillation
100 mA to 200 mA	Certain ventricular fibrillation
More than 200 mA	Severe burns and muscular contractions; heart more apt to stop than to go into fibrillation
More than a few amperes	Irreparable damage to body tissue

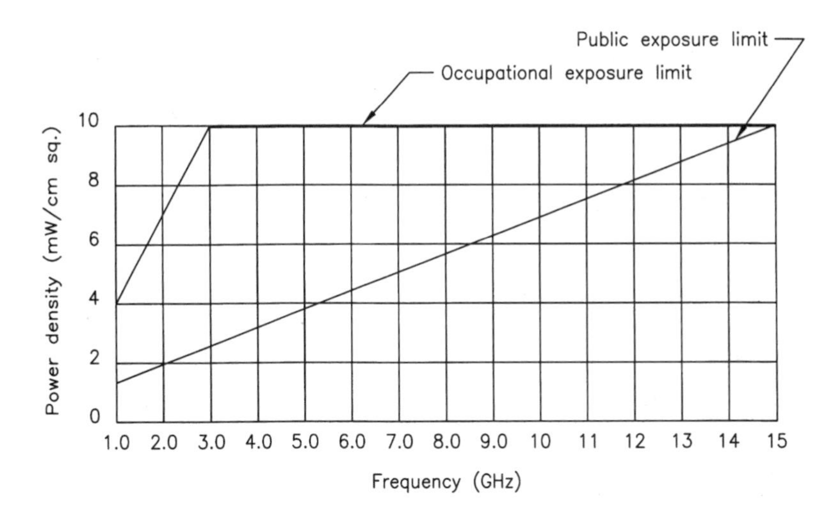

Figure 2.24.1 The power density limits for nonionizing radiation exposure for humans

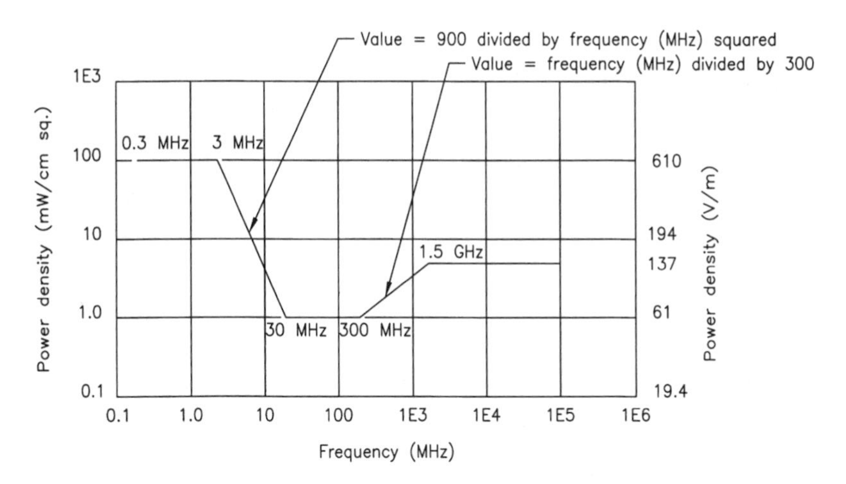

Figure 2.24.2 ANSI/IEEE exposure guidelines for microwave frequencies.

Table 2.24.2 Definition of PCB Terms as Identified by the EPA

Term	Definition	Examples
PCB	Any chemical substance that is limited to the biphenyl molecule that has been chlorinated to varying degrees, or any combination of substances that contain such substances.	PCB dielectric fluids, PCB heat-transfer fluids, PCB hydraulic fluids, 2,2',4-trichlorobiphenyl
PCB article	Any manufactured article, other than a PCB container, that contains PCBs and whose surface has been in direct contact with PCBs.	Capacitors, transformers, electric motors, pumps, pipes
PCB container	A device used to contain PCBs or PCB articles, and whose surface has been in direct contact with PCBs.	Packages, cans, bottles, bags, barrels, drums, tanks
PCB article container	A device used to contain PCB articles or equipment, and whose surface has not been in direct contact with PCBs.	Packages, cans, bottles, bags, barrels, drums, tanks
PCB equipment	Any manufactured item, other than a PCB container or PCB article container, that contains a PCB article or other PCB equipment.	Microwave systems, fluorescent light ballasts, electronic equipment
PCB item	Any PCB article, PCB article container, PCB container, or PCB equipment that deliberately or unintentionally contains, or has as a part of it, any PCBs.	
PCB transformer	Any transformer that contains PCBs in concentrations of 500 ppm or greater.	
PCB contaminated	Any electric equipment that contains more than 50, but less than 500 ppm of PCBs. (Oil-filled electric equipment other than circuit breakers, reclosers, and cable whose PCB concentration is unknown must be assumed to be PCB-contaminated electric equipment.)	Transformers, capacitors, contaminated circuit breakers, reclosers, voltage regulators, switches, cable, electromagnets

Table 2.24.3 Common Trade Names for PCB Insulating Material

Apirolio	Abestol	Askarel	Aroclor B	Chlorexto	Chlophen
Chlorinol	Clorphon	Diaclor	DK	Dykanol	EEC-18
Elemex	Eucarel	Fenclor	Hyvol	Inclor	Inerteen
Kanechlor	No-Flamol	Phenodlor	Pydraul	Pyralene	Pyranol
Pyroclor	Sal-T-Kuhl	Santothern FR	Santovac	Solvol	Thermin

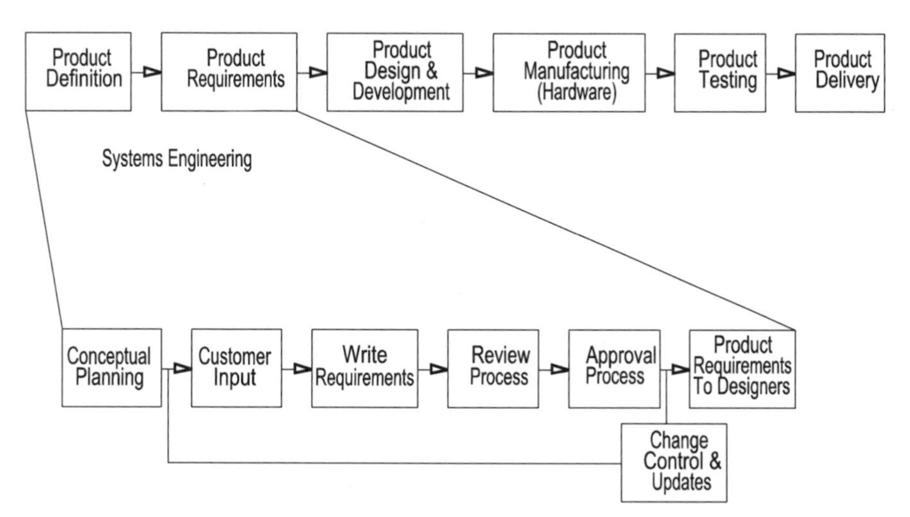

Figure 2.24.3 The product development and documentation process.

Chapter

3

Standard Units and Conversion Ratios

3.1 Standard Electrical Units

Name	Symbol	Quantity
ampere	A	electric current
ampere per meter	A/m	magnetic field strength
ampere per square meter	A/m^2	current density
becquerel	Bg	activity (of a radionuclide)
candela	cd	luminous intensity
coulomb	C	electric charge
coulomb per kilogram	C/kg	exposure (x and gamma rays)
coulomb per sq. meter	C/m^2	electric flux density
cubic meter	m^3	volume
cubic meter per kilogram	m^3/kg	specific volume
degree Celsius	°C	Celsius temperature
farad	F	capacitance
farad per meter	F/m	permittivity
henry	H	inductance
henry per meter	H/m	permeability
hertz	Hz	frequency
joule	J	energy, work, quantity of heat
joule per cubic meter	J/m^3	energy density
joule per kelvin	J/K	heat capacity
joule per kilogram K	J/(kg•K)	specific heat capacity
joule per mole	J/mol	molar energy

kelvin	K	thermodynamic temperature
kilogram	kg	mass
kilogram per cubic meter	kg/m^3	density, mass density
lumen	lm	luminous flux
lux	lx	luminance
meter	m	length
meter per second	m/s	speed, velocity
meter per second sq.	m/s^2	acceleration
mole	mol	amount of substance
newton	N	force
newton per meter	N/m	surface tension
ohm	Ω	electrical resistance
pascal	Pa	pressure, stress
pascal second	Pa•s	dynamic viscosity
radian	rad	plane angle
radian per second	rad/s	angular velocity
radian per second squared	rad/s^2	angular acceleration
second	s	time
siemens	S	electrical conductance
square meter	m^2	area
steradian	sr	solid angle
tesla	T	magnetic flux density
volt	V	electrical potential
volt per meter	V/m	electric field strength
watt	W	power, radiant flux
watt per meter kelvin	W/(m•K)	thermal conductivity
watt per square meter	W/m^2	heat (power) flux density
weber	Wb	magnetic flux

3.2 Standard Prefixes

Multiple	Prefix	Symbol
10^{18}	exa	E
10^{15}	peta	P
10^{12}	tera	T
10^{9}	giga	G
10^{6}	mega	M
10^{3}	kilo	k
10^{2}	hecto	h
10	deka	da
10^{-1}	deci	d
10^{-2}	centi	c
10^{-3}	milli	m
10^{-6}	micro	μ
10^{-9}	nano	n
10^{-12}	pico	p
10^{-15}	femto	f
10^{-18}	atto	a

3.3 Common Standard Units

Unit	Symbol
centimeter	cm
cubic centimeter	cm^3
cubic meter per sec.	m^3/s
gigahertz	GHz
gram	g
kilohertz	kHz
kilohm	kΩ
kilojoule	kJ
kilometer	km
kilovolt	kV

kilovoltampere	kVA
kilowatt	kW
megahertz	MHz
megavolt	MV
megawatt	MW
megohm	MΩ
microampere	μA
microfarad	μF
microgram	μg
microhenry	μH
microsecond	μs
microwatt	μW
milliampere	mA
milligram	mg
millihenry	mH
millimeter	mm
millisecond	ms
millivolt	mV
milliwatt	mW
nanoampere	nA
nanofarad	nF
nanometer	nm
nanosecond	ns
nanowatt	nW
picoampere	pA
picofarad	pF
picosecond	ps
picowatt	pW

3.4 Conversion Reference Data

To Convert	Into	Multiply By
abcoulomb	statcoulombs	2.998×10^{10}
acre	sq. chain (Gunters)	10
acre	rods	160
acre	square links (Gunters)	1×10^{5}

acre	Hectare or sq. hectometer	0.4047
acre-feet	cubic feet	43,560.0
acre-feet	gallons	3.259×10^5
acres	sq. feet	43,560.0
acres	sq. meters	4,047
acres	sq. miles	1.562×10^{-3}
acres	sq. yards	4,840
ampere-hours	coulombs	3,600.0
ampere-hours	faradays	0.03731
amperes/sq. cm	amps/sq. in	6.452
amperes/sq. cm	amps/sq. meter	10^4
amperes/sq. in	amps/sq. cm	0.1550
amperes/sq. in	amps/sq. meter	1,550.0
amperes/sq. meter	amps/sq. cm	10^{-4}
amperes/sq. meter	amps/sq. in	6.452×10^{-4}
ampere-turns	gilberts	1.257
ampere-turns/cm	amp-turns/in	2.540
ampere-turns/cm	amp-turns/meter	100.0
ampere-turns/cm	gilberts/cm	1.257
ampere-turns/in	amp-turns/cm	0.3937
ampere-turns/in	amp-turns/m	39.37
ampere-turns/in	gilberts/cm	0.4950
ampere-turns/meter	amp-turns/cm	0.01
ampere-turns/meter	amp-turns/in	0.0254
ampere-turns/meter	gilberts/cm	0.01257
Angstrom unit	inch	3937×10^{-9}
Angstrom unit	meter	1×10^{-10}
Angstrom unit	micron or (Mu)	1×10^{-4}
are	acre (U.S.)	0.02471
ares	sq. yards	119.60
ares	acres	0.02471
ares	sq. meters	100.0
astronomical unit	kilometers	1.495×10^8
atmospheres	ton/sq. in	0.007348
atmospheres	cm of mercury	76.0
atmospheres	ft of water (at 4°C)	33.90

atmospheres	in of mercury (at 0°C)	29.92
atmospheres	kg/sq. cm	1.0333
atmospheres	kg/sq. m	10,332
atmospheres	pounds/sq. in	14.70
atmospheres	tons/sq. ft	1.058
barrels (U.S., dry)	cubic inches	7056
barrels (U.S., dry)	quarts (dry)	105.0
barrels (U.S., liquid)	gallons	31.5
barrels (oil)	gallons (oil)	42.0
bars	atmospheres	0.9869
bars	dynes/sq. cm	10^4
bars	kg/sq. m	1.020×10^4
bars	pounds/sq. ft	2,089
bars	pounds/sq. in	14.50
baryl	dyne/sq. cm	1.000
bolt (U.S. cloth)	meters	36.576
Btu	liter-atmosphere	10.409
Btu	ergs	1.0550×10^{10}
Btu	foot-lb	778.3
Btu	gram-calories	252.0
Btu	horsepower-hr	3.931×10^{-4}
Btu	joules	1,054.8
Btu	kilogram-calories	0.2520
Btu	kilogram-meters	107.5
Btu	kilowatt-hr	2.928×10^{-4}
Btu/hr	foot-pounds/s	0.2162
Btu/hr	gram-calories/s	0.0700
Btu/hr	horsepower-hr	3.929×10^{-4}
Btu/hr	watts	0.2931
Btu/min	foot-lbs/s	12.96
Btu/min	horsepower	0.02356
Btu/min	kilowatts	0.01757
Btu/min	watts	17.57
Btu/sq. ft/min	watts/sq. in	0.1221
bucket (br. dry)	cubic cm	1.818×10^4
bushels	cubic ft	1.2445
bushels	cubic in	2,150.4

bushels	cubic m	0.03524
bushels	liters	35.24
bushels	pecks	4.0
bushels	pints (dry)	64.0
bushels	quarts (dry)	32.0
calories, gram (mean)	Btu (mean)	3.9685×10^{-3}
candle/sq. cm	Lamberts	3.142
candle/sq. in	Lamberts	0.4870
centares (centiares)	sq. meters	1.0
Centigrade	Fahrenheit	(C° x 9/5) + 32
centigrams	grams	0.01
centiliter	ounce fluid (U.S.)	0.3382
centiliter	cubic inch	0.6103
centiliter	drams	2.705
centiliter	liters	0.01
centimeter	feet	3.281×10^{-2}
centimeter	inches	0.3937
centimeter	kilometers	10^{-5}
centimeter	meters	0.01
centimeter	miles	6.214×10^{-6}
centimeter	millimeters	10.0
centimeter	mils	393.7
centimeter	yards	1.094×10^{-2}
centimeter-dynes	cm-grams	1.020×10^{-3}
centimeter-dynes	meter-kg	1.020×10^{-8}
centimeter-dynes	pound-ft	7.376×10^{-8}
centimeter-grams	cm-dynes	980.7
centimeter-grams	meter-kg	10^{-5}
centimeter-grams	pound-ft	7.233×10^{-5}
centimeters of mercury	atmospheres	0.01316
centimeters of mercury	feet of water	0.4461
centimeters of mercury	kg/sq. meter	136.0
centimeters of mercury	pounds/sq. ft	27.85
centimeters of mercury	pounds/sq. in	0.1934
centimeters/sec	feet/min	1.9686
centimeters/sec	feet/sec	0.03281

centimeters/sec	kilometers/hr	0.036
centimeters/sec	knots	0.1943
centimeters/sec	meters/min	0.6
centimeters/sec	miles/hr	0.02237
centimeters/sec	miles/min	3.728×10^{-4}
centimeters/sec/sec	feet/sec/sec	0.03281
centimeters/sec/sec	km/hr/sec	0.036
centimeters/sec/sec	meters/sec/sec	0.01
centimeters/sec/sec	miles/hr/sec	0.02237
chain	inches	792.00
chain	meters	20.12
chains (surveyor's or Gunter's)	yards	22.00
circular mils	sq. cm	5.067×10^{-6}
circular mils	sq. mils	0.7854
circular mils	sq. inches	7.854×10^{-7}
circumference	Radians	6.283
cord feet	cubic feet	16
cords	cord feet	8
coulomb	statcoulombs	2.998×10^{9}
coulombs	faradays	1.036×10^{-5}
coulombs/sq. cm	coulombs/sq. in	64.52
coulombs/sq. cm	coulombs/sq. meter	10^{4}
coulombs/sq. in	coulombs/sq. cm	0.1550
coulombs/sq. in	coulombs/sq. meter	1,550
coulombs/sq. meter	coulombs/sq. cm	10^{-4}
coulombs/sq. meter	coulombs/sq. in	6.452×10^{-4}
cubic centimeters	cubic feet	3.531×10^{-5}
cubic centimeters	cubic inches	0.06102
cubic centimeters	cubic meters	10^{-6}
cubic centimeters	cubic yards	1.308×10^{-6}
cubic centimeters	gallons (U.S. liq.)	2.642×10^{-4}
cubic centimeters	liters	0.001
cubic centimeters	pints (U.S. liq.)	2.113×10^{-3}
cubic centimeters	quarts (U.S. liq.)	1.057×10^{-3}
cubic feet	bushels (dry)	0.8036
cubic feet	cubic cm	28,320.0

cubic feet	cubic inches	1,728.0
cubic feet	cubic meters	0.02832
cubic feet	cubic yards	0.03704
cubic feet	gallons (U.S. liq.)	7.48052
cubic feet	liters	28.32
cubic feet	pints (U.S. liq.)	59.84
cubic feet	quarts (U.S. liq.)	29.92
cubic feet/min	cubic cm/sec	472.0
cubic feet/min	gallons/sec	0.1247
cubic feet/min	liters/sec	0.4720
cubic feet/min	pounds of water/min	62.43
cubic feet/sec	million gal/day	0.646317
cubic feet/sec	gallons/min	448.831
cubic inches	cubic cm	16.39
cubic inches	cubic feet	5.787×10^{-4}
cubic inches	cubic meters	1.639×10^{-5}
cubic inches	cubic yards	2.143×10^{-5}
cubic inches	gallons	4.329×10^{-3}
cubic inches	liters	0.01639
cubic inches	mil-feet	1.061×10^{5}
cubic inches	pints (U.S. liq.)	0.03463
cubic inches	quarts (U.S. liq.)	0.01732
cubic meters	bushels (dry)	28.38
cubic meters	cubic cm	10^{6}
cubic meters	cubic feet	35.31
cubic meters	cubic inches	61,023.0
cubic meters	cubic yards	1.308
cubic meters	gallons (U.S. liq.)	264.2
cubic meters	liters	1,000.0
cubic meters	pints (U.S. liq.)	2,113.0
cubic meters	quarts (U.S. liq.)	1,057.
cubic yards	cubic cm	7.646×10^{5}
cubic yards	cubic feet	27.0
cubic yards	cubic inches	46,656.0
cubic yards	cubic meters	0.7646
cubic yards	gallons (U.S. liq.)	202.0
cubic yards	liters	764.6

cubic yards	pints (U.S. liq.)	1,615.9
cubic yards	quarts (U.S. liq.)	807.9
cubic yards/min	cubic ft/sec	0.45
cubic yards/min	gallons/sec	3.367
cubic yards/min	liters/sec	12.74
Dalton	gram	1.650×10^{-24}
days	seconds	86,400.0
decigrams	grams	0.1
deciliters	liters	0.1
decimeters	meters	0.1
degrees (angle)	quadrants	0.01111
degrees (angle)	radians	0.01745
degrees (angle)	seconds	3,600.0
degrees/sec	radians/sec	0.01745
degrees/sec	revolutions/min	0.1667
degrees/sec	revolutions/sec	2.778×10^{-3}
dekagrams	grams	10.0
dekaliters	liters	10.0
dekameters	meters	10.0
drams (apothecaries or troy)	ounces (avoirdupois)	0.1371429
drams (apothecaries or troy)	ounces (troy)	0.125
drams (U.S., fluid or apothecaries)	cubic cm	3.6967
drams	grams	1.7718
drams	grains	27.3437
drams	ounces	0.0625
dyne/cm	erg/sq. millimeter	0.01
dyne/sq. cm	atmospheres	9.869×10^{-7}
dyne/sq. cm	inch of mercury at 0°C	2.953×10^{-5}
dyne/sq. cm	inch of water at 4°C	4.015×10^{-4}
dynes	grams	1.020×10^{-3}
dynes	joules/cm	10^{-7}
dynes	joules/meter (newtons)	10^{-5}
dynes	kilograms	1.020×10^{-6}
dynes	poundals	7.233×10^{-5}
dynes	pounds	2.248×10^{-6}

dynes/sq. cm	bars	10^{-6}
ell	cm	114.30
ell	inches	45
em, pica	inch	0.167
em, pica	cm	0.4233
erg/sec	Dyne-cm/sec	1.000
ergs	Btu	9.480×10^{-11}
ergs	dyne-centimeters	1.0
ergs	foot-pounds	7.367×10^{-8}
ergs	gram-calories	0.2389×10^{-7}
ergs	gram-cm	1.020×10^{-3}
ergs	horsepower-hr	3.7250×10^{-14}
ergs	joules	10^{-7}
ergs	kg-calories	2.389×10^{-11}
ergs	kg-meters	1.020×10^{-8}
ergs	kilowatt-hr	0.2778×10^{-13}
ergs	watt-hours	0.2778×10^{-10}
ergs/sec	Btu/min	$5,688 \times 10^{-9}$
ergs/sec	ft-lb/min	4.427×10^{-6}
ergs/sec	ft-lb/sec	7.3756×10^{-8}
ergs/sec	horsepower	1.341×10^{-10}
ergs/sec	kg-calories/min	1.433×10^{-9}
ergs/sec	kilowatts	10^{-10}
farad	microfarads	10^{6}
Faraday/sec	ampere (absolute)	9.6500×10^{4}
faradays	ampere-hours	26.80
faradays	coulombs	9.649×10^{4}
fathom	meter	1.828804
fathoms	feet	6.0
feet	centimeters	30.48
feet	kilometers	3.048×10^{-4}
feet	meters	0.3048
feet	miles (naut.)	1.645×10^{-4}
feet	miles (stat.)	1.894×10^{-4}

feet	millimeters	304.8
feet	mils	1.2×10^4
feet of water	atmospheres	0.02950
feet of water	in of mercury	0.8826
feet of water	kg/sq. cm	0.03048
feet of water	kg/sq. meter	304.8
feet of water	pounds/sq. ft	62.43
feet of water	pounds/sq. in	0.4335
feet/min	cm/sec	0.5080
feet/min	feet/sec	0.01667
feet/min	km/hr	0.01829
feet/min	meters/min	0.3048
feet/min	miles/hr	0.01136
feet/sec	cm/sec	30.48
feet/sec	km/hr	1.097
feet/sec	knots	0.5921
feet/sec	meters/min	18.29
feet/sec	miles/hr	0.6818
feet/sec	miles/min	0.01136
feet/sec/sec	cm/sec/sec	30.48
feet/sec/sec	km/hr/sec	1.097
feet/sec/sec	meters/sec/sec	0.3048
feet/sec/sec	miles/hr/sec	0.6818
feet/100 feet	per centigrade	1.0
foot-candle	lumen/sq. meter	10.764
foot-pounds	Btu	1.286×10^{-3}
foot-pounds	ergs	1.356×10^{7}
foot-pounds	gram-calories	0.3238
foot-pounds	hp-hr	5.050×10^{-7}
foot-pounds	joules	1.356
foot-pounds	kg-calories	3.24×10^{-4}
foot-pounds	kg-meters	0.1383
foot-pounds	kilowatt-hr	3.766×10^{-7}
foot-pounds/min	Btu/min	1.286×10^{-3}
foot-pounds/min	foot-pounds/sec	0.01667
foot-pounds/min	horsepower	3.030×10^{-5}

foot-pounds/min	kg-calories/min	3.24×10^{-4}
foot-pounds/min	kilowatts	2.260×10^{-5}
foot-pounds/sec	Btu/hr	4.6263
foot-pounds/sec	Btu/min	0.07717
foot-pounds/sec	horsepower	1.818×10^{-3}
foot-pounds/sec	kg-calories/min	0.01945
foot-pounds/sec	kilowatts	1.356×10^{-3}
Furlongs	miles (U.S.)	0.125
furlongs	rods	40.0
furlongs	feet	660.0
gallons	cubic cm	3,785.0
gallons	cubic feet	0.1337
gallons	cubic inches	231.0
gallons	cubic meters	3.785×10^{-3}
gallons	cubic yards	4.951×10^{-3}
gallons	liters	3.785
gallons (liq. Br. Imp.)	gallons (U.S. liq.)	1.20095
gallons (U.S.)	gallons (Imp.)	0.83267
gallons of water	pounds of water	8.3453
gallons/min	cubic ft/sec	2.228×10^{-3}
gallons/min	liters/sec	0.06308
gallons/min	cubic ft/hr	8.0208
gausses	lines/sq. in	6.452
gausses	webers/sq. cm	10^{-8}
gausses	webers/sq. in	6.452×10^{-8}
gausses	webers/sq. meter	10^{-4}
gilberts	ampere-turns	0.7958
gilberts/cm	amp-turns/cm	0.7958
gilberts/cm	amp-turns/in	2.021
gilberts/cm	amp-turns/meter	79.58
gills	liters	0.1183
gills	pints (liq.)	0.25
gills (British)	cubic cm	142.07
grade	radian	0.01571
grains	drams (avoirdupois)	0.03657143
grains (troy)	grains (avdp.)	1.0

grains (troy)	grams	0.06480
grains (troy)	ounces (avdp.)	2.0833×10^{-3}
grains (troy)	pennyweight (troy)	0.04167
grains/Imp. gal	parts/million	14.286
grains/U.S. gal	parts/million	17.118
grains/U.S. gal	pounds/million gal	142.86
gram-calories	Btu	3.9683×10^{-3}
gram-calories	ergs	4.1868×10^{7}
gram-calories	foot-pounds	3.0880
gram-calories	horsepower-hr	1.5596×10^{-6}
gram-calories	kilowatt-hr	1.1630×10^{-6}
gram-calories	watt-hr	1.1630×10^{-3}
gram-calories/sec	Btu/hr	14.286
gram-centimeters	Btu	9.297×10^{-8}
gram-centimeters	ergs	980.7
gram-centimeters	joules	9.807×10^{-5}
gram-centimeters	kg-calories	2.343×10^{-8}
gram-centimeters	kg-meters	10^{-5}
grams	dynes	980.7
grams	grains	15.43
grams	joules/cm	9.807×10^{-5}
grams	joules/meter (newtons)	9.807×10^{-3}
grams	kilograms	0.001
grams	milligrams	1,000
grams	ounces (avdp.)	0.03527
grams	ounces (troy)	0.03215
grams	poundals	0.07093
grams	pounds	2.205×10^{-3}
grams/cm	pounds/inch	5.600×10^{-3}
grams/cubic cm	pounds/cubic ft	62.43
grams/cubic cm	pounds/cubic in	0.03613
grams/cubic cm	pounds/mil-foot	3.405×10^{-7}
grams/liter	grains/gal	58.417
grams/liter	pounds/1,000 gal	8.345
grams/liter	pounds/cubic ft	0.062427

grams/liter	parts/million	1,000.0
grams/sq. cm	pounds/sq. ft	2.0481
hand	cm	10.16
hectares	acres	2.471
hectares	sq. feet	1.076×10^5
hectograms	grams	100.0
hectoliters	liters	100.0
hectometers	meters	100.0
hectowatts	watts	100.0
henries	millihenries	1,000.0
horsepower	Btu/min	42.44
horsepower	foot-lb/min	33,000
horsepower	foot-lb/sec	550.0
horsepower	kg-calories/min	10.68
horsepower	kilowatts	0.7457
horsepower	watts	745.7
horsepower (boiler)	Btu/hr	33.479
horsepower (boiler)	kilowatts	9.803
horsepower, metric (542.5 ft lb./sec)	horsepower (550 ft lb./sec)	0.9863
horsepower (550 ft lb./sec)	horsepower, metric (542.5 ft lb./sec)	1.014
horsepower-hr	Btu	2,547
horsepower-hr	ergs	2.6845×10^{13}
horsepower-hr	foot-lb	1.98×10^6
horsepower-hr	gram-calories	641,190
horsepower-hr	joules	2.684×10^6
horsepower-hr	kg-calories	641.1
horsepower-hr	kg-meters	2.737×10^5
horsepower-hr	kilowatt-hr	0.7457
hours	days	4.167×10^{-2}
hours	weeks	5.952×10^{-3}
hundredweights (long)	pounds	112
hundredweights (long)	tons (long)	0.05
hundredweights (short)	ounces (avoirdupois)	1,600
hundredweights (short)	pounds	100
hundredweights (short)	tons (metric)	0.0453592

hundredweights (short)	tons (long)	0.0446429
inches	centimeters	2.540
inches	meters	2.540×10^{-2}
inches	miles	1.578×10^{-5}
inches	millimeters	25.40
inches	mils	1,000.0
inches	yards	2.778×10^{-2}
inches of mercury	atmospheres	0.03342
inches of mercury	feet of water	1.133
inches of mercury	kg/sq. cm	0.03453
inches of mercury	kg/sq. meter	345.3
inches of mercury	pounds/sq. ft	70.73
inches of mercury	pounds/sq. in	0.4912
inches of water (at 4°C)	atmospheres	2.458×10^{-3}
inches of water (at 4°C)	inches of mercury	0.07355
inches of water (at 4°C)	kg/sq. cm	2.540×10^{-3}
inches of water (at 4°C)	ounces/sq. in	0.5781
inches of water (at 4°C)	pounds/sq. ft	5.204
inches of water (at 4°C)	pounds/sq. in	0.03613
international ampere	ampere (absolute)	0.9998
international Volt	volts (absolute)	1.0003
international volt	joules (absolute)	1.593×10^{-19}
international volt	joules	9.654×10^{4}
joules	Btu	9.480×10^{-4}
joules	ergs	10^{7}
joules	foot-pounds	0.7376
joules	kg-calories	2.389×10^{-4}
joules	kg-meters	0.1020
joules	watt-hr	2.778×10^{-4}
joules/cm	grams	1.020×10^{4}
joules/cm	dynes	10^{7}
joules/cm	joules/meter (newtons)	100.0
joules/cm	poundals	723.3
joules/cm	pounds	22.48
kilogram-calories	Btu	3.968
kilogram-calories	foot-pounds	3,088

kilogram-calories	hp-hr	1.560×10^{-3}
kilogram-calories	joules	4,186
kilogram-calories	kg-meters	426.9
kilogram-calories	kilojoules	4.186
kilogram-calories	kilowatt-hr	1.163×10^{-3}
kilogram meters	Btu	9.294×10^{-3}
kilogram meters	ergs	9.804×10^{7}
kilogram meters	foot-pounds	7.233
kilogram meters	joules	9.804
kilogram meters	kg-calories	2.342×10^{-3}
kilogram meters	kilowatt-hr	2.723×10^{-6}
kilograms	dynes	980,665
kilograms	grams	1,000.0
kilograms	joules/cm	0.09807
kilograms	joules/meter (newtons)	9.807
kilograms	poundals	70.93
kilograms	pounds	2.205
kilograms	tons (long)	9.842×10^{-4}
kilograms	tons (short)	1.102×10^{-3}
kilograms/cubic meter	grams/cubic cm	0.001
kilograms/cubic meter	pounds/cubic ft	0.06243
kilograms/cubic meter	pounds/cubic in	3.613×10^{-5}
kilograms/cubic meter	pounds/mil-foot	3.405×10^{-10}
kilograms/meter	pounds/ft	0.6720
kilograms/sq. cm	dynes	980,665
kilograms/sq. cm	atmospheres	0.9678
kilograms/sq. cm	feet of water	32.81
kilograms/sq. cm	inches of mercury	28.96
kilograms/sq. cm	pounds/sq. ft	2,048
kilograms/sq. cm	pounds/sq. in	14.22
kilograms/sq. meter	atmospheres	9.678×10^{-5}
kilograms/sq. meter	bars	98.07×10^{-6}
kilograms/sq. meter	feet of water	3.281×10^{-3}
kilograms/sq. meter	inches of mercury	2.896×10^{-3}
kilograms/sq. meter	pounds/sq. ft	0.2048

kilograms/sq. meter	pounds/sq. in	1.422×10^{-3}
kilograms/sq. mm	kg/sq. meter	10^{6}
kilolines	maxwells	1,000.0
kiloliters	liters	1,000.0
kilometers	centimeters	10^{5}
kilometers	feet	3,281
kilometers	inches	3.937×10^{4}
kilometers	meters	1,000.0
kilometers	miles	0.6214
kilometers	millimeters	10^{4}
kilometers	yards	1,094
kilometers/hr	cm/sec	27.78
kilometers/hr	feet/min	54.68
kilometers/hr	feet/sec	0.9113
kilometers/hr	knots	0.5396
kilometers/hr	meters/min	16.67
kilometers/hr	miles/hr	0.6214
kilometers/hr/sec	cm/sec/sec	27.78
kilometers/hr/sec	feet/sec/sec	0.9113
kilometers/hr/sec	meters/sec/sec	0.2778
kilometers/hr/sec	miles/hr/sec	0.6214
kilowatt-hr	Btu	3,413
kilowatt-hr	ergs	3.600×10^{13}
kilowatt-hr	foot-lb	2.655×10^{6}
kilowatt-hr	gram-calories	859,850
kilowatt-hr	horsepower-hr	1.341
kilowatt-hr	joules	3.6×10^{6}
kilowatt-hr	kg-calories	860.5
kilowatt-hr	kg-meters	3.671×10^{5}
kilowatt-hr	pounds of water raised from 62° to 212°F	22.75
kilowatts	Btu/min	56.92
kilowatts	foot-lb/min	4.426×10^{4}
kilowatts	foot-lb/sec	737.6
kilowatts	horsepower	1.341
kilowatts	kg-calories/min	14.34

kilowatts	watts	1,000.0
knots	feet/hr	6,080
knots	kilometers/hr	1.8532
knots	nautical miles/hr	1.0
knots	statute miles/hr	1.151
knots	yards/hr	2,027
knots	feet/sec	1.689
league	miles (approx.)	3.0
light year	miles	5.9×10^{12}
light year	kilometers	9.4637×10^{12}
lines/sq. cm	gausses	1.0
lines/sq. in	gausses	0.1550
lines/sq. in	webers/sq. cm	1.550×10^{-9}
lines/sq. in	webers/sq. in	10^{-8}
lines/sq. in	webers/sq. meter	1.550×10^{-5}
links (engineer's)	inches	12.0
links (surveyor's)	inches	7.92
liters	bushels (U.S. dry)	0.02838
liters	cubic cm	1,000.0
liters	cubic feet	0.03531
liters	cubic inches	61.02
liters	cubic meters	0.001
liters	cubic yards	1.308×10^{-3}
liters	gallons (U.S. liq.)	0.2642
liters	pints (U.S. liq.)	2.113
liters	quarts (U.S. liq.)	1.057
liters/min	cubic ft/sec	5.886×10^{-4}
liters/min	gal/sec	4.403×10^{-3}
lumen	spherical candle power	0.07958
lumen	watt	0.001496
lumens/sq. ft	foot-candles	1.0
lumens/sq. ft	lumen/sq. meter	10.76
lux	foot-candles	0.0929
maxwells	kilolines	0.001
maxwells	webers	10^{-8}
megalines	maxwells	10^{6}

megohms	microhms	10^{12}
megohms	ohms	10^{6}
meter-kilograms	cm-dynes	9.807×10^{7}
meter-kilograms	cm-grams	10^{5}
meter-kilograms	pound-feet	7.233
meters	centimeters	100.0
meters	feet	3.281
meters	inches	39.37
meters	kilometers	0.001
meters	miles (naut.)	5.396×10^{-4}
meters	miles (stat.)	6.214×10^{-4}
meters	millimeters	1,000.0
meters	yards	1.094
meters	varas	1.179
meters/min	cm/sec	1,667
meters/min	feet/min	3.281
meters/min	feet/sec	0.05468
meters/min	km/hr	0.06
meters/min	knots	0.03238
meters/min	miles/hr	0.03728
meters/sec	feet/min	196.8
meters/sec	feet/sec	3.281
meters/sec	kilometers/hr	3.6
meters/sec	kilometers/min	0.06
meters/sec	miles/hr	2.237
meters/sec	miles/min	0.03728
meters/sec/sec	cm/sec/sec	100.0
meters/sec/sec	ft/sec/sec	3.281
meters/sec/sec	km/hr/sec	3.6
meters/sec/sec	miles/hr/sec	2.237
microfarad	farads	10^{-6}
micrograms	grams	10^{-6}
microhms	megohms	10^{-12}
microhms	ohms	10^{-6}
microliters	liters	10^{-6}
microns	meters	1×10^{-6}

miles (naut.)	feet	6,080.27
miles (naut.)	kilometers	1.853
miles (naut.)	meters	1,853
miles (naut.)	miles (statute)	1.1516
miles (naut.)	yards	2,027
miles (statute)	centimeters	1.609×10^5
miles (statute)	feet	5,280
miles (statute)	inches	6.336×10^4
miles (statute)	kilometers	1.609
miles (statute)	meters	1,609
miles (statute)	miles (naut.)	0.8684
miles (statute)	yards	1,760
miles/hr	cm/sec	44.70
miles/hr	feet/min	88
miles/hr	feet/sec	1.467
miles/hr	km/hr	1.609
miles/hr	km/min	0.02682
miles/hr	knots	0.8684
miles/hr	meters/min	26.82
miles/hr	miles/min	0.1667
miles/hr/sec	cm/sec/sec	44.70
miles/hr/sec	feet/sec/sec	1.467
miles/hr/sec	km/hr/sec	1.609
miles/hr/sec	meters/sec/sec	0.4470
miles/min	cm/sec	2,682
miles/min	feet/sec	88
miles/min	km/min	1.609
miles/min	knots/min	0.8684
miles/min	miles/hr	60
mil-feet	cubic inches	9.425×10^{-6}
milliers	kilograms	1,000
milligrams	grains	0.01543236
milligrams	grams	0.001
milligrams/liter	parts/million	1.0
millihenries	henries	0.001
milliliters	liters	0.001
millimeters	centimeters	0.1

millimeters	feet	3.281×10^{-3}
millimeters	inches	0.03937
millimeters	kilometers	10^{-6}
millimeters	meters	0.001
millimeters	miles	6.214×10^{-7}
millimeters	mils	39.37
millimeters	yards	1.094×10^{-3}
millimicrons	meters	1×10^{-9}
million gal/day	cubic ft/sec	1.54723
mils	centimeters	2.540×10^{-3}
mils	feet	8.333×10^{-5}
mils	inches	0.001
mils	kilometers	2.540×10^{-8}
mils	yards	2.778×10^{-5}
miner's inches	cubic ft/min	1.5
minims (British)	cubic cm	0.059192
minims (U.S., fluid)	cubic cm	0.061612
minutes (angles)	degrees	0.01667
minutes (angles)	quadrants	1.852×10^{-4}
minutes (angles)	radians	2.909×10^{-4}
minutes (angles)	seconds	60.0
myriagrams	kilograms	10.0
myriameters	kilometers	10.0
myriawatts	kilowatts	10.0
nepers	decibels	8.686
Newton	dynes	1 x 105
ohm (international)	ohm (absolute)	1.0005
ohms	megohms	10^{-6}
ohms	microhms	10^{6}
ounces	drams	16.0
ounces	grains	437.5
ounces	grams	28.349527
ounces	pounds	0.0625
ounces	ounces (troy)	0.9115
ounces	tons (long)	2.790×10^{-5}

ounces	tons (metric)	2.835×10^{-5}
ounces (fluid)	cubic inches	1.805
ounces (fluid)	liters	0.02957
ounces (troy)	grains	480.0
ounces (troy)	grams	31.103481
ounces (troy)	ounces (avdp.)	1.09714
ounces (troy)	pennyweights (troy)	20.0
ounces (troy)	pounds (troy)	0.08333
ounces/sq. inch	dynes/sq. cm	4,309
ounces/sq. in	pounds/sq. in	0.0625
parsec	miles	19×10^{12}
parsec	kilometers	3.084×10^{13}
parts/million	grains/U.S. gal	0.0584
parts/million	grains/Imp. gal	0.07016
parts/million	pounds/million gal	8.345
pecks (British)	cubic inches	554.6
pecks (British)	liters	9.091901
pecks (U.S.)	bushels	0.25
pecks (U.S.)	cubic inches	537.605
pecks (U.S.)	liters	8.809582
pecks (U.S.)	quarts (dry)	8
pennyweights (troy)	grains	24.0
pennyweights (troy)	ounces (troy)	0.05
pennyweights (troy)	grams	1.55517
pennyweights (troy)	pounds (troy)	4.1667×10^{-3}
pints (dry)	cubic inches	33.60
pints (liq.)	cubic cm	473.2
pints (liq.)	cubic feet	0.01671
pints (liq.)	cubic inches	28.87
pints (liq.)	cubic meters	4.732×10^{-4}
pints (liq.)	cubic yards	6.189×10^{-4}
pints (liq.)	gallons	0.125
pints (liq.)	liters	0.4732
pints (liq.)	quarts (liq.)	0.5
Planck's quantum	erg - second	6.624×10^{-27}
poise	gram/cm sec	1.00
poundals	dynes	13,826

poundals	grams	14.10
poundals	joules/cm	1.383×10^{-3}
poundals	joules/meter (newtons)	0.1383
poundals	kilograms	0.01410
poundals	pounds	0.03108
pound-feet	cm-dynes	1.356×10^{7}
pound-feet	cm-grams	13,825
pound-feet	meter-kg	0.1383
pounds	drams	256
pounds	dynes	44.4823×10^{4}
pounds	grains	7,000
pounds	grams	453.5924
pounds	joules/cm	0.04448
pounds	joules/meter (newtons)	4.448
pounds	kilograms	0.4536
pounds	ounces	16.0
pounds	ounces (troy)	14.5833
pounds	poundals	32.17
pounds	pounds (troy)	1.21528
pounds	tons (short)	0.0005
pounds (avoirdupois)	ounces (troy)	14.5833
pounds (troy)	grains	5,760
pounds (troy)	grams	373.24177
pounds (troy)	ounces (avdp.)	13.1657
pounds (troy)	ounces (troy)	12.0
pounds (troy)	pennyweights (troy)	240.0
pounds (troy)	pounds (avdp.)	0.822857
pounds (troy)	tons (long)	3.6735×10^{-4}
pounds (troy)	tons (metric)	3.7324×10^{-4}
pounds (troy)	tons (short)	4.1143×10^{-4}
pounds of water	cubic ft	0.01602
pounds of water	cubic inches	27.68
pounds of water	gallons	0.1198
pounds of water/min	cubic ft/sec	2.670×10^{-4}
pounds/cubic ft	grams/cubic cm	0.01602
pounds/cubic ft	kg/cubic meter	16.02

pounds/cubic ft	pounds/cubic in	5.787×10^{-4}
pounds/cubic ft	pounds/mil-foot	5.456×10^{-9}
pounds/cubic in	gm/cubic cm	27.68
pounds/cubic in	kg/cubic meter	2.768×10^{4}
pounds/cubic in	pounds/cubic ft	1,728
pounds/cubic in	pounds/mil-foot	9.425×10^{-6}
pounds/ft	kg/meter	1.488
pounds/in	gm/cm	178.6
pounds/mil-foot	gm/cubic cm	2.306×10^{6}
pounds/sq. ft	atmospheres	4.725×10^{-4}
pounds/sq. ft	feet of water	0.01602
pounds/sq. ft	inches of mercury	0.01414
pounds/sq. ft	kg/sq. meter	4.882
pounds/sq. ft	pounds/sq. in	6.944×10^{-3}
pounds/sq. in	atmospheres	0.06804
pounds/sq. in	feet of water	2.307
pounds/sq. in	inches of mercury	2.036
pounds/sq. in	kg/sq. meter	703.1
pounds/sq. in	pounds/sq. ft	144.0
quadrants (angle)	degrees	90.0
quadrants (angle)	minutes	5,400.0
quadrants (angle)	radians	1.571
quadrants (angle)	seconds	3.24×10^{5}
quarts (dry)	cubic inches	67.20
quarts (liq.)	cubic cm	946.4
quarts (liq.)	cubic feet	0.03342
quarts (liq.)	cubic inches	57.75
quarts (liq.)	cubic meters	9.464×10^{-4}
quarts (liq.)	cubic yards	1.238×10^{-3}
quarts (liq.)	gallons	0.25
quarts (liq.)	liters	0.9463
radians	degrees	57.30
radians	minutes	3,438
radians	quadrants	0.6366
radians	seconds	2.063×10^{5}
radians/sec	degrees/sec	57.30

radians/sec	revolutions/min	9.549
radians/sec	revolutions/sec	0.1592
radians/sec/sec	revolutions/min/min	573.0
radians/sec/sec	revolutions/min/sec	9.549
radians/sec/sec	revolutions/sec/sec	0.1592
revolutions	degrees	360.0
revolutions	quadrants	4.0
revolutions	radians	6.283
revolutions/min	degrees/sec	6.0
revolutions/min	radians/sec	0.1047
revolutions/min	revolutions/sec	0.01667
revolutions/min/min	radians/sec/sec	1.745×10^{-3}
revolutions/min/min	revolutions/min/sec	0.01667
revolutions/min/min	revolutions/sec/sec	2.778×10^{-4}
revolutions/sec	degrees/sec	360.0
revolutions/sec	radians/sec	6.283
revolutions/sec	revolutions/min	60.0
revolutions/sec/sec	radians/sec/sec	6.283
revolutions/sec/sec	revolutions/min/min	3,600.0
revolutions/sec/sec	revolutions/min/sec	60.0
rod	chain (Gunters)	0.25
rod	meters	5.029
rods	feet	16.5
rods (surveyors' meas.)	yards	5.5
scruples	grains	20
seconds (angle)	degrees	2.778×10^{-4}
seconds (angle)	minutes	0.01667
seconds (angle)	quadrants	3.087×10^{-6}
seconds (angle)	radians	4.848×10^{-6}
slug	kilogram	14.59
slug	pounds	32.17
sphere	steradians	12.57
square centimeters	circular mils	1.973×10^{5}
square centimeters	sq. feet	1.076×10^{-3}
square centimeters	sq. inches	0.1550
square centimeters	sq. meters	0.0001

square centimeters	sq. miles	3.861×10^{-11}
square centimeters	sq. millimeters	100.0
square centimeters	sq. yards	1.196×10^{-4}
square feet	acres	2.296×10^{-5}
square feet	circular mils	1.833×10^{8}
square feet	sq. cm	929.0
square feet	sq. inches	144.0
square feet	sq. meters	0.09290
square feet	sq. miles	3.587×10^{-8}
square feet	sq. millimeters	9.290×10^{4}
square feet	sq. yards	0.1111
square inches	circular mils	1.273×10^{6}
square inches	sq. cm	6.452
square inches	sq. feet	6.944×10^{-3}
square inches	sq. millimeters	645.2
square inches	sq. mils	10^{6}
square inches	sq. yards	7.716×10^{-4}
square kilometers	acres	247.1
square kilometers	sq. cm	10^{10}
square kilometers	sq. ft	10.76×10^{6}
square kilometers	sq. inches	1.550×10^{9}
square kilometers	sq. meters	10^{6}
square kilometers	sq. miles	0.3861
square kilometers	sq. yards	1.196×10^{6}
square meters	acres	2.471×10^{-4}
square meters	sq. cm	10^{4}
square meters	sq. feet	10.76
square meters	sq. inches	1,550
square meters	sq. miles	3.861×10^{-7}
square meters	sq. millimeters	10^{6}
square meters	sq. yards	1.196
square miles	acres	640.0
square miles	sq. feet	27.88×10^{6}
square miles	sq. km	2.590

square miles	sq. meters	2.590×10^6
square miles	sq. yards	3.098×10^6
square millimeters	circular mils	1,973
square millimeters	sq. cm	0.01
square millimeters	sq. feet	1.076×10^{-5}
square millimeters	sq. inches	1.550×10^{-3}
square mils	circular mils	1.273
square mils	sq. cm	6.452×10^{-6}
square mils	sq. inches	10^{-6}
square yards	acres	2.066×10^{-4}
square yards	sq. cm	8,361
square yards	sq. feet	9.0
square yards	sq. inches	1,296
square yards	sq. meters	0.8361
square yards	sq. miles	3.228×10^{-7}
square yards	sq. millimeters	8.361×10^5
temperature (°C)+273	absolute temperature (°C)	1.0
temperature (°C)+17.78	temperature (°F)	1.8
temperature (°F)+460	absolute temperature (°F)	1.0
temperature (°F)–32	temperature (°C)	5/9
tons (long)	kilograms	1,016
tons (long)	pounds	2,240
tons (long)	tons (short)	1.120
tons (metric)	kilograms	1,000
tons (metric)	pounds	2,205
tons (short)	kilograms	907.1848
tons (short)	ounces	32,000
tons (short)	ounces (troy)	29,166.66
tons (short)	pounds	2,000
tons (short)	pounds (troy)	2,430.56
tons (short)	tons (long)	0.89287
tons (short)	tons (metric)	0.9078
tons (short)/sq. ft	kg/sq. meter	9,765
tons (short)/sq. ft	pounds/sq. in	2,000
tons of water/24 hr	pounds of water/hr	83.333
tons of water/24 hr	gallons/min	0.16643

tons of water/24 hr	cubic ft/hr	1.3349
volt (absolute)	statvolts	0.003336
volt/inch	volt/cm	0.39370
watt-hours	Btu	3.413
watt-hours	ergs	3.60×10^{10}
watt-hours	foot-pounds	2,656
watt-hours	gram-calories	859.85
watt-hours	horsepower-hr	1.341×10^{-3}
watt-hours	kilogram-calories	0.8605
watt-hours	kilogram-meters	367.2
watt-hours	kilowatt-hr	0.001
watt (international)	watt (absolute)	1.0002
watts	Btu/hr	3.4129
watts	Btu/min	0.05688
watts	ergs/sec	107
watts	foot-lb/min	44.27
watts	foot-lb/sec	0.7378
watts	horsepower	1.341×10^{-3}
watts	horsepower (metric)	1.360×10^{-3}
watts	kg-calories/min	0.01433
watts	kilowatts	0.001
watts (Abs.)	Btu (mean)/min	0.056884
watts (Abs.)	joules/sec	1
webers	maxwells	10^{8}
webers	kilolines	10^{5}
webers/sq. in	gausses	1.550×10^{7}
webers/sq. in	lines/sq. in	10^{8}
webers/sq. in	webers/sq. cm	0.1550
webers/sq. in	webers/sq. meter	1,550
webers/sq. meter	gausses	10^{4}
webers/sq. meter	lines/sq. in	6.452×10^{4}
webers/sq. meter	webers/sq. cm	10^{-4}
webers/sq. meter	webers/sq. in	6.452×10^{-4}
yards	centimeters	91.44
yards	kilometers	9.144×10^{-4}
yards	meters	0.9144

yards	miles (naut.)	4.934×10^{-4}
yards	miles (stat.)	5.682×10^{-4}
yards	millimeters	914.4

Chapter

4

Dictionary of Audio/Video Terms

4.1 Common Terms

A

absolute delay The amount of time a signal is delayed. The delay may be expressed in time or number of pulse events.

absolute zero The lowest temperature theoretically possible, –273.16°C. *Absolute zero* is equal to zero degrees Kelvin.

absorption The transference of some or all of the energy contained in an electromagnetic wave to the substance or medium in which it is propagating or upon which it is incident.

absorption auroral The loss of energy in a radio wave passing through an area affected by solar auroral activity.

ac coupling A method of coupling one circuit to another through a capacitor or transformer so as to transmit the varying (ac) characteristics of the signal while blocking the static (dc) characteristics.

ac/dc coupling Coupling between circuits that accommodates the passing of both ac and dc signals (may also be referred to as simply dc coupling).

accelerated life test A special form of reliability testing performed by an equipment manufacturer. The unit under test is subjected to stresses that exceed those typically experienced in normal operation. The goal of an *accelerated life test* is to improve the reliability of products shipped by forcing latent failures in components to become evident before the unit leaves the factory.

accelerating electrode The electrode that causes electrons emitted from an electron gun to accelerate in their journey to the screen of a cathode ray tube.

accelerating voltage The voltage applied to an electrode that accelerates a beam of electrons or other charged particles.

acceptable reliability level The maximum number of failures allowed per thousand operating hours of a given component or system.

acceptance test The process of testing newly purchased equipment to ensure that it is fully compliant with contractual specifications.

access The point at which entry is gained to a circuit or facility.

acquisition time In a communication system, the amount of time required to attain synchronism.

active Any device or circuit that introduces gain or uses a source of energy other than that inherent in the signal to perform its function.

adapter A fitting or electrical connector that links equipment that cannot be connected directly.

adaptive A device able to adjust or react to a condition or application, as an *adaptive circuit*. This term usually refers to filter circuits.

adaptive system A general name for a system that is capable of reconfiguring itself to meet new requirements.

adder A device whose output represents the sum of its inputs.

adjacent channel interference Interference to communications caused by a transmitter operating on an adjacent radio channel. The sidebands of the transmitter mix with the carrier being received on the desired channel, resulting in noise.

admittance A measure of how well alternating current flows in a conductor. It is the reciprocal of *impedance* and is expressed in *siemens*. The real part of admittance is *conductance*; the imaginary part is *susceptance*.

AFC (automatic frequency control) A circuit that automatically keeps an oscillator on frequency by comparing the output of the oscillator with a standard frequency source or signal.

air core An inductor with no magnetic material in its core.

algorithm A prescribed finite set of well-defined rules or processes for the solution of a problem in a finite number of steps.

alignment The adjustment of circuit components so that an entire system meets minimum performance values. For example, the stages in a radio are aligned to ensure proper reception.

allocation The planned use of certain facilities and equipment to meet current, pending, and/or forecasted circuit- and carrier-system requirements.

alternating current (ac) A continuously variable current, rising to a maximum in one direction, falling to zero, then reversing direction and rising to a maximum in the other direction, then falling to zero and repeating the cycle. Alternating current usually follows a sinusoidal growth and decay curve. Note that the correct usage of the term *ac* is lower case.

alternator A generator that produces alternating current electric power.

ambient electromagnetic environment The radiated or conducted electromagnetic signals and noise at a specific location and time.

ambient level The magnitude of radiated or conducted electromagnetic signals and noise at a specific test location when equipment-under-test is not powered.

ambient temperature The temperature of the surrounding medium, typically air, that comes into contact with an apparatus. Ambient temperature may also refer simply to room temperature.

American National Standards Institute (ANSI) A nonprofit organization that coordinates voluntary standards activities in the U.S.

American Wire Gauge (AWG) The standard American method of classifying wire diameter.

ammeter An instrument that measures and records the amount of current in amperes flowing in a circuit.

amp (A) An abbreviation of the term *ampere.*

ampacity A measure of the current carrying capacity of a power cable. *Ampacity* is determined by the maximum continuous-performance temperature of the insulation, by the heat generated in the cable (as a result of conductor and insulation losses), and by the heat-dissipating properties of the cable and its environment.

ampere (amp) The standard unit of electric current.

ampere per meter The standard unit of magnetic field strength.

ampere-hour The energy that is consumed when a current of one ampere flows for a period of one hour.

ampere-turns The product of the number of turns of a coil and the current in amperes flowing through the coil.

amplification The process that results when the output of a circuit is an enlarged reproduction of the input signal. Amplifiers may be designed to provide amplification of voltage, current, or power, or a combination of these quantities.

amplification factor In a vacuum tube, the ratio of the change in plate voltage to the change in grid voltage that causes a corresponding change in plate current. Amplification factor is expressed by the Greek letter μ (*mu*).

amplifier (1—general) A device that receives an input signal and provides as an output a magnified replica of the input waveform. **(2—audio)** An amplifier designed to cover the normal audio frequency range (20 Hz to 20 kHz). **(3—balanced)** A circuit with two identical connected signal branches that operate in phase opposition, with input and output connections each balanced to ground. **(4—bridging)** An amplifying circuit featuring high input impedance to prevent loading of the source. **(5—broadband)** An amplifier capable of operating over a specified broad band of frequencies with acceptably small amplitude variations as a function of frequency. **(6—buffer)** An amplifier stage used to isolate a frequency-sensitive circuit from variations in the load presented by following stages. **(7—linear)** An amplifier in which the instantaneous output signal is a linear function of the corresponding input signal. **(8—magnetic)** An amplifier incorporating a control device dependent on magnetic saturation. A small dc signal applied to a control circuit triggers a large change in operating impedance and, hence, in the output of the circuit. **(9—micro-**

phone) A circuit that amplifies the low level output from a microphone to make it sufficient to be used as an input signal to a power amplifier or another stage in a modulation circuit. Such a circuit is commonly known as a *preamplifier.* **(10—push-pull)** A balanced amplifier with two similar amplifying units connected in phase opposition in order to cancel undesired harmonics and minimize distortion. **(11—tuned radio frequency)** An amplifier tuned to a particular radio frequency or band so that only selected frequencies are amplified.

amplifier operating class (1—general) The operating point of an amplifying stage. The operating point, termed the operating *class*, determines the period during which current flows in the output. **(2—class A)** An amplifier in which output current flows during the whole of the input current cycle. **(3—class AB)** An amplifier in which the output current flows for more than half but less than the whole of the input cycle. **(4—class B)** An amplifier in which output current is cut off at zero input signal; a half-wave rectified output is produced. **(5—class C)** An amplifier in which output current flows for less than half the input cycle. **(6—class D)** An amplifier operating in a pulse-only mode.

amplitude The magnitude of a signal in voltage or current, frequently expressed in terms of *peak*, *peak-to-peak*, or *root-mean-square* (RMS). The actual amplitude of a quantity at a particular instant often varies in a sinusoidal manner.

amplitude distortion A distortion mechanism occurring in an amplifier or other device when the output amplitude is not a linear function of the input amplitude under specified conditions.

amplitude equalizer A corrective network that is designed to modify the amplitude characteristics of a circuit or system over a desired frequency range.

amplitude-versus-frequency distortion The distortion in a transmission system caused by the nonuniform attenuation or gain of the system with respect to frequency under specified conditions.

analog carrier system A carrier system whose signal amplitude, frequency, or phase is varied continuously as a function of a modulating input.

anode (1 — general) A positive pole or element. **(2—vacuum tube)** The outermost positive element in a vacuum tube, also called the *plate.* **(3—battery)** The positive element of a battery or cell.

anodize The formation of a thin film of oxide on a metallic surface, usually to produce an insulating layer.

antenna (1—general) A device used to transmit or receive a radio signal. An antenna is usually designed for a specified frequency range and serves to couple electromagnetic energy from a transmission line to and/or from the free space through which it travels. Directional antennas concentrate the energy in a particular horizontal or vertical direction. **(2—aperiodic)** An antenna that is not periodic or resonant at particular frequencies, and so can be used over a wide band of frequencies. **(3—artificial)** A device that behaves, so far as the transmitter is concerned, like a proper antenna, but does not radiate any power at radio frequencies. **(4—broadband)** An antenna that operates within specified performance limits over a wide band of frequencies, without requiring retuning for each individual frequency.

(5—Cassegrain) A double reflecting antenna, often used for ground stations in satellite systems. **(6—coaxial)** A dipole antenna made by folding back on itself a quarter wavelength of the outer conductor of a coaxial line, leaving a quarter wavelength of the inner conductor exposed. **(7—corner)** An antenna within the angle formed by two plane-reflecting surfaces. **(8—dipole)** A center-fed antenna, one half-wavelength long. **(9—directional)** An antenna designed to receive or emit radiation more efficiently in a particular direction. **(10—dummy)** An artificial antenna, designed to accept power from the transmitter but not to radiate it. **(11—ferrite)** A common AM broadcast receive antenna that uses a small coil mounted on a short rod of ferrite material. **(12—flat top)** An antenna in which all the horizontal components are in the same horizontal plane. **(13—folded dipole)** A radiating device consisting of two ordinary half-wave dipoles joined at their outer ends and fed at the center of one of the dipoles. **(14—horn reflector)** A radiator in which the feed horn extends into a parabolic reflector, and the power is radiated through a window in the horn. **(15—isotropic)** A theoretical antenna in free space that transmits or receives with the same efficiency in all directions. **(16—log-periodic)** A broadband directional antenna incorporating an array of dipoles of different lengths, the length and spacing between dipoles increasing logarithmically away from the feeder element. **(17—long wire)** An antenna made up of one or more conductors in a straight line pointing in the required direction with a total length of several wavelengths at the operating frequency. **(18—loop)** An antenna consisting of one or more turns of wire in the same or parallel planes. **(19—nested rhombic)** An assembly of two rhombic antennas, one smaller than the other, so that the complete diamond-shaped antenna fits inside the area occupied by the larger unit. **(20—omnidirectional)** An antenna whose radiating or receiving properties are the same in all horizontal plane directions. **(21—periodic)** A resonant antenna designed for use at a particular frequency. **(22—quarter-wave)** A dipole antenna whose length is equal to one quarter of a wavelength at the operating frequency. **(23—rhombic)** A large diamond-shaped antenna, with sides of the diamond several wavelengths long. The rhombic antenna is fed at one of the corners, with directional efficiency in the direction of the diagonal. **(24—series fed)** A vertical antenna that is fed at its lower end. **(25—shunt fed)** A vertical antenna whose base is grounded, and is fed at a specified point above ground. The point at which the antenna is fed above ground determines the operating impedance. **(26—steerable)** An antenna so constructed that its major lobe may readily be changed in direction. **(27—top-loaded)** A vertical antenna capacitively loaded at its upper end, often by simple enlargement or the attachment of a disc or plate. **(28—turnstile)** An antenna with one or more tiers of horizontal dipoles, crossed at right angles to each other and with excitation of the dipoles in phase quadrature. **(29—whip)** An antenna constructed of a thin semiflexible metal rod or tube, fed at its base. **(30—Yagi)** A directional antenna constructed of a series of dipoles cut to specific lengths. *Director* elements are placed in front of the active dipole and *reflector* elements are placed behind the active element.

antenna array A group of several antennas coupled together to yield a required degree of directivity.

antenna beamwidth The angle between the *half-power* points (3 dB points) of the main lobe of the antenna pattern when referenced to the peak power point of the antenna pattern.

Antenna beamwidth is measured in degrees and normally refers to the horizontal radiation pattern.

antenna directivity factor The ratio of the power flux density in the desired direction to the average value of power flux density at crests in the antenna directivity pattern in the interference section.

antenna factor A factor that, when applied to the voltage appearing at the terminals of measurement equipment, yields the electrical field strength at an antenna. The unit of antenna factor is volts per meter per measured volt.

antenna gain The ratio of the power required at the input of a theoretically perfect omnidirectional reference antenna to the power supplied to the input of the given antenna to produce the same field at the same distance. When not specified otherwise, the figure expressing the gain of an antenna refers to the gain in the direction of the radiation main lobe. In services using *scattering* modes of propagation, the full gain of an antenna may not be realizable in practice and the apparent gain may vary with time.

antenna gain-to-noise temperature For a satellite earth terminal receiving system, a figure of merit that equals G/T, where G is the gain in dB of the earth terminal antenna at the receive frequency, and T is the equivalent noise temperature of the receiving system in Kelvins.

antenna matching The process of adjusting an antenna matching circuit (or the antenna itself) so that the input impedance of the antenna is equal to the characteristic impedance of the transmission line.

antenna monitor A device used to measure the ratio and phase between the currents flowing in the towers of a directional AM broadcast station.

antenna noise temperature The temperature of a resistor having an available noise power per unit bandwidth equal to that at the antenna output at a specified frequency.

antenna pattern A diagram showing the efficiency of radiation in all directions from the antenna.

antenna power rating The maximum continuous-wave power that can be applied to an antenna without degrading its performance.

antenna preamplifier A small amplifier, usually mast-mounted, for amplifying weak signals to a level sufficient to compensate for down-lead losses.

apparent power The product of the root-mean-square values of the voltage and current in an alternating-current circuit without a correction for the phase difference between the voltage and current.

arc A sustained luminous discharge between two or more electrodes.

arithmetic mean The sum of the values of several quantities divided by the number of quantities, also referred to as the *average*.

armature winding The winding of an electrical machine, either a motor or generator, in which current is induced.

array (1—antenna) An assembly of several directional antennas so placed and interconnected that directivity may be enhanced. **(2—broadside)** An antenna array whose elements are all in the same plane, producing a major lobe perpendicular to the plane. **(3—colinear)** An antenna array whose elements are in the same line, either horizontal or vertical. **(4—end-fire)** An antenna array whose elements are in parallel rows, one behind the other, producing a major lobe perpendicular to the plane in which individual elements are placed. **(5—linear)** An antenna array whose elements are arranged end-to-end. **(6—stacked)** An antenna array whose elements are stacked, one above the other.

artificial line An assembly of resistors, inductors, and capacitors that simulates the electrical characteristics of a transmission line.

assembly A manufactured part made by combining several other parts or subassemblies.

assumed values A range of values, parameters, levels, and other elements assumed for a mathematical model, hypothetical circuit, or network, from which analysis, additional estimates, or calculations will be made. The range of values, while not measured, represents the best engineering judgment and is generally derived from values found or measured in real circuits or networks of the same generic type, and includes projected improvements.

atmosphere The gaseous envelope surrounding the earth, composed largely of oxygen, carbon dioxide, and water vapor. The atmosphere is divided into four primary layers: *troposphere*, *stratosphere*, *ionosphere*, and *exosphere*.

atmospheric noise Radio noise caused by natural atmospheric processes, such as lightning.

attack time The time interval in seconds required for a device to respond to a control stimulus.

attenuation The decrease in amplitude of an electrical signal traveling through a transmission medium caused by dielectric and conductor losses.

attenuation coefficient The rate of decrease in the amplitude of an electrical signal caused by attenuation. The *attenuation coefficient* can be expressed in decibels or nepers per unit length. It may also be referred to as the *attenuation constant*.

attenuation distortion The distortion caused by attenuation that varies over the frequency range of a signal.

attenuation-limited operation The condition prevailing when the received signal amplitude (rather than distortion) limits overall system performance.

attenuator A fixed or adjustable component that reduces the amplitude of an electrical signal without causing distortion.

atto A prefix meaning one *quintillionth*.

attraction The attractive force between two unlike magnetic poles (N/S) or electrically charged bodies (+/-).

attributes The characteristics of equipment that aid planning and circuit design.

automatic frequency control (AFC) A system designed to maintain the correct operating frequency of a receiver. Any drift in tuning results in the production of a control voltage, which is used to adjust the frequency of a local oscillator so as to minimize the tuning error.

automatic gain control (AGC) An electronic circuit that compares the level of an incoming signal with a previously defined standard and automatically amplifies or attenuates the signal so it arrives at its destination at the correct level.

autotransformer A transformer in which both the primary and secondary currents flow through one common part of the coil.

auxiliary power An alternate source of electric power, serving as a back-up for the primary utility company ac power.

availability A measure of the degree to which a system, subsystem, or equipment is operable and not in a stage of congestion or failure at any given point in time.

avalanche effect The effect obtained when the electric field across a barrier region is sufficiently strong for electrons to collide with *valence electrons*, thereby releasing more electrons and giving a cumulative multiplication effect in a semiconductor.

average life The mean value for a normal distribution of product or component lives, generally applied to mechanical failures resulting from "wear-out."

B

back emf A voltage induced in the reverse direction when current flows through an inductance. *Back emf* is also known as *counter-emf.*

back scattering A form of wave scattering in which at least one component of the scattered wave is deflected opposite to the direction of propagation of the incident wave.

background noise The total system noise in the absence of information transmission, independent of the presence or absence of a signal.

backscatter The deflection or reflection of radiant energy through angles greater than 90° with respect to the original angle of travel.

backscatter range The maximum distance from which backscattered radiant energy can be measured.

backup A circuit element or facility used to replace an element that has failed.

backup supply A redundant power supply that takes over if the primary power supply fails.

balance The process of equalizing the voltage, current, or other parameter between two or more circuits or systems.

balanced A circuit having two sides (conductors) carrying voltages that are symmetrical about a common reference point, typically ground.

balanced circuit A circuit whose two sides are electrically equal in all transmission respects.

balanced line A transmission line consisting of two conductors in the presence of ground capable of being operated in such a way that when the voltages of the two conductors at all transverse planes are equal in magnitude and opposite in polarity with respect to ground, the currents in the two conductors are equal in magnitude and opposite in direction.

balanced modulator A modulator that combines the information signal and the carrier so that the output contains the two sidebands without the carrier.

balanced three-wire system A power distribution system using three conductors, one of which is balanced to have a potential midway between the potentials of the other two.

balanced-to-ground The condition when the impedance to ground on one wire of a two-wire circuit is equal to the impedance to ground on the other wire.

balun (balanced/unbalanced) A device used to connect balanced circuits with unbalanced circuits.

band A range of frequencies between a specified upper and lower limit.

band elimination filter A filter having a single continuous attenuation band, with neither the upper nor lower cut-off frequencies being zero or infinite. A *band elimination filter* may also be referred to as a *band-stop*, *notch*, or *band reject* filter.

bandpass filter A filter having a single continuous transmission band with neither the upper nor the lower cut-off frequencies being zero or infinite. A bandpass filter permits only a specific band of frequencies to pass; frequencies above or below are attenuated.

bandwidth The range of signal frequencies that can be transmitted by a communications channel with a defined maximum loss or distortion. Bandwidth indicates the information-carrying capacity of a channel.

bandwidth expansion ratio The ratio of the necessary bandwidth to the baseband bandwidth.

bandwidth-limited operation The condition prevailing when the frequency spectrum or bandwidth, rather than the amplitude (or power) of the signal, is the limiting factor in communication capability. This condition is reached when the system distorts the shape of the waveform beyond tolerable limits.

bank A group of similar items connected together in a specified manner and used in conjunction with one another.

bare A wire conductor that is not enameled or enclosed in an insulating sheath.

baseband The band of frequencies occupied by a signal before it modulates a carrier wave to form a transmitted radio or line signal.

baseband channel A channel that carries a signal without modulation, in contrast to a *passband* channel.

baseband signal The original form of a signal, unchanged by modulation.

bath tub The shape of a typical graph of component failure rates: high during an initial period of operation, falling to an acceptable low level during the normal usage period, and then rising again as the components become time-expired.

battery A group of several cells connected together to furnish current by conversion of chemical, thermal, solar, or nuclear energy into electrical energy. A single cell is itself sometimes also called a battery.

bay A row or suite of racks on which transmission, switching, and/or processing equipment is mounted.

Bel A unit of power measurement, named in honor of Alexander Graham Bell. The commonly used unit is one tenth of a Bel, or a decibel (dB). One Bel is defined as a tenfold increase in power. If an amplifier increases the power of a signal by 10 times, the power gain of the amplifier is equal to 1 Bel or 10 *decibels* (dB). If power is increased by 100 times, the power gain is 2 Bels or 20 decibels.

bend A transition component between two elements of a transmission waveguide.

bending radius The smallest bend that may be put into a cable under a stated pulling force. The bending radius is typically expressed in inches.

bias A dc voltage difference applied between two elements of an active electronic device, such as a vacuum tube, transistor, or integrated circuit. Bias currents may or may not be drawn, depending on the device and circuit type.

bidirectional An operational qualification which implies that the transmission of information occurs in both directions.

bifilar winding A type of winding in which two insulated wires are placed side by side. In some components, bifilar winding is used to produce balanced circuits.

bipolar A signal that contains both positive-going and negative-going amplitude components. A bipolar signal may also contain a zero amplitude state.

bleeder A high resistance connected in parallel with one or more filter capacitors in a high voltage dc system. If the power supply load is disconnected, the capacitors discharge through the bleeder.

block diagram An overview diagram that uses geometric figures to represent the principal divisions or sections of a circuit, and lines and arrows to show the path of a signal, or to show program functionalities. It is not a *schematic*, which provides greater detail.

blocking capacitor A capacitor included in a circuit to stop the passage of direct current.

BNC An abbreviation for *bayonet Neill-Concelman*, a type of cable connector used extensively in RF applications (named for its inventor).

Boltzmann's constant 1.38×10^{-23} joules.

bridge A type of network circuit used to match different circuits to each other, ensuring minimum transmission impairment.

bridging The shunting or paralleling of one circuit with another.

broadband The quality of a communications link having essentially uniform response over a given range of frequencies. A communications link is said to be *broadband* if it offers no perceptible degradation to the signal being transported.

buffer A circuit or component that isolates one electrical circuit from another.

burn-in The operation of a device, sometimes under extreme conditions, to stabilize its characteristics and identify latent component failures before bringing the device into normal service.

bus A central conductor for the primary signal path. The term bus may also refer to a signal path to which a number of inputs may be connected for feed to one or more outputs.

busbar A main dc power bus.

bypass capacitor A capacitor that provides a signal path that effectively shunts or bypasses other components.

bypass relay A switch used to bypass the normal electrical route of a signal or current in the event of power, signal, or equipment failure.

C

cable An electrically and/or optically conductive interconnecting device.

cable loss Signal loss caused by passing a signal through a coaxial cable. Losses are the result of resistance, capacitance, and inductance in the cable.

cable splice The connection of two pieces of cable by joining them mechanically and closing the joint with a weather-tight case or sleeve.

cabling The wiring used to interconnect electronic equipment.

calibrate The process of checking, and adjusting if necessary, a test instrument against one known to be set correctly.

calibration The process of identifying and measuring errors in instruments and/or procedures.

capacitance The property of a device or component that enables it to store energy in an electrostatic field and to release it later. A capacitor consists of two conductors separated by an insulating material. When the conductors have a voltage difference between them, a charge will be stored in the electrostatic field between the conductors.

capacitor A device that stores electrical energy. A capacitor allows the apparent flow of alternating current, while blocking the flow of direct current. The degree to which the device permits ac current flow depends on the frequency of the signal and the size of the capacitor. Capacitors are used in filters, delay-line components, couplers, frequency selectors, timing elements, voltage transient suppression, and other applications.

carrier A single frequency wave that, prior to transmission, is modulated by another wave containing information. A carrier may be modulated by manipulating its amplitude and/or frequency in direct relation to one or more applied signals.

carrier frequency The frequency of an unmodulated oscillator or transmitter. Also, the average frequency of a transmitter when a signal is frequency modulated by a symmetrical signal.

cascade connection A tandem arrangement of two or more similar component devices or circuits, with the output of one connected to the input of the next.

cascaded An arrangement of two or more circuits in which the output of one circuit is connected to the input of the next circuit.

cathode ray tube (CRT) A vacuum tube device, usually glass, that is narrow at one end and widens at the other to create a surface onto which images can be projected. The narrow end contains the necessary circuits to generate and focus an electron beam on the luminescent screen at the other end. CRTs are used to display pictures in TV receivers, video monitors, oscilloscopes, computers, and other systems.

cell An elementary unit of communication, of power supply, or of equipment.

Celsius A temperature measurement scale, expressed in degrees C, in which water freezes at 0°C and boils at 100°C. To convert to degrees Fahrenheit, multiply by 0.555 and add 32. To convert to Kelvins add 273 (approximately).

center frequency In frequency modulation, the resting frequency or initial frequency of the carrier before modulation.

center tap A connection made at the electrical center of a coil.

channel The smallest subdivision of a circuit that provides a single type of communication service.

channel decoder A device that converts an incoming modulated signal on a given channel back into the source-encoded signal.

channel encoder A device that takes a given signal and converts it into a form suitable for transmission over the communications channel.

channel noise level The ratio of the channel noise at any point in a transmission system to some arbitrary amount of circuit noise chosen as a reference. This ratio is usually expressed in *decibels above reference noise*, abbreviated *dBrn*.

channel reliability The percent of time a channel is available for use in a specific direction during a specified period.

channelization The allocation of communication circuits to channels and the forming of these channels into groups for higher order multiplexing.

characteristic The property of a circuit or component.

characteristic impedance The impedance of a transmission line, as measured at the driving point, if the line were of infinite length. In such a line, there would be no standing waves. The *characteristic impedance* may also be referred to as the *surge impedance*.

charge The process of replenishing or replacing the electrical charge in a secondary cell or storage battery.

charger A device used to recharge a battery. Types of charging include: (1) *constant voltage charge*, (2) *equalizing charge*, and (3) *trickle charge*.

chassis ground A connection to the metal frame of an electronic system that holds the components in a place. The chassis ground connection serves as the ground return or electrical common for the system.

circuit Any closed path through which an electrical current can flow. In a *parallel circuit*, components are connected between common inputs and outputs such that all paths are parallel to each other. The same voltage appears across all paths. In a *series circuit*, the same current flows through all components.

circuit noise level The ratio of the circuit noise at some given point in a transmission system to an established reference, usually expressed in decibels above the reference.

circuit reliability The percentage of time a circuit is available to the user during a specified period of scheduled availability.

circular mil The measurement unit of the cross-sectional area of a circular conductor. A *circular mil* is the area of a circle whose diameter is one mil, or 0.001 inch.

clear channel A transmission path wherein the full bandwidth is available to the user, with no portions of the channel used for control, framing, or signaling. Can also refer to a classification of AM broadcast station.

clipper A limiting circuit which ensures that a specified output level is not exceeded by restricting the output waveform to a maximum peak amplitude.

clipping The distortion of a signal caused by removing a portion of the waveform through restriction of the amplitude of the signal by a circuit or device.

coax A short-hand expression for *coaxial cable*, which is used to transport high-frequency signals.

coaxial cable A transmission line consisting of an inner conductor surrounded first by an insulating material and then by an outer conductor, either solid or braided.The mechanical dimensions of the cable determine its *characteristic impedance*.

coherence The correlation between the phases of two or more waves.

coherent The condition characterized by a fixed phase relationship among points on an electromagnetic wave.

coherent pulse The condition in which a fixed phase relationship is maintained between consecutive pulses during pulse transmission.

cold joint A soldered connection that was inadequately heated, with the result that the wire is held in place by rosin flux, not solder. A cold joint is sometimes referred to as a *dry joint*.

comb filter An electrical filter circuit that passes a series of frequencies and rejects the frequencies in between, producing a frequency response similar to the teeth of a comb.

common A point that acts as a reference for circuits, often equal in potential to the local ground.

common mode Signals identical with respect to amplitude, frequency, and phase that are applied to both terminals of a cable and/or both the input and reference of an amplifier.

common return A return path that is common to two or more circuits, and returns currents to their source or to ground.

common return offset The dc common return potential difference of a line.

communications system A collection of individual communications networks, transmission systems, relay stations, tributary stations, and terminal equipment capable of interconnection and interoperation to form an integral whole. The individual components must serve a common purpose, be technically compatible, employ common procedures, respond to some form of control, and, in general, operate in unison.

commutation A successive switching process carried out by a commutator.

commutator A circular assembly of contacts, insulated one from another, each leading to a different portion of the circuit or machine.

compatibility The ability of diverse systems to exchange necessary information at appropriate levels of command directly and in usable form. Communications equipment items are compatible if signals can be exchanged between them without the addition of buffering or translation for the specific purpose of achieving workable interface connections, and if the equipment or systems being interconnected possess comparable performance characteristics, including the suppression of undesired radiation.

complex wave A waveform consisting of two or more sinewave components. At any instant of time, a complex wave is the algebraic sum of all its sinewave components.

compliance For mechanical systems, a property which is the reciprocal of stiffness.

component An assembly, or part thereof, that is essential to the operation of some larger circuit or system. A *component* is an immediate subdivision of the assembly to which it belongs.

COMSAT The *Communications Satellite Corporation*, an organization established by an act of Congress in 1962. COMSAT launches and operates the international satellites for the INTELSAT consortium of countries.

concentricity A measure of the deviation of the center conductor position relative to its ideal location in the exact center of the dielectric cross-section of a coaxial cable.

conditioning The adjustment of a channel in order to provide the appropriate transmission characteristics needed for data or other special services.

conditioning equipment The equipment used to match transmission levels and impedances, and to provide equalization between facilities.

conductance A measure of the capability of a material to conduct electricity. It is the reciprocal of *resistance* (ohm) and is expressed in *siemens*. (Formerly expressed as *mho*.)

conducted emission An electromagnetic energy propagated along a conductor.

conduction The transfer of energy through a medium, such as the conduction of electricity by a wire, or of heat by a metallic frame.

conduction band A partially filled or empty atomic energy band in which electrons are free to move easily, allowing the material to carry an electric current.

conductivity The conductance per unit length.

conductor Any material that is capable of carrying an electric current.

configuration A relative arrangement of parts.

connection A point at which a junction of two or more conductors is made.

connector A device mounted on the end of a wire or fiber optic cable that mates to a similar device on a specific piece of equipment or another cable.

constant-current source A source with infinitely high output impedance so that output current is independent of voltage, for a specified range of output voltages.

constant-voltage charge A method of charging a secondary cell or storage battery during which the terminal voltage is kept at a constant value.

constant-voltage source A source with low, ideally zero, internal impedance, so that voltage will remain constant, independent of current supplied.

contact The points that are brought together or separated to complete or break an electrical circuit.

contact bounce The rebound of a contact, which temporarily opens the circuit after its initial *make*.

contact form The configuration of a contact assembly on a relay. Many different configurations are possible from simple *single-make* contacts to complex arrangements involving *breaks* and *makes*.

contact noise A noise resulting from current flow through an electrical contact that has a rapidly varying resistance, as when the contacts are corroded or dirty.

contact resistance The resistance at the surface when two conductors make contact.

continuity A continuous path for the flow of current in an electrical circuit.

continuous wave An electromagnetic signal in which successive oscillations of the waves are identical.

control The supervision that an operator or device exercises over a circuit or system.

control grid The grid in an electron tube that controls the flow of current from the cathode to the anode.

convention A generally acceptable symbol, sign, or practice in a given industry.

Coordinated Universal Time (UTC) The time scale, maintained by the BIH (Bureau International de l'Heure) that forms the basis of a coordinated dissemination of standard frequencies and time signals.

copper loss The loss resulting from the heating effect of current.

corona A bluish luminous discharge resulting from ionization of the air near a conductor carrying a voltage gradient above a certain *critical level*.

corrective maintenance The necessary tests, measurements, and adjustments required to remove or correct a fault.

cosmic noise The random noise originating outside the earth's atmosphere.

coulomb The standard unit of electric quantity or charge. One *coulomb* is equal to the quantity of electricity transported in 1 second by a current of 1 ampere.

Coulomb's Law The attraction and repulsion of electric charges act on a line between them. The charges are inversely proportional to the square of the distance between them, and proportional to the product of their magnitudes. (Named for the French physicist Charles-Augustine de Coulomb, 1736–1806.)

counter-electromotive force The effective electromotive force within a system that opposes the passage of current in a specified direction.

couple The process of linking two circuits by inductance, so that energy is transferred from one circuit to another.

coupled mode The selection of either ac or dc coupling.

coupling The relationship between two components that enables the transfer of energy between them. Included are *direct coupling* through a direct electrical connection, such as a wire; *capacitive coupling* through the capacitance formed by two adjacent conductors; and *inductive coupling* in which energy is transferred through a magnetic field. Capacitive coupling is also called *electrostatic coupling*. Inductive coupling is often referred to as *electromagnetic coupling*.

coupling coefficient A measure of the electrical coupling that exists between two circuits. The *coupling coefficient* is equal to the ratio of the mutual impedance to the square root of the product of the self impedances of the coupled circuits.

cross coupling The coupling of a signal from one channel, circuit, or conductor to another, where it becomes an undesired signal.

crossover distortion A distortion that results in an amplifier when an irregularity is introduced into the signal as it crosses through a zero reference point. If an amplifier is properly designed and biased, the upper half cycle and lower half cycle of the signal coincide at the zero crossover reference.

crossover frequency The frequency at which output signals pass from one channel to the other in a *crossover network*. At the *crossover frequency* itself, the outputs to each side are equal.

crossover network A type of filter that divides an incoming signal into two or more outputs, with higher frequencies directed to one output, and lower frequencies to another.

crosstalk Undesired transmission of signals from one circuit into another circuit in the same system. Crosstalk is usually caused by unintentional capacitive (ac) coupling.

crosstalk coupling The ratio of the power in a disturbing circuit to the induced power in the disturbed circuit, observed at a particular point under specified conditions. Crosstalk coupling is typically expressed in dB.

crowbar A short-circuit or low resistance path placed across the input to a circuit, usually for protective purposes.

CRT (cathode ray tube) A vacuum tube device that produces light when energized by the electron beam generated inside the tube. A CRT includes an electron gun, deflection mechanism, and phosphor-covered faceplate.

crystal A solidified form of a substance that has atoms and molecules arranged in a symmetrical pattern.

crystal filter A filter that uses piezoelectric crystals to create resonant or antiresonant circuits.

crystal oscillator An oscillator using a piezoelectric crystal as the tuned circuit that controls the resonant frequency.

crystal-controlled oscillator An oscillator in which a piezoelectric-effect crystal is coupled to a tuned oscillator circuit in such a way that the crystal pulls the oscillator frequency to its own natural frequency and does not allow frequency drift.

current (1—general) A general term for the transfer of electricity, or the movement of electrons or *holes*. **(2—alternating)** An electric current that is constantly varying in amplitude and periodically reversing direction. **(3—average)** The arithmetic mean of the instantaneous values of current, averaged over one complete half cycle. **(4—charging)** The current that flows in to charge a capacitor when it is first connected to a source of electric potential. **(5—direct)** Electric current that flows in one direction only. **(6—eddy)** A wasteful current that flows in the core of a transformer and produces heat. *Eddy currents* are largely eliminated through the use of laminated cores. **(7—effective)** The ac current that will produce the same effective heat in a resistor as is produced by dc. If the ac is sinusoidal, the *effective current* value is 0.707 times the peak ac value. **(8—fault)** The current that flows between conductors or to ground during a fault condition. **(9—ground fault)** A fault current that flows to ground. **(10—ground return)** A current that returns through the earth. **(11—lagging)** A phenomenon observed in an inductive circuit where alternating current lags behind the voltage that produces it. **(12—leading)** A phenomenon observed in a capacitive circuit where alternating current leads the voltage that produces it. **(13—magnetizing)** The current in a transformer primary winding that is just sufficient to magnetize the core and offset iron losses. **(14—neutral)** The current that flows in the neutral conductor of an unbalanced polyphase power circuit. If correctly balanced, the neutral would carry no net current. **(15—peak)** The maximum value reached by a varying current during one cycle. **(16—pick-up)** The minimum current at which a relay just begins to operate. **(17—plate)** The anode current of an electron tube. **(18—residual)** The vector sum of the currents in the phase wires of an unbalanced polyphase power circuit. **(19—space)** The total current flowing through an electron tube.

current amplifier A low output impedance amplifier capable of providing high current output.

current probe A sensor, clamped around an electrical conductor, in which an induced current is developed from the magnetic field surrounding the conductor. For measurements, the current probe is connected to a suitable test instrument.

current transformer A transformer-type of instrument in which the primary carries the current to be measured and the secondary is in series with a low current ammeter. A current transformer is used to measure high values of alternating current.

current-carrying capacity A measure of the maximum current that can be carried continuously without damage to components or devices in a circuit.

cut-off frequency The frequency above or below which the output current in a circuit is reduced to a specified level.

cycle The interval of time or space required for a periodic signal to complete one period.

cycles per second The standard unit of frequency, expressed in Hertz (one cycle per second).

D

damped oscillation An oscillation exhibiting a progressive diminution of amplitude with time.

damping The dissipation and resultant reduction of any type of energy, such as electromagnetic waves.

dB (decibel) A measure of voltage, current, or power gain equal to 0.1 Bel. Decibels are given by the equations 20 log V_{out}/V_{in}, 20 log I_{out}/I_{in}, or 10 log P_{out}/P_{in}.

dBk A measure of power relative to 1 kilowatt. 0 dBk equals 1 kW.

dBm (decibels above 1 milliwatt) A logarithmic measure of power with respect to a reference power of one milliwatt.

dBmv A measure of voltage gain relative to 1 millivolt at 75 ohms.

dBr The power difference expressed in dB between any point and a reference point selected as the *zero relative transmission level* point. A power expressed in *dBr* does not specify the absolute power; it is a relative measurement only.

dBu A term that reflects comparison between a measured value of voltage and a reference value of 0.775 V, expressed under conditions in which the impedance at the point of measurement (and of the reference source) are not considered.

dbV A measure of voltage gain relative to 1 V.

dBW A measure of power relative to 1 watt. 0 dBW equals 1 W.

dc An abbreviation for *direct current*. Note that the preferred usage of the term *dc* is lower case.

dc amplifier A circuit capable of amplifying dc and slowly varying alternating current signals.

dc component The portion of a signal that consists of direct current. This term may also refer to the average value of a signal.

dc coupled A connection configured so that both the signal (ac component) and the constant voltage on which it is riding (dc component) are passed from one stage to the next.

dc coupling A method of coupling one circuit to another so as to transmit the static (dc) characteristics of the signal as well as the varying (ac) characteristics. Any dc offset present on the input signal is maintained and will be present in the output.

dc offset The amount that the dc component of a given signal has shifted from its correct level.

dc signal bounce Overshoot of the proper dc voltage level resulting from multiple ac couplings in a signal path.

de-energized A system from which sources of power have been disconnected.

deca A prefix meaning *ten*.

decay The reduction in amplitude of a signal on an exponential basis.

decay time The time required for a signal to fall to a certain fraction of its original value.

decibel (dB) One tenth of a Bel. The decibel is a logarithmic measure of the ratio between two powers.

decode The process of recovering information from a signal into which the information has been encoded.

decoder A device capable of deciphering encoded signals. A decoder interprets input instructions and initiates the appropriate control operations as a result.

decoupling The reduction or removal of undesired coupling between two circuits or stages.

deemphasis The reduction of the high-frequency components of a received signal to reverse the preemphasis that was placed on them to overcome attenuation and noise in the transmission process.

defect An error made during initial planning that is normally detected and corrected during the development phase. Note that a *fault* is an error that occurs in an in-service system.

deflection The control placed on electron direction and motion in CRTs and other vacuum tube devices by varying the strengths of electrostatic (electrical) or electromagnetic fields.

degradation In susceptibility testing, any undesirable change in the operational performance of a test specimen. This term does not necessarily mean malfunction or catastrophic failure.

degradation failure A failure that results from a gradual change in performance characteristics of a system or part with time.

delay The amount of time by which a signal is delayed or an event is retarded.

delay circuit A circuit designed to delay a signal passing through it by a specified amount.

delay distortion The distortion resulting from the difference in phase delays at two frequencies of interest.

delay equalizer A network that adjusts the velocity of propagation of the frequency components of a complex signal to counteract the delay distortion characteristics of a transmission channel.

delay line A transmission network that increases the propagation time of a signal traveling through it.

delta connection A common method of joining together a three-phase power supply, with each phase across a different pair of the three wires used.

delta-connected system A 3-phase power distribution system where a single-phase output can be derived from each of the adjacent pairs of an equilateral triangle formed by the service drop transformer secondary windings.

demodulator Any device that recovers the original signal after it has modulated a high-frequency carrier. The output from the unit may be in baseband composite form.

demultiplexer (demux) A device used to separate two or more signals that were previously combined by a compatible multiplexer and are transmitted over a single channel.

derating factor An operating safety margin provided for a component or system to ensure reliable performance. A *derating allowance* also is typically provided for operation under extreme environmental conditions, or under stringent reliability requirements.

desiccant A drying agent used for drying out cable splices or sensitive equipment.

design A layout of all the necessary equipment and facilities required to make a special circuit, piece of equipment, or system work.

design objective The desired electrical or mechanical performance characteristic for electronic circuits and equipment.

detection The rectification process that results in the modulating signal being separated from a modulated wave.

detectivity The reciprocal of *noise equivalent power*.

detector A device that converts one type of energy into another.

device A functional circuit, component, or network unit, such as a vacuum tube or transistor.

dewpoint The temperature at which moisture will condense out.

diagnosis The process of locating errors in software, or equipment faults in hardware.

diagnostic routine A software program designed to trace errors in software, locate hardware faults, or identify the cause of a breakdown.

dielectric An insulating material that separates the elements of various components, including capacitors and transmission lines. Dielectric materials include air, plastic, mica, ceramic, and Teflon. A dielectric material must be an insulator. (*Teflon* is a registered trademark of Du Pont.)

dielectric constant The ratio of the capacitance of a capacitor with a certain dielectric material to the capacitance with a vacuum as the dielectric. The *dielectric constant* is considered a measure of the capability of a dielectric material to store an electrostatic charge.

dielectric strength The potential gradient at which electrical breakdown occurs.

differential amplifier An input circuit that rejects voltages that are the same at both input terminals but amplifies any voltage difference between the inputs. Use of a differential amplifier causes any signal present on both terminals, such as common mode hum, to cancel itself.

differential dc The maximum dc voltage that can be applied between the differential inputs of an amplifier while maintaining linear operation.

differential gain The difference in output amplitude (expressed in percent or dB) of a small high frequency sinewave signal at two stated levels of a low frequency signal on which it is superimposed.

differential phase The difference in output phase of a small high frequency sinewave signal at two stated levels of a low frequency signal on which it is superimposed.

differential-mode interference An interference source that causes a change in potential of one side of a signal transmission path relative to the other side.

diffuse reflection The scattering effect that occurs when light, radio, or sound waves strike a rough surface.

diffusion The spreading or scattering of a wave, such as a radio wave.

diode A semiconductor or vacuum tube with two electrodes that passes electric current in one direction only. Diodes are used in rectifiers, gates, modulators, and detectors.

direct coupling A coupling method between stages that permits dc current to flow between the stages.

direct current An electrical signal in which the direction of current flow remains constant.

discharge The conversion of stored energy, as in a battery or capacitor, into an electric current.

discontinuity An abrupt nonuniform point of change in a transmission circuit that causes a disruption of normal operation.

discrete An individual circuit component.

discrete component A separately contained circuit element with its own external connections.

discriminator A device or circuit whose output amplitude and polarity vary according to how much the input signal varies from a standard or from another signal. A discriminator can be used to recover the modulating waveform in a frequency modulated signal.

dish An antenna system consisting of a parabolic shaped reflector with a signal feed element at the focal point. Dish antennas commonly are used for transmission and reception from microwave stations and communications satellites.

dispersion The wavelength dependence of a parameter.

display The representation of text and images on a cathode-ray tube, an array of light-emitting diodes, a liquid-crystal readout, or another similar device.

display device An output unit that provides a visual representation of data.

distortion The difference between the wave shape of an original signal and the signal after it has traversed a transmission circuit.

distortion-limited operation The condition prevailing when the shape of the signal, rather than the amplitude (or power), is the limiting factor in communication capability. This condition is reached when the system distorts the shape of the waveform beyond tolerable limits. For linear systems, *distortion-limited* operation is equivalent to *bandwidth-limited* operation.

disturbance The interference with normal conditions and communications by some external energy source.

disturbance current The unwanted current of any irregular phenomenon associated with transmission that tends to limit or interfere with the interchange of information.

disturbance power The unwanted power of any irregular phenomenon associated with transmission that tends to limit or interfere with the interchange of information.

disturbance voltage The unwanted voltage of any irregular phenomenon associated with transmission that tends to limit or interfere with the interchange of information.

diversity receiver A receiver using two antennas connected through circuitry that senses which antenna is receiving the stronger signal. Electronic gating permits the stronger source to be routed to the receiving system.

documentation A written description of a program. *Documentation* can be considered as any record that has permanence and can be read by humans or machines.

down-lead A lead-in wire from an antenna to a receiver.

downlink The portion of a communication link used for transmission of signals from a satellite or airborne platform to a surface terminal.

downstream A specified signal modification occurring after other given devices in a signal path.

downtime The time during which equipment is not capable of doing useful work because of malfunction. This does not include preventive maintenance time. In other words, *downtime* is measured from the occurrence of a malfunction to the correction of that malfunction.

drift A slow change in a nominally constant signal characteristic, such as frequency.

drift-space The area in a klystron tube in which electrons drift at their entering velocities and form electron *bunches*.

drive The input signal to a circuit, particularly to an amplifier.

driver An electronic circuit that supplies an isolated output to drive the input of another circuit.

drop-out value The value of current or voltage at which a relay will cease to be operated.

dropout The momentary loss of a signal.

dropping resistor A resistor designed to carry current that will make a required voltage available.

duplex separation The frequency spacing required in a communications system between the *forward* and *return* channels to maintain interference at an acceptably low level.

duplex signaling A configuration permitting signaling in both transmission directions simultaneously.

duty cycle The ratio of operating time to total elapsed time of a device that operates intermittently, expressed in percent.

dynamic A situation in which the operating parameters and/or requirements of a given system are continually changing.

dynamic range The maximum range or extremes in amplitude, from the lowest to the highest (noise floor to system clipping), that a system is capable of reproducing. The dynamic range is expressed in dB against a reference level.

dynamo A rotating machine, normally a dc generator.

dynamotor A rotating machine used to convert dc into ac.

E

earth A large conducting body with no electrical potential, also called *ground.*

earth capacitance The capacitance between a given circuit or component and a point at ground potential.

earth current A current that flows to earth/ground, especially one that follows from a fault in the system. *Earth current* may also refer to a current that flows in the earth, resulting from ionospheric disturbances, lightning, or faults on power lines.

earth fault A fault that occurs when a conductor is accidentally grounded/earthed, or when the resistance to earth of an insulator falls below a specified value.

earth ground A large conducting body that represents *zero level* in the scale of electrical potential. An *earth ground* is a connection made either accidentally or by design between a conductor and earth.

earth potential The potential taken to be the arbitrary zero in a scale of electric potential.

effective ground A connection to ground through a medium of sufficiently low impedance and adequate current-carrying capacity to prevent the buildup of voltages that might be hazardous to equipment or personnel.

effective resistance The increased resistance of a conductor to an alternating current resulting from the *skin effect*, relative to the direct-current resistance of the conductor. Higher frequencies tend to travel only on the outer skin of the conductor, whereas dc flows uniformly through the entire area.

efficiency The useful power output of an electrical device or circuit divided by the total power input, expressed in percent.

electric Any device or circuit that produces, operates on, transmits, or uses electricity.

electric charge An excess of either electrons or protons within a given space or material.

electric field strength The magnitude, measured in volts per meter, of the electric field in an electromagnetic wave.

electric flux The amount of electric charge, measured in coulombs, across a dielectric of specified area. *Electric flux* may also refer simply to electric lines of force.

electricity An energy force derived from the movement of negative and positive electric charges.

electrode An electrical terminal that emits, collects, or controls an electric current.

electrolysis A chemical change induced in a substance resulting from the passage of electric current through an electrolyte.

electrolyte A nonmetallic conductor of electricity in which current is carried by the physical movement of ions.

electromagnet An iron or steel core surrounded by a wire coil. The core becomes magnetized when current flows through the coil but loses its magnetism when the current flow is stopped.

electromagnetic compatibility The capability of electronic equipment or systems to operate in a specific electromagnetic environment, at designated levels of efficiency and within a defined margin of safety, without interfering with itself or other systems.

electromagnetic field The electric and magnetic fields associated with radio and light waves.

electromagnetic induction An electromotive force created with a conductor by the relative motion between the conductor and a nearby magnetic field.

electromagnetism The study of phenomena associated with varying magnetic fields, electromagnetic radiation, and moving electric charges.

electromotive force (EMF) An electrical potential, measured in volts, that can produce the movement of electrical charges.

electron A stable elementary particle with a negative charge that is mainly responsible for electrical conduction. Electrons move when under the influence of an electric field. This movement constitutes an *electric current.*

electron beam A stream of emitted electrons, usually in a vacuum.

electron gun A hot cathode that produces a finely focused stream of fast electrons, which are necessary for the operation of a vacuum tube, such as a cathode ray tube. The gun is made up of a hot cathode electron source, a control grid, accelerating anodes, and (usually) focusing electrodes.

electron lens A device used for focusing an electron beam in a cathode ray tube. Such focusing can be accomplished by either magnetic forces, in which external coils are used to create the proper magnetic field within the tube, or electrostatic forces, where metallic plates within the tube are charged electrically in such a way as to control the movement of electrons in the beam.

electron volt The energy acquired by an electron in passing through a potential difference of one volt in a vacuum.

electronic A description of devices (or systems) that are dependent on the flow of electrons in electron tubes, semiconductors, and other devices, and not solely on electron flow in ordinary wires, inductors, capacitors, and similar passive components.

Electronic Industries Association (EIA) A trade organization, based in Washington, DC, representing the manufacturers of electronic systems and parts, including communications systems. The association develops standards for electronic components and systems.

electronic switch A transistor, semiconductor diode, or a vacuum tube used as an on/off switch in an electrical circuit. Electronic switches can be controlled manually, by other circuits, or by computers.

electronics The field of science and engineering that deals with electron devices and their utilization.

electroplate The process of coating a given material with a deposit of metal by electrolytic action.

electrostatic The condition pertaining to electric charges that are at rest.

electrostatic field The space in which there is electric stress produced by static electric charges.

electrostatic induction The process of inducing static electric charges on a body by bringing it near other bodies that carry high electrostatic charges.

element A substance that consists of atoms of the same atomic number. Elements are the basic units in all chemical changes other than those in which *atomic changes*, such as fusion and fission, are involved.

EMI (electromagnetic interference) Undesirable electromagnetic waves that are radiated unintentionally from an electronic circuit or device into other circuits or devices, disrupting their operation.

emission (1—radiation) The radiation produced, or the production of radiation by a radio transmitting system. The emission is considered to be a *single emission* if the modulating signal and other characteristics are the same for every transmitter of the radio transmitting system and the spacing between antennas is not more than a few wavelengths. **(2—cathode)** The release of electrons from the cathode of a vacuum tube. **(3—parasitic)** A spurious radio frequency emission unintentionally generated at frequencies that are independent of the carrier frequency being amplified or modulated. **(4—secondary)** In an electron tube, emission of electrons by a plate or grid because of bombardment by *primary emission* electrons from the cathode of the tube. **(5—spurious)** An emission outside the radio frequency band authorized for a transmitter. **(6—thermonic)** An emission from a cathode resulting from high temperature.

emphasis The intentional alteration of the frequency-amplitude characteristics of a signal to reduce the adverse effects of noise in a communication system.

empirical A conclusion not based on pure theory, but on practical and experimental work.

emulation The use of one system to imitate the capabilities of another system.

enable To prepare a circuit for operation or to allow an item to function.

enabling signal A signal that permits the occurrence of a specified event.

encode The conversion of information from one form into another to obtain characteristics required by a transmission or storage system.

encoder A device that processes one or more input signals into a specified form for transmission and/or storage.

energized The condition when a circuit is switched on, or powered up.

energy spectral density A frequency-domain description of the energy in each of the frequency components of a pulse.

envelope The boundary of the family of curves obtained by varying a parameter of a wave.

envelope delay The difference in absolute delay between the fastest and slowest propagating frequencies within a specified bandwidth.

envelope delay distortion The maximum difference or deviation of the envelope-delay characteristic between any two specified frequencies.

envelope detection A demodulation process that senses the shape of the modulated RF envelope. A diode detector is one type of envelop detection device.

environmental An equipment specification category relating to temperature and humidity.

EQ (equalization) network A network connected to a circuit to correct or control its transmission frequency characteristics.

equalization (EQ) The reduction of frequency distortion and/or phase distortion of a circuit through the introduction of one or more networks to compensate for the difference in attenuation, time delay, or both, at the various frequencies in the transmission band.

equalize The process of inserting in a line a network with complementary transmission characteristics to those of the line, so that when the loss or delay in the line and that in the equalizer are combined, the overall loss or delay is approximately equal at all frequencies.

equalizer A network that corrects the transmission-frequency characteristics of a circuit to allow it to transmit selected frequencies in a uniform manner.

equatorial orbit The plane of a satellite orbit which coincides with that of the equator of the primary body.

equipment A general term for electrical apparatus and hardware, switching systems, and transmission components.

equipment failure The condition when a hardware fault stops the successful completion of a task.

equipment ground A protective ground consisting of a conducting path to ground of noncurrent carrying metal parts.

equivalent circuit A simplified network that emulates the characteristics of the real circuit it replaces. An equivalent circuit is typically used for mathematical analysis.

equivalent noise resistance A quantitative representation in resistance units of the spectral density of a noise voltage generator at a specified frequency.

error A collective term that includes all types of inconsistencies, transmission deviations, and control failures.

excitation The current that energizes field coils in a generator.

expandor A device with a nonlinear gain characteristic that acts to increase the gain more on larger input signals than it does on smaller input signals.

extremely high frequency (EHF) The band of microwave frequencies between the limits of 30 GHz and 300 GHz (wavelengths between 1 cm and 1 mm).

extremely low frequency The radio signals with operating frequencies below 300 Hz (wavelengths longer than 1000 km).

F

fail-safe operation A type of control architecture for a system that prevents improper functioning in the event of circuit or operator failure.

failure A detected cessation of ability to perform a specified function or functions within previously established limits. A *failure* is beyond adjustment by the operator by means of controls normally accessible during routine operation of the system. (This requires that measurable limits be established to define "satisfactory performance".)

failure effect The result of the malfunction or failure of a device or component.

failure in time (FIT) A unit value that indicates the reliability of a component or device. One failure in time corresponds to a failure rate of 10^{-9} per hour.

failure mode and effects analysis (FMEA) An iterative documented process performed to identify basic faults at the component level and determine their effects at higher levels of assembly.

failure rate The ratio of the number of actual failures to the number of times each item has been subjected to a set of specified stress conditions.

fall time The length of time during which a pulse decreases from 90 percent to 10 percent of its maximum amplitude.

farad The standard unit of capacitance equal to the value of a capacitor with a potential of one volt between its plates when the charge on one plate is one coulomb and there is an equal and opposite charge on the other plate. The farad is a large value and is more commonly expressed in *microfarads* or *picofarads*. The *farad* is named for the English chemist and physicist Michael Faraday (179–1867).

fast frequency shift keying (FFSK) A system of digital modulation where the digits are represented by different frequencies that are related to the baud rate, and where transitions occur at the zero crossings.

fatigue The reduction in strength of a metal caused by the formation of crystals resulting from repeated flexing of the part in question.

fault A condition that causes a device, a component, or an element to fail to perform in a required manner. Examples include a short-circuit, broken wire, or intermittent connection.

fault to ground A fault caused by the failure of insulation and the consequent establishment of a direct path to ground from a part of the circuit that should not normally be grounded.

fault tree analysis (FTA) An iterative documented process of a systematic nature performed to identify basic faults, determine their causes and effects, and establish their probabilities of occurrence.

feature A distinctive characteristic or part of a system or piece of equipment, usually visible to end users and designed for their convenience.

Federal Communications Commission (FCC) The federal agency empowered by law to regulate all interstate radio and wireline communications services originating in the United States, including radio, television, facsimile, telegraph, data transmission, and telephone systems. The agency was established by the Communications Act of 1934.

feedback The return of a portion of the output of a device to the input. *Positive feedback* adds to the input, *negative feedback* subtracts from the input.

feedback amplifier An amplifier with the components required to feed a portion of the output back into the input to alter the characteristics of the output signal.

feedline A transmission line, typically coaxial cable, that connects a high frequency energy source to its load.

femto A prefix meaning *one quadrillionth* (10^{-15}).

ferrite A ceramic material made of powdered and compressed ferric oxide, plus other oxides (mainly cobalt, nickel, zinc, yttrium-iron, and manganese). These materials have low eddy current losses at high frequencies.

ferromagnetic material A material with low relative permeability and high coercive force so that it is difficult to magnetize and demagnetize. Hard ferromagnetic materials retain magnetism well, and are commonly used in permanent magnets.

fidelity The degree to which a system, or a portion of a system, accurately reproduces at its output the essential characteristics of the signal impressed upon its input.

field strength The strength of an electric, magnetic, or electromagnetic field.

filament A wire that becomes hot when current is passed through it, used either to emit light (for a light bulb) or to heat a cathode to enable it to emit electrons (for an electron tube).

film resistor A type of resistor made by depositing a thin layer of resistive material on an insulating core.

filter A network that passes desired frequencies but greatly attenuates other frequencies.

filtered noise White noise that has been passed through a filter. The power spectral density of filtered white noise has the same shape as the transfer function of the filter.

fitting A coupling or other mechanical device that joins one component with another.

fixed A system or device that is not changeable or movable.

flashover An arc or spark between two conductors.

flashover voltage The voltage between conductors at which flashover just occurs.

flat face tube The design of CRT tube with almost a flat face, giving improved legibility of text and reduced reflection of ambient light.

flat level A signal that has an equal amplitude response for all frequencies within a stated range.

flat loss A circuit, device, or channel that attenuates all frequencies of interest by the same amount, also called *flat slope*.

flat noise A noise whose power per unit of frequency is essentially independent of frequency over a specified frequency range.

flat response The performance parameter of a system in which the output signal amplitude of the system is a faithful reproduction of the input amplitude over some range of specified input frequencies.

floating A circuit or device that is not connected to any source of potential or to ground.

fluorescence The characteristic of a material to produce light when excited by an external energy source. Minimal or no heat results from the process.

flux The electric or magnetic lines of force resulting from an applied energy source.

flywheel effect The characteristic of an oscillator that enables it to sustain oscillations after removal of the control stimulus. This characteristic may be desirable, as in the case of a phase-locked loop employed in a synchronous system, or undesirable, as in the case of a voltage-controlled oscillator.

focusing A method of making beams of radiation converge on a target, such as the face of a CRT.

Fourier analysis A mathematical process for transforming values between the frequency domain and the time domain. This term also refers to the decomposition of a time-domain signal into its frequency components.

Fourier transform An integral that performs an actual transformation between the frequency domain and the time domain in Fourier analysis.

frame A segment of an analog or digital signal that has a repetitive characteristic, in that corresponding elements of successive *frames* represent the same things.

free electron An electron that is not attached to an atom and is, thus, mobile when an electromotive force is applied.

free running An oscillator that is not controlled by an external synchronizing signal.

free-running oscillator An oscillator that is not synchronized with an external timing source.

frequency The number of complete cycles of a periodic waveform that occur within a given length of time. Frequency is usually specified in cycles per second (*Hertz*). Frequency is the reciprocal of wavelength. The higher the frequency, the shorter the wavelength. In general, the higher the frequency of a signal, the more capacity it has to carry information, the smaller an antenna is required, and the more susceptible the signal is to absorption by the atmosphere and by physical structures. At microwave frequencies, radio signals take on a *line-of-sight* characteristic and require highly directional and focused antennas to be used successfully.

frequency accuracy The degree of conformity of a given signal to the specified value of a frequency.

frequency allocation The designation of radio-frequency bands for use by specific radio services.

frequency content The band of frequencies or specific frequency components contained in a signal.

frequency converter A circuit or device used to change a signal of one frequency into another of a different frequency.

frequency coordination The process of analyzing frequencies in use in various bands of the spectrum to achieve reliable performance for current and new services.

frequency counter An instrument or test set used to measure the frequency of a radio signal or any other alternating waveform.

frequency departure An unintentional deviation from the nominal frequency value.

frequency difference The algebraic difference between two frequencies. The two frequencies can be of identical or different nominal values.

frequency displacement The end-to-end shift in frequency that may result from independent frequency translation errors in a circuit.

frequency distortion The distortion of a multifrequency signal caused by unequal attenuation or amplification at the different frequencies of the signal. This term may also be referred to as *amplitude distortion*.

frequency domain A representation of signals as a function of frequency, rather than of time.

frequency modulation (FM) The modulation of a carrier signal so that its instantaneous frequency is proportional to the instantaneous value of the modulating wave.

frequency multiplier A circuit that provides as an output an exact multiple of the input frequency.

frequency offset A frequency shift that occurs when a signal is sent over an analog transmission facility in which the modulating and demodulating frequencies are not identical. A channel with frequency offset does not preserve the waveform of a transmitted signal.

frequency response The measure of system linearity in reproducing signals across a specified bandwidth. Frequency response is expressed as a frequency range with a specified amplitude tolerance in dB.

frequency response characteristic The variation in the transmission performance (gain or loss) of a system with respect to variations in frequency.

frequency reuse A technique used to expand the capacity of a given set of frequencies or channels by separating the signals either geographically or through the use of different polarization techniques. Frequency reuse is a common element of the *frequency coordination* process.

frequency selectivity The ability of equipment to separate or differentiate between signals at different frequencies.

frequency shift The difference between the frequency of a signal applied at the input of a circuit and the frequency of that signal at the output.

frequency shift keying (FSK) A commonly-used method of digital modulation in which a one and a zero (the two possible states) are each transmitted as separate frequencies.

frequency stability A measure of the variations of the frequency of an oscillator from its mean frequency over a specified period of time.

frequency standard An oscillator with an output frequency sufficiently stable and accurate that it is used as a reference.

frequency-division multiple access (FDMA) The provision of multiple access to a transmission facility, such as an earth satellite, by assigning each transmitter its own frequency band.

frequency-division multiplexing (FDM) The process of transmitting multiple analog signals by an orderly assignment of frequency slots, that is, by dividing transmission bandwidth into several narrow bands, each of which carries a single communication and is sent simultaneously with others over a common transmission path.

full duplex A communications system capable of transmission simultaneously in two directions.

full-wave rectifier A circuit configuration in which both positive and negative half-cycles of the incoming ac signal are rectified to produce a unidirectional (dc) current through the load.

functional block diagram A diagram illustrating the definition of a device, system, or problem on a logical and functional basis.

functional unit An entity of hardware and/or software capable of accomplishing a given purpose.

fundamental frequency The lowest frequency component of a complex signal.

fuse A protective device used to limit current flow in a circuit to a specified level. The fuse consists of a metallic link that melts and opens the circuit at a specified current level.

fuse wire A fine-gauge wire made of an alloy that overheats and melts at the relatively low temperatures produced when the wire carries overload currents. When used in a fuse, the wire is called a fuse (or fusible) link.

G

gain An increase or decrease in the level of an electrical signal. Gain is measured in terms of decibels or number-of-times of magnification. Strictly speaking, *gain* refers to an increase in level. Negative numbers, however, are commonly used to denote a decrease in level.

gain-bandwidth The gain times the frequency of measurement when a device is biased for maximum obtainable gain.

gain/frequency characteristic The gain-versus-frequency characteristic of a channel over the bandwidth provided, also referred to as *frequency response*.

gain/frequency distortion A circuit defect in which a change in frequency causes a change in signal amplitude.

galvanic A device that produces direct current by chemical action.

gang The mechanical connection of two or more circuit devices so that they can all be adjusted simultaneously.

gang capacitor A variable capacitor with more than one set of moving plates linked together.

gang tuning The simultaneous tuning of several different circuits by turning a single shaft on which ganged capacitors are mounted.

ganged One or more devices that are mechanically coupled, normally through the use of a shared shaft.

gas breakdown The ionization of a gas between two electrodes caused by the application of a voltage that exceeds a threshold value. The ionized path has a low impedance. Certain types of circuit and line protectors rely on gas breakdown to divert hazardous currents away from protected equipment.

gas tube A protection device in which a sufficient voltage across two electrodes causes a gas to ionize, creating a low impedance path for the discharge of dangerous voltages.

gas-discharge tube A gas-filled tube designed to carry current during gas breakdown. The gas-discharge tube is commonly used as a protective device, preventing high voltages from damaging sensitive equipment.

gauge A measure of wire diameter. In measuring wire gauge, the lower the number, the thicker the wire.

Gaussian distribution A statistical distribution, also called the *normal* distribution. The graph of a Gaussian distribution is a bell-shaped curve.

Gaussian noise Noise in which the distribution of amplitude follows a Gaussian model, that is, the noise is random but distributed about a reference voltage of zero.

Gaussian pulse A pulse that has the same form as its own Fourier transform.

generator A machine that converts mechanical energy into electrical energy, or one form of electrical energy into another form.

geosynchronous The attribute of a satellite in which the relative position of the satellite as viewed from the surface of a given planet is stationary. For earth, the geosynchronous position is 22,300 miles above the planet.

getter A metal used in vaporized form to remove residual gas from inside an electron tube during manufacture.

giga A prefix meaning one billion.

gigahertz (GHz) A measure of frequency equal to one billion cycles per second. Signals operating above 1 gigahertz are commonly known as *microwaves*, and begin to take on the characteristics of visible light.

glitch A general term used to describe a wide variety of momentary signal discontinuities.

graceful degradation An equipment failure mode in which the system suffers reduced capability, but does not fail altogether.

graticule A fixed pattern of reference markings used with oscilloscope CRTs to simplify measurements. The graticule may be etched on a transparent plate covering the front of the CRT or, for greater accuracy in readings, may be electrically generated within the CRT itself.

grid (1—general) A mesh electrode within an electron tube that controls the flow of electrons between the cathode and plate of the tube. **(2—bias)** The potential applied to a grid in an electron tube to control its center operating point. **(3—control)** The grid in an electron tube to which the input signal is usually applied. **(4—screen)** The grid in an electron tube, typically held at a steady potential, that screens the control grid from changes in anode potential. **(5—suppressor)** The grid in an electron tube near the anode (plate) that suppresses the emission of secondary electrons from the plate.

ground An electrical connection to earth or to a common conductor usually connected to earth.

ground clamp A clamp used to connect a ground wire to a ground rod or system.

ground loop An undesirable circulating ground current in a circuit grounded via multiple connections or at multiple points.

ground plane A conducting material at ground potential, physically close to other equipment, so that connections may be made readily to ground the equipment at the required points.

ground potential The point at zero electric potential.

ground return A conductor used as a path for one or more circuits back to the ground plane or central facility ground point.

ground rod A metal rod driven into the earth and connected into a mesh of interconnected rods so as to provide a low resistance link to ground.

ground window A single-point interface between the integrated ground plane of a building and an isolated ground plane.

ground wire A copper conductor used to extend a good low-resistance earth ground to protective devices in a facility.

grounded The connection of a piece of equipment to earth via a low resistance path.

grounding The act of connecting a device or circuit to ground or to a conductor that is grounded.

group delay A condition where the different frequency elements of a given signal suffer differing propagation delays through a circuit or a system. The delay at a lower frequency is different from the delay at a higher frequency, resulting in a time-related distortion of the signal at the receiving point.

group delay time The rate of change of the total phase shift of a waveform with angular frequency through a device or transmission facility.

group velocity The speed of a pulse on a transmission line.

guard band A narrow bandwidth between adjacent channels intended to reduce interference or crosstalk.

H

half-wave rectifier A circuit or device that changes only positive or negative half-cycle inputs of alternating current into direct current.

Hall effect The phenomenon by which a voltage develops between the edges of a current-carrying metal strip whose faces are perpendicular to an external magnetic field.

hard-wired Electrical devices connected through physical wiring.

harden The process of constructing military telecommunications facilities so as to protect them from damage by enemy action, especially *electromagnetic pulse* (EMP) radiation.

hardware Physical equipment, such as mechanical, magnetic, electrical, or electronic devices or components.

harmonic A periodic wave having a frequency that is an integral multiple of the fundamental frequency. For example, a wave with twice the frequency of the fundamental is called the *second harmonic*.

harmonic analyzer A test set capable of identifying the frequencies of the individual signals that make up a complex wave.

harmonic distortion The production of harmonics at the output of a circuit when a periodic wave is applied to its input. The level of the distortion is usually expressed as a percentage of the level of the input.

hazard A condition that could lead to danger for operating personnel.

headroom The difference, in decibels, between the typical operating signal level and a peak overload level.

heat loss The loss of useful electrical energy resulting from conversion into unwanted heat.

heat sink A device that conducts heat away from a heat-producing component so that it stays within a safe working temperature range.

heater In an electron tube, the filament that heats the cathode to enable it to emit electrons.

hecto A prefix meaning 100.

henry The standard unit of electrical inductance, equal to the self-inductance of a circuit or the mutual inductance of two circuits when there is an induced electromotive force of one volt and a current change of one ampere per second. The symbol for inductance is *H*, named for the American physicist Joseph Henry (1797–1878).

hertz (Hz) The unit of frequency that is equal to one cycle per second. Hertz is the reciprocal of the *period*, the interval after which the same portion of a periodic waveform recurs. Hertz was named for the German physicist Heinrich R. Hertz (1857–1894).

heterodyne The mixing of two signals in a nonlinear device in order to produce two additional signals at frequencies that are the sum and difference of the original frequencies.

heterodyne frequency The sum of, or the difference between, two frequencies, produced by combining the two signals together in a modulator or similar device.

heterodyne wavemeter A test set that uses the heterodyne principle to measure the frequencies of incoming signals.

high-frequency loss Loss of signal amplitude at higher frequencies through a given circuit or medium. For example, high frequency loss could be caused by passing a signal through a coaxial cable.

high *Q* An inductance or capacitance whose ratio of reactance to resistance is high.

high tension A high voltage circuit.

high-pass filter A network that passes signals of higher than a specified frequency but attenuates signals of all lower frequencies.

homochronous Signals whose corresponding significant instants have a constant but uncontrolled phase relationship with each other.

horn gap A lightning arrester utilizing a gap between two horns. When lightning causes a discharge between the horns, the heat produced lengthens the arc and breaks it.

horsepower The basic unit of mechanical power. One horsepower (hp) equals 550 foot-pounds per second or 746 watts.

hot A charged electrical circuit or device.

hot dip galvanized The process of galvanizing steel by dipping it into a bath of molten zinc.

hot standby System equipment that is fully powered but not in service. A *hot standby* can rapidly replace a primary system in the event of a failure.

hum Undesirable coupling of the 60 Hz power sine wave into other electrical signals and/or circuits.

HVAC An abbreviation for *heating, ventilation, and air conditioning* system.

hybrid system A communication system that accommodates both digital and analog signals.

hydrometer A testing device used to measure specific gravity, particularly the specific gravity of the dilute sulphuric acid in a lead-acid storage battery, to learn the state of charge of the battery.

hygrometer An instrument that measures the relative humidity of the atmosphere.

hygroscopic The ability of a substance to absorb moisture from the air.

hysteresis The property of an element evidenced by the dependence of the value of the output, for a given excursion of the input, upon the history of prior excursions and direction of the input. Originally, *hysteresis* was the name for magnetic phenomena only—the lagging of flux density behind the change in value of the magnetizing flux—but now, the term is also used to describe other inelastic behavior.

hysteresis loop The plot of magnetizing current against magnetic flux density (or of other similarly related pairs of parameters), which appears as a loop. The area within the loop is proportional to the power loss resulting from hysteresis.

hysteresis loss The loss in a magnetic core resulting from hysteresis.

I

I^2R loss The power lost as a result of the heating effect of current passing through resistance.

idling current The current drawn by a circuit, such as an amplifier, when no signal is present at its input.

image frequency A frequency on which a carrier signal, when heterodyned with the local oscillator in a superheterodyne receiver, will cause a sum or difference frequency that is the same as the intermediate frequency of the receiver. Thus, a signal on an *image frequency* will be demodulated along with the desired signal and will interfere with it.

impact ionization The ionization of an atom or molecule as a result of a high energy collision.

impedance The total passive opposition offered to the flow of an alternating current. *Impedance* consists of a combination of resistance, inductive reactance, and capacitive reactance. It is the vector sum of resistance and reactance ($R + jX$) or the vector of magnitude Z at an angle θ.

impedance characteristic A graph of the impedance of a circuit showing how it varies with frequency.

impedance irregularity A discontinuity in an impedance characteristic caused, for example, by the use of different coaxial cable types.

impedance matching The adjustment of the impedances of adjoining circuit components to a common value so as to minimize reflected energy from the junction and to maximize energy transfer across it. Incorrect adjustment results in an *impedance mismatch*.

impedance matching transformer A transformer used between two circuits of different impedances with a turns ratio that provides for maximum power transfer and minimum loss by reflection.

impulse A short high energy surge of electrical current in a circuit or on a line.

impulse current A current that rises rapidly to a peak then decays to zero without oscillating.

impulse excitation The production of an oscillatory current in a circuit by impressing a voltage for a relatively short period compared with the duration of the current produced.

impulse noise A noise signal consisting of random occurrences of energy spikes, having random amplitude and bandwidth.

impulse response The amplitude-versus-time output of a transmission facility or device in response to an impulse.

impulse voltage A unidirectional voltage that rises rapidly to a peak and then falls to zero, without any appreciable oscillation.

in-phase The property of alternating current signals of the same frequency that achieve their peak positive, peak negative, and zero amplitude values simultaneously.

incidence angle The angle between the perpendicular to a surface and the direction of arrival of a signal.

increment A small change in the value of a quantity.

induce To produce an electrical or magnetic effect in one conductor by changing the condition or position of another conductor.

induced current The current that flows in a conductor because a voltage has been induced across two points in, or connected to, the conductor.

induced voltage A voltage developed in a conductor when the conductor passes through magnetic lines of force.

inductance The property of an inductor that opposes any change in a current that flows through it. The standard unit of inductance is the *Henry.*

induction The electrical and magnetic interaction process by which a changing current in one circuit produces a voltage change not only in its own circuit (*self inductance*) but also in other circuits to which it is linked magnetically.

inductive A circuit element exhibiting inductive reactance.

inductive kick A voltage surge produced when a current flowing through an inductance is interrupted.

inductive load A load that possesses a net inductive reactance.

inductive reactance The reactance of a circuit resulting from the presence of inductance and the phenomenon of induction.

inductor A coil of wire, usually wound on a core of high permeability, that provides high inductance without necessarily exhibiting high resistance.

inert An inactive unit, or a unit that has no power requirements.

infinite line A transmission line that appears to be of infinite length. There are no reflections back from the far end because it is terminated in its characteristic impedance.

infra low frequency (ILF) The frequency band from 300 Hz to 3000 Hz.

inhibit A control signal that prevents a device or circuit from operating.

injection The application of a signal to an electronic device.

input The waveform fed into a circuit, or the terminals that receive the input waveform.

insertion gain The gain resulting from the insertion of a transducer in a transmission system, expressed as the ratio of the power delivered to that part of the system following the transducer to the power delivered to that same part before insertion. If more than one component is involved in the input or output, the particular component used must be specified. This ratio is usually expressed in decibels. If the resulting number is negative, an *insertion loss* is indicated.

insertion loss The signal loss within a circuit, usually expressed in decibels as the ratio of input power to output power.

insertion loss-vs.-frequency characteristic The amplitude transfer characteristic of a system or component as a function of frequency. The amplitude response may be stated as actual gain, loss, amplification, or attenuation, or as a ratio of any one of these quantities at a particular frequency, with respect to that at a specified reference frequency.

inspection lot A collection of units of product from which a sample is drawn and inspected to determine conformance with acceptability criteria.

instantaneous value The value of a varying waveform at a given instant of time. The value can be in volts, amperes, or phase angle.

Institute of Electrical and Electronics Engineers (IEEE) The organization of electrical and electronics scientists and engineers formed in 1963 by the merger of the Institute of Radio Engineers (IRE) and the American Institute of Electrical Engineers (AIEE).

instrument multiplier A measuring device that enables a high voltage to be measured using a meter with only a low voltage range.

instrument rating The range within which an instrument has been designed to operate without damage.

insulate The process of separating one conducting body from another conductor.

insulation The material that surrounds and insulates an electrical wire from other wires or circuits. *Insulation* may also refer to any material that does not ionize easily and thus presents a large impedance to the flow of electrical current.

insulator A material or device used to separate one conducting body from another.

intelligence signal A signal containing information.

intensity The strength of a given signal under specified conditions.

interconnect cable A short distance cable intended for use between equipment (generally less than 3 m in length).

interface A device or circuit used to interconnect two pieces of electronic equipment.

interface device A unit that joins two interconnecting systems.

interference emission An emission that results in an electrical signal being propagated into and interfering with the proper operation of electrical or electronic equipment.

interlock A protection device or system designed to remove all dangerous voltages from a machine or piece of equipment when access doors or panels are opened or removed.

intermediate frequency A frequency that results from combining a signal of interest with a signal generated within a radio receiver. In superheterodyne receivers, all incoming signals are converted to a single intermediate frequency for which the amplifiers and filters of the receiver have been optimized.

intermittent A noncontinuous recurring event, often used to denote a problem that is difficult to find because of its unpredictable nature.

intermodulation The production, in a nonlinear transducer element, of frequencies corresponding to the sums and differences of the fundamentals and harmonics of two or more frequencies that are transmitted through the transducer.

intermodulation distortion (IMD) The distortion that results from the mixing of two input signals in a nonlinear system. The resulting output contains new frequencies that represent the sum and difference of the input signals and the sums and differences of their harmonics. IMD is also called *intermodulation noise*.

intermodulation noise In a transmission path or device, the noise signal that is contingent upon modulation and demodulation, resulting from nonlinear characteristics in the path or device.

internal resistance The actual resistance of a source of electric power. The total electromotive force produced by a power source is not available for external use; some of the energy is used in driving current through the source itself.

International Standards Organization (ISO) An international body concerned with worldwide standardization for a broad range of industrial products, including telecommunications equipment. Members are represented by national standards organizations, such as ANSI (American National Standards Institute) in the United States. ISO was established in 1947 as a specialized agency of the United Nations.

International Telecommunications Union (ITU) A specialized agency of the United Nations established to maintain and extend international cooperation for the maintenance, development, and efficient use of telecommunications. The union does this through standards and recommended regulations, and through technical and telecommunications studies.

International Telecommunications Satellite Consortium (Intelsat) A nonprofit cooperative of member nations that owns and operates a satellite system for international and, in many instances, domestic communications.

interoperability The condition achieved among communications and electronics systems or equipment when information or services can be exchanged directly between them or their users, or both.

interpolate The process of estimating unknown values based on a knowledge of comparable data that falls on both sides of the point in question.

interrupting capacity The rating of a circuit breaker or fuse that specifies the maximum current the device is designed to interrupt at its rated voltage.

interval The points or numbers lying between two specified endpoints.

inverse voltage The effective value of voltage across a rectifying device, which conducts a current in one direction during one half cycle of the alternating input, during the half cycle when current is not flowing.

inversion The change in the polarity of a pulse, such as from positive to negative.

inverter A circuit or device that converts a direct current into an alternating current.

ionizing radiation The form of electromagnetic radiation that can turn an atom into an ion by knocking one or more of its electrons loose. Examples of ionizing radiation include X rays, gamma rays, and cosmic rays

***IR* drop** A drop in voltage because of the flow of current (*I*) through a resistance (*R*), also called *resistance drop.*

***IR* loss** The conversion of electrical power to heat caused by the flow of electrical current through a resistance.

isochronous A signal in which the time interval separating any two significant instants is theoretically equal to a specified unit interval or to an integral multiple of the unit interval.

isolated ground A ground circuit that is isolated from all equipment framework and any other grounds, except for a single-point external connection.

isolated ground plane A set of connected frames that are grounded through a single connection to a ground reference point. That point and all parts of the frames are insulated from any other ground system in a building.

isolated pulse A pulse uninfluenced by other pulses in the same signal.

isophasing amplifier A timing device that corrects for small timing errors.

isotropic A quantity exhibiting the same properties in all planes and directions.

J

jack A receptacle or connector that makes electrical contact with the mating contacts of a plug. In combination, the plug and jack provide a ready means for making connections in electrical circuits.

jacket An insulating layer of material surrounding a wire in a cable.

jitter Small, rapid variations in a waveform resulting from fluctuations in a supply voltage or other causes.

joule The standard unit of work that is equal to the work done by one newton of force when the point at which the force is applied is displaced a distance of one meter in the direction of the force. The *joule* is named for the English physicist James Prescott Joule (1818-1889).

Julian date A chronological date in which days of the year are numbered in sequence. For example, the first day is 001, the second is 002, and the last is 365 (or 366 in a leap year).

K

Kelvin (K) The standard unit of thermodynamic temperature. Zero degrees Kelvin represents *absolute zero*. Water freezes at 273 K and water boils at 373 K under standard pressure conditions.

kilo A prefix meaning one thousand.

kilohertz (kHz) A unit of measure of frequency equal to 1,000 Hz.

kilovar A unit equal to one thousand volt-amperes.

kilovolt (kV) A unit of measure of electrical voltage equal to 1,000 V.

kilowatt A unit equal to one thousand watts.

Kirchoff's Law At any point in a circuit, there is as much current flowing into the point as there is flowing away from it.

klystron (1—general) A family of electron tubes that function as microwave amplifiers and oscillators. Simplest in form are two-cavity klystrons in which an electron beam passes through a cavity that is excited by a microwave input, producing a velocity-modulated beam which passes through a second cavity a precise distance away that is coupled to a tuned circuit, thereby producing an amplified output of the original input signal frequency. If part of the output is fed back to the input, an oscillator can be the result. **(2—multi-cavity)** An amplifier device for UHF and microwave signals based on velocity modulation of an electron beam. The beam is directed through an input cavity, where the input RF signal polarity initializes a *bunching effect* on electrons in the beam. The bunching effect excites subsequent cavities, which increase the bunching through an energy flywheel concept. Finally, the beam passes to an output cavity that couples the amplified signal to the load (antenna system). The beam falls onto a collector element that forms the return path for the current and dissipates the heat resulting from electron beam bombardment. **(3—reflex)** A klystron with only one cavity. The action is the same as in a two-cavity klystron but the beam is reflected back into the cavity in which it was first excited, after being sent out to a reflector. The one cavity, therefore, acts both as the original exciter (or buncher) and as the collector from which the output is taken.

knee In a response curve, the region of maximum curvature.

ku band Radio frequencies in the range of 15.35 GHz to 17.25 GHz, typically used for satellite telecommunications.

L

ladder network A type of filter with components alternately across the line and in the line.

lag The difference in phase between a current and the voltage that produced it, expressed in electrical degrees.

lagging current A current that lags behind the alternating electromotive force that produced it. A circuit that produces a *lagging current* is one containing inductance alone, or whose effective impedance is inductive.

lagging load A load whose combined inductive reactance exceeds its capacitive reactance. When an alternating voltage is applied, the current lags behind the voltage.

laminate A material consisting of layers of the same or different materials bonded together and built up to the required thickness.

latitude An angular measurement of a point on the earth above or below the equator. The equator represents 0°, the north pole +90°, and the south pole –90°.

layout A proposed or actual arrangement or allocation of equipment.

LC circuit An electrical circuit with both inductance (*L*) and capacitance (*C*) that is resonant at a particular frequency.

LC ratio The ratio of inductance to capacitance in a given circuit.

lead An electrical wire, usually insulated.

leading edge The initial portion of a pulse or wave in which voltage or current rise rapidly from zero to a final value.

leading load A reactive load in which the reactance of capacitance is greater than that of inductance. Current through such a load *leads* the applied voltage causing the current.

leakage The loss of energy resulting from the flow of electricity past an insulating material, the escape of electromagnetic radiation beyond its shielding, or the extension of magnetic lines of force beyond their intended working area.

leakage resistance The resistance of a path through which leakage current flows.

level The strength or intensity of a given signal.

level alignment The adjustment of transmission levels of single links and links in tandem to prevent overloading of transmission subsystems.

life cycle The predicted useful life of a class of equipment, operating under normal (specified) working conditions.

life safety system A system designed to protect life and property, such as emergency lighting, fire alarms, smoke exhaust and ventilating fans, and site security.

life test A test in which random samples of a product are checked to see how long they can continue to perform their functions satisfactorily. A form of *stress testing* is used, including temperature, current, voltage, and/or vibration effects, cycled at many times the rate that would apply in normal usage.

limiter An electronic device in which some characteristic of the output is automatically prevented from exceeding a predetermined value.

limiter circuit A circuit of nonlinear elements that restricts the electrical excursion of a variable in accordance with some specified criteria.

limiting A process by which some characteristic at the output of a device is prevented from exceeding a predetermined value.

line loss The total end-to-end loss in decibels in a transmission line.

line-up The process of adjusting transmission parameters to bring a circuit to its specified values.

linear A circuit, device, or channel whose output is directly proportional to its input.

linear distortion A distortion mechanism that is independent of signal amplitude.

linearity A constant relationship, over a designated range, between the input and output characteristics of a circuit or device.

lines of force A group of imaginary lines indicating the direction of the electric or magnetic field at all points along it.

lissajous pattern The looping patterns generated by a CRT spot when the horizontal (X) and vertical (Y) deflection signals are sinusoids. The lissajous pattern is useful for evaluating the delay or phase of two sinusoids of the same frequency.

live A device or system connected to a source of electric potential.

load The work required of an electrical or mechanical system.

load factor The ratio of the average load over a designated period of time to the peak load occurring during the same period.

load line A straight line drawn across a grouping of plate current/plate voltage characteristic curves showing the relationship between grid voltage and plate current for a particular plate load resistance of an electron tube.

logarithm The power to which a base must be raised to produce a given number. Common logarithms are to base 10.

logarithmic scale A meter scale with displacement proportional to the logarithm of the quantity represented.

long persistence The quality of a cathode ray tube that has phosphorescent compounds on its screen (in addition to fluorescent compounds) so that the image continues to glow after the original electron beam has ceased to create it by producing the usual fluorescence effect. Long persistence is often used in radar screens or where photographic evidence is needed of a display. Most such applications, however, have been superseded through the use of digital storage techniques.

longitude The angular measurement of a point on the surface of the earth in relation to the meridian of Greenwich (London). The earth is divided into 360° of longitude, beginning at the Greenwich mean. As one travels west around the globe, the longitude increases.

longitudinal current A current that travels in the same direction on both wires of a pair. The return current either flows in another pair or via a ground return path.

loss The power dissipated in a circuit, usually expressed in decibels, that performs no useful work.

loss deviation The change of actual loss in a circuit or system from a designed value.

loss variation The change in actual measured loss over time.

lossy The condition when the line loss per unit length is significantly greater than some defined normal parameter.

lossy cable A coaxial cable constructed to have high transmission loss so it can be used as an artificial load or as an attenuator.

lot size A specific quantity of similar material or a collection of similar units from a common source; in inspection work, the quantity offered for inspection and acceptance at any one time. The **lot size** may be a collection of raw material, parts, subassemblies inspected during production, or a consignment of finished products to be sent out for service.

low tension A low voltage circuit.

low-pass filter A filter network that passes all frequencies below a specified frequency with little or no loss, but that significantly attenuates higher frequencies.

lug A tag or projecting terminal onto which a wire may be connected by wrapping, soldering, or crimping.

lumped constant A resistance, inductance, or capacitance connected at a point, and not distributed uniformly throughout the length of a route or circuit.

M

mA An abbreviation for *milliamperes* (0.001 A).

magnet A device that produces a magnetic field and can attract iron, and attract or repel other magnets.

magnetic field An energy field that exists around magnetic materials and current-carrying conductors. Magnetic fields combine with electric fields in light and radio waves.

magnetic flux The field produced in the area surrounding a magnet or electric current. The standard unit of flux is the *Weber*.

magnetic flux density A vector quantity measured by a standard unit called the *Tesla*. The *magnetic flux density* is the number of magnetic lines of force per unit area, at right angles to the lines.

magnetic leakage The magnetic flux that does not follow a useful path.

magnetic pole A point that appears from the outside to be the center of magnetic attraction or repulsion at or near one end of a magnet.

magnetic storm A violent local variation in the earth's magnetic field, usually the result of sunspot activity.

magnetism A property of iron and some other materials by which external magnetic fields are maintained, other magnets being thereby attracted or repelled.

magnetization The exposure of a magnetic material to a magnetizing current, field, or force.

magnetizing force The force producing magnetization.

magnetomotive force The force that tends to produce lines of force in a magnetic circuit. The *magnetomotive force* bears the same relationship to a magnetic circuit that voltage does to an electrical circuit.

magnetron A high-power, ultra high frequency electron tube oscillator that employs the interaction of a strong electric field between an anode and cathode with the field of a strong permanent magnet to cause oscillatory electron flow through multiple internal cavity resonators. The magnetron may operate in a continuous or pulsed mode.

maintainability The probability that a failure will be repaired within a specified time after the failure occurs.

maintenance Any activity intended to keep a functional unit in satisfactory working condition. The term includes the tests, measurements, replacements, adjustments, and repairs necessary to keep a device or system operating properly.

malfunction An equipment failure or a fault.

manometer A test device for measuring gas pressure.

margin The difference between the value of an operating parameter and the value that would result in unsatisfactory operation. Typical *margin* parameters include signal level, signal-to-noise ratio, distortion, crosstalk coupling, and/or undesired emission level.

Markov model A statistical model of the behavior of a complex system over time in which the probabilities of the occurrence of various future states depend only on the present state of the system, and not on the path by which the present state was achieved. This term was named for the Russian mathematician Andrei Andreevich Markov (1856–1922).

master clock An accurate timing device that generates a synchronous signal to control other clocks or equipment.

master oscillator A stable oscillator that provides a standard frequency signal for other hardware and/or systems.

matched termination A termination that absorbs all the incident power and so produces no reflected waves or mismatch loss.

matching The connection of channels, circuits, or devices in a manner that results in minimal reflected energy.

matrix A logical network configured in a rectangular array of intersections of input/output signals.

Maxwell's equations Four differential equations that relate electric and magnetic fields to electromagnetic waves. The equations are a basis of electrical and electronic engineering.

mean An arithmetic average in which values are added and divided by the number of such values.

mean time between failures (MTBF) For a particular interval, the total functioning life of a population of an item divided by the total number of failures within the population during the measurement interval.

mean time to failure (MTTF) The measured operating time of a single piece of equipment divided by the total number of failures during the measured period of time. This measurement is normally made during that period between early life and wear-out failures.

mean time to repair (MTTR) The total corrective maintenance time on a component or system divided by the total number of corrective maintenance actions during a given period of time.

measurement A procedure for determining the amount of a quantity.

median A value in a series that has as many readings or values above it as below.

medium An electronic pathway or mechanism for passing information from one point to another.

mega A prefix meaning one million.

megahertz (MHz) A quantity equal to one million Hertz (cycles per second).

megohm A quantity equal to one million ohms.

metric system A decimal system of measurement based on the meter, the kilogram, and the second.

micro A prefix meaning one millionth.

micron A unit of length equal to one millionth of a meter (1/25,000 of an inch).

microphonic(s) Unintended noise introduced into an electronic system by mechanical vibration of electrical components.

microsecond One millionth of a second (0.000001 s).

microvolt A quantity equal to one-millionth of a volt.

milli A prefix meaning one thousandth.

milliammeter A test instrument for measuring electrical current, often part of a *multimeter.*

millihenry A quantity equal to one-thousandth of a henry.

milliwatt A quantity equal to one thousandth of a watt.

minimum discernible signal The smallest input that will produce a discernible change in the output of a circuit or device.

mixer A circuit used to combine two or more signals to produce a third signal that is a function of the input waveforms.

mixing ratio The ratio of the mass of water vapor to the mass of dry air in a given volume of air. The *mixing ratio* affects radio propagation.

mode An electromagnetic field distribution that satisfies theoretical requirements for propagation in a waveguide or oscillation in a cavity.

modified refractive index The sum of the refractive index of the air at a given height above sea level, and the ratio of this height to the radius of the earth.

modular An equipment design in which major elements are readily separable, and which the user may replace, reducing the mean-time-to-repair.

modulation The process whereby the amplitude, frequency, or phase of a single-frequency wave (the *carrier*) is varied in step with the instantaneous value of, or samples of, a complex wave (the *modulating wave*).

modulator A device that enables the intelligence in an information-carrying modulating wave to be conveyed by a signal at a higher frequency. A *modulator* modifies a carrier wave by amplitude, phase, and/or frequency as a function of a control signal that carries intelligence. Signals are *modulated* in this way to permit more efficient and/or reliable transmission over any of several media.

module An assembly replaceable as an entity, often as an interchangeable plug-in item. A *module* is not normally capable of being disassembled.

monostable A device that is stable in one state only. An input pulse causes the device to change state, but it reverts immediately to its stable state.

motor A machine that converts electrical energy into mechanical energy.

motor effect The repulsion force exerted between adjacent conductors carrying currents in opposite directions.

moving coil Any device that utilizes a coil of wire in a magnetic field in such a way that the coil is made to move by varying the applied current, or itself produces a varying voltage because of its movement.

ms An abbreviation for *millisecond* (0.001 s).

multimeter A test instrument fitted with several ranges for measuring voltage, resistance, and current, and equipped with an analog meter or digital display readout. The *multimeter* is also known as a *volt-ohm-milliammeter*, or *VOM*.

multiplex (MUX) The use of a common channel to convey two or more channels. This is done either by splitting of the common channel frequency band into narrower bands, each of which is used to constitute a distinct channel (*frequency division multiplex*), or by allotting this common channel to multiple users in turn to constitute different intermittent channels (*time division multiplex*).

multiplexer A device or circuit that combines several signals onto a single signal.

multiplexing A technique that uses a single transmission path to carry multiple channels. In *time division multiplexing* (TDM), path time is shared. For *frequency division multiplexing* (FDM) or *wavelength division multiplexing* (WDM), signals are divided into individual channels sent along the same path but at different frequencies.

multiplication Signal mixing that occurs within a multiplier circuit.

multiplier A circuit in which one or more input signals are mixed under the direction of one or more control signals. The resulting output is a composite of the input signals, the characteristics of which are determined by the scaling specified for the circuit.

mutual induction The property of the magnetic flux around a conductor that induces a voltage in a nearby conductor. The voltage generated in the secondary conductor in turn induces a voltage in the primary conductor. The inductance of two conductors so coupled is referred to as *mutual inductance*.

mV An abbreviation for *millivolt* (0.001 V).

mW An abbreviation for *milliwatt* (0.001 W).

N

nano A prefix meaning one billionth.

nanometer 1×10^{-9} meter.

nanosecond (ns) One billionth of a second (1×10^{-9} s).

narrowband A communications channel of restricted bandwidth, often resulting in degradation of the transmitted signal.

narrowband emission An emission having a spectrum exhibiting one or more sharp peaks that are narrow in width compared to the nominal bandwidth of the measuring instrument, and are far enough apart in frequency to be resolvable by the instrument.

National Electrical Code (NEC) A document providing rules for the installation of electric wiring and equipment in public and private buildings, published by the National Fire Protection Association. The NEC has been adopted as law by many states and municipalities in the U.S.

National Institute of Standards and Technology (NIST) A nonregulatory agency of the Department of Commerce that serves as a national reference and measurement laboratory for the physical and engineering sciences. Formerly called the *National Bureau of Standards*, the agency was renamed in 1988 and given the additional responsibility of aiding U.S. companies in adopting new technologies to increase their international competitiveness.

negative In a conductor or semiconductor material, an excess of electrons or a deficiency of positive charge.

negative feedback The return of a portion of the output signal from a circuit to the input but 180° out of phase. This type of feedback decreases signal amplitude but stabilizes the amplifier and reduces distortion and noise.

negative impedance An impedance characterized by a decrease in voltage drop across a device as the current through the device is increased, or a decrease in current through the device as the voltage across it is increased.

neutral A device or object having no electrical charge.

neutral conductor A conductor in a power distribution system connected to a point in the system that is designed to be at neutral potential. In a balanced system, the neutral conductor carries no current.

neutral ground An intentional ground applied to the neutral conductor or neutral point of a circuit, transformer, machine, apparatus, or system.

newton The standard unit of force. One *newton* is the force that, when applied to a body having a mass of 1 kg, gives it an acceleration of 1 m/s^2.

nitrogen A gas widely used to pressurize radio frequency transmission lines. If a small puncture occurs in the cable sheath, the nitrogen keeps moisture out so that service is not adversely affected.

node The points at which the current is at minimum in a transmission system in which standing waves are present.

noise Any random disturbance or unwanted signal in a communication system that tends to obscure the clarity or usefulness of a signal in relation to its intended use.

noise factor (NF) The ratio of the noise power measured at the output of a receiver to the noise power that would be present at the output if the thermal noise resulting from the resistive component of the source impedance were the only source of noise in the system.

noise figure A measure of the noise in dB generated at the input of an amplifier, compared with the noise generated by an impedance-method resistor at a specified temperature.

noise filter A network that attenuates noise frequencies.

noise generator A generator of wideband random noise.

noise power ratio (NPR) The ratio, expressed in decibels, of signal power to intermodulation product power plus residual noise power, measured at the baseband level.

noise suppressor A filter or digital signal processing circuit in a receiver or transmitter that automatically reduces or eliminates noise.

noise temperature The temperature, expressed in Kelvin, at which a resistor will develop a particular noise voltage. The noise temperature of a radio receiver is the value by which the temperature of the resistive component of the source impedance should be increased—if it were the only source of noise in the system—to cause the noise power at the output of the receiver to be the same as in the real system.

nominal The most common value for a component or parameter that falls between the maximum and minimum limits of a tolerance range.

nominal value A specified or intended value independent of any uncertainty in its realization.

nomogram A chart showing three or more scales across which a straight edge may be held in order to read off a graphical solution to a three-variable equation.

nonionizing radiation Electromagnetic radiation that does not turn an atom into an ion. Examples of nonionizing radiation include visible light and radio waves.

nonconductor A material that does not conduct energy, such as electricity, heat, or sound.

noncritical technical load That part of the technical power load for a facility not required for minimum acceptable operation.

noninductive A device or circuit without significant inductance.

nonlinearity A distortion in which the output of a circuit or system does not rise or fall in direct proportion to the input.

nontechnical load The part of the total operational load of a facility used for such purposes as general lighting, air conditioning, and ventilating equipment during normal operation.

normal A line perpendicular to another line or to a surface.

normal-mode noise Unwanted signals in the form of voltages appearing in line-to-line and line-to-neutral signals.

normalized frequency The ratio between the actual frequency and its nominal value.

normalized frequency departure The frequency departure divided by the nominal frequency value.

normalized frequency difference The algebraic difference between two normalized frequencies.

normalized frequency drift The frequency drift divided by the nominal frequency value.

normally closed Switch contacts that are closed in their nonoperated state, or relay contacts that are closed when the relay is de-energized.

normally open Switch contacts that are open in their nonoperated state, or relay contacts that are open when the relay is de-energized.

north pole The pole of a magnet that seeks the north magnetic pole of the earth.

notch filter A circuit designed to attenuate a specific frequency band; also known as a *band stop filter.*

notched noise A noise signal in which a narrow band of frequencies has been removed.

ns An abbreviation for *nanosecond.*

null A zero or minimum amount or position.

O

octave Any frequency band in which the highest frequency is twice the lowest frequency.

off-line A condition wherein devices or subsystems are not connected into, do not form a part of, and are not subject to the same controls as an operational system.

offset An intentional difference between the realized value and the nominal value.

ohm The unit of electric resistance through which one ampere of current will flow when there is a difference of one volt. The quantity is named for the German physicist Georg Simon Ohm (1787–1854).

Ohm's law A law that sets forth the relationship between voltage (E), current (I), and resistance (R). The law states that $E = I^2 R$. *Ohm's Law* is named for the German physicist Georg Simon Ohm (1787–1854).

ohmic loss The power dissipation in a line or circuit caused by electrical resistance.

ohmmeter A test instrument used for measuring resistance, often part of a *multimeter.*

ohms-per-volt A measure of the sensitivity of a voltmeter.

on-line A device or system that is energized and operational, and ready to perform useful work.

open An interruption in the flow of electrical current, as caused by a broken wire or connection.

open-circuit A defined loop or path that closes on itself and contains an infinite impedance.

open-circuit impedance The input impedance of a circuit when its output terminals are open, that is, not terminated.

open-circuit voltage The voltage measured at the terminals of a circuit when there is no load and, hence, no current flowing.

operating lifetime The period of time during which the principal parameters of a component or system remain within a prescribed range.

optimize The process of adjusting for the best output or maximum response from a circuit or system.

orbit The path, relative to a specified frame of reference, described by the center of mass of a satellite or other object in space, subjected solely to natural forces (mainly gravitational attraction).

order of diversity The number of independently fading propagation paths or frequencies, or both, used in a diversity reception system.

original equipment manufacturer (OEM) A manufacturer of equipment that is used in systems assembled and sold by others.

oscillation A variation with time of the magnitude of a quantity with respect to a specified reference when the magnitude is alternately greater than and smaller than the reference.

oscillator A nonrotating device for producing alternating current, the output frequency of which is determined by the characteristics of the circuit.

oscilloscope A test instrument that uses a display, usually a cathode-ray tube, to show the instantaneous values and waveforms of a signal that varies with time or some other parameter.

out-of-band energy Energy emitted by a transmission system that falls outside the frequency spectrum of the intended transmission.

outage duration The average elapsed time between the start and the end of an outage period.

outage probability The probability that an outage state will occur within a specified time period. In the absence of specific known causes of outages, the *outage probability* is the sum of all outage durations divided by the time period of measurement.

outage threshold A defined value for a supported performance parameter that establishes the minimum operational service performance level for that parameter.

output impedance The impedance presented at the output terminals of a circuit, device, or channel.

output stage The final driving circuit in a piece of electronic equipment.

ovenized crystal oscillator (OXO) A crystal oscillator enclosed within a temperature regulated heater (oven) to maintain a stable frequency despite external temperature variations.

overcoupling A degree of coupling greater than the *critical coupling* between two resonant circuits. *Overcoupling* results in a wide bandwidth circuit with two peaks in the response curve.

overload In a transmission system, a power greater than the amount the system was designed to carry. In a power system, an overload could cause excessive heating. In a communications system, distortion of a signal could result.

overshoot The first maximum excursion of a pulse beyond the 100% level. Overshoot is the portion of the pulse that exceeds its defined level temporarily before settling to the correct level. Overshoot amplitude is expressed as a percentage of the defined level.

P

pentode An electron tube with five electrodes, the cathode, control grid, screen grid, suppressor grid, and plate.

photocathode An electrode in an electron tube that will emit electrons when bombarded by photons of light.

picture tube A cathode-ray tube used to produce an image by variation of the intensity of a scanning beam on a phosphor screen.

pin A terminal on the base of a component, such as an electron tube.

plasma (1—arc) An ionized gas in an arc-discharge tube that provides a conducting path for the discharge. **(2—solar)** The ionized gas at extremely high temperature found in the sun.

plate (1—electron tube) The anode of an electron tube. **(2—battery)** An electrode in a storage battery. **(3—capacitor)** One of the surfaces in a capacitor. **(4—chassis)** A mounting surface to which equipment may be fastened.

propagation time delay The time required for a signal to travel from one point to another.

protector A device or circuit that prevents damage to lines or equipment by conducting dangerously high voltages or currents to ground. Protector types include spark gaps, semiconductors, varistors, and gas tubes.

proximity effect A nonuniform current distribution in a conductor, caused by current flow in a nearby conductor.

pseudonoise In a spread-spectrum system, a seemingly random series of pulses whose frequency spectrum resembles that of continuous noise.

pseudorandom A sequence of signals that appears to be completely random but have, in fact, been carefully drawn up and repeat after a significant time interval.

pseudorandom noise A noise signal that satisfies one or more of the standard tests for statistical randomness. Although it seems to lack any definite pattern, there is a sequence of pulses that repeats after a long time interval.

pseudorandom number sequence A sequence of numbers that satisfies one or more of the standard tests for statistical randomness. Although it seems to lack any definite pattern, there is a sequence that repeats after a long time interval.

pulsating direct current A current changing in value at regular or irregular intervals but which has the same direction at all times.

pulse One of the elements of a repetitive signal characterized by the rise and decay in time of its magnitude. A *pulse* is usually short in relation to the time span of interest.

pulse decay time The time required for the trailing edge of a pulse to decrease from 90 percent to 10 percent of its peak amplitude.

pulse duration The time interval between the points on the leading and trailing edges of a pulse at which the instantaneous value bears a specified relation to the peak pulse amplitude.

pulse duration modulation (PDM) The modulation of a pulse carrier by varying the width of the pulses according to the instantaneous values of the voltage samples of the modulating signal (also called *pulse width modulation*).

pulse edge The leading or trailing edge of a pulse, defined as the 50 percent point of the pulse rise or fall time.

pulse fall time The interval of time required for the edge of a pulse to fall from 90 percent to 10 percent of its peak amplitude.

pulse interval The time between the start of one pulse and the start of the next.

pulse length The duration of a pulse (also called *pulse width*).

pulse level The voltage amplitude of a pulse.

pulse period The time between the start of one pulse and the start of the next.

pulse ratio The ratio of the length of any pulse to the total pulse period.

pulse repetition period The time interval from the beginning of one pulse to the beginning of the next pulse.

pulse repetition rate The number of times each second that pulses are transmitted.

pulse rise time The time required for the leading edge of a pulse to rise from 10 percent to 90 percent of its peak amplitude.

pulse train A series of pulses having similar characteristics.

pulse width The measured interval between the 50 percent amplitude points of the leading and trailing edges of a pulse.

puncture A breakdown of insulation or of a dielectric, such as in a cable sheath or in the insulant around a conductor.

pW An abbreviation for picowatt, a unit of power equal to 10^{-12} W (–90 dBm).

Q

Q (quality factor) A figure of merit that defines how close a coil comes to functioning as a pure inductor. *High Q* describes an inductor with little energy loss resulting from resistance. *Q* is found by dividing the inductive reactance of a device by its resistance.

quadrature A state of alternating current signals separated by one quarter of a cycle (90°).

quadrature amplitude modulation (QAM) A process that allows two different signals to modulate a single carrier frequency. The two signals of interest amplitude modulate two samples of the carrier that are of the same frequency, but differ in phase by 90°. The two resultant signals can be added and transmitted. Both signals may be recovered at a decoder when they are demodulated 90° apart.

quadrature component The component of a voltage or current at an angle of 90° to a reference signal, resulting from inductive or capacitive reactance.

quadrature phase shift keying (QPSK) A type of phase shift keying using four phase states.

quality The absence of objectionable distortion.

quality assurance (QA) All those activities, including surveillance, inspection, control, and documentation, aimed at ensuring that a given product will meet its performance specifications.

quality control (QC) A function whereby management exercises control over the quality of raw material or intermediate products in order to prevent the production of defective devices or systems.

quantum noise Any noise attributable to the discrete nature of electromagnetic radiation. Examples include shot noise, photon noise, and recombination noise.

quantum-limited operation An operation wherein the minimum detectable signal is limited by quantum noise.

quartz A crystalline mineral that when electrically excited vibrates with a stable period. Quartz is typically used as the frequency-determining element in oscillators and filters.

quasi-peak detector A detector that delivers an output voltage that is some fraction of the peak value of the regularly repeated pulses applied to it. The fraction increases toward unity as the pulse repetition rate increases.

quick-break fuse A fuse in which the fusible link is under tension, providing for rapid operation.

quiescent An inactive device, signal, or system.

quiescent current The current that flows in a device in the absence of an applied signal.

R

rack An equipment rack, usually measuring 19 in (48.26 cm) wide at the front mounting rails.

rack unit (RU) A unit of measure of vertical space in an equipment enclosure. One rack unit is equal to 1.75 in (4.45 cm).

radiate The process of emitting electromagnetic energy.

radiation The emission and propagation of electromagnetic energy in the form of waves. *Radiation* is also called *radiant energy.*

radiation scattering The diversion of thermal, electromagnetic, or nuclear radiation from its original path as a result of interactions or collisions with atoms, molecules, or large particles in the atmosphere or other media between the source of radiation and a point some distance away. As a result of scattering, radiation (especially gamma rays and neutrons) will be received at such a point from many directions, rather than only from the direction of the source.

radio The transmission of signals over a distance by means of electromagnetic waves in the approximate frequency range of 150 kHz to 300 GHz. The term may also be used to describe the equipment used to transmit or receive electromagnetic waves.

radio detection The detection of the presence of an object by radio location without precise determination of its position.

radio frequency interference (RFI) The intrusion of unwanted signals or electromagnetic noise into various types of equipment resulting from radio frequency transmission equipment or other devices using radio frequencies.

radio frequency spectrum Those frequency bands in the electromagnetic spectrum that range from several hundred thousand cycles per second (*very low frequency*) to several billion cycles per second (*microwave frequencies*).

radio recognition In military communications, the determination by radio means of the "friendly" or "unfriendly" character of an aircraft or ship.

random noise Electromagnetic signals that originate in transient electrical disturbances and have random time and amplitude patterns. Random noise is generally undesirable; however, it may also be generated for testing purposes.

rated output power The power available from an amplifier or other device under specified conditions of operation.

RC constant The time constant of a resistor-capacitor circuit. The *RC constant* is the time in seconds required for current in an RC circuit to rise to 63 percent of its final steady value or fall to 37 percent of its original steady value, obtained by multiplying the resistance value in ohms by the capacitance value in farads.

RC network A circuit that contains resistors and capacitors, normally connected in series.

reactance The part of the impedance of a network resulting from inductance or capacitance. The *reactance* of a component varies with the frequency of the applied signal.

reactive power The power circulating in an ac circuit. It is delivered to the circuit during part of the cycle and is returned during the other half of the cycle. The *reactive power* is obtained by multiplying the voltage, current, and the sine of the phase angle between them.

reactor A component with inductive reactance.

received signal level (RSL) The value of a specified bandwidth of signals at the receiver input terminals relative to an established reference.

receiver Any device for receiving electrical signals and converting them to audible sound, visible light, data, or some combination of these elements.

receptacle An electrical socket designed to receive a mating plug.

reception The act of receiving, listening to, or watching information-carrying signals.

rectification The conversion of alternating current into direct current.

rectifier A device for converting alternating current into direct current. A *rectifier* normally includes filters so that the output is, within specified limits, smooth and free of ac components.

rectify The process of converting alternating current into direct current.

redundancy A system design that provides a back-up for key circuits or components in the event of a failure. Redundancy improves the overall reliability of a system.

redundant A configuration when two complete systems are available at one time. If the online system fails, the backup will take over with no loss of service.

reference voltage A voltage used for control or comparison purposes.

reflectance The ratio of reflected power to incident power.

reflection An abrupt change, resulting from an impedance mismatch, in the direction of propagation of an electromagnetic wave. For light, at the interface of two dissimilar materials, the incident wave is returned to its medium of origin.

reflection coefficient The ratio between the amplitude of a reflected wave and the amplitude of the incident wave. For large smooth surfaces, the reflection coefficient may be near unity.

reflection gain The increase in signal strength that results when a reflected wave combines, in phase, with an incident wave.

reflection loss The apparent loss of signal strength caused by an impedance mismatch in a transmission line or circuit. The loss results from the reflection of part of the signal back toward the source from the point of the impedance discontinuity. The greater the mismatch, the greater the loss.

reflectometer A device that measures energy traveling in each direction in a waveguide, used in determining the standing wave ratio.

refraction The bending of a sound, radio, or light wave as it passes obliquely from a medium of one density to a medium of another density that varies its speed.

regulation The process of adjusting the level of some quantity, such as circuit gain, by means of an electronic system that monitors an output and feeds back a controlling signal to constantly maintain a desired level.

regulator A device that maintains its output voltage at a constant level.

relative envelope delay The difference in envelope delay at various frequencies when compared with a reference frequency that is chosen as having zero delay.

relative humidity The ratio of the quantity of water vapor in the atmosphere to the quantity that would cause saturation at the ambient temperature.

relative transmission level The ratio of the signal power in a transmission system to the signal power at some point chosen as a reference. The ratio is usually determined by applying a standard test signal at the input to the system and measuring the gain or loss at the location of interest.

relay A device by which current flowing in one circuit causes contacts to operate that control the flow of current in another circuit.

relay armature The movable part of an electromechanical relay, usually coupled to spring sets on which contacts are mounted.

relay bypass A device that, in the event of a loss of power or other failure, routes a critical signal around the equipment that has failed.

release time The time required for a pulse to drop from steady-state level to zero, also referred to as the *decay time*.

reliability The ability of a system or subsystem to perform within the prescribed parameters of quality of service. *Reliability* is often expressed as the probability that a system or subsystem will perform its intended function for a specified interval under stated conditions.

reliability growth The action taken to move a hardware item toward its reliability potential, during development or subsequent manufacturing or operation.

reliability predictions The compiled failure rates for parts, components, subassemblies, assemblies, and systems. These generic failure rates are used as basic data to predict the reliability of a given device or system.

remote control A system used to control a device from a distance.

remote station A station or terminal that is physically remote from a main station or center but can gain access through a communication channel.

repeater The equipment between two circuits that receives a signal degraded by normal factors during transmission and amplifies the signal to its original level for retransmission.

repetition rate The rate at which regularly recurring pulses are repeated.

reply A transmitted message that is a direct response to an original message.

repulsion The mechanical force that tends to separate like magnetic poles, like electric charges, or conductors carrying currents in opposite directions.

reset The act of restoring a device to its default or original state.

residual flux The magnetic flux that remains after a magnetomotive force has been removed.

residual magnetism The magnetism or flux that remains in a core after current ceases to flow in the coil producing the magnetomotive force.

residual voltage The vector sum of the voltages in all the phase wires of an unbalanced polyphase power system.

resistance The opposition of a material to the flow of electrical current. Resistance is equal to the voltage drop through a given material divided by the current flow through it. The standard unit of resistance is the *ohm*, named for the German physicist Georg Simon Ohm (1787–1854).

resistance drop The fall in potential (volts) between two points, the product of the current and resistance.

resistance-grounded A circuit or system grounded for safety through a resistance, which limits the value of the current flowing through the circuit in the event of a fault.

resistive load A load in which the voltage is in phase with the current.

resistivity The resistance per unit volume or per unit area.

resistor A device the primary function of which is to introduce a specified resistance into an electrical circuit.

resonance A tuned condition conducive to oscillation, when the reactance resulting from capacitance in a circuit is equal in value to the reactance resulting from inductance.

resonant frequency The frequency at which the inductive reactance and capacitive reactance of a circuit are equal.

resonator A resonant cavity.

return A return path for current, sometimes through ground.

reversal A change in magnetic polarity, in the direction of current flow.

reverse current A small current that flows through a diode when the voltage across it is such that normal forward current does not flow.

reverse voltage A voltage in the reverse direction from that normally applied.

rheostat A two-terminal variable resistor, usually constructed with a sliding or rotating shaft that can be used to vary the resistance value of the device.

ripple An ac voltage superimposed on the output of a dc power supply, usually resulting from imperfect filtering.

rise time The time required for a pulse to rise from 10 percent to 90 percent of its peak value.

roll-off A gradual attenuation of gain-frequency response at either or both ends of a transmission pass band.

root-mean-square (RMS) The square root of the average value of the squares of all the instantaneous values of current or voltage during one half-cycle of an alternating current. For an alternating current, the RMS voltage or current is equal to the amount of direct current or voltage that would produce the same heating effect in a purely resistive circuit. For a sinewave, the root-mean-square value is equal to 0.707 times the peak value. RMS is also called the *effective value*.

rotor The rotating part of an electric generator or motor.

RU An abbreviation for *rack unit*.

S

scan One sweep of the target area in a camera tube, or of the screen in a picture tube.

screen grid A grid in an electron tube that improves performance of the device by shielding the control grid from the plate.

self-bias The provision of bias in an electron tube through a voltage drop in the cathode circuit.

shot noise The noise developed in a vacuum tube or photoconductor resulting from the random number and velocity of emitted charge carriers.

slope The rate of change, with respect to frequency, of transmission line attenuation over a given frequency spectrum.

slope equalizer A device or circuit used to achieve a specified slope in a transmission line.

smoothing circuit A filter designed to reduce the amount of ripple in a circuit, usually a dc power supply.

snubber An electronic circuit used to suppress high frequency noise.

solar wind Charged particles from the sun that continuously bombard the surface of the earth.

solid A single wire conductor, as contrasted with a stranded, braided, or rope-type wire.

solid-state The use of semiconductors rather than electron tubes in a circuit or system.

source The part of a system from which signals or messages are considered to originate.

source terminated A circuit whose output is terminated for correct impedance matching with standard cable.

spare A system that is available but not presently in use.

spark gap A gap between two electrodes designed to produce a spark under given conditions.

specific gravity The ratio of the weight of a volume, liquid, or solid to the weight of the same volume of water at a specified temperature.

spectrum A continuous band of frequencies within which waves have some common characteristics.

spectrum analyzer A test instrument that presents a graphic display of signals over a selected frequency bandwidth. A cathode-ray tube is often used for the display.

spectrum designation of frequency A method of referring to a range of communication frequencies. In American practice, the designation is a two or three letter acronym for the name. The ranges are: below 300 Hz, ELF (extremely low frequency); 300 Hz–3000 Hz, ILF (infra low frequency); 3 kHz–30 kHz, VLF (very low frequency); 30 kHz–300 kHz, LF (low frequency); 300 kHz–3000 kHz, MF (medium frequency); 3 MHz–30 MHz, HF (high frequency); 30 MHz–300 MHz, VHF (very high frequency); 300 MHz–3000 MHz, UHF (ultra high frequency); 3 GHz–30 GHz, SHF (super high frequency); 30 GHz–300 GHz, EHF (extremely high frequency); 300 GHz–3000 GHz, THF (tremendously high frequency).

spherical antenna A type of satellite receiving antenna that permits more than one satellite to be accessed at any given time. A spherical antenna has a broader angle of acceptance than a parabolic antenna.

spike A high amplitude, short duration pulse superimposed on an otherwise regular waveform.

split-phase A device that derives a second phase from a single phase power supply by passing it through a capacitive or inductive reactor.

splitter A circuit or device that accepts one input signal and distributes it to several outputs.

splitting ratio The ratio of the power emerging from the output ports of a coupler.

sporadic An event occurring at random and infrequent intervals.

spread spectrum A communications technique in which the frequency components of a narrowband signal are spread over a wide band. The resulting signal resembles white noise. The technique is used to achieve signal security and privacy, and to enable the use of a common band by many users.

spurious signal Any portion of a given signal that is not part of the fundamental waveform. Spurious signals include transients, noise, and hum.

square wave A square or rectangular-shaped periodic wave that alternately assumes two fixed values for equal lengths of time, the transition being negligible in comparison with the duration of each fixed value.

square wave testing The use of a square wave containing many odd harmonics of the fundamental frequency as an input signal to a device. Visual examination of the output signal on an oscilloscope indicates the amount of distortion introduced.

stability The ability of a device or circuit to remain stable in frequency, power level, and/or other specified parameters.

standard The specific signal configuration, reference pulses, voltage levels, and other parameters that describe the input/output requirements for a particular type of equipment.

standard time and frequency signal A time-controlled radio signal broadcast at scheduled intervals on a number of different frequencies by government-operated radio stations to provide a method for calibrating instruments.

standing wave ratio (SWR) The ratio of the maximum to the minimum value of a component of a wave in a transmission line or waveguide, such as the maximum voltage to the minimum voltage.

static charge An electric charge on the surface of an object, particularly a dielectric.

station One of the input or output points in a communications system.

stator The stationary part of a rotating electric machine.

status The present condition of a device.

statute mile A unit of distance equal to 1,609 km or 5,280 ft.

steady-state A condition in which circuit values remain essentially constant, occurring after all initial transients or fluctuating conditions have passed.

steady-state condition A condition occurring after all initial transient or fluctuating conditions have damped out in which currents, voltages, or fields remain essentially constant or oscillate uniformly without changes in characteristics such as amplitude, frequency, or wave shape.

steep wavefront A rapid rise in voltage of a given signal, indicating the presence of high frequency odd harmonics of a fundamental wave frequency.

step up (or down) The process of increasing (or decreasing) the voltage of an electrical signal, as in a step-up (or step-down) transformer.

straight-line capacitance A capacitance employing a variable capacitor with plates so shaped that capacitance varies directly with the angle of rotation.

stray capacitance An unintended—and usually undesired—capacitance between wires and components in a circuit or system.

stray current A current through a path other than the intended one.

stress The force per unit of cross-sectional area on a given object or structure.

subassembly A functional unit of a system.

subcarrier (SC) A carrier applied as modulation on another carrier, or on an intermediate subcarrier.

subharmonic A frequency equal to the fundamental frequency of a given signal divided by a whole number.

submodule A small circuit board or device that mounts on a larger module or device.

subrefraction A refraction for which the refractivity gradient is greater than standard.

subsystem A functional unit of a system.

superheterodyne receiver A radio receiver in which all signals are first converted to a common frequency for which the intermediate stages of the receiver have been optimized, both for tuning and filtering. Signals are converted by mixing them with the output of a local oscillator whose output is varied in accordance with the frequency of the received signals so as to maintain the desired *intermediate frequency.*

suppressor grid The fifth grid of a pentode electron tube, which provides screening between plate and screen grid.

surface leakage A leakage current from line to ground over the face of an insulator supporting an open wire route.

surface refractivity The refractive index, calculated from observations of pressure, temperature, and humidity at the surface of the earth.

surge A rapid rise in current or voltage, usually followed by a fall back to the normal value.

susceptance The reciprocal of reactance, and the imaginary component of admittance, expressed in siemens.

sweep The process of varying the frequency of a signal over a specified bandwidth.

sweep generator A test oscillator, the frequency of which is constantly varied over a specified bandwidth.

switching The process of making and breaking (connecting and disconnecting) two or more electrical circuits.

synchronization The process of adjusting the corresponding significant instants of signals—for example, the zero-crossings—to make them synchronous. The term *synchronization* is often abbreviated as *sync.*

synchronize The process of causing two systems to operate at the same speed.

synchronous In step or in phase, as applied to two or more devices; a system in which all events occur in a predetermined timed sequence.

synchronous detection A demodulation process in which the original signal is recovered by multiplying the modulated signal by the output of a synchronous oscillator locked to the carrier.

synchronous system A system in which the transmitter and receiver are operating in a fixed time relationship.

system standards The minimum required electrical performance characteristics of a specific collection of hardware and/or software.

systems analysis An analysis of a given activity to determine precisely what must be accomplished and how it is to be done.

T

tetrode A four element electron tube consisting of a cathode, control grid, screen grid, and plate.

thyratron A gas-filled electron tube in which plate current flows when the grid voltage reaches a predetermined level. At that point, the grid has no further control over the current, which continues to flow until it is interrupted or reversed.

tolerance The permissible variation from a standard.

torque A moment of force acting on a body and tending to produce rotation about an axis.

total harmonic distortion (THD) The ratio of the sum of the amplitudes of all signals harmonically related to the fundamental versus the amplitude of the fundamental signal. THD is expressed in percent.

trace The pattern on an oscilloscope screen when displaying a signal.

tracking The locking of tuned stages in a radio receiver so that all stages are changed appropriately as the receiver tuning is changed.

trade-off The process of weighing conflicting requirements and reaching a compromise decision in the design of a component or a subsystem.

transceiver Any circuit or device that receives and transmits signals.

transconductance The mutual conductance of an electron tube expressed as the change in plate current divided by the change in control grid voltage that produced it.

transducer A device that converts energy from one form to another.

transfer characteristics The intrinsic parameters of a system, subsystem, or unit of equipment which, when applied to the input of the system, subsystem, or unit of equipment, will fully describe its output.

transformer A device consisting of two or more windings wrapped around a single core or linked by a common magnetic circuit.

transformer ratio The ratio of the number of turns in the secondary winding of a transformer to the number of turns in the primary winding, also known as the *turns ratio*.

transient A sudden variance of current or voltage from a steady-state value. A transient normally results from changes in load or effects related to switching action.

transient disturbance A voltage pulse of high energy and short duration impressed upon the ac waveform. The overvoltage pulse can be one to 100 times the normal ac potential (or more) and can last up to 15 ms. Rise times measure in the nanosecond range.

transient response The time response of a system under test to a stated input stimulus.

transition A sequence of actions that occurs when a process changes from one state to another in response to an input.

transmission The transfer of electrical power, signals, or an intelligence from one location to another by wire, fiber optic, or radio means.

transmission facility A transmission medium and all the associated equipment required to transmit information.

transmission loss The ratio, in decibels, of the power of a signal at a point along a transmission path to the power of the same signal at a more distant point along the same path. This value is often used as a measure of the quality of the transmission medium for conveying signals. Changes in power level are normally expressed in decibels by calculating ten times the logarithm (base 10) of the ratio of the two powers.

transmission mode One of the field patterns in a waveguide in a plane transverse to the direction of propagation.

transmission system The set of equipment that provides single or multichannel communications facilities capable of carrying audio, video, or data signals.

transmitter The device or circuit that launches a signal into a passive medium, such as the atmosphere.

transparency The property of a communications system that enables it to carry a signal without altering or otherwise affecting the electrical characteristics of the signal.

tray The metal cabinet that holds circuit boards.

tremendously high frequency (THF) The frequency band from 300 GHz to 3000 GHz.

triangular wave An oscillation, the values of which rise and fall linearly, and immediately change upon reaching their peak maximum and minimum. A graphical representation of a triangular wave resembles a triangle.

trim The process of making fine adjustments to a circuit or a circuit element.

trimmer A small mechanically-adjustable component connected in parallel or series with a major component so that the net value of the two can be finely adjusted for tuning purposes.

triode A three-element electron tube, consisting of a cathode, control grid, and plate.

triple beat A third-order beat whose three beating carriers all have different frequencies, but are spaced at equal frequency separations.

troposphere The layer of the earth's atmosphere, between the surface and the stratosphere, in which about 80 percent of the total mass of atmospheric air is concentrated and in which temperature normally decreases with altitude.

trouble A failure or fault affecting the service provided by a system.

troubleshoot The process of investigating, localizing, and (if possible) correcting a fault.

tube (1—electron) An evacuated or gas-filled tube enclosed in a glass or metal case in which the electrodes are maintained at different voltages, giving rise to a controlled flow of electrons from the cathode to the anode. **(2—cathode ray, CRT)** An electron beam tube used for the display of changing electrical phenomena, generally similar to a television picture tube. **(3—cold-cathode)** An electron tube whose cathode emits electrons without the need of a heating filament. **(4—gas)** A gas-filled electron tube in which the gas plays an essential role in operation of the device. **(5—mercury-vapor)** A tube filled with mercury vapor at low pressure, used as a rectifying device. **(6—metal)** An electron tube enclosed in a metal case. **(7—traveling wave, TWT)** A wide band microwave amplifier in which a stream of electrons interacts with a guided electromagnetic wave moving substantially in synchronism with the electron stream, resulting in a net transfer of energy from the electron stream to the wave. **(8—velocity-modulated)** An electron tube in which the velocity of the electron stream is continually changing, as in a klystron.

tune The process of adjusting the frequency of a device or circuit, such as for resonance or for maximum response to an input signal.

tuned trap A series resonant network bridged across a circuit that eliminates ("traps") the frequency of the resonant network.

tuner The radio frequency and intermediate frequency parts of a radio receiver that produce a low level output signal.

tuning The process of adjusting a given frequency; in particular, to adjust for resonance or for maximum response to a particular incoming signal.

turns ratio In a transformer, the ratio of the number of turns on the secondary to the number of turns on the primary.

tweaking The process of adjusting an electronic circuit to optimize its performance.

twin-line A feeder cable with two parallel, insulated conductors.

two-phase A source of alternating current circuit with two sinusoidal voltages that are 90° apart.

U

ultra high frequency (UHF) The frequency range from 300 MHz to 3000 MHz.

ultraviolet radiation Electromagnetic radiation in a frequency range between visible light and high-frequency X-rays.

unattended A device or system designed to operate without a human attendant.

unattended operation A system that permits a station to receive and transmit messages without the presence of an attendant or operator.

unavailability A measure of the degree to which a system, subsystem, or piece of equipment is not operable and not in a committable state at the start of a mission, when the mission is called for at a random point in time.

unbalanced circuit A two-wire circuit with legs that differ from one another in resistance, capacity to earth or to other conductors, leakage, or inductance.

unbalanced line A transmission line in which the magnitudes of the voltages on the two conductors are not equal with respect to ground. A coaxial cable is an example of an unbalanced line.

unbalanced modulator A modulator whose output includes the carrier signal.

unbalanced output An output with one leg at ground potential.

unbalanced wire circuit A circuit whose two sides are inherently electrically unlike.

uncertainty An expression of the magnitude of a possible deviation of a measured value from the true value. Frequently, it is possible to distinguish two components: the *systematic uncertainty* and the *random uncertainty*. The random uncertainty is expressed by the standard deviation or by a multiple of the standard deviation. The systematic uncertainty is generally estimated on the basis of the parameter characteristics.

undamped wave A signal with constant amplitude.

underbunching A condition in a traveling wave tube wherein the tube is not operating at its optimum bunching rate.

Underwriters Laboratories, Inc. A laboratory established by the National Board of Fire Underwriters which tests equipment, materials, and systems that may affect insurance risks, with special attention to fire dangers and other hazards to life.

ungrounded A circuit or line not connected to ground.

unicoupler A device used to couple a balanced circuit to an unbalanced circuit.

unidirectional A signal or current flowing in one direction only.

uniform transmission line A transmission line with electrical characteristics that are identical, per unit length, over its entire length.

unit An assembly of equipment and associated wiring that together forms a complete system or independent subsystem.

unity coupling In a theoretically perfect transformer, complete electromagnetic coupling between the primary and secondary windings with no loss of power.

unity gain An amplifier or active circuit in which the output amplitude is the same as the input amplitude.

unity power factor A power factor of 1.0, which means that the load is—in effect—a pure resistance, with ac voltage and current completely in phase.

unterminated A device or system that is not terminated.

up-converter A frequency translation device in which the frequency of the output signal is greater than that of the input signal. Such devices are commonly found in microwave radio and satellite systems.

uplink A transmission system for sending radio signals from the ground to a satellite or aircraft.

upstream A device or system placed ahead of other devices or systems in a signal path.

useful life The period during which a low, constant failure rate can be expected for a given device or system. The *useful life* is the portion of a product life cycle between break-in and wear out.

user A person, organization, or group that employs the services of a system for the transfer of information or other purposes.

V

VA An abbreviation for *volt-amperes*, volts times amperes.

vacuum relay A relay whose contacts are enclosed in an evacuated space, usually to provide reliable long-term operation.

vacuum switch A switch whose contacts are enclosed in an evacuated container so that spark formation is discouraged.

vacuum tube An electron tube. The most common vacuum tubes include the diode, triode, tetrode, and pentode.

validity check A test designed to ensure that the quality of transmission is maintained over a given system.

varactor A semiconductor that behaves like a capacitor under the influence of an external control voltage.

varactor diode A semiconductor device whose capacitance is a function of the applied voltage. A varactor diode, also called a *variable reactance diode* or simply a *varactor*, is often used to tune the operating frequency of a radio circuit.

variable frequency oscillator (VFO) An oscillator whose frequency can be set to any required value in a given range of frequencies.

variable impedance A capacitor, inductor, or resistor that is adjustable in value.

variable-gain amplifier An amplifier whose gain can be controlled by an external signal source.

variable-reluctance A transducer in which the input (usually a mechanical movement) varies the magnetic reluctance of a device.

variation monitor A device used for sensing a deviation in voltage, current, or frequency, which is capable of providing an alarm and/or initiating transfer to another power source when programmed limits of voltage, frequency, current, or time are exceeded.

varicap A diode used as a variable capacitor.

VCXO (voltage controlled crystal oscillator) A device whose output frequency is determined by an input control voltage.

vector A quantity having both magnitude and direction.

vector diagram A diagram using vectors to indicate the relationship between voltage and current in a circuit.

vector sum The sum of two vectors which, when they are at right angles to each other, equal the length of the hypotenuse of the right triangle so formed. In the general case, the vector sum of the two vectors equals the diagonal of the parallelogram formed on the two vectors.

velocity of light The speed of propagation of electromagnetic waves in a vacuum, equal to 299,792,458 m/s, or approximately 186,000 mi/s. For rough calculations, the figure of 300,000 km/s is used.

velocity of propagation The velocity of signal transmission. In free space, electromagnetic waves travel at the speed of light. In a cable, the velocity is substantially lower.

vernier A device that enables precision reading of a measuring set or gauge, or the setting of a dial with precision.

very low frequency (VLF) A radio frequency in the band 3 kHz to 30 kHz.

vestigial sideband A form of transmission in which one sideband is significantly attenuated. The carrier and the other sideband are transmitted without attenuation.

vibration testing A testing procedure whereby subsystems are mounted on a test base that vibrates, thereby revealing any faults resulting from badly soldered joints or other poor mechanical design features.

volt The standard unit of electromotive force, equal to the potential difference between two points on a conductor that is carrying a constant current of one ampere when the power dissipated between the two points is equal to one watt. One *volt* is equivalent to the potential difference across a resistance of one ohm when one ampere is flowing through it. The volt is named for the Italian physicist Alessandro Volta (1745–1827).

volt-ampere (VA) The apparent power in an ac circuit (volts times amperes).

volt-ohm-milliammeter (VOM) A general purpose multirange test meter used to measure voltage, resistance, and current.

voltage The potential difference between two points.

voltage drop A decrease in electrical potential resulting from current flow through a resistance.

voltage gradient The continuous drop in electrical potential, per unit length, along a uniform conductor or thickness of a uniform dielectric.

voltage level The ratio of the voltage at a given point to the voltage at an arbitrary reference point.

voltage reference circuit A stable voltage reference source.

voltage regulation The deviation from a nominal voltage, expressed as a percentage of the nominal voltage.

voltage regulator A circuit used for controlling and maintaining a voltage at a constant level.

voltage stabilizer A device that produces a constant or substantially constant output voltage despite variations in input voltage or output load current.

voltage to ground The voltage between any given portion of a piece of equipment and the ground potential.

voltmeter An instrument used to measure differences in electrical potential.

vox A voice-operated relay circuit that permits the equivalent of push-to-talk operation of a transmitter by the operator.

VSAT (very small aperture terminal) A satellite Ku-band earth station intended for fixed or portable use. The antenna diameter of a VSAT is on the order of 1.5 m or less.

W

watt The unit of power equal to the work done at one joule per second, or the rate of work measured as a current of one ampere under an electric potential of one volt. Designated by the symbol *W*, the watt is named after the Scottish inventor James Watt (1736–1819).

watt meter A meter indicating in watts the rate of consumption of electrical energy.

watt-hour The work performed by one watt over a one hour period.

wave A disturbance that is a function of time or space, or both, and is propagated in a medium or through space.

wave number The reciprocal of wavelength; the number of wave lengths per unit distance in the direction of propagation of a wave.

waveband A band of wavelengths defined for some given purpose.

waveform The characteristic shape of a periodic wave, determined by the frequencies present and their amplitudes and relative phases.

wavefront A continuous surface that is a locus of points having the same phase at a given instant. A *wavefront* is a surface at right angles to rays that proceed from the wave source. The surface passes through those parts of the wave that are in the same phase and travel in the same direction. For parallel rays the wavefront is a plane; for rays that radiate from a point, the wavefront is spherical.

waveguide Generally, a rectangular or circular pipe that constrains the propagation of an acoustic or electromagnetic wave along a path between two locations. The dimensions of a waveguide determine the frequencies for optimum transmission.

wavelength For a sinusoidal wave, the distance between points of corresponding phase of two consecutive cycles.

weber The unit of magnetic flux equal to the flux that, when linked to a circuit of one turn, produces an electromotive force of one volt as the flux is reduced at a uniform rate to zero in one second. The *weber* is named for the German physicist Wilhelm Eduard Weber (1804–1891).

weighted The condition when a correction factor is applied to a measurement.

weighting The adjustment of a measured value to account for conditions that would otherwise be different or appropriate during a measurement.

weighting network A circuit, used with a test instrument, that has a specified amplitude-versus-frequency characteristic.

wideband The passing or processing of a wide range of frequencies. The meaning varies with the context.

Wien bridge An ac bridge used to measure capacitance or inductance.

winding A coil of wire used to form an inductor.

wire A single metallic conductor, usually solid-drawn and circular in cross section.

working range The permitted range of values of an analog signal over which transmitting or other processing equipment can operate.

working voltage The rated voltage that may safely be applied continuously to a given circuit or device.

X

x-band A microwave frequency band from 5.2 GHz to 10.9 GHz.

x-cut A method of cutting a quartz plate for an oscillator, with the x-axis of the crystal perpendicular to the faces of the plate.

X ray An electromagnetic radiation of approximately 100 nm to 0.1 nm, capable of penetrating nonmetallic materials.

Y

y-cut A method of cutting a quartz plate for an oscillator, with the y-axis of the crystal perpendicular to the faces of the plate.

yield strength The magnitude of mechanical stress at which a material will begin to deform. Beyond the *yield strength* point, extension is no longer proportional to stress and rupture is possible.

yoke A material that interconnects magnetic cores. *Yoke* can also refer to the deflection windings of a CRT.

yttrium-iron garnet (YIG) A crystalline material used in microwave devices.

14.5.5 Bibliography

Whitaker, Jerry C.: *Power Vacuum Tubes Handbook*, CRC Press, Boca Raton, Fla., 1999.

General Services Administration, Information Technology Service, National Communications System: *Glossary of Telecommunication Terms*, Technology and Standards Division, Federal Standard 1037C, General Services Administration, Washington, D.C., August, 7, 1996.

Chapter

5

Acronyms and Abbreviations

5.1 General Electronics Terms[1]

A

a	atto (10^{-18})
Å	angstrom
A	ampere
AAR	automatic alternate routing
AARTS	automatic audio remote test set
ac	alternating current
ACA	automatic circuit assurance
ACC	automatic callback calling
ACD	automatic call distributor
ac-dc	alternating current - direct current
ACK	acknowledge character
ACTS	Advanced Communications Technology Satellite
ACU	automatic calling unit
A-D	analog-to-digital
ADC	analog-to-digital converter; analog-to-digital conversion
ADCCP	Advanced Data Communication Control Procedures
ADH	automatic data handling

1. This chapter adapted from: General Services Administration, Information Technology Service, National Communications System: *Glossary of Telecommunication Terms*, Technology and Standards Division, Federal Standard 1037C, General Services Administration, Washington, D.C., August, 7, 1996.

ADP	automatic data processing
ADPCM	adaptive differential pulse-code modulation
ADPE	automatic data processing equipment
ADU	automatic dialing unit
ADX	automatic data exchange
AECS	Aeronautical Emergency Communications System [Plan]
AF	audio frequency
AFC	area frequency coordinator; automatic frequency control
AFRS	Armed Forces Radio Service
AGC	automatic gain control
AGE	aerospace ground equipment
AI	artificial intelligence
AIG	address indicator group; address indicating group
AIM	amplitude intensity modulation
AIN	advanced intelligent network
AIOD	automatic identified outward dialing
AIS	automated information system
AJ	anti-jamming
ALC	automatic level control; automatic load control
ALE	automatic link establishment
ALU	arithmetic and logic unit
AM	amplitude modulation
AMA	automatic message accounting
AMC	administrative management complex
AME	amplitude modulation equivalent; automatic message exchange
AMI	alternate mark inversion [signal]
AM/PM/VSB	amplitude modulation/phase modulation/vestigial sideband
AMPS	automatic message processing system
AMPSSO	automated message processing system security officer
AMSC	American Mobile Satellite Corporation
AMTS	automated maritime telecommunications system
ANI	automatic number identification

ANL	automatic noise limiter
ANMCC	Alternate National Military Command Center
ANS	American National Standard
ANSI	American National Standards Institute
AP	anomalous propagation
APC	adaptive predictive coding
API	application program interface
APK	amplitude phase-shift keying
APL	average picture level
ARP	address resolution protocol
ARPA	Advanced Research Projects Agency [now DARPA]
ARPANET	Advanced Research Projects Agency Network
ARQ	automatic repeat-request
ARS	automatic route selection
ARSR	air route surveillance radar
ARU	audio response unit
ASCII	American Standard Code for Information Interchange
ASIC	application-specific integrated circuits
ASP	Aggregated Switch Procurement; adjunct service point
ASR	automatic send and receive; airport surveillance radar
AT	access tandem
ATACS	Army Tactical Communications System
ATB	all trunks busy
ATCRBS	air traffic control radar beacon system
ATDM	asynchronous time-division multiplexing
ATE	automatic test equipment
ATM	asynchronous transfer mode
ATV	advanced television
au	astronomical unit
AUI	attachment unit interface
AUTODIN	Automatic Digital Network
AUTOVON	Automatic Voice Network

AVD	alternate voice/data
AWG	American wire gauge
AWGN	additive white Gaussian noise
AZ	azimuth

B

b	bit
B	bel; byte
balun	balanced to unbalanced
basecom	base communications
BASIC	beginners' all-purpose symbolic instruction code
BCC	block check character
BCD	binary coded decimal; binary-coded decimal notation
BCI	bit-count integrity
Bd	baud
B8ZS	bipolar with eight-zero substitution
bell	BEL character
BER	bit error ratio
BERT	bit error ratio tester
BETRS	basic exchange telecommunications radio service
BEX	broadband exchange
BIH	International Time Bureau
B-ISDN	broadband ISDN
bi-sync	binary synchronous [communication]
bit	binary digit
BIT	built-in test
BITE	built-in test equipment
BIU	bus interface unit
BNF	Backus Naur form
BOC	Bell Operating Company
bpi	bits per inch; bytes per inch
BPSK	binary phase-shift keying
b/s	bits per second

b/in	bits per inch
BPSK	binary phase-shift keying
BR	bit rate
BRI	basic rate interface
BSA	basic serving arrangement
BSE	basic service element
BSI	British Standards Institution
B6ZS	bipolar with six-zero substitution
B3ZS	bipolar with three-zero substitution
BTN	billing telephone number
BW	bandwidth

C

c	centi (10^{-2})
CACS	centralized alarm control system
CAM	computer-aided manufacturing
CAMA	centralized automatic message accounting
CAN	cancel character
CAP	competitive access provider; customer administration panel
CARS	cable television relay service [station]
CAS	centralized attendant services
CASE	computer-aided software engineering; computer aided system engineering; computer-assisted software engineering
CATV	cable TV; cable television; community antenna television
CBX	computer branch exchange
C^2	command and control
C^3	command, control, and communications
C^3CM	C^3 countermeasures
C^3I	command, control, communications and intelligence
CCA	carrier-controlled approach
CCD	charge-coupled device
CCH	connections per circuit hour
CCIF	International Telephone Consultative Committee

CCIR	International Radio Consultative Committee
CCIS	common-channel interoffice signaling
CCIT	International Telegraph Consultative Committee
CCITT	International Telegraph and Telephone Consultative Committee
CCL	continuous communications link
CCS	hundred call-seconds
CCSA	common control switching arrangement
CCTV	closed-circuit television
CCW	cable cutoff wavelength
cd	candela
CD	collision detection; compact disk
CDF	combined distribution frame; cumulative distribution function
CDMA	code-division multiple access
CDPSK	coherent differential phase-shift keying
CDR	call detail recording
CD ROM	compact disk read-only memory
CDT	control data terminal
CDU	central display unit
C-E	communications-electronics
CEI	comparably efficient interconnection
CELP	code-excited linear prediction
CEP	circular error probable
CFE	contractor-furnished equipment
cgs	centimeter-gram-second
ChR	channel reliability
CIAS	circuit inventory and analysis system
CIC	content indicator code
CIF	common intermediate format
CIFAX	ciphered facsimile
CiR	circuit reliability
C/kT	carrier-to-receiver noise density
CLASS	custom local area signaling service

cm	centimeter
CMI	coded mark inversion
CMIP	Common Management Information Protocol
CMIS	common management information service
CMOS	complementary metal oxide substrate
CMRR	common-mode rejection ratio
CNR	carrier-to-noise ratio; combat-net radio
CNS	complementary network service
C.O.	central office
COAM	customer owned and maintained equipment
COBOL	common business oriented language
codec	coder-decoder
COG	centralized ordering group
COMINT	communications intelligence
COMJAM	communications jamming
compandor	compressor-expander
COMPUSEC	computer security
COMSAT	Communications Satellite Corporation
COMSEC	communications security
CONEX	connectivity exchange
CONUS	Continental United States
COP	Committee of Principals
COR	Council of Representatives
COT	customer office terminal
CPAS	cellular priority access services
CPE	customer premises equipment
cpi	characters per inch
cpm	counts per minute
cps	characters per second
CPU	central processing unit; communications processor unit
CR	channel reliability; circuit reliability
CRC	cyclic redundancy check

CRITICOM	Critical Intelligence Communications
CROM	control read-only memory
CRT	cathode ray tube
c/s	cycles per second
CSA	Canadian Standards Association
CSC	circuit-switching center; common signaling channel
CSMA	carrier sense multiple access
CSMA/CA	carrier sense multiple access with collision avoidance
CSMA/CD	carrier sense multiple access with collision detection
CSU	channel service unit; circuit switching unit; customer service unit
CTS	clear to send
CTX	Centrex® [service]; clear to transmit
CVD	chemical vapor deposition
CVSD	continuously variable slope delta [modulation]
cw	carrier wave; composite wave; continuous wave
CX	composite signaling
cxr	carrier

D

d	deci (10^{-1})
da	deka (10)
D-A	digital-to-analog; digital-to-analog converter
D/L	downlink
DACS	digital access and cross-connect system
DAMA	demand assignment multiple access
DARPA	Defense Advanced Research Projects Agency
dB	decibel
dBa	decibels adjusted
dBa0	noise power measured at zero transmission level point
dBc	dB relative to carrier power
dBm	dB referred to 1 milliwatt
dBm(psoph)	noise power in dBm measured by a set with psophometric weighting
DBMS	database management system

dBmV	dB referred to 1 millivolt across 75 ohms
dBm0	noise power in dBm referred to or measured at 0TLP
dBm0p	noise power in dBm0 measured by a psophometric or noise measuring set having psophometric weighting
dBr	power difference in dB between any point and a reference point
dBrn	dB above reference noise
dBrnC	noise power in dBrn measured by a set with C-message weighting
dBrnC0	noise power in dBrnC referred to or measured at 0TLP
dBrn(f_1–f_2)	flat noise power in dBrn
dBrn(144)	noise power in dBrn measured by a set with 144-line weighting
dBv	dB relative to 1 V (volt) peak-to-peak
dBW	dB referred to 1 W (watt)
dBx	dB above reference coupling
dc	direct current
DCA	Defense Communications Agency
DCE	data circuit-terminating equipment
DCL	direct communications link
DCPSK	differentially coherent phase-shift keying
DCS	Defense Communications System
DCTN	Defense Commercial Telecommunications Network
DCWV	direct-current working volts
DDD	direct distance dialing
DDN	Defense Data Network
DDS	digital data service
DEL	delete character
demarc	demarcation point
demux	demultiplex; demultiplexer; demultiplexing
dequeue	double-ended queue
DES	Data Encryption Standard
detem	detector/emitter
DFSK	double-frequency shift keying
DIA	Defense Intelligence Agency
DID	direct inward dialing

DIN	Deutsches Institut für Normung
DIP	dual in-line package
DISA	Defense Information Systems Agency
DISC	disconnect command
DISN	Defense Information System Network
DISNET	Defense Integrated Secure Network
DLA	Defense Logistic Agency
DLC	digital loop carrier
DLE	data link escape character
DM	delta modulation
DMA	Defense Mapping Agency; direct memory access
DME	distance measuring equipment
DMS	Defense Message System
DNA	Defense Nuclear Agency
DNIC	data network identification code
DNPA	data numbering plan area
DNS	Domain Name System
DO	design objective
DoC	Department of Commerce
DOD	Department of Defense; direct outward dialing
DODD	Department of Defense Directive
DODISS	Department of Defense Index of Specifications and Standards
DOD-STD	Department of Defense Standard
DOS	Department of State
DPCM	differential pulse-code modulation
DPSK	differential phase-shift keying
DQDB	distributed-queue dual-bus [network]
DRAM	dynamic random access memory
DRSI	destination station routing indicator
DS	digital signal; direct support
DS0	digital signal 0
DS1	digital signal 1

DS1C	digital signal 1C
DS2	digital signal 2
DS3	digital signal 3
DS4	digital signal 4
DSA	dial service assistance
DSB	double sideband (transmission); Defense Science Board
DSB-RC	double-sideband reduced carrier transmission
DSB-SC	double-sideband suppressed-carrier transmission
DSC	digital selective calling
DSCS	Defense Satellite Communications System
DSE	data switching exchange
DSI	digital speech interpolation
DSL	digital subscriber line
DSN	Defense Switched Network
DSR	data signaling rate
DSS	direct station selection
DSSCS	Defense Special Service Communications System
DSTE	data subscriber terminal equipment
DSU	data service unit
DTE	data terminal equipment
DTG	date-time group
DTMF	dual-tone multifrequency (signaling)
DTN	data transmission network
DTS	Diplomatic Telecommunications Service
DTU	data transfer unit; data tape unit; digital transmission unit; direct to user
DVL	direct voice link
DX signaling	direct current signaling; duplex signaling

E

E	exa (10^{18})
E-MAIL	electronic mail
EAS	extended area service
EBCDIC	extended binary coded decimal interchange code

E_b/N_0	signal energy per bit per hertz of thermal noise
EBO	embedded base organization
EBS	Emergency Broadcast System
EBX	electronic branch exchange
EC	Earth coverage; Earth curvature
ECC	electronically controlled coupling; enhance call completion
ECCM	electronic counter-countermeasures
ECM	electronic countermeasures
EDC	error detection and correction
EDI	electronic data interchange
EDTV	extended-definition television
EHF	extremely high frequency
EIA	Electronic Industries Association
eirp	effective isotropically radiated power; equivalent isotopically radiated power
EIS	Emergency Information System
el	elevation
ELF	extremely low frequency
ELINT	electronics intelligence; electromagnetic intelligence
ELSEC	electronics security
ELT	emergency locator transmitter
EMC	electromagnetic compatibility
EMCON	emission control
EMD	equilibrium mode distribution
EME	electromagnetic environment
emf	electromotive force
EMI	electromagnetic interference; electromagnetic interference control
EMP	electromagnetic pulse
EMR	electromagnetic radiation
e.m.r.p.	effective monopole radiated power; equivalent monopole radiated power
EMS	electronic message system
EMSEC	emanations security
emu	electromagnetic unit

EMV	electromagnetic vulnerability
EMW	electromagnetic warfare; electromagnetic wave
ENQ	enquiry character
EO	end office
E.O.	Executive Order
EOD	end of data
EOF	end of file
EOL	end of line
EOM	end of message
EOP	end of program; end output
EOS	end-of-selection character
EOT	end-of-transmission character; end of tape
EOW	engineering orderwire
EPROM	erasable programmable read-only memory
EPSCS	enhanced private switched communications system
ERL	echo return loss
ERLINK	emergency response link
ERP, e.r.p.	effective radiated power
ES	end system; expert system
ESC	escape character; enhanced satellite capability
ESF	extended superframe
ESM	electronic warfare support measures
ESP	enhanced service provider
ESS	electronic switching system
ETB	end-of-transmission-block character
ETX	end-of-text character
EW	electronic warfare
EXCSA	Exchange Carriers Standards Association

F

f	femto (10^{-15})
f	frequency
FAA	Federal Aviation Administration

FAQ file	Frequently Asked Questions file
FAX	facsimile
FC	functional component
FCC	Federal Communications Commission
FCS	frame check sequence
FDDI	fiber distributed data interface
FDDI-2	fiber distributed data interface-2
FDHM	full duration at half maximum
FDM	frequency-division multiplexing
FDMA	frequency-division multiple access
FDX	full duplex
FEC	forward error correction
FECC	Federal Emergency Communications Coordinators
FED-STD	Federal Standard
FEMA	Federal Emergency Management Agency
FEP	front-end processor
FET	field effect transistor
FIFO	first-in first-out
FIP	Federal Information Processing
FIPS	Federal Information Processing Standards
FIR	finite impulse response
FIRMR	Federal Information Resources Management Regulations
FISINT	foreign instrumentation signals intelligence
flops	floating-point operations per second
FM	frequency modulation
FO	fiber optics
FOC	final operational capability; full operational capability
FOT	frequency of optimum traffic; frequency of optimum transmission
FPIS	forward propagation ionospheric scatter
fps	foot-pound-second
FPS	frames per second; focus projection and scanning
FRP	Federal Response Plan

FSDPSK	filtered symmetric differential phase-shift keying
FSK	frequency-shift keying
FSS	fully separate subsidiary
FT	fiber optic T-carrier
FTAM	file transfer, access, and management
FTF	Federal Telecommunications Fund
ft/min	feet per minute
FTP	file transfer protocol
ft/s	feet per second
FTS	Federal Telecommunications System
FTS2000	Federal Telecommunications System 2000
FTSC	Federal Telecommunications Standards Committee
FWHM	full width at half maximum
FX	fixed service; foreign exchange service
FYDP	Five Year Defense Plan

G

g	profile parameter
G	giga (10^9)
GBH	group busy hour
GCT	Greenwich Civil Time
GDF	group distribution frame
GETS	Government Emergency Telecommunications Service
GFE	Government-furnished equipment
GGCL	government-to-government communications link
GHz	gigahertz
GII	Global Information Infrastructure
GMT	Greenwich Mean Time
GOS	grade of service
GOSIP	Government Open Systems Interconnection Profile
GSA	General Services Administration
GSTN	general switched telephone network
G/T	antenna gain-to-noise-temperature

GTP Government Telecommunications Program

GTS Government Telecommunications System

GUI graphical user interface

H

h hecto (10^2); hour; Planck's constant

HCS hard clad silica (fiber)

HDLC high-level data link control

HDTV high-definition television

HDX half-duplex (operation)

HE_{11} mode the fundamental hybrid mode (of an optical fiber)

HEMP high-altitude electromagnetic pulse

HERF hazards of electromagnetic radiation to fuel

HERO hazards of electromagnetic radiation to ordnance

HERP hazards of electromagnetic radiation to personnel

HF high frequency

HFDF high-frequency distribution frame

HLL high-level language

HPC high probability of completion

HV high voltage

Hz hertz

I

IA International Alphabet

I&C installation and checkout

IC integrated circuit

ICI incoming call identification

ICNI Integrated Communications, Navigation, and Identification

ICW interrupted continuous wave

IDDD International Direct Distance Dialing

IDF intermediate distribution frame

IDN integrated digital network

IDTV improved-definition television

IEC	International Electrotechnical Commission
IEEE	Institute of Electrical and Electronics Engineers
IES	Industry Executive Subcommittee
IF	intermediate frequency
I/F	interface
IFF	identification, friend or foe
IFRB	International Frequency Registration Board
IFS	ionospheric forward scatter
IIR	infinite impulse response
IITF	Information Infrastructure Task Force
ILD	injection laser diode
ILS	instrument landing system
IM	intensity modulation; intermodulation
I&M	installation and maintenance
IMD	intermodulation distortion
IMP	interface message processor
IN	intelligent network
INFOSEC	information systems security
INS	inertial navigation system
INTELSAT	International Telecommunications Satellite Consortium
INWATS	Inward Wide-Area Telephone Service
I/O	input/output (device)
IOC	integrated optical circuit; initial operational capability; input-output controller
IP	Internet protocol; intelligent peripheral
IPA	intermediate power amplifier
IPC	information processing center
IPM	impulses per minute; interference prediction model; internal polarization modulation; interruptions per minute
in/s	inches per second
ips	interruptions per second
IPX	Internet Packet Exchange
IQF	intrinsic quality factor
IR	infrared

IRAC	Interdepartment Radio Advisory Committee
IRC	international record carrier; Interagency Radio Committee
ISB	independent-sideband (transmission)
ISDN	Integrated Services Digital Network
ISM	industrial, scientific, and medical (applications)
ISO	International Organization for Standardization
ITA	International Telegraph Alphabet
ITA-5	International Telegraph Alphabet Number 5
ITC	International Teletraffic Congress
ITS	Institute for Telecommunication Sciences
ITSO	International Telecommunications Satellite Organization
ITU	International Telecommunication Union
IVDT	integrated voice data terminal
IXC	interexchange carrier

J

JANAP	Joint Army-Navy-Air Force Publication(s)
JCS	Joint Chiefs of Staff
JPL	Jet Propulsion Laboratory
JSC	Joint Steering Committee; Joint Spectrum Center
JTC^3A	Joint Tactical Command, Control and Communications Agency
JTIDS	Joint Tactical Information Distribution System
JTRB	Joint Telecommunications Resources Board
JTSSG	Joint Telecommunications Standards Steering Group
JWID	Joint Warrior Interoperability Demonstration

K

k	kilo (10^3); Boltzmann's constant
K	coefficient of absorption; kelvin
KDC	key distribution center
KDR	keyboard data recorder
KDT	keyboard display terminal
kg	kilogram

kg·m·s	kilogram-meter-second
kHz	kilohertz
km	kilometer
kΩ, k	kilohm
KSR	keyboard send/receive device
kT	noise power density
KTS	key telephone system
KTU	key telephone unit

L

LAN	local area network
LAP-B	Data Link Layer protocol (CCITT Recommendation X.25 [1989])
LAP-D	link access procedure D
laser	light amplification by stimulated emission of radiation
LASINT	laser intelligence
LATA	local access and transport area
LBO	line buildout
LC	limited capability
LCD	liquid crystal display
LD	long distance
LDM	limited distance modem
LEC	local exchange carrier
LED	light-emitting diode
LF	low frequency
LFB	look-ahead-for-busy (information)
LIFO	last-in first-out
LLC	logical link control (sublayer)
l/m	lines per minute
LMF	language media format
LMR	land mobile radio
LNA	launch numerical aperture
LOF	lowest operating frequency

loran	long-range aid to navigation system; long-range radio navigation; long-range radio aid to navigation system
LOS	line of sight, loss of signal
LP	linearly polarized (mode); linear programming; linking protection; log-periodic (antenna); log-periodic (array)
LPA	linear power amplifier
LPC	linear predictive coding
LPD	low probability of detection
LPI	low probability of interception
lpi	lines per inch
lpm	lines per minute
LP_{01}	the fundamental mode (of an optical fiber)
LQA	link quality analysis
LRC	longitudinal redundancy check
LSB	lower sideband, least significant bit
LSI	large scale integrated (circuit); large scale integration; line status indication
LTC	line traffic coordinator
LUF	lowest usable high frequency
LULT	line-unit-line termination
LUNT	line-unit-network termination
LV	low voltage

M

m	meter
M	mega (10^6)
MAC	medium access control [sublayer]
MACOM	major command
MAN	metropolitan area network
MAP	manufacturers' automation protocol
maser	microwave amplification by the stimulated emission of radiation
MAU	medium access unit
MCC	maintenance control circuit
MCEB	Military Communications-Electronics Board

MCM	multicarrier modulation
MCS	Master Control System
MCW	modulated continuous wave
MCXO	microcomputer compensated crystal oscillator
MDF	main distribution frame
MDT	mean downtime
MEECN	Minimum Essential Emergency Communications Network
MERCAST	merchant-ship broadcast system
MF	medium frequency; multifrequency (signaling)
MFD	mode field diameter
MFJ	Modification of Final Judgment
MFSK	multiple frequency-shift keying
MHF	medium high frequency
MHS	message handling service; message handling system
MHz	megahertz
mi	mile
MIC	medium interface connector; microphone; microwave integrated circuit; minimum ignition current; monolithic integrated circuit; mutual interface chart
MILNET	military network
MIL-STD	Military Standard
min	minute
MIP	medium interface point
MIPS, mips	million instructions per second
MIS	management information system
MKS	meter-kilogram-second
MLPP	multilevel precedence and preemption
MMW	millimeter wave
modem	modulator-demodulator
mol	mole
ms	millisecond (10^{-3} second)
MSB	most significant bit
MSK	minimum-shift keying
MTBF	mean time between failures

MTBM mean time between maintenance

MTBO mean time between outages

MTBPM mean time between preventive maintenance

MTF modulation transfer function

MTSO mobile telephone switching office

MTSR mean time to service restoration

MTTR mean time to repair

μ micro (10^{-6})

μs microsecond

MUF maximum usable frequency

MUX multiplex; multiplexer

MUXing multiplexing

mw microwave

MWI message waiting indicator

MWV maximum working voltage

N

n nano (10^{-9}); refractive index

N_0 sea level refractivity; spectral noise density

NA numerical aperture

NACSEM National Communications Security Emanation Memorandum

NACSIM National Communications Security Information Memorandum

NAK negative-acknowledge character

NASA National Aeronautics and Space Administration

NATA North American Telecommunications Association

NATO North Atlantic Treaty Organization

NAVSTAR Navigational Satellite Timing and Ranging

NBFM narrowband frequency modulation

NBH network busy hour

NBRVF narrowband radio voice frequency

NBS National Bureau of Standards

NBSV narrowband secure voice

NCA National Command Authorities

NCC	National Coordinating Center for Telecommunications
NCS	National Communications System; net control station
NCSC	National Communications Security Committee
NDCS	network data control system
NDER	National Defense Executive Reserve
NEACP	National Emergency Airborne Command Post
NEC	National Electric Code®
NEP	noise equivalent power
NES	noise equivalent signal
NF	noise figure
NFS	Network File System
NIC	network interface card
NICS	NATO Integrated Communications System
NID	network interface device; network inward dialing; network information database
NII	National Information Infrastructure
NIOD	network inward/outward dialing
NIST	National Institute of Standards and Technology
NIU	network interface unit
NLP	National Level Program
nm	nanometer
NMCS	National Military Command System
nmi	nautical mile
NOD	network outward dialing
Np	neper
NPA	numbering plan area
NPR	noise power ratio
NRI	net radio interface
NRM	network resource manager
NRRC	Nuclear Risk Reduction Center
NRZ	non-return-to-zero
NRZI	non-return-to-zero inverted
NRZ-M	non-return-to-zero mark

NRZ-S	non-return-to-zero space
NRZ1	non-return-to-zero, change on ones
NRZ-1	non-return-to-zero mark
ns	nanosecond
NSA	National Security Agency
NSC	National Security Council
NS/EP	National Security or Emergency Preparedness telecommunications
NSTAC	National Security Telecommunications Advisory Committee
NTCN	National Telecommunications Coordinating Network
NTDS	Naval Tactical Data System
NT1	Network termination 1
NT2	Network termination 2
NTI	network terminating interface
NTIA	National Telecommunications and Information Administration
NTMS	National Telecommunications Management Structure
NTN	network terminal number
NTSC	National Television Standards Committee; National Television Standards Committee (standard)
NUL	null character
NVIS	near vertical incidence skywave

O

O&M	operations and maintenance
OC	operations center
OCC	other common carrier
OCR	optical character reader; optical character recognition
OCU	orderwire control unit
OCVCXO	oven controlled-voltage controlled crystal oscillator
OCXO	oven controlled crystal oscillator
OD	optical density; outside diameter
OFC	optical fiber, conductive
OFCP	optical fiber, conductive, plenum
OFCR	optical fiber, conductive, riser

OFN	optical fiber, nonconductive
OFNP	optical fiber, nonconductive, plenum
OFNR	optical fiber, nonconductive, riser
OMB	Office of Management and Budget
ONA	open network architecture
opm	operations per minute
OPMODEL	operations model
OPSEC	operations security
OPX	off-premises extension
OR	off-route service; off-route aeronautical mobile service
OSHA	Occupational Safety and Health Administration
OSI	open switching interval; Open Systems Interconnection
OSI-RM	Open Systems Interconnection—Reference Model
OSRI	originating stations routing indicator
OSSN	originating stations serial number
OTAM	over-the-air management of automated HF network nodes
OTAR	over-the-air rekeying
OTDR	optical time domain reflectometer; optical time domain reflectometry
OW	orderwire [circuit]

P

p	pico (10^{-12})
P	peta (10^{15})
PABX	private automatic branch exchange
PAD	packet assembler/disassembler
PAL	phase alternation by line
PAL-M	phase alternation by line—modified
PAM	pulse-amplitude modulation
PAMA	pulse-address multiple access
p/a r	peak-to-average ratio
PAR	performance analysis and review
PARAMP	parametric amplifier
par meter	peak-to-average ratio meter

PAX	private automatic exchange
PBER	pseudo-bit-error-ratio
PBX	private branch exchange
PC	carrier power (of a radio transmitter); personal computer
PCB	power circuit breaker; printed circuit board
PCM	pulse-code modulation; plug compatible module; process control module
PCS	Personal Communications Services; personal communications system; plastic-clad silica (fiber)
PCSR	parallel channels signaling rate
PD	photodetector
PDM	pulse delta modulation; pulse-duration modulation
PDN	public data network
PDS	protected distribution system; power distribution system; program data source
PDT	programmable data terminal
PDU	protocol data unit
PE	phase-encoded (recording)
PEP	peak envelope power (of a radio transmitter)
pF	picofarad
PF	power factor
PFM	pulse-frequency modulation
PI	protection interval
PIC	plastic insulated cable
ping	packet Internet groper
PIV	peak inverse voltage
PLA	programmable logic array
PL/I	programming language 1
PLL	phase-locked loop
PLN	private line network
PLR	pulse link repeater
PLS	physical signaling sublayer
pm	phase modulation
PM	mean power; polarization-maintaining (optical fiber); preventive maintenance; pulse modulation

PMB	pilot-make-busy (circuit)
PMO	program management office
POI	point of interface
POP	point of presence
POSIX	portable operating system interface for computer environments
POTS	plain old telephone service
PP	polarization-preserving (optical fiber)
P-P	peak-to-peak (value)
P/P	point-to-point
PPM	pulse-position modulation
pps	pulses per second
PR	pulse rate
PRF	pulse-repetition frequency
PRI	primary rate interface
PRM	pulse-rate modulation
PROM	programmable read-only memory
PRR	pulse repetition rate
PRSL	primary area switch locator
PS	permanent signal
psi	pounds (force) per square inch
PSK	phase-shift keying
PSN	public switched network
p-static	precipitation static
PSTN	public switched telephone network
PTF	patch and test facility
PTM	pulse-time modulation
PTT	postal, telephone, and telegraph; push-to-talk (operation)
PTTC	paper tape transmission code
PTTI	precise time and time interval
PU	power unit
PUC	public utility commission; public utilities commission
PVC	permanent virtual circuit; polyvinyl chloride (insulation)

pW	picowatt
PWM	pulse-width modulation
PX	private exchange

Q

QA	quality assurance
QAM	quadrature amplitude modulation
QC	quality control
QCIF	quarter common intermediate format
QMR	qualitative material requirement
QOS	quality of service
QPSK	quadrature phase-shift keying
QRC	quick reaction capability

R

racon	radar beacon
rad	radian; radiation absorbed dose
radar	radio detection and ranging
RADHAZ	electromagnetic radiation hazards
RADINT	radar intelligence
RAM	random access memory; reliability, availability, and maintainability
R&D	research and development
RATT	radio teletypewriter system
RBOC	Regional Bell Operating Company
RbXO	rubidium-crystal oscillator
RC	reflection coefficient; resource controller
RCC	radio common carrier
RCVR	receiver
RDF	radio-direction finding
REA	Rural Electrification Administration
REN	ringer equivalency number
RF	radio frequency; range finder
RFI	radio frequency interference

RFP	request for proposal
RFQ	request for quotation
RGB	red-green-blue
RH	relative humidity
RHR	radio horizon range
RI	routing indicator
RISC	reduced instruction set chip
RJ	registered jack
RJE	remote job entry
rms	root-mean-square (deviation)
RO	read only; receive only
ROA	recognized operating agency
ROC	required operational capability
ROM	read-only memory
ROSE	remote operations service element protocol
rpm	revolutions per minute
RPM	rate per minute
RPOA	recognized private operating agency
rps	revolutions per second
RQ	repeat-request
RR	repetition rate
RSL	received signal level
rss	root-sum-square
R/T	real time
RTA	remote trunk arrangement
RTS	request to send
RTTY	radio teletypewriter
RTU	remote terminal unit
RTX	request to transmit
RVA	reactive volt-ampere
RVWG	Reliability and Vulnerability Working Group
RWI	radio and wire integration

RX	receive; receiver
RZ	return-to-zero

S

s	second
SCC	specialized common carrier
SCE	service creation environment
SCF	service control facility
SCP	service control point
SCPC	single channel per carrier
SCR	semiconductor-controlled rectifier; silicon-controlled rectifier
SCSR	single channel signaling rate
SDLC	synchronous data link control
SDM	space-division multiplexing
SDN	software-defined network
SECDEF	Secretary of Defense
SECORD	secure voice cord board
SECTEL	secure telephone
SETAMS	systems engineering, technical assistance, and management services
SEVAS	Secure Voice Access System
S-F	store-and-forward
SF	single-frequency (signaling)
SGDF	supergroup distribution frame
S/H	sample and hold
SHA	sidereal hour angle
SHARES	Shared Resources (SHARES) HF Radio Program
SHF	super high frequency
SI	International System of Units
SID	sudden ionospheric disturbance
SIGINT	signals intelligence
SINAD	signal-plus-noise-plus-distortion to noise-plus-distortion ratio
SLD	superluminescent diode
SLI	service logic interpreter

SLP	service logic program
SMDR	station message-detail recording
SMSA	standard metropolitan statistical area
SNR	signal-to-noise ratio
SOH	start-of-heading character
SOM	start of message
sonar	sound navigation and ranging
SONET	synchronous optical network
SOP	standard operating procedure
SOR	start of record
SOW	statement of work
(S+N)/N	signal-plus-noise-to-noise ratio
sr	steradian
S/R	send and receive
SSB	single-sideband (transmission)
SSB-SC	single-sideband suppressed carrier (transmission)
SSN	station serial number
SSP	service switching point
SSUPS	solid-state uninterruptible power system
STALO	stabilized local oscillator
STD	subscriber trunk dialing
STFS	standard time and frequency signal (service); standard time and frequency service
STL	standard telegraph level; studio-to-transmitter link
STP	standard temperature and pressure; signal transfer point
STU	secure telephone unit
STX	start-of-text character
SUB	substitute character
SWR	standing wave ratio
SX	simplex signaling
SXS	step-by-step switching system
SYN	synchronous idle character
SYSGEN	system generation

T

T	tera (10^{12})
TADIL	tactical data information link
TADIL-A	tactical data information link-A
TADIL-B	tactical data information link-B
TADS	teletypewriter automatic dispatch system
TADSS	Tactical Automatic Digital Switching System
TAI	International Atomic Time
TASI	time-assignment speech interpolation
TAT	trans-Atlantic telecommunication (cable)
TC	toll center
TCB	trusted computing base
TCC	telecommunications center
TCCF	Tactical Communications Control Facility
TCF	technical control facility
TCP	transmission control protocol
TCS	trusted computer system
TCU	teletypewriter control unit
TCVXO	temperature compensated-voltage controlled crystal oscillator
TCXO	temperature controlled crystal oscillator
TD	time delay; transmitter distributor
TDD	Telecommunications Device for the Deaf
TDM	time-division multiplexing
TDMA	time-division multiple access
TE	transverse electric [mode]
TED	trunk encryption device
TEK	traffic encryption key
TEM	transverse electric and magnetic [mode]
TEMPEST	compromising emanations
TEMS	telecommunications management system
TGM	trunk group multiplexer
THD	total harmonic distortion

THF tremendously high frequency

THz terahertz

TIA Telecommunications Industry Association

TIE time interval error

TIFF tag image file format

TIP terminal interface processor

T_K response timer

TLP transmission level point

TM transverse magnetic [mode]

TP toll point

TRANSEC transmission security

TRC transverse redundancy check

TRF tuned radio frequency

TRI-TAC tri-services tactical [equipment]

TSK transmission security key

TSP Telecommunications Service Priority [system]

TSPS traffic service position system

TSR telecommunications service request; terminate and stay resident

TTL transistor-transistor logic

TTTN tandem tie trunk network

TTY teletypewriter

TTY/TDD Telecommunications Device for the Deaf

TV television

TW traveling wave

TWT traveling wave tube

TWTA traveling wave tube amplifier

TWX® teletypewriter exchange service

TX transmit; transmitter

U

UDP User Datagram Protocol

UHF ultra high frequency

U/L uplink

ULF	ultra low frequency
UPS	uninterruptible power supply
UPT	Universal Personal Telecommunications service
USB	upper sideband
USDA	U.S. Department of Agriculture
USFJ	U.S. Forces, Japan
USFK	U.S. Forces, Korea
USNO	U.S. Naval observatory
USTA	U.S. Telephone Association
UT	Universal Time
UTC	Coordinated Universal Time
uv	ultraviolet

V

V	volt
VA	volt-ampere
VAN	value-added network
VAR	value added reseller
VARISTAR	variable resistor
vars	volt-amperes reactive
VC	virtual circuit
VCO	voltage-controlled oscillator
VCXO	voltage-controlled crystal oscillator
V/D	voice/data
Vdc	volts direct current
VDU	video display unit; visual display unit
VF	voice frequency
VFCT	voice frequency carrier telegraph
VFDF	voice frequency distribution frame
VFO	variable-frequency oscillator
VFTG	voice-frequency telegraph
VHF	very high frequency
VLF	very low frequency

V/m volts per meter

VNL via net loss

VNLF via net loss factor

vocoder voice-coder

vodas voice-operated device anti-sing

vogad voice-operated gain-adjusting device

volcas voice-operated loss control and echo/signaling suppression

vox voice-operated relay circuit; voice operated transmit

VRC vertical redundancy check

VSAT very small aperture terminal

VSB vestigial sideband [transmission]

VSM vestigial sideband modulation

VSWR voltage standing wave ratio

VT virtual terminal

VTU video teleconferencing unit

vu volume unit

W

WADS wide area data service

WAIS Wide Area Information Servers

WAN wide area network

WARC World Administrative Radio Conference

WATS Wide Area Telecommunications Service; Wide Area Telephone Service

WAWS Washington Area Wideband System

WDM wavelength-division multiplexing

WHSR White House Situation Room

WIN WWMCCS Intercomputer Network

WITS Washington Integrated Telecommunications System

WORM write once, read many times

wpm words per minute

wps words per second

wv working voltage

WVDC working voltage direct current

WWDSA	worldwide digital system architecture
WWMCCS	Worldwide Military Command and Control System
WWW	World Wide Web

X

XMIT	transmit
XMSN	transmission
XMTD	transmitted
XMTR	transmitter
XO	crystal oscillator
XOFF	transmitter off
XON	transmitter on
XT	crosstalk
XTAL	crystal

Z

Z	Zulu time
ZD	zero defects
Z_0	characteristic impedance
0TLP	zero transmission level point

14.5.6 Bibliography

Whitaker, Jerry C.: *Power Vacuum Tubes Handbook*, CRC Press, Boca Raton, Fla., 1999.

General Services Administration, Information Technology Service, National Communications System: *Glossary of Telecommunication Terms*, Technology and Standards Division, Federal Standard 1037C, General Services Administration, Washington, D.C., August, 7, 1996.

Chapter 6

Reference Documents by Subject

6.1 Introduction

The following documents are listed as a means of finding additional information on specific aspects of audio and video engineering.

6.2 Audio

6.2a Principles and Sound and Hearing

Backus, John: *The Acoustical Foundations of Music,* Norton, New York, N.Y., 1969.

Batteau, D. W.: "The Role of the Pinna in Human Localization," *Proc. R. Soc. London*, B168, pp. 158–180, 1967.

Benade, A. H.: *Fundamentals of Musical Acoustics,* Oxford University Press, New York, N.Y., 1976.

Beranek, Leo L: *Acoustics,* McGraw-Hill, New York, N.Y., 1954.

Blauert, J., and W. Lindemann: "Auditory Spaciousness: Some Further Psychoacoustic Studies," *J. Acoust. Soc. Am.*, vol. 80, 533–542, 1986.

Blauert, J: *Spatial Hearing*, translation by J. S. Allen, M.I.T., Cambridge. Mass., 1983.

Bloom, P. J.: "Creating Source Elevation Illusions by Spectral Manipulations," *J. Audio Eng. Soc.*, vol. 25, pp. 560–565, 1977.

Bose, A. G.: "On the Design, Measurement and Evaluation of Loudspeakers," presented at the 35th convention of the Audio Engineering Society, preprint 622, 1962.

Buchlein, R.: "The Audibility of Frequency Response Irregularities" (1962), reprinted in English translation in *J. Audio Eng. Soc.*, vol. 29, pp. 126–131, 1981.

Denes, Peter B., and E. N. Pinson: *The Speech Chain,* Bell Telephone Laboratories, Waverly, 1963.

Durlach, N. I., and H. S. Colburn: "Binaural Phenemena," in *Handbook of Perception*, E. C. Carterette and M. P. Friedman (eds.), vol. 4, Academic, New York, N.Y., 1978.

Ehara, Shiro: "Instantaneous Pressure Distributions of Orchestra Sounds," *J. Acoust. Soc. Japan*, vol. 22, pp. 276–289, 1966.

Fletcher, H., and W. A. Munson: "Loudness, Its Definition, Measurement and Calculation," *J. Acoust. Soc. Am.*, vol. 5, pp. 82–108, 1933.

Fryer, P.: "Loudspeaker Distortions—Can We Rear Them?," *Hi-Fi News Record Rev.*, vol. 22, pp. 51–56, 1977.

Gabrielsson, A., and B. Lindstrom: "Perceived Sound Quality of High-Fidelity Loudspeakers." *J. Audio Eng. Soc.*, vol. 33, pp. 33–53, 1985.

Gabrielsson, A., and H. Siogren: "Perceived Sound Quality of Sound-Reproducing Systems," *J. Aoust. Soc. Am.*, vol. 65, pp. 1019–1033, 1979.

Haas, H.: "The Influence of a Single Echo on the Audibility of Speech," *Acustica*, vol. I, pp. 49– 58, 1951; English translation reprinted in *J. Audio Eng. Soc.*, vol. 20, pp. 146–159, 1972.

Hall, Donald: *Musical Acoustics—An Introduction*, Wadsworth, Belmont, Calif., 1980.

International Electrotechnical Commission: *Sound System Equipment*, part 10, *Programme Level Meters*, Publication 268-1 0A, 1978.

International Organization for Standardization: *Normal Equal-Loudness Contours for Pure Tones and Normal Threshold for Hearing under Free Field Listening Conditions*, Recommendation R226, December 1961.

Jones, B. L., and E. L. Torick: "A New Loudness Indicator for Use in Broadcasting," *J. SMPTE*, Society of Motion Picture and Television Engineers, White Plains, N.Y., vol. 90, pp. 772– 777, 1981.

Kuhl, W., and R. Plantz: "The Significance of the Diffuse Sound Radiated from Loudspeakers for the Subjective Hearing Event," *Acustica*, vol. 40, pp. 182–190, 1978.

Kuhn, G. F.: "Model for the Interaural Time Differences in the Azimuthal Plane," *J. Acoust. Soc. Am.*, vol. 62, pp. 157–167, 1977.

Kurozumi, K., and K. Ohgushi: "The Relationship between the Cross-Correlation Coefficient of Two-Channel Acoustic Signals and Sound Image Quality," *J. Acoust. Soc. Am.*, vol. 74, pp. 1726–1733, 1983.

Main, Ian G.: *Vibrations and Waves in Physics,* Cambridge, London, 1978.

Mankovsky, V. S.: *Acoustics of Studios and Auditoria,* Focal Press, London, 1971.

Meyer, J.: *Acoustics and the Performance of Music*, Verlag das Musikinstrument, Frankfurt am Main, 1987.

Morse, Philip M.: *Vibrations and Sound,* 1964, reprinted by the Acoustical Society of America, New York, N.Y., 1976.

Olson, Harry F.: *Acoustical Engineering*, Van Nostrand, New York, N.Y., 1957.

Pickett, J. M.: *The Sounds of Speech Communications*, University Park Press, Baltimore, MD, 1980.

Pierce, John R.: *The Science of Musical Sound*, Scientific American Library, New York, N.Y., 1983.

Piercy, J. E., and T. F. W. Embleton: "Sound Propagation in the Open Air," in *Handbook* of *Noise Control*, 2d ed., C. M. Harris (ed.), McGraw-Hill, New York, N.Y., 1979.

Plomp, R.: *Aspects* of *Tone Sensation—A Psychophysical Study*," Academic, New York, N.Y., 1976.

Rakerd, B., and W. M. Hartmann: "Localization of Sound in Rooms, II—The Effects of a Single Reflecting Surface," *J. Acoust. Soc. Am.*, vol. 78, pp. 524–533, 1985.

Rasch, R. A., and R. Plomp: "The Listener and the Acoustic Environment," in D. Deutsch (ed.), *The Psychology of Music*, Academic, New York, N.Y., 1982.

Robinson, D. W., and R. S. Dadson: "A Redetermination of the Equal-Loudness Relations for Pure Tones," *Br. J. Appl. Physics*, vol. 7, pp. 166–181, 1956.

Scharf, B.: "Loudness," in E. C. Carterette and M. P. Friedman (eds.), *Handbook* of *Perception*, vol. 4, *Hearing*, chapter 6, Academic, New York, N.Y., 1978.

Shaw, E. A. G., and M. M. Vaillancourt: "Transformation of Sound-Pressure Level from the Free Field to the Eardrum Presented in Numerical Form," *J. Acoust. Soc. Am.*, vol. 78, pp. 1120–1123, 1985.

Shaw, E. A. G., and R. Teranishi: "Sound Pressure Generated in an External-Ear Replica and Real Human Ears by a Nearby Sound Source," *J. Acoust. Soc. Am.*, vol. 44, pp. 240–249, 1968.

Shaw, E. A. G.: "Aural Reception," in A. Lara Saenz and R. W. B. Stevens (eds.), *Noise Pollution*, Wiley, New York, N.Y., 1986.

Shaw, E. A. G.: "External Ear Response and Sound Localization," in R. W. Gatehouse (ed.), *Localization of Sound: Theory and Applications*, Amphora Press, Groton, Conn., 1982.

Shaw, E. A. G.: "Noise Pollution—What Can be Done?" *Phys. Today*, vol. 28, no. 1, pp. 46–58, 1975.

Shaw, E. A. G.: "The Acoustics of the External Ear," in W. D. Keidel and W. D. Neff (eds.), *Handbook* of *Sensory Physiology*, vol. V/I, *Auditory System,* Springer-Verlag, Berlin, 1974.

Shaw, E. A. G.: "Transformation of Sound Pressure Level from the Free Field to the Eardrum in the Horizontal Plane," *J. Acoust. Soc. Am.*, vol. 56, pp. 1848–1861, 1974.

Stephens, R. W. B., and A. E. Bate: *Acoustics and Vibrational Physics,* 2nd ed., E. Arnold (ed.), London, 1966.

Stevens, W. R.: "Loudspeakers—Cabinet Effects," *Hi-Fi News Record Rev.*, vol. 21, pp. 87–93, 1976.

Sundberg, Johan: "The Acoustics of the Singing Voice," in *The Physics of Music,* introduction by C. M. Hutchins, Scientific American/Freeman, San Francisco, Calif., 1978.

Tonic, F. E.: "Loudness—Applications and Implications to Audio," *dB,* Part 1, vol. 7, no. 5, pp. 27–30; Part 2, vol. 7, no. 6, pp. 25–28, 1973.

Toole, F. E., and B. McA. Sayers: "Lateralization Judgments and the Nature of Binaural Acoustic Images," *J. Acoust. Soc. Am.*, vol. 37, pp. 319–324, 1965.

Toole, F. E.: "Loudspeaker Measurements and Their Relationship to Listener Preferences," *J. Audio Eng. Soc.*, vol. 34, part 1, pp. 227–235, part 2, pp. 323–348, 1986.

Toole, F. E.: "Subjective Measurements of Loudspeaker Sound Quality and Listener Performance," *J. Audio Eng. Soc.*, vol. 33, pp. 2–32, 1985.

Voelker, E. J.: "Control Rooms for Music Monitoring," *J. Audio Eng. Soc.*, vol. 33, pp. 452–462, 1985.

Ward, W. D.: "Subjective Musical Pitch," *J. Acoust. Soc. Am.*, vol. 26, pp. 369–380, 1954.

Waterhouse, R. V., and C. M. Harris: "Sound in Enclosed Spaces," in *Handbook of Noise Control*, 2d ed., C. M. Harris (ed.), McGraw-Hill, New York, N.Y., 1979.

Wong, G. S. K.: "Speed of Sound in Standard Air," *J. Acoust. Soc. Am.*, vol. 79, pp. 1359–1366, 1986.

Zurek, P. M.: "Measurements of Binaural Echo Suppression," *J. Acoust. Soc. Am.*, vol. 66, pp. 1750–1757, 1979.

Zwislocki, J. J.: "Masking—Experimental and Theoretical Aspects of Simultaneous, Forward, Backward and Central Masking," in E. C. Carterette and M. P. Friedman (eds.), *Handbook of Perception,* vol. 4, *Hearing*, chapter 8, Academic, New York, N.Y., 1978.

6.2b The Audio Spectrum

Bendat, J. S., and A. G. Riersol: *Engineering Applications of Correlation and Spectral Analysis*, Wiley, New York, 1980.

Bendat, J. S., and A. G. Piersol: *Random Data: Analysis and Measurement Procedures,* Wiley-Interscience, New York, N.Y., 1971.

Blinchikoff, H. J., and A. I. Zverev: *Filtering in the Time and Frequency Domains*, Wiley, New York, N.Y., 1976.

Bloom, P. J., and D. Preis: "Perceptual Identification and Discrimination of Phase Distortions," *IEEE ICASSP Proc.*, pp. 1396–1399, April 1983.

Bode, H. W.: *Network Analysis and Feedback Amplifier Design,* Van Nostrand, New York, N.Y., 1945.

Bracewell, R.: *The Fourier Integral and Its Applications,* McGraw-Hill, New York, N.Y., 1965.

Cheng, D. K.: *Analysis of Linear Systems*, Addison-Wesley, Reading, Mass., 1961.

Childers, D. G.: *Modern Spectral Analysis,* IEEE, New York, N.Y., 1978.

Connor, F. R.: *Signals,* Arnold, London, 1972.

Deer, J. A., P. J. Bloom, and D. Preis: "Perception of Phase Distortion in All-Pass Filters," *J. Audio Eng. Soc.*, vol. 33, no. 10, pp. 782–786, October 1985.

Di Toro. M. J.: "Phase and Amplitude Distortion in Linear Networks," *Proc. IRE*, vol. 36, pp. 24–36, January 1948.

Guillemin, E. A.: *Communication Networks*, vol. 11, Wiley, New York, N.Y., 1935.

Henderson. K. W., and W. H. Kautz: "Transient Response of Conventional Filters," *IRE Trans. Circuit Theory*, CT-5, pp. 333–347, December 1958.

Hewlett-Packard: "Application Note 63—Section II, Appendix A, "Table of Important Transforms," Hewlett-Packard, Palo Alto, Calif, pp. 37, 38, 1954.

Jenkins, G. M., and D. G. Watts: *Spectral Analysis and Its Applications,* Holden-Day, San Francisco, Calif., 1968.

Kharkevich, A. A.: *Spectra and Analysis,* English translation, Consultants Bureau, New York, N.Y., 1960.

Kupfmuller, K.: *Die Systemtheorie der elektrischen Nachrichtenuhertragung,* S. Hirzel Verlag, Stuttgart, 1968.

Lane, C. E.: "Phase Distortion in Telephone Apparatus," *Bell* Syst. *Tech. J.*, vol. 9, pp. 493–521, July 1930.

Lathi, B. P.: *Signals, Systems and Communications,* Wiley, New York, N.Y., 1965.

Lynn, P. A.: *An Introduction to the Analysis and Processing of Signals,* 2nd ed. Macmillan, London, 1982.

Mallinson, J. C.: "Tutorial Review of Magnetic Recording." *Proc. IEEE*, vol. 62, pp. 196–208, February 1976.

Members of the Technical Staff of Bell Telephone Laboratories: *Transmission Systems for Communications,* 4th ed., Western Electric Company, Technical Publications, Winston-Salem, N.C., 197 I.

Oppenheim, A. V., and R. W. Schafer: *Digital Signal Processing*, Prentice-Hall, Englewood Cliffs, N.J., 1975.

Panter, P. F.: *Modulation, Noise and Spectral Analysis,* McGraw-Hill, New York, N.Y., 1965.

Papoulis, A.: *Signal Analysis*, McGraw-Hill, New York, N.Y., 1977.

Papoulis, A.: *The Fourier Integral and Its Applications,* McGraw-Hill, New York, N.Y., 1962.

Peus, S.: "Microphones and Transients," *db Mag.*, translated from *Radio Mentor* by S. Temmer, vol. 11, pp. 35–38, May 1977.

Preis, D: "A Catalog of Frequency and Transient Responses," *J. Audio Eng. Soc.*, vol. 25, no. 12, pp. 990–1007, December 1977.

Pries, D.: "Audio Signal Processing with Transversal Filters," *IEEE Conf. Proc.*, 1979 ICASSP, pp. 310–313, April 1979.

Preis, D.: "Hilbert-Transformer Side-Chain Phase Equalizer for Analogue Magnetic Recording," *Electron. Lett.*, vol. 13, pp. 616–617, September 1977.

Preis, D.: "Impulse Testing and Peak Clipping," *J. Audio Eng. Soc.*, vol. 25, no. 1, pp. 2–14, January 1977.

Preis, D.: "Least-Squares Time-Domain Deconvolution for Transversal-Filter Equalizers," *Electron. Lett.*, vol. 13, no. 12, pp. 356–357, June 1977.

Preis, D.: "Linear Distortion," *J. Audio Eng. Soc.*, vol. 24, no. 5, pp. 346–367, June 1976.

Pries, D.: "Measures and Perception of Phase Distortion in Electroacoustical Systems," *IEEE Conf. Proc.*, 1980 ICASSP, pp. 490–493, 1980.

Pries, D.: "Phase Equalization for Analogue Magnetic Recorders by Transversal Filtering," *Electron. Lett.*, vol. 13, pp. 127–128, March 1977.

Pries, D.: "Phase Equalization for Magnetic Recording," *IEEE Conf. Proc.*, 1981 ICASSP, pp. 790–795, March 1981.

Preis, D.: "Phase Distortion and Phase Equalization in Audio Signal Processing—A Tutorial Review," *J. Audio Eng Soc.*, vol. 30, no. 11, pp. 774–794, November 1982.

Pries, D., and C. Bunks: "Three Algorithms for the Design of Transversal-Filter Equalizers," *Proc. 1981 IEEE Int. Symp. Circuits Sys.*, pp. 536–539, 1981.

Pries, D., and P. J. Bloom: "Perception of Phase Distortion in Anti-Alias Filters," *J. Audio Eng. Soc.*, vol. 32, no. 11, pp. 842–848, November 1984.

Preis, D., F. Hlawatsch, P. J. Bloom, and J. A. Deer: "Wigner Distribution Analysis of Filters with Perceptible Phase Distortion," *J. Audio Eng. Soc.*, December 1987.

Rabiner, L. R., and C. M. Rader (eds.): *Digital Signal Processing,* IEEE, New York, N.Y., 1972.

Schwartz, M.: *Information Transmission, Modulation and Noise,* McGraw-Hill, New York, N.Y., 1970.

Small, R. H.: "Closed-Box Loudspeaker Systems, Part 1: Analysis," *J. Audio Eng. Soc.*, vol. 20, pp. 798–808, December 1972.

Totzek, U., and D. Press: "How to Measure and Interpret Coherence Loss in Magnetic Recording," *J. Audio Eng. Soc.*, December 1987.

Totzek, U., D. Preis, and J. F. Boebme: "A Spectral Model for Time-Base Distortions and Magnetic Recording," *Archiv. fur Elektronik und Ubertragungstechnik*, vol. 41, no. 4, pp. 223–231, July-August 1987.

Westman, H. P. (ed.): *ITT Reference Data for Radio Engineers,* Howard W. Sams, New York, N.Y., 1973.

Wheeler, H. A.: "The Interpretation of Amplitude and Phase Distortion in Terms of Paired Echoes," *Proc. IRE*, vol. 27, pp. 359–385, June 1939.

Williams, A. B.: *Active Filter Design,* Artech House. Dedham, Mass., 1975.

Zverev, A. 1.: *Handbook of Filter Synthesis*, Wiley, New York, N.Y., 1967.

6.2c Architectural Acoustic Principles and Design Techniques

ANSI: *American National Standard.for Rating Noise with Respect to Speech Interference,* ANSI S3.14-1977, American National Standards Institute, New York, N.Y., 1977.

ANSI: *Method for the Measurement of Monosyllabic Word Intelligibility,* ANSI S3.2-1960, rev. 1977, American National Standards Institute, New York, N.Y., 1976.

ASA Standards Index 2, Acoustical Society of America, New York, N.Y., 1980.

ASHRAE: *ASHRAE Handbook—1984 Systems,* American Society of Heating, Refrigerating and Air-Conditioning Engineers, Atlanta, Ga., 1984.7.

Beranek, L. L.: *Acoustics,* McGraw-Hill, New York, N.Y., 1954.

Beranek, L. L.: *Noise and Vibration Control,* McGraw-Hill. New York, N.Y., 1971.

Catalogue of STC and IIC Ratings for Wall and Floor/Ceiling Assemblies, Office of Noise Control, Berkeley, Calif.

Egan, M. D.: Co*ncepts in Architectural Acoustics,* McGraw-Hill, New York, N.Y., 1972.

Huntington, W. C., R. A. Mickadeit, and W. Cavanaugh: *Building Construction Materials,* 5th ed., Wiley, New York, N.Y., 1981.

Jones, Robert S.: *Noise and Vibration Control in Buildings,* McGraw-Hill, New York, N.Y., 1980.

Kryter, K. D.: *The Effects of Noise on Man,* Academic, New York, N.Y., 1985.

Lyon R. H., and R. G. Cann: *Acoustical Scale Modeling,* Grozier Technical Systems, Inc., Brookline, Mass.

Marris, Cyril M.: *Handbook of Noise Control*, 2nd ed., McGraw-Hill, New York, N.Y., 1979.

Marshall, Harold, and M. Barron: "Spatial Impression Due to Early Lateral Reflections in Concert Halls: The Derivation of the Physical Measure," *JSV,* vol.77, no. 2, pp. 211–232, 1981.

Morse, P. M.: *Vibration and Sound*, American Institute of Physics, New York, N.Y., 1981.

Siebein, Gary W.: *Prolect Design Phase Analysis Techniques for Predicting the Acoustical Qualities of Buildings,* research report to the National Science Foundation, grant CEE8307948, Florida Architecture and Building Research Center, Gainesville, Fla., 1986.

Talaske, Richard H., Ewart A. Wetherill, and William J. Cavanaugh (eds.): *Halls for Music Performance Two Decades of Experience, 1962-1982*, American Institute of Physics for the Acoustical Society of America, New York, N.Y., 1982.

6.2d Microphone Devices and Systems

"A Phased Array," *Hi Fi News Record Rev.,* July 1981.

Bevan, W. R., R. B. Schulein, and C. E. Seeler: "Design of a Studio-Quality Condenser Microphone Using Electret Technology," *J. Audio Eng Soc.* (*Engineering Reports*), vol. 26, pp. 947–957, December 1978.

Black, H. S.: U.S. Patent 2,102,671.

Blumlein, A.: British Patent 394,325, December 14, 1931; reprinted in *J. Audio Eng. Soc.*, vol. 6, April 1958.

Dooley, W. L., and R. D. Streicher: "M-S Stereo: A Powerful Technique for Working in Stereo," *J. Audio Eng. Soc.*, vol. 30, pp. 707–718, October 1982.

Eargle, J.: *Sound Recording,* Van Nostrand Reinhold, New York, 1976.

Eargle, J.: *The Microphone Handbook,* Elar Publishing, Plainview, N.Y., 1981.

Fewer, D. R.: "Design Principles for Junction Transistor Audio Power Amplifiers," *Trans. IRE PGA*, AU-3(6), November–December 1955.

Fredericksen, E., N. Firby, and H. Mathiasen: "Prepolarized Condenser Microphones for Measurement Purposes," *Tech. Rev.,* Bruel & Kjaer, Copenhagen, no.4, 1979.

Garner, L. H.: "High-Power Solid State Amplifiers," *Trans. IRE PGA*, 15(4), December 1967.

Gordon, J.: "Recording in 2 and 4 Channels," *Audio*, pp. 36–38, December 1973.

Harper, C. A. (ed.): *Handbook of Components for Electronics,* McGraw-Hill, New York, N.Y., 1977.

Instruction Book for RCA BK-16A Dynamic Microphone, IB-24898, Radio Corporation of America, Camden, N.J.

Killion, M. C., and E. V. Carlson: "A Subminiature Electret-Condenser Microphone of New Design," *J. Audio Eng. Soc.*, vol. 22, pg. 237–243, May 1974.

Kirchner, R. J.: "Properties of Junction Transistors," *Trans. IRE PGA*, AU-3(4), July-August 1955.

Kishi, K., N. Tsuchiya, and K. Shimura: "Unidirectional Microphone," U.S. Patent 3,581,012, May 25, 1971.

Kubota, H.: "Back Electret Microphones," presented at the 55th Convention of the Audio Engineering Society, *J. Audio Eng. Soc.* (*Abstracts*), vol. 24, no. 862, preprint 1157, December 1976.

Lipshitz, S. P.: "Stereo Microphone Techniques: Are the Purists Wrong?" presented at the 78th Convention of the Audio Engineering Society, *J. Audio Eng. Soc.* (*Abstracts*), vol. 33, pg. 594, preprint 2261, July-August 1985.

Long, E. M., and R. J. Wickersham: "Pressure Recording Process and Device," U.S. Patent 4,361,736, November 30, 1982.

Lynn, D. K., C. S. Meyer, and D. C. Hamilton (eds.): *Analysis and Design of Integrated Circuits,* McGraw-Hill, New York, N.Y., 1967.

Microphones—Anthology, Audio Engineering Society, New York, 1979.

"Miking with the 3-Point System," *Audio*, pp. 28–36, December 1975.

Nisbett, A.: *The Technique of the Sound Studio,* Hastings House, New York, N.Y., 1974.

Olson, H. F.: *Acoustical Engineering,* Van Nostrand, Princeton, N.J., 1957.

Olson, H. F.: "Directional Electrostatic Microphone," U.S. Patent 3,007,012, October 31, 1961.

Olson, H. F. (ed.): *McGraw-Hill Encyclopedia of Science and Technology,* 5th ed., vol. 18, McGraw-Hill, New York, N.Y., pg. 506, 1982.

Olson, H. F.: *Music, Physics, and Engineering,* 2d ed., Dover, New York, N.Y., 1967.

Olson, H. F.: "Ribbon Velocity Microphones," *J. Audio Eng. Soc.*, vol. 18, pp. 263–268, June 1970.

Petersen, A., and D. B. Sinclair: "A Singled-Ended Push-Pull Audio Amplifier," *Proc. IRE*, vol. 40, January 1952.

Rasmussen, G.: "A New Condenser Microphone," *Tech. Rev.,* Bruel & Kjaer, Copenhagen, no.1, 1959.

Sank, J. R.: "Equipment Profile-Nakamichi CM-700 Electret Condenser Microphone System." *Audio*, September 1978.

Shockley, W: "A Unipolar Field-Effect Transistor," *Proc. IRE*, vol. 40, November 1952.

Shockley, W: "The Theory of P-N Junctions in Semiconductors and P-N Junction Transistors," *Proc. JRE*, vol. 41, June 1953.

Trent, R. L.: "Design Principles for Transistor Audio Amplifiers," *Trans. IRE PGA*, AU-3(5), September–October 1955.

Walker, P. J.: "A Current Dumping Audio Power Amplifier," *Wireless World,* December 1975.

Weinberg, L.: *Network Analysis and Synthesis,* McGraw-Hill, New York, N.Y., 1967.

Widlar, R. J.: "A Unique Current Design for a High Performance Operational Amplifier Especially Suited to Monolithic Construction," *Proc. NEC*, 1965.

Woszczyk, W. R.: "A Microphone Technique Employing the Principle of Second-Order Gradient Unidirectionality," presented at the 69th Convention of the Audio Engineering Society., *J. Audio Eng. Soc.* (*Abstracts*), vol. 29, pg. 550, preprint 1800, July-August 1981.

6.2e Sound Reproduction Devices and Systems

Allison, R., et at.: "On the Magnitude and Audibility of FM Distortion in Loudspeakers," *J. Audio Eng Soc.*, vol. 30, no. 10, pg. 694, 1982.

Beranek, L. L.: *Acoustics*, McGraw-Hill, New York, N.Y., pg. 183–185, 1954.

Hayasaka, T., et al.: *Onkyo-Kogaku Gairon* (*An Introduction to Sound and Vibration*), Nikkan Kogyo Shinbunshya, pg. 67, 1973 (in Japanese).

Hayasaka, T., et al.: *Onkyo-Shindo Ron* (*Sound and Vibration*), Maruzen Kabushikigaishya, pg. 201, 1974 (in Japanese).

Hirata, Y.: "Study of Nonlinear Distortion in Audio Instruments," *J. Audio Eng. Soc.*, vol. 29, no. 9, pg. 607, 1981.

Kinsler, L. E., et al: *Fundamentals of Acoustics*, Wiley, New York, N.Y., 1982.

Melillo, L., et al.: "Ferrolluids as a Means of Controlling Woofer Design Parameters," presented at the 63d Convention of the Audio Engineering Society, vol. 1, pg. 177, 1979.

Morse, P. M.: *Vibration and Sound*, McGraw-Hill, New York, N.Y., pg. 326, 1948.

Morse, P. M., and K. U. Ingard: Theoretical Acoustics, McGraw-Hill, New York, N.Y., pg. 366, 1968.

Niguchi, H., et al.: "Reinforced Olefin Polymer Diaphragm for Loudspeakers," *J. Audio Eng. Soc.*, vol. 29, no. 11, pg. 808, 1981.

Okahara, M., et al: *Audio Handbook*, Ohm Sya, pg. 285, 1978 (in Japanese).

Olson, H. F.: *Elements of Acoustical Engineering*, Van Nostrand, Princeton, N.J., 1957.

Rayleigh, J. W. S.: *The Theory of Sound*, Dover, New York, N.Y., pg 162, 1945.

Sakamoto, N.: *Loudspeaker and Loudspeaker Systems*, Nikkan Kogyo Shinbunshya, pg. 36, 1967 (in Japanese).

Sakamotoet, N., et. al.: "Loudspeaker with Honeycomb Disk Diaphragm," *J. Audio Eng. Soc.*, vol. 29, no. 10, pg. 711, 1981.

Shindo, T., et al: "Effect of Voice-Coil and Surround on Vibration and Sound Pressure Response of Loudspeaker Cones," *J. Audio Eng. Soc.*, vol. 28, no 7–8, pg. 490, 1980.

Suwa, H., et al.: "Heat Pipe Cooling Enables Loudspeakers to Handle Higher Power," presented at the 63d Convention of the Audio Engineering Society, vol. 1, pg. 213, 1979.

Suzuki, H., et al.: "Radiation and Diffraction Effects by Convex and Concave Domes," *J. Audio Eng Soc.*, vol. 29, no. 12, pg. 873, 1981.

Takahashi, S., et al.: "Glass-Fiber and Graphite-Flake Reinforced Polyimide Composite Diaphragm for Loudspeakers," *J. Audio Eng. Soc.*, vol. 31, no. 10, pg. 723, 1983.

Tsuchiya, H., et al.: "Reducing Harmonic Distortion in Loudspeakers," presented at the 63d Convention of the Audio Engineering Society, vol. 2, pg. 1, 1979.

Yamamoto, T., et al.: "High-Fidelity Loudspeakers with Boronized Titanium Diaphragm," *J. Audio Eng. Soc.*, vol. 28, no. 12, pg. 868, 1980.

Yoshihisa, N., et al.: "Nonlinear Distortion in Cone Loudspeakers," Chuyu-Ou Univ. Rep., vol. 23, pg. 271, 1980.

6.2f Digital Coding of Audio Signals

Alkin, Oktay: "Digital Coding Schemes," *The Electronics Handbook*, Jerry C. Whitaker (ed.), CRC Press, Boca Raton, Fla., pp. 1252–1258, 1996.

Benson, K. B., and D. G. Fink: "Digital Operations in Video Systems," *HDTV: Advanced Television for the 1990s*, McGraw-Hill, New York, pp. 4.1–4.8, 1990.

Chambers, J. A., S. Tantaratana, and B. W. Bomar: "Digital Filters," *The Electronics Handbook*, Jerry C. Whitaker (ed.), CRC Press, Boca Raton, Fla., pp. 749–772, 1996.

Garrod, Susan A. R.: "D/A and A/D Converters," *The Electronics Handbook*, Jerry C. Whitaker (ed.), CRC Press, Boca Raton, Fla., pp. 723–730, 1996.

Garrod, Susan, and R. Borns: *Digital Logic: Analysis, Application, and Design*, Saunders College Publishing, Philadelphia, 1991.

Lee, E. A., and D. G. Messerschmitt: *Digital Communications*, 2nd ed., Kluwer, Norell, Mass., 1994.

Nyquist, H.: "Certain Factors Affecting Telegraph Speed," *Bell System Tech. J.*, vol. 3, pp. 324–346, March 1924.

Parks, T. W., and J. H. McClellan: "A Program for the Design of Linear Phase Infinite Impulse Response Filters," *IEEE Trans. Audio Electroacoustics*, AU-20(3), pp. 195–199, 1972.

Peterson, R., R. Ziemer, and D. Borth: *Introduction to Spread Spectrum Communications*, Prentice-Hall, Englewood Cliffs, N. J., 1995.

Pohlmann, Ken: *Principles of Digital Audio*, McGraw-Hill, New York, N.Y., 2000.

Sklar, B.: *Digital Communications: Fundamentals and Applications*, Prentice-Hall, Englewood Cliffs, N. J., 1988.

TMS320C55x DSP Functional Overview, Texas Instruments, Dallas, TX, literature No. SRPU312, June 2000.

Ungerboeck, G.: "Trellis-Coded Modulation with Redundant Signal Sets," parts I and II, *IEEE Comm. Mag.*, vol. 25 (Feb.), pp. 5-11 and 12-21, 1987.

Ziemer, R., and W. Tranter: *Principles of Communications: Systems, Modulation, and Noise*, 4th ed., Wiley, New York, 1995.

Ziemer, Rodger E.: "Digital Modulation," *The Electronics Handbook*, Jerry C. Whitaker (ed.), CRC Press, Boca Raton, Fla., pp. 1213–1236, 1996.

6.2g Compression Technologies for Audio

Brandenburg, K., and Gerhard Stoll: "ISO-MPEG-1 Audio: A Generic Standard for Coding of High Quality Digital Audio," *92nd AES Convention Proceedings*, Audio Engineering Society, New York, N.Y., 1992, revised 1994.

Ehmer, R. H.: "Masking Patterns of Tones," J. Acoust. Soc. Am., vol. 31, pp. 1115–1120, August 1959.

Fibush, David K.: "Testing MPEG-Compressed Signals," *Broadcast Engineering*, Overland Park, Kan., pp. 76–86, February 1996.

Herre, J., and B. Grill: "MPEG-4 Audio—Overview and Perspectives for the Broadcaster," *IBC 2000 Proceedings*, International Broadcast Convention, Amsterdam, September 2000.

IEEE Standard Dictionary of Electrical and Electronics Terms, ANSI/IEEE Standard 100-1984, Institute of Electrical and Electronics Engineers, New York, 1984.

ITU-R Recommendation BS-775, "Multi-channel Stereophonic Sound System with and Without Accompanying Picture."

Lyman, Stephen, "A Multichannel Audio Infrastructure Based on Dolby E Coding," *Proceedings of the NAB Broadcast Engineering Conference*, National Association of Broadcasters, Washington, D.C., 1999.

Moore, B. C. J., and B. R. Glasberg: "Formulae Describing Frequency Selectivity as a Function of Frequency and Level, and Their Use in Calculating Excitation Patterns," Hearing Research, vol. 28, pp. 209–225, 1987.

Robin, Michael, and Michel Poulin: *Digital Television Fundamentals*, McGraw-Hill, New York, N.Y., 1998.

SMPTE Standard for Television: "12-Channel Serial Interface for Digital Audio and Auxiliary Data," SMPTE 324M, SMPTE, White Plains, N.Y., 1999.

SMPTE Standard for Television: "Channel Assignments and Levels on Multichannel Audio Media," SMPTE 320M-1999, SMPTE, White Plains, N.Y., 1999.

Smyth, Stephen: "Digital Audio Data Compression," *Broadcast Engineering*, Intertec Publishing, Overland Park, Kan., February 1992.

Terry, K. B., and S. B. Lyman: "Dolby E—A New Audio Distribution Format for Digital Broadcast Applications," *International Broadcasting Convention Proceedings*, IBC, London, England, pp. 204–209, September 1999.

Todd, C., et. al.: "AC-3: Flexible Perceptual Coding for Audio Transmission and Storage," AES 96th Convention, Preprint 3796, Audio Engineering Society, New York, February 1994.

Vernon, S., and T. Spath: "Carrying Multichannel Audio in a Stereo Production and Distribution Infrastructure," *Proceedings of IBC 2000*, International Broadcasting Convention, Amsterdam, September 2000.

Wylie, Fred: "Audio Compression Techniques," *The Electronics Handbook*, Jerry C. Whitaker (ed.), CRC Press, Boca Raton, Fla., pp. 1260–1272, 1996.

Wylie, Fred: "Audio Compression Technologies," *NAB Engineering Handbook*, 9th ed., Jerry C. Whitaker (ed.), National Association of Broadcasters, Washington, D.C., 1998.

Zwicker, E.: "Subdivision of the Audible Frequency Range Into Critical Bands (Frequenzgruppen)," J. Acoust. Soc. of Am., vol. 33, p. 248, February 1961.

6.2h Audio Networking

ATSC, "Guide to the Use of the Digital Television Standard," Advanced Television Systems Committee, Washington, D.C., Doc. A/54, Oct. 4, 1995.

Craig, Donald: "Network Architectures: What does Isochronous Mean?," *IBC Daily News*, IBC, Amsterdam, September 1999.

Dahlgren, Michael W.: "Servicing Local Area Networks," *Broadcast Engineering*, Intertec Publishing, Overland Park, Kan., November 1989.

Fibush, David: *A Guide to Digital Television Systems and Measurement*, Tektronix, Beaverton, OR, 1994.

Gaggioni, H., M. Ueda, F. Saga, K. Tomita, and N. Kobayashi, "The Development of a High-Definition Serial Digital Interface," Sony Technical Paper, Sony Broadcast Group, San Jose, Calif., 1998.

Gallo and Hancock: *Networking Explained*, Digital Press, pp. 191–235, 1999.

Goldman, J: *Applied Data Communications: A Business Oriented Approach*, 2md ed., Wiley, New York, N.Y., 1998.

Goldman, J: *Local Area Networks: A Business Oriented Approach*, 2nd ed., Wiley, New York, N.Y., 2000.

Goldman, James E.: "Network Communication," in *The Electronics Handbook*, Jerry C. Whitaker (ed.), CRC Press, Boca Raton, Fla., 1996.

Held, G.: *Ethernet Networks: Design Implementation, Operation and Management*, Wiley, New York, N.Y., 1994.

Held, G.: *Internetworking LANs and WANs*, Wiley, New York, N.Y., 1993.

Held, G.: *Local Area Network Performance Issues and Answers*, Wiley, New York, N.Y., 1994.

Held, G.: *The Complete Modem Reference*, Wiley, New York, N.Y., 1994.

International Organization for Standardization: "Information Processing Systems—Open Systems Interconnection—Basic Reference Model," ISO 7498, 1984.

Legault, Alain, and Janet Matey: "Interconnectivity in the DTV Era—The Emergence of SDTI," *Proceedings of Digital Television '98*, Intertec Publishing, Overland Park, Kan., 1998.

Miller, Mark A.: "Servicing Local Area Networks," *Microservice Management*, Intertec Publishing, Overland Park, Kan., February 1990.

Miller, Mark A.: *LAN Troubleshooting Handbook*, M&T Books, Redwood City, Calif., 1990.

"Networking and Internet Broadcasting," Omneon Video Networks, Campbell, Calif, 1999.

"Networking and Production," Omneon Video Networks, Campbell, Calif., 1999.

Owen, Peter: "Gigabit Ethernet for Broadcast and Beyond," *Proceedings of DTV99*, Intertec Publishing, Overland Park, Kan., November 1999.

Piercy, John: "ATM Networked Video: Moving From Leased-Lines to Packetized Transmission," *Proceedings of the Transition to Digital Conference*, Intertec Publishing, Overland Park, Kan., 1996.

"SMPTE Recommended Practice—Error Detection Checkwords and Status Flags for Use in Bit-Serial Digital Interfaces for Television," RP 165-1994, SMPTE, White Plains, N.Y., 1994.

"SMPTE Recommended Practice—SDTI-CP MPEG Decoder Templates," RP 204, SMPTE, White Plains, N.Y., 1999.

"SMPTE Standard for Television—24-Bit Digital Audio Format for HDTV Bit-Serial Interface," SMPTE 299M-1997, SMPTE, White Plains, N.Y., 1997.

"SMPTE Standard for Television—Ancillary Data Packet and Space Formatting," SMPTE 291M-1998, SMPTE, White Plains, N.Y., 1998.

"SMPTE Standard for Television—Bit-Serial Digital Interface for High-Definition Television Systems," SMPTE 292M-1998, SMPTE, White Plains, N.Y., 1998.

"SMPTE Standard for Television—Element and Metadata Definitions for the SDTI-CP," SMPTE 331M-2000, SMPTE, White Plains, N.Y., 2000.

"SMPTE Standard for Television—Encapsulation of Data Packet Streams over SDTI (SDTI-PF)," SMPTE 332M-2000, SMPTE, White Plains, N.Y., 2000.

"SMPTE Standard for Television—Mapping of AES3 Data into MPEG-2 Transport Stream," SMPTE 302M-1998, SMPTE, White Plains, N.Y., 1998.

"SMPTE Standard for Television—SDTI Content Package Format (SDTI-CP)," SMPTE 326M-2000, SMPTE, White Plains, N.Y., 2000.

"SMPTE Standard for Television—Serial Data Transport Interface," SMPTE 305M-1998, SMPTE, White Plains, N.Y., 1998.

SMPTE 344M, "540 Mb/s Serial Digital Interface," SMPTE, White Plains, N.Y., 2000.

"SMPTE Standard for Television—High Data-Rate Serial Data Transport Interface (HD-SDTI)," SMPTE 348M, SMPTE, White Plains, N.Y., 2000.

"SMPTE Standard for Television—Signals and Generic Data over High-Definition Interfaces," SMPTE 346M, SMPTE, White Plains, N.Y., 2000.

"SMPTE Standard for Television—Vertical Ancillary Data Mapping for Bit-Serial Interface, SMPTE 334M, SMPTE, White Plains, N.Y., 2000.

"Technology Brief—Networking and Storage Strategies," Omneon Video Networks, Campbell, Calif., 1999.

Turow, Dan: "SDTI and the Evolution of Studio Interconnect," *International Broadcasting Convention Proceedings*, IBC, Amsterdam, September 1998.

Wilkinson, J. H., H. Sakamoto, and P. Horne: "SDDI as a Video Data Network Solution," *International Broadcasting Convention Proceedings*, IBC, Amsterdam, September 1997.

Wu, Tsong-Ho: "Network Switching Concepts," *The Electronics Handbook*, Jerry C. Whitaker (ed.), CRC Press, Boca Raton, Fla., p. 1513, 1996.

6.2i Audio Recording Systems

Anderson, D: "Fibre Channel-Arbitrated Loop: The Preferred Path to Higher I/O Performance, Flexibility in Design," Seagate Technology Paper #MN-24, Seagate, Scotts Valley, Calif., 1995.

Bate, G.: "Recent Developments in Magnetic Recording Materials," *J. Appl. Phys*. pg. 2447, 1981.

Bertram, H. N.: "Long Wavelength ac Bias Recording Theory," *IEEE Trans. Magnetics*, vol. MAG-10, pp. 1039–1048, 1974.

Bozorth, Richard M.: *Ferromagnetism*, Van Nostrand, Princeton, N.J., 1961.

Chikazumi, Soshin: *Physics of Magnetism*, Wiley, New York, N.Y., 1964.

Goldberg, Thomas: "New Storage Technology," *Proceedings of the Advanced Television Summit*, Intertec Publishing, Overland Park, Kan., 1996.

Grega, Joe: "Magnetic and Optical Recording Media," in *NAB Engineering Handbook*, 9th ed., Jerry C. Whitaker (ed.), National Association of Broadcasters, Washington, D.C., pp. 893–906, 1999.

Hawthorne, J. M., and C. J. Hefielinger: "Polyester Films," in *Encyclopedia of Polymer Science and Technology*, N. M. Bikales (ed.), vol. 11, Wiley, New York, N.Y., pg. 42, 1969.

Heyn, T.: "The RAID Advantage," Seagate Technology Paper, Seagate, Scotts Valley, Calif., 1995.

Jorgensen, F.: *The Complete Handbook of Magnetic Recording*, Tab Books, Blue Ridge Summit, Pa., 1980.

Kalil, F. (ed): *Magnetic Tape Recording for the Eighties*, NASA References Publication 1975, April 1982.

Kraus, John D.: *Electromagnetics*, McGraw-Hill, New York, N.Y., 1953.

Lehtinen, Rick, "Editing Systems," *Broadcast Engineering*, Intertec Publishing, Overland Park. Kan., pp. 26–36, May 1996.

Lueck, L. B. (ed): *Symposium Proceedings Textbook*, Symposium on Magnetic Media Manufacturing Methods, Honolulu, May 25–27, 1983.

McConathy, Charles F.: "A Digital Video Disk Array Primer," *SMPTE Journal*, SMPTE, New York, N.Y., pp. 220–223, April 1998.

McKnight, John G.: "Erasure of Magnetic Tape," *J. Audio Eng. Soc.*, Audio Engineering Society, New York, N.Y., vol. 11, no.3, pp. 223–232, 1963.

Nylen, P., and E. Sunderland: *Modern Surface Coatings*, Interscience Publishers Division, Wiley, London, 1965.

Pear, C. B.: *Magnetic Recording in Science and Industry*, Reinhold, New York, N.Y., 1967.

Perry, R. H., and A. A. Nishimura: "Magnetic Tape," in *Encyclopedia of Chemical Technology*, 3d ed., Kirk Othmer (ed.), vol. 14, Wiley, New York, N.Y., pp. 732–753, 1981.

Plank, Bob: "Video Disk and Server Operation," *International Broadcast Engineer*, September 1995.

Robin, Michael, and Michel Poulin: "Multimedia and Television," in *Digital Television Fundamentals*, McGraw-Hill, New York, N.Y., pp. 455–488, 1997.

Sharrock, Michael P., and D. P. Stubs: "Perpendicular Magnetic Recording Technology: A Review," *SMPTE J.*, SMPTE, White Plains, N.Y., vol. 93, pp. 1127–1133, December 1984.

Smit, J., and H. P. J. Wijn: *Ferrite*, Wiley, New York, N.Y., 1959.

Tochihara, S.: "Magnetic Coatings and Their Applications in Japan," *Prog. Organic Coatings*, vol. 10, pp. 195–204, 1982.

Tyson, H: "Barracuda and Elite: Disk Drive Storage for Professional Audio/Video," Seagate Technology Paper #SV-25, Seagate, Scotts Valley, Calif., 1995.

Whitaker, Jerry C.: "Data Storage Systems," in *The Electronics Handbook*, Jerry C. Whitaker (ed.), CRC Press, Boca Raton, Fla., pp. 1445–1459, 1996.

Whitaker, Jerry C.: "Tape Recording Technology," *Broadcast Engineering*, Intertec Publishing, Overland Park, Kan., vol. 31, no. 11, pp. 78–108, 1989.

6.2j Audio Production Facility Design

DeSantis, Gene, Jerry C. Whitaker, and C. Robert Paulson: *Interconnecting Electronic Systems*, CRC Press, Boca Raton, Fla., 1992.

Whitaker, Jerry C.: *Facility Design Handbook*, CRC Press, Boca Raton, Fla., 2000.

6.2k Radio Broadcast Transmission Systems

Bean, B. R., and E. J. Dutton: "Radio Meteorology," National Bureau of Standards Monograph 92, March 1, 1966.

Benson, B., and Whitaker, J.: *Television and Audio Handbook for Technicians and Engineers*, McGraw-Hill, New York, N.Y., 1989.

Bingeman, Grant: "AM Tower Impedance Matching," *Broadcast Engineering*, Intertec Publishing, Overland Park, Kan., July 1985.

Bixby, Jeffrey: "AM DAs—Doing it Right," *Broadcast Engineering*, Intertec Publishing, Overland Park, Kan., February 1984.

Bullington, K.: "Radio Propagation at Frequencies above 30 Mc," *Proc. IRE*, pg. 1122, October 1947.

Bullington, K.: "Radio Propagation Variations at VHF and UHF," *Proc. IRE*, pg. 27, January 1950.

Burrows, C. R., and M. C. Gray: "The Effect of the Earth's Curvature on Groundwave Propagation," *Proc. IRE*, pg. 16, January 1941.

Chick, Elton B.: "Monitoring Directional Antennas," *Broadcast Engineering*, Intertec Publishing, Overland Park, Kan., July 1985.

Collocott, T. C., A. B. Dobson, and W. R. Chambers (eds.): *Dictionary of Science & Technology.*

DeComier, Bill: "Inside FM Multiplexer Systems," *Broadcast Engineering*, Intertec Publishing, Overland Park, Kan., May 1988.

de Lisle, E. W.: "Computations of VHF and UHF Propagation for Radio Relay Applications," RCA, Report by International Division, New York, N.Y.

Dickson, F. H., J. J. Egli, J. W. Herbstreit, and G. S. Wickizer: "Large Reductions of VHF Transmission Loss and Fading by the Presence of a Mountain Obstacle in Beyond-Line-of-Sight Paths," *Proc. IRE*, vol. 41, no. 8, pg. 96, August 1953.

"Documents of the XVth Plenary Assembly," CCIR Report 238, vol. 5, Geneva, 1982.

"Documents of the XVth Plenary Assembly," CCIR Report 563, vol. 5, Geneva, 1982.

"Documents of the XVth Plenary Assembly," CCIR Report 881, vol. 5, Geneva, 1982.

Dougherty, H. T., and E. J. Dutton: "The Role of Elevated Ducting for Radio Service and Interference Fields," NTIA Report 81–69, March 1981.

Eckersley, T. L.: "Ultra-Short-Wave Refraction and Diffraction," *J. Inst. Elec. Engrs*., pg. 286, March 1937.

Epstein, J., and D. Peterson: "An Experimental Study of Wave Propagation at 850 Mc," *Proc. IRE*, pg. 595, May 1953.

Fink, D. G., (ed.): *Television Engineering Handbook*, McGraw-Hill, New York, N.Y., 1957.

Fink, D., and D. Christiansen (eds.): Electronics Engineer's Handbook, 3rd ed., McGraw-Hill, New York, N.Y., 1989.

Handbook of Physics, McGraw-Hill, New York, N.Y., 1958.

Harrison, Cecil: "Passive Filters," in *The Electronics Handbook*, Jerry C. Whitaker (ed.), CRC Press, Boca Raton, Fla., pp. 279–290, 1996.

Hauptstuek, Jim: "Interconnecting the Digital Chain," *NAB 1996 Broadcast Engineering Conference Proceedings*, National Association of Broadcasters, Washington, D.C., pp. 360–358, 1996.

Heymans, Dennis: "Hot Switches and Combiners," *Broadcast Engineering*, Overland Park, Kan., December 1987.

Jordan, Edward C. (ed.): "*Reference Data for Engineers—Radio, Electronics, Computer and Communications*, 7th ed., Howard W. Sams, Indianapolis, Ind., 1985.

Judd, D. B., and G. Wyszecki: *Color in Business, Science and Industry*, 3rd ed., John Wiley and Sons, New York, N.Y.

Kaufman, Ed: *IES Illumination Handbook*, Illumination Engineering Society.

Lapedes, D. N. (ed.): *The McGraw-Hill Encyclopedia of Science & Technology*, 2nd ed., McGraw-Hill, New York, N.Y.

Longley, A. G., and P. L. Rice: "Prediction of Tropospheric Radio Transmission over Irregular Terrain—A Computer Method," ESSA (Environmental Science Services Administration), U.S. Dept. of Commerce, Report ERL (Environment Research Laboratories) 79-ITS 67, July 1968.

McClanahan, M. E.: "Aural Broadcast Auxiliary Links," in *NAB Engineering Handbook*, 8th ed., E. B. Cructhfield (ed.), National Association of Broadcasters, Washington, D.C, pp. 671–678, 1992.

McPetrie, J. S., and L. H. Ford: "An Experimental Investigation on the Propagation of Radio Waves over Bare Ridges in the Wavelength Range 10 cm to 10 m," J. *Inst. Elec. Engrs*., pt. 3, vol. 93, pg. 527, 1946.

Megaw, E. C. S.: "Some Effects of Obstacles on the Propagation of Very Short Radio Waves," J. *Inst. Elec. Engrs*., pt. 3, vol. 95, no. 34, pg. 97, March 1948.

Mullaney, John H.: "The Folded Unipole Antenna for AM Broadcast," *Broadcast Engineering*, Intertec Publishing, Overland Park, Kan., January 1960.

Mullaney, John H.: "The Folded Unipole Antenna," *Broadcast Engineering*, Intertec Publishing, Overland Park, Kan., July 1986.

National Bureau of Standards Circular 462, "Ionospheric Radio Propagation," June 1948.

NIST: *Manual of Regulations and Procedures for Federal Radio Frequency Management*, September 1995 edition, revisions for September 1996, January and May 1997, NTIA, Washington, D.C., 1997.

Norgard, John: "Electromagnetic Spectrum," *NAB Engineering Handbook*, 9th ed., Jerry C. Whitaker (ed.), National Association of Broadcasters, Washington, D.C., 1999.

Norgard, John: "Electromagnetic Spectrum," *The Electronics Handbook*, Jerry C. Whitaker (ed.), CRC Press, Boca Raton, Fla., 1996.

Norton, K. A.: "Ground Wave Intensity over a Finitely Conducting Spherical Earth," *Proc. IRE*, pg. 622, December 1941.

Norton, K. A.: "The Propagation of Radio Waves over a Finitely Conducting Spherical Earth," *Phil. Mag*., June 1938.

Parker, Darryl: "TFT DMM92 Meets STL Requirements," *Radio World*, Falls Church, VA, October 21, 1992.

"Radio Wave Propagation," Summary Technical Report of the Committee on Propagation of the National Defense Research Committee, Academic Press, New York, N.Y., 1949.

"Report of the Ad Hoc Committee, Federal Communications Commission," vol. 1, May 1949; vol. 2, July 1950.

Rollins, William W., and Robert L. Band: "T1 Digital STL: Discrete vs. Composite Transmission," *NAB 1996 Broadcast Engineering Conference Proceedings*, National Association of Broadcasters, Washington, D.C., pp. 356–359, 1996

Salek, Stanley: "Analysis of FM Booster System Configurations," *Proceedings of the 1992 NAB Broadcast Engineering Conference*, National Association of Broadcasters, Washington, DC, April 1992.

Selvidge, H.:"Diffraction Measurements at Ultra High Frequencies," *Proc. IRE*, pg. 10, January 1941.

Smith, E. E., and E. W. Davis: "Wind-induced Ions Thwart TV Reception," *IEEE Spectrum*, pp. 52—55, February 1981.

Stemson, A: *Photometry and Radiometry for Engineers*, John Wiley and Sons, New York, N.Y.

Stenberg, James T.: "Using Super Power Isolators in the Broadcast Plant," *Proceedings of the Broadcast Engineering Conference*, Society of Broadcast Engineers, Indianapolis, IN, 1988.

Surette, Robert A.: "Combiners and Combining Networks," in *The Electronics Handbook*, Jerry C. Whitaker (ed.), CRC Press, Boca Raton, Fla., pp. 1368–1381, 1996.

The Cambridge Encyclopedia, Cambridge University Press, 1990.

The Columbia Encyclopedia, Columbia University Press, 1993.

"The Propagation of Radio Waves through the Standard Atmosphere," *Summary Technical Report of the Committee on Propagation*, vol. 3, National Defense Research Council, Washington, D.C., 1946, published by Academic Press, New York, N.Y.

van der Pol, Balth, and H. Bremmer: "The Diffraction of Electromagnetic Waves from an Electrical Point Source Round a Finitely Conducting Sphere, with Applications to Radiotelegraphy and to Theory of the Rainbow," pt. 1, *Phil. Mag.*, July, 1937; pt. 2, *Phil. Mag.*, November 1937.

Vaughan, T., and E. Pivit: "High Power Isolator for UHF Television," *Proceedings of the NAB Engineering Conference*, National Association of Broadcasters, Washington, D.C., 1989.

Webster's New World Encyclopedia, Prentice Hall, 1992.

Westberg, J. M.: "Effect of 90° Stub on Medium Wave Antennas," *NAB Engineering Handbook*, 7the ed., National Association of Broadcasters, Washington, D.C., 1985.

Whitaker, Jerry C., (ed.): *A Primer: Digital Aural Studio to Transmitter Links*, TFT, Santa Clara, CA, 1994.

Whitaker, Jerry C., and Skip. Pizzi: "Radio Electronic News Gathering and Field Production," in *NAB Engineering Handbook*, 8th ed., E. B. Cructhfield (ed.), National Association of Broadcasters, Washington, D.C, pp. 1051–1072, 1992.

Wyszecki, G., and W. S. Stiles: *Color Science, Concepts and Methods, Quantitative Data and Formulae*, 2nd ed., John Wiley and Sons, New York, N.Y.

6.21 Radio Receivers

Amos, S. W.: "FM Detectors," *Wireless World*, vol. 87, no. 1540, pg. 77, January 1981.

Benson, K. Blair, and Jerry C. Whitaker: *Television and Audio Handbook for Engineers and Technicians*, McGraw-Hill, New York, N.Y., 1990.

Engelson, M., and J. Herbert: "Effective Characterization of CDMA Signals," *Microwave Journal*, pg. 90, January 1995.

Howald, R.: "Understand the Mathematics of Phase Noise," *Microwaves & RF*, pg. 97, December 1993.

Johnson, J. B:, "Thermal Agitation of Electricity in Conduction," *Phys. Rev.*, vol. 32, pg. 97, July 1928.

Nyquist, H.: "Thermal Agitation of Electrical Charge in Conductors," *Phys. Rev.*, vol. 32, pg. 110, July 1928.

Pleasant, D.: "Practical Simulation of Bit Error Rates," *Applied Microwave and Wireless*, pg. 65, Spring 1995.

Rohde, Ulrich L.: *Digital PLL Frequency Synthesizers*, Prentice-Hall, Englewood Cliffs, N.J., 1983.

Rohde, Ulrich L.: "Key Components of Modern Receiver Design—Part 1," *QST*, pg. 29, May 1994.

Rohde, Ulrich L. Rohde and David P. Newkirk: *RF/Microwave Circuit Design for Wireless Applications*, John Wiley & Sons, New York, N.Y., 2000.

Rohde, Ulrich L, and Jerry C. Whitaker: *Communications Receivers*, 3rd ed., McGraw-Hill, New York, N.Y., 2000.

"Standards Testing: Bit Error Rate," application note 3SW-8136-2, Tektronix, Beaverton, OR, July 1993.

Using Vector Modulation Analysis in the Integration, Troubleshooting and Design of Digital RF Communications Systems, Product Note HP89400-8, Hewlett-Packard, Palo Alto, Calif., 1994.

Watson, R.: "Receiver Dynamic Range; Pt. 1, Guidelines for Receiver Analysis," *Microwaves & RF*, vol. 25, pg. 113, December 1986.

"Waveform Analysis: Noise and Jitter," application note 3SW8142-2, Tektronix, Beaverton, OR, March 1993.

Wilson, E.: "Evaluate the Distortion of Modular Cascades," *Microwaves*, vol. 20, March 1981.

Whitaker, Jerry C. (ed.): *NAB Engineering Handbook*, 9th ed., National Association of Broadcasters, Washington, D.C., 1999.

6.2m Standards and Practices

Baumgartner, Fred, and Terrence Baun: "Broadcast Engineering Documentation," in *NAB Engineering Handbook*, 9th ed., Jerry C. Whitaker (ed.), National Association of Broadcasters, Washington, D.C., 1999.

Baumgartner, Fred, and Terrence Baun: "Engineering Documentation," in *The Electronics Handbook*, Jerry C. Whitaker (ed.), CRC Press, Boca Raton, Fla., 1996.

"Current Intelligence Bulletin #45," National Institute for Occupational Safety and Health, Division of Standards Development and Technology Transfer, February 24, 1986.

Code of Federal Regulations, 40, Part 761.

Delatore, J. P., E. M. Prell, and M. K. Vora: "Translating Customer Needs Into Product Specifications", *Quality Progress*, January 1989.

DeSantis, Gene: "Systems Engineering Concepts," in *NAB Engineering Handbook*, 9th ed., Jerry C. Whitaker (ed.), National Association of Broadcasters, Washington, D.C., 1999.

DeSantis, Gene: "Systems Engineering," in *The Electronics Handbook*, Jerry C. Whitaker (ed.), CRC Press, Boca Raton, Fla., 1996.

"Electrical Standards Reference Manual," U.S. Department of Labor, Washington, D.C.

Finkelstein, L.: "Systems Theory", *IEE Proceedings*, vol. 135, Part A, no. 6, July 1988.

Hammett, William F.: "Meeting IEEE C95.1-1991 Requirements," *NAB 1993 Broadcast Engineering Conference Proceedings*, National Association of Broadcasters, Washington, D.C., pp. 471–476, April 1993.

Hammar, Willie: *Occupational Safety Management and Engineering*, Prentice Hall, New York, N.Y.

Hoban, F. T., and W. M. Lawbaugh: *Readings In Systems Engineering*, NASA, Washington, D.C., 1993.

Markley, Donald: "Complying with RF Emission Standards," *Broadcast Engineering*, Intertec Publishing, Overland Park, Kan., May 1986.

"Occupational Injuries and Illnesses in the United States by Industry," OSHA Bulletin 2278, U.S. Department of Labor, Washington, D.C, 1985.

OSHA, "Electrical Hazard Fact Sheets," U.S. Department of Labor, Washington, D.C, January 1987.

OSHA, "Handbook for Small Business," U.S. Department of Labor, Washington, D.C.

Pfrimmer, Jack, "Identifying and Managing PCBs in Broadcast Facilities," *1987 NAB Engineering Conference Proceedings*, National Association of Broadcasters, Washington, D.C, 1987.

"Safety Precautions," Publication no. 3386A, Varian Associates, Palo Alto, Calif., March 1985.

Shinners, S. M.: *A Guide to Systems Engineering and Management*, Lexington, 1976.

Smith, Milford K., Jr., "RF Radiation Compliance," *Proceedings of the Broadcast Engineering Conference*, Society of Broadcast Engineers, Indianapolis, IN, 1989.

System Engineering Management Guide, Defense Systems Management College, Virginia, 1983.

"Toxics Information Series," Office of Toxic Substances, July 1983.

Tuxal, J. G.: *Introductory System Engineering*, McGraw-Hill, New York, N.Y., 1972.

Whitaker, Jerry C.: *AC Power Systems*, 2nd Ed., CRC Press, Boca Raton, Fla., 1998.

Whitaker, Jerry C.: G. DeSantis, and C. Paulson: *Interconnecting Electronic Systems*, CRC Press, Boca Raton, Fla., 1993.

Whitaker, Jerry C.: *Maintaining Electronic Systems*, CRC Press, Boca Raton, Fla. 1991.

Whitaker, Jerry C.: *Power Vacuum Tubes Handbook*, 2nd Ed., CRC Press, Boca Raton, Fla., 1999.

Whitaker, Jerry C.: *Radio Frequency Transmission Systems: Design and Operation*, McGraw-Hill, New York, N.Y., 1990.

6.3 Video

6.3a Light, Vision, and Photometry

Barten, Peter G. J.: "Physical Model for the Contrast Sensitivity of the Human Eye," *Human Vision, Visual Processing, and Digital Display III*, Bernice E. Rogowitz ed., Proc. SPIE 1666, SPIE, Bellingham, Wash., pp. 57–72, 1992.

Boynton, R. M.: *Human Color Vision*, Holt, New York, 1979.

Committee on Colorimetry, Optical Society of America: *The Science of Color*, Optical Society of America, New York, N.Y., 1953.

Daly, Scott: "The Visible Differences Predictor: An Algorithm for the Assessment of Image Fidelity," *Human Vision, Visual Processing, and Digital Display III*, Bernice E. Rogowitz ed., Proc. SPIE 1666, SPIE, Bellingham, Wash., pp. 2–15, 1992.

Davson, H.: *Physiology of the Eye*, 4th ed., Academic, New York, N.Y., 1980.

Evans, R. M., W. T. Hanson, Jr., and W. L. Brewer: *Principles of Color Photography*, Wiley, New York, N.Y., 1953.

Fink, D. G.: *Television Engineering Handbook*, McGraw-Hill, New York, N.Y., 1957.

Fink, D. G: *Television Engineering*, 2nd ed., McGraw-Hill, New York, N.Y., 1952.

Grogan, T. A.: "Image Evaluation with a Contour-Based Perceptual Model," *Human Vision, Visual Processing, and Digital Display III*, Bernice E. Rogowitz ed., Proc. SPIE 1666, SPIE, Bellingham, Wash., pp. 188–197, 1992.

Grogan, Timothy A.: "Image Evaluation with a Contour-Based Perceptual Model," *Human Vision, Visual Processing, and Digital Display III*, Bernice E. Rogowitz ed., Proc. SPIE 1666, SPIE, Bellingham, Wash., pp. 188–197, 1992.

Hecht, S., S. Shiaer, and E. L. Smith: "Intermittent Light Stimulation and the Duplicity Theory of Vision," Cold Spring Harbor Symposia on Quantitative Biology, vol. 3, pg. 241, 1935.

Hecht, S.: "The Visual Discrimination of Intensity and the Weber-Fechner Law," *J. Gen Physiol.*, vol. 7, pg. 241, 1924.

IES Lighting Handbook, Illuminating Engineering Society of North America, New York, N.Y., 1981.

Kingslake, R. (ed.): *Applied Optics and Optical Engineering*, vol. 1, Academic, New York, N.Y., 1965.

Martin, Russel A., Albert J. Ahumanda, Jr., and James O. Larimer: "Color Matrix Display Simulation Based Upon Luminance and Chromatic Contrast Sensitivity of Early Vision," in *Human Vision, Visual Processing, and Digital Display III*, Bernice E. Rogowitz ed., Proc. SPIE 1666, SPIE, Bellingham, Wash., pp. 336–342, 1992.

Polysak, S. L.: *The Retina*, University of Chicago Press, Chicago, Ill., 1941.

Reese, G.: "Enhancing Images with Intensity-Dependent Spread Functions," *Human Vision, Visual Processing, and Digital Display III*, Bernice E. Rogowitz ed., Proc. SPIE 1666, SPIE, Bellingham, Wash., pp. 253–261, 1992.

Reese, Greg: "Enhancing Images with Intensity-Dependent Spread Functions," *Human Vision, Visual Processing, and Digital Display III*, Bernice E. Rogowitz ed., Proc. SPIE 1666, SPIE, Bellingham, Wash., pp. 253–261, 1992.

Schade, O. H.: "Electro-optical Characteristics of Television Systems," *RCA Review*, vol. 9, pp. 5–37, 245–286, 490–530, 653–686, 1948.

Wright, W. D.: *The Measurement of Colour*, 4th ed., Adam Hilger, London, 1969.

6.3b Color Vision, Representation, and Reproduction

Baldwin, M., Jr.: "The Subjective Sharpness of Simulated Television Images," *Proceedings of the IRE*, vol. 28, July 1940.

Belton, J.: "The Development of the CinemaScope by Twentieth Century Fox," *SMPTE Journal*, vol. 97, SMPTE, White Plains, N.Y., September 1988.

Benson, K. B., and D. G. Fink: *HDTV: Advanced Television for the 1990s*, McGraw-Hill, New York, N.Y., 1990.

Bingley, F. J.: "Colorimetry in Color Television—Pt. I," *Proc. IRE*, vol. 41, pp. 838–851, 1953.

Bingley, F. J.: "Colorimetry in Color Television—Pts. II and III," *Proc. IRE*, vol. 42, pp. 48–57, 1954.

Bingley, F. J.: "The Application of Projective Geometry to the Theory of Color Mixture," *Proc. IRE*, vol. 36, pp. 709–723, 1948.

Boynton, R.M.: *Human Color Vision*, Holt, New York, N.Y., p. 404, 1979.

"Colorimetry," Publication no. 15, Commission Internationale de l'Eclairage, Paris, 1971.

DeMarsh, L. E.: "Colorimetric Standards in US Color Television," *J. SMPTE*, vol. 83, pp. 1–5, 1974.

Epstein, D. W.: "Colorimetric Analysis of RCA Color Television System," *RCA Review*, vol. 14, pp. 227–258, 1953.

Fink, D. G., et. al.: "The Future of High Definition Television," *SMPTE Journal*, vol. 89, SMPTE, White Plains, N.Y., February/March 1980.

Fink, D. G.: "Perspectives on Television: The Role Played by the Two NTSCs in Preparing Television Service for the American Public," *Proceedings of the IEEE*, vol. 64, IEEE, New York, N.Y., September 1976.

Fink, D. G.: *Color Television Standards*, McGraw-Hill, New York, N.Y., 1986.

Foley, James D., et al.: *Computer Graphics: Principles and Practice*, 2nd ed., Addison-Wesley, Reading, Mass., pp. 584–592, 1991.

Fujio, T., J. Ishida, T. Komoto and T. Nishizawa: "High Definition Television Systems—Signal Standards and Transmission," *SMPTE Journal*, vol. 89, SMPTE, White Plains, N.Y., August 1980.

Guild, J.: "The Colorimetric Properties of the Spectrum," *Phil. Trans. Roy. Soc. A.*, vol. 230, pp. 149–187, 1931.

Herman, S.: "The Design of Television Color Rendition," *J. SMPTE*, SMPTE, White Plains, N.Y., vol. 84, pp. 267–273, 1975.

Hubel, David H.: *Eye, Brain and Vision*, Scientific American Library, New York, N.Y., 1988.

Hunt, R. W. G.: *The Reproduction of Colour*, 3d ed., Fountain Press, England, 1975.

Isnardi, M. A.: "Exploring and Exploiting Subchannels in the NTSC Spectrum," *SMPTE J.*, SMPTE, White Plains, N.Y., vol. 97, pp. 526–532, July 1988.

Isnardi, M. A.: "Multidimensional Interpretation of NTSC Encoding and Decoding," *IEEE Transactions on Consumer Electronics*, vol. 34, pp. 179–193, February 1988.

Judd, D. B., and G. Wyszencki: *Color in Business, Science, and Industry,*. 3rd ed., Wiley, New York, N.Y., pp. 44-45, 1975.

Judd, D. B.: "The 1931 C.I.E. Standard Observer and Coordinate System for Colorimetry," *Journal of the Optical Society of America*, vol. 23, 1933.

Kaufman, J. E. (ed.): *IES Lighting Handbook-1981 Reference Volume*, Illuminating Engineering Society of North America, New York, N.Y., 1981.

Kelly, K. L.: "Color Designation of Lights," *Journal of the Optical Society of America*, vol. 33, 1943.

Kelly, R. D., A. V. Bedbord and M. Trainer: "Scanning Sequence and Repetition of Television Images," *Proceedings of the IRE*, vol. 24, April 1936.

Miller, Howard: "Options in Advanced Television Broadcasting in North America," *Proceedings of the ITS*, International Television Symposium, Montreux, Switzerland, 1991.

Morizono, M.: "Technological Trends in High-Resolution Displays Including HDTV," *SID International Symposium Digest*, paper 3.1, May 1990.

Munsell Book of Color, Munsell Color Co., 2441 No. Calvert Street, Baltimore, MD 21218.

Neal, C. B.: "Television Colorimetry for Receiver Engineers," *IEEE Trans. BTR*, vol. 19, pp. 149–162, 1973.

Newhall, S. M., D. Nickerson, and D. B. Judd: “Final Report of the OSA Subcommittee on the Spacing of the Munsell Colors,” *Journal of the Optical Society of America*, vol. 33, pp. 385–418, 1943.

Nickerson, D.: “History of the Munsell Color System, Company and Foundation, I,” *Color Res. Appl.*, vol. 1, pp. 7–10, 1976.

Nickerson, D.: “History of the Munsell Color System, Company and Foundation, II: Its Scientific Application,” *Color Res. Appl.*, vol. 1, pp. 69–77, 1976.

Nickerson, D.: “History of the Munsell Color System, Company and Foundation, III,” *Color Res. Appl.,* vol. 1, pp. 121–130, 1976.

Nyquist, H.: “Certain Factors Affecting Telegraph Speed,” *Bell System Tech. J.*, vol. 3, pp. 324–346, March 1924.

Pearson, M. (ed.): Proc. ISCC Conf. on Optimum Reproduction of Color, Williamsburg, Va., 1971, Graphic Arts Research Center, Rochester, N.Y., 1971.

Pitts, K. and N. Hurst: “How Much Do People Prefer Widescreen (16 × 9) to Standard NTSC (4 × 3)?,” *IEEE Transactions on Consumer Electronics*, IEEE, New York, N.Y., August 1989.

Pointer, M. R.: “The Gamut of Real Surface Colours,” *Color Res. Appl.*, vol. 5, pp. 145–155, 1980.

Pointer, R. M.: “The Gamut of Real Surface Colors, *Color Res. App.*, vol. 5, 1945.

Pritchard, D. H.: “US Color Television Fundamentals—A Review,” *IEEE Trans. CE*, vol. 23, pp. 467–478, 1977.

Smith, A. R.: “Color Gamut Transform Pairs,” *SIGGRAPH 78*, 12–19, 1978.

Smith, V. C., and J. Pokorny: “Spectral Sensitivity of the Foveal Cone Pigments Between 400 and 500 nm,” *Vision Res.*, vol. 15, pp. 161–171, 1975.

Sproson, W. N.: *Colour Science in Television and Display Systems*, Adam Hilger, Bristol, England, 1983.

Tektronix application note #21W-7165: “Colorimetry and Television Camera Color Measurement,” Tektronix, Beaverton, Ore., 1992.

Uba, T., K. Omae, R. Ashiya, and K. Saita: “16:9 Aspect Ratio 38V-High Resolution Trinitron for HDTV,” *IEEE Transactions on Consumer Electronics*, IEEE, New York, N.Y., February 1988.

van Raalte, John A.: “CRT Technologies for HDTV Applications,” *1991 HDTV World Conference Proceedings*, National Association of Broadcasters, Washington, D.C., April 1991.

Wentworth, J. W.: *Color Television Engineering*, McGraw-Hill, New York, N.Y., 1955.

Wintringham, W. T.: “Color Television and Colorimetry,” *Proc. IRE*, vol. 39, pp. 1135– 1172, 1951.

Wright, W. D.: “A Redetermination of the Trichromatic Coefficients of the Spectral Colours,” *Trans. Opt. Soc.*, vol. 30, pp. 141–164, 1928–1929.

Wright, W. D.: *The Measurement of Colour*, 4th ed., Adam Hilger, London, 1969.

Wyszecki, G., and W. S. Stiles: *Color Science*, 2nd ed., Wiley, New York, N.Y., 1982.

6.3c Optical Components and Systems

Fink, D. G. (ed.): *Television Engineering Handbook*, McGraw-Hill, New York, N.Y., 1957.

Hardy, A. C., and F. H. Perrin: *The Principles of Optics*, McGraw-Hill, New York, N.Y., 1932.

Kingslake, Rudolf (ed.): *Applied Optics and Optical Engineering*, vol. 1, Chapter 6, Academic, New York, N.Y., 1965.

Sears, F. W.: *Principles of Physics*, III, Optics, Addison-Wesley, Cambridge, Mass., 1946.

Williams, Charles S., and Becklund, Orville A.: *Optics: A Short Course for Engineers and Scientists*, Wiley Interscience, New York, N.Y., 1972.

6.3d Digital Coding of Video Signals

Alkin, Oktay: "Digital Coding Schemes," *The Electronics Handbook*, Jerry C. Whitaker (ed.), CRC Press, Boca Raton, Fla., pp. 1252–1258, 1996.

Benson, K. B., and D. G. Fink: "Digital Operations in Video Systems," *HDTV: Advanced Television for the 1990s*, McGraw-Hill, New York, pp. 4.1–4.8, 1990.

Chambers, J. A., S. Tantaratana, and B. W. Bomar: "Digital Filters," *The Electronics Handbook*, Jerry C. Whitaker (ed.), CRC Press, Boca Raton, Fla., pp. 749–772, 1996.

DeMarsh, LeRoy E.: "Displays and Colorimetry for Future Television," *SMPTE Journal*, SMPTE, White Plains, N.Y., pp. 666–672, October 1994.

Garrod, Susan A. R.: "D/A and A/D Converters," *The Electronics Handbook*, Jerry C. Whitaker (ed.), CRC Press, Boca Raton, Fla., pp. 723–730, 1996.

Garrod, Susan, and R. Borns: *Digital Logic: Analysis, Application, and Design*, Saunders College Publishing, Philadelphia, 1991.

Hunold, Kenneth: "4:2:2 or 4:1:1—What are the Differences?," *Broadcast Engineering*, Intertec Publishing, Overland Park, Kan., pp. 62–74, October 1997.

Lee, E. A., and D. G. Messerschmitt: *Digital Communications*, 2nd ed., Kluwer, Norell, Mass., 1994.

Mazur, Jeff: "Video Special Effects Systems," *NAB Engineering Handbook*, 9th ed., Jerry C. Whitaker (ed.), National Association of Broadcasters, Washington, D.C., to be published 1998.

Nyquist, H.: "Certain Factors Affecting Telegraph Speed," *Bell System Tech. J.*, vol. 3, pp. 324–346, March 1924.

Parks, T. W., and J. H. McClellan: "A Program for the Design of Linear Phase Infinite Impulse Response Filters," *IEEE Trans. Audio Electroacoustics*, AU-20(3), pp. 195–199, 1972.

Peterson, R., R. Ziemer, and D. Borth: *Introduction to Spread Spectrum Communications*, Prentice-Hall, Englewood Cliffs, N. J., 1995.

Pohlmann, Ken: *Principles of Digital Audio*, McGraw-Hill, New York, N.Y., 2000.

Sklar, B.: *Digital Communications: Fundamentals and Applications*, Prentice-Hall, Englewood Cliffs, N. J., 1988.

TMS320C55x DSP Functional Overview, Texas Instruments, Dallas, TX, literature No. SRPU312, June 2000.

"SMPTE C Color Monitor Colorimetry," SMPTE Recommended Practice RP 145-1994, SMPTE, White Plains, N.Y., June 1, 1994.

Ungerboeck, G.: "Trellis-Coded Modulation with Redundant Signal Sets," parts I and II, *IEEE Comm. Mag*., vol. 25 (Feb.), pp. 5-11 and 12-21, 1987.

Ziemer, R., and W. Tranter: *Principles of Communications: Systems, Modulation, and Noise*, 4th ed., Wiley, New York, 1995.

Ziemer, Rodger E.: "Digital Modulation," *The Electronics Handbook*, Jerry C. Whitaker (ed.), CRC Press, Boca Raton, Fla., pp. 1213–1236, 1996.

6.3e Electron Optics and Deflection

Aiken, W. R.: "A Thin Cathode Ray Tube," *Proc. IRE*, vol. 45, no. 12, pp. 1599–1604, December 1957.

Barkow, W. H., and J. Gross: "The RCA Large Screen 110° Precision In-Line System," ST-5015, RCA Entertainment, Lancaster, Pa.

Boers, J.: "Computer Simulation of Space Charge Flows," Rome Air Development Command RADC-TR-68-175, University of Michigan, 1968.

Casteloano, Joseph A.: *Handbook of Display Technology*, Academic, New York, N.Y., 1992.

Cathode Ray Tube Displays, MIT Radiation Laboratory Series, vol. 22, McGraw-Hill, New York, N.Y., 1953.

Cloz, R., et al.: "Mechanism of Thin Film Electroluminescence," Conference Record, *SID Proceedings*, Society for Information Display, San Jose, Calif., vol. 20, no. 3, 1979.

Dasgupta, B. B.: "Recent Advances in Deflection Yoke Design," *SID International Symposium Digest of Technical Papers*, Society for Information Display, San Jose, Calif., pp. 248–252, May 1999.

Fink, Donald, (ed.): *Television Engineering Handbook*, McGraw-Hill, New York, N.Y., 1957.

Fink, Donald, and Donald Christiansen (eds.): *Electronics Engineers Handbook*, 3rd ed., McGraw-Hill, New York, N.Y., 1989.

Hutter, Rudolph G. E., "The Deflection of Electron Beams," in *Advances in Image Pickup and Display*, B. Kazan (ed.), vol. 1, pp. 212–215, Academic, New York, N.Y., 1974.

IEEE Standard Dictionary of Electrical and Electronics Terms, 2nd ed., Wiley, New York, N.Y., 1977.

Jordan, Edward C. (ed.): *Reference Data for Engineers: Radio, Electronics, Computer, and Communications*, 7th ed., Howard W. Sams, Indianapolis, IN, 1985.

Langmuir, D.: "Limitations of Cathode Ray Tubes," *Proc. IRE*, vol. 25, pp. 977–991, 1937.

Luxenberg, H. R., and R. L. Kuehn (eds.): *Display Systems Engineering*, McGraw-Hill, New York, N.Y. 1968.

Morell, A. M., et al.: "Color Television Picture Tubes," in *Advances in Image Pickup and Display*, vol. 1, B. Kazan (ed.), pg. 136, Academic, New York,N.Y., 1974.

Moss, Hilary: *Narrow Angle Electron Guns and Cathode Ray Tubes*, Academic, New York, N.Y., 1968.

Nix, L.: "Spot Growth Reduction in Bright, Wide Deflection Angle CRTs," *SID Proc.*, Society for Information Display, San Jose, Calif., vol. 21, no. 4, pg. 315, 1980.

Pender, H., and K. McIlwain (eds.), *Electrical Engineers Handbook*, Wiley, New York, N.Y., 1950.

Poole, H. H.: *Fundamentals of Display Systems*, Spartan, Washington, D.C., 1966.

Popodi, A. E., "Linearity Correction for Magnetically Deflected Cathode Ray Tubes," *Elect. Design News*, vol. 9, no. 1, January 1964.

Sadowski, M.: *RCA Review*, vol 95, 1957.

Sherr, S.: *Electronic Displays*, Wiley, New York, N.Y., 1979.

Sherr, S.: *Fundamentals of Display Systems Design*, Wiley, New York, N.Y., 1970.

Sinclair, Clive, "Small Flat Cathode Ray Tube," *SID Digest*, Society for Information Display, San Jose, Calif., pp. 138–139, 1981.

Spangenberg, K. R., *Vacuum Tubes*, McGraw-Hill, New York, N.Y., 1948.

True, R.: "Space Charge Limited Beam Forming Systems Analyzed by the Method of Self-Consistent Fields with Solution of Poisson's Equation on a Deformable Relaxation Mesh," Ph.D. thesis, University of Connecticut, Storrs, 1968.

Zworykin, V. K., and G. Morton: *Television*, 2d ed., Wiley, New York, N.Y., 1954.

6.3f Video Cameras

Bendel, Sidney L., and C. A. Johnson: "Matching the Performance of a New Pickup Tube to the TK-47 Camera," *SMPTE J.*, SMPTE, White Plains, N.Y., vol. 86, no. 11, pp. 838–841, November 1980.

Crutchfield, E. B., (ed.): *NAB Engineering Handbook*, 8th ed., National Association of Broadcasters, Washington, D.C., 1993.

Crutchfield, E. B., (ed.); *NAB Engineering Handbook*, 7th ed., National Association of Broadcasters, Washington, D.C., 1988.

Favreau, M., S. Soca, J. Bajon, and M. Cattoen: "Adaptive Contrast Corrector Using Real-Time Histogram Modification," *SMPTE J.*, SMPTE, White Plains, N.Y., vol. 93, pp. 488–491, May 1984.

Fink, D. G., and D. Christiansen (eds.): *Electronics Engineers' Handbook*, 2nd ed., McGraw-Hill, New York, N.Y., pp. 20–30, 1982.

Gloeggler, Peter: "Video Pickup Devices and Systems," in *NAB Engineering Handbook*, 9th Ed., Jerry C. Whitaker (ed.), National Association of Broadcasters, Washington, D.C., 1999.

Inglis, A.F.: *Behind the Tube*, Focal Press, London, 1990.

Levitt, R.S.: *Operating Characteristics of the Plumbicon*,1968.

Mathias, H.: "Gamma and Dynamic Range Needs for an HDTV Electronic Cinematography System," *SMPTE J.*, SMPTE, White Plains, N.Y., vol. 96, pp. 840–845, September 1987.

Ogomo, M., T. Yamada, K. Ando, and E. Yamazaki: "Considerations on Required Property for HDTV Displays," *Proc. of HDTV 90 Colloquium*, vols. 1, 2B, 1990.

Philips Components: "Plumbicon Application Bulletin 43," Philips, Slatersville, R.I., January 1985.

Rao, N.V.: "Development of High-Resolution Camera Tube for 2000 line TV System," 1968.

"SMPTE Standard for Television—Broadcast Cameras: Hybrid Electrical and Fiber-Optic Connector," SMPTE 304M-1998, SMPTE, White Plains, N.Y., 1998.

"SMPTE Standard for Television—Camera Positioning Information Conveyed by Ancillary Data Packets," SMPTE 315M-1999, SMPTE, White Plains, N.Y., 1999.

"SMPTE Standard for Television—Hybrid Electrical and Fiber-Optic Camera Cable," SMPTE 311M-1998, SMPTE, White Plains, N.Y., 1998.

SPIE: *Electron Image Tubes and Image Intensifiers*, SPIE, Bellingham, Wash., vol. 1243, pp. 80–86.

Steen, R.: "CCDs vs. Camera Tubes: A Comparison," *Broadcast Engineering*, Intertec Publishing, Overland Park, Kan., May,1991.

Stupp, E.H.: Physical Properties of the Plumbicon, 1968.

Tanaka, H., and L. J. Thorpe: "The Sony PCL HDVS Production Facility," *SMPTE J.*, SMPTE, White Plains, N.Y., vol. 100, pp. 404–415, June 1991.

Thorpe, L. J., E. Tamura, and T. Iwasaki: "New Advances in CCD Imaging," *SMPTE J.*, SMPTE, White Plains, N.Y., vol. 97, pp. 378–387, May 1988.

Thorpe, L., et. al.: "New High Resolution CCD Imager," *NAB Engineering Conference Proceedings*, National Association of Broadcasters, Washington, D.C., pp. 334– 345, 1988.

Thorpe, Laurence J.: "HDTV and Film—Digitization and Extended Dynamic Range," *133rd SMPTE Technical Conference*, Paper no. 133-100, SMPTE, White Plains, N.Y., October 1991.

Thorpe, Laurence J.: "Television Cameras," in *Electronic Engineers' Handbook*, 4th ed., Donald Christiansen (ed.), McGraw-Hill, New York, N.Y., pp. 24.58–24.74, 1997.

Thorpe, Laurence J.: "The HDTV Camcorder and the March to Marketplace Reality," *SMPTE Journal*, SMPTE, White Plains, N.Y., pp. 164–177, March 1998.

6.3g Monochrome and Color Image Display Devices

Aiken, J. A.: "A Thin Cathode Ray Tube," *Proc. IRE*, vol. 45, pg. 1599, 1957.

Allison, J.: *Electronic Engineering Semiconductors and Devices*, 2nd ed., McGraw-Hill, London, pg. 308–309, 1990.

Amm, D. T., and R. W. Corrigan: "Optical Performance of the Grating Light Valve Technology," Projection Displays V Symposium, SPIE Proceedings, SPIE, San Jose, Calif., vol. EI 3634-10, February 1999.

Amm, D. T., and R.W. Corrigan: "Grating Light Valve Technology: Update and Novel Applications," SID Symposium—Anaheim, SID, San Jose, Calif., May 1998.

Ashizaki, S., Y. Suzuki, K. Mitsuda, and H. Omae: "Direct-View and Projection CRTs for HDTV," *IEEE Transactions on Consumer Electronics*, vol. 34, no. 1, pp. 91–98, February 1988.

Barbin, R., and R. Hughes: "New Color Picture Tube System for Portable TV Receivers," *IEEE Trans. Broadcast TV Receivers*, vol. BTR-18, no. 3, pp. 193–200, August 1972.

Barkow, W. H., and J. Gross: "The RCA Large Screen 110° Precision In-Line System," ST-5015, RCA Entertainment, Lancaster, Pa.

Bates, W., P. Gelinas, and P. Recuay: "Light Valve Projection System for HDTV," *Proceedings of the ITS*, International Television Symposium, Montreux, Switzerland, 1991.

Baur, G.: *The Physics and Chemistry of Liquid Crystal Devices*, G. J. Sprokel (ed.), Plenum, New York, N.Y., pg. 62, 1980.

Bauman, E.: "The Fischer Large-Screen Projection System," *SMPTE Journal*, SMPTE, White Plains, N.Y., vol. 60, pg. 351, 1953.

Benson, K. B., and D. G. Fink: *HDTV: Advanced Television for the 1990s*, McGraw-Hill, New York, N.Y., 1990.

Blacker, A., et al.: "A New Form of Extended Field Lens for Use in Color Television Picture Tube Guns," *IEEE Trans. Consumer Electronics*, pp. 238–246, August 1966.

Blaha, Richard J.: "Degaussing Circuits for Color TV Receivers," *IEEE Trans. Broadcast TV Receivers*, vol. BTR-18, no. 1, pp. 7–10, February 1972.

Blaha, Richard J.: "Large Screen Display Technology Assessment for Military Applications," *Large-Screen Projection Displays II*, William P. Bleha, Jr., (ed.), Proc. SPIE 1255, SPIE, Bellingham, Wash., pp. 80–92, 1990.

Bleha, Wiliam. P.: "Image Light Amplifier (ILA) Technology for Large-Screen Projection," *SMPTE Journal*, SMPTE, White Plains, N.Y., pp. 710–717, October 1997.

Bleha, William P., Jr., (ed.): *Large-Screen Projection Displays II*, Proc. SPIE 1255, SPIE, Bellingham, Wash., 1990.

Burgmans, A., et. al.: *Information Display*, pg. 14, April/May 1998.

Buzak, Thomas S.: "Recent Advances in PALC Technology," in *Proceedings of the 18th International Display Research Conference*, Society for Information Display, San Jose, Calif., Asia Display '98, pp. 273–276, 1998.

Carpenter, C.: et al., "An Analysis of Focusing and Deflection in the Post-Deflection Focus Color Kinescope," *IRE Trans. Electron Devices*, vol. 2, pp. 1–7, 1955.

Casteloano, Joseph A.: *Handbook of Display Technology*, Academic, New York, N.Y., 1992.

Chang, I.: "Recent Advances in Display Technologies," *Proc. SID*, Society for Information Display, San Jose, Calif., vol. 21, no. 2, pg. 45, 1980.

Chen, H., and R. Hughes: "A High Performance Color CRT Gun with an Asymmetrical Beam Forming Region," *IEEE Trans. Consumer Electronics*, vol. CE-26, pp. 459–465, August, 1980.

Chen, K. C., W. Y. Ho and C. H. Tseng: "Invar Mask for Color Cathode Ray Tubes," in *Display Technologies*, Shu-Hsia Chen and Shin-Tson Wu (eds.), Proc. SPIE 1815, SPIE, Bellingham, Wash., pp.42–48, 1992.

Cheng, J. B., and Q. H. Wang: "Studies on YAG Phosphor Screen for HDTV Projector," *Proc SPIE 2892*, SPIE, Bellingham, Wash., pg. 36, 1996.

Cheng, Jia-Shyong, et. al.: "The Optimum Design of LCD Parameters in Projection and Direct View Applications," *Display Technologies*, Shu-Hsia Chen and Shin-Tson Wu (eds.), Proc. SPIE 1815, SPIE, Bellingham, Wash., pp. 69–80, 1992.

Clapp, R., et al.: "A New Beam Indexing Color Television Display System," *Proc. IRE*, vol. 44, no. 9, pp. 1108–1114, September 1956.

Cohen, C.: "Sony's Pocket TV Slims Down CRT Technology," *Electronics*, pg. 81, February 10, 1982.

Corrigan, R. W., B. R. Lang, D.A. LeHoty, and P.A. Alioshin: "An Alternative Architecture for High Performance Display," Silicon Light Machines, Sunnyvale, Calif., 1999. Presented at the 141st SMPTE Technical Conference (paper 141-25).

Credelle, T. L., et al.: "Cathodoluminescent Flat Panel TV Using Electron Beam Guides," *SID Int. Symp. Digest*, Society for Information Display, San Jose, Calif., pg. 26, 1980.

Credelle, T. L., et al.: *Japan Display '83*, pg. 26, 1983.

Credelle, T. L.: "Modular Flat Display Device with Beam Convergence," U.S. Patent 4,131,823.

"CRT Control Grid Having Orthogonal Openings on Opposite Sides," U.S. Patent 4,242,613, Dec. 30, 1980.

"CRTs: Glossary of Terms and Definitions," Publication TEP92, Electronic Industries Association, Washington, 1975.

Davis, C., and D. Say: "High Performance Guns for Color TV—A Comparison of Recent Designs," *IEEE Trans. Consumer Electronics*, vol. CE-25, August 1979.

Donofrio, R.: "Image Sharpness of a Color Picture Tube by MTF Techniques," *IEEE Trans. Broadcast TV Receivers*, vol. BTR-18, no. 1, pp. 1–6, February 1972.

Dressler, R.: "The PDF Chromatron—A Single or Multi-Gun CRT," *Proc. IRE*, vol. 41, no. 7, July 1953.

Dworsky, Lawrence N., and Babu R. Chalamala: *Field-Emission Displays*, Society for Information Display, San Jose, Calif., pp. F-1/3–F1/66, 1999.

Eccles, D. A., and Y. Zhang: "Digital-Television Signal Processing and Display Technology," *SID 99 Digest*, Society for Information Display, San Jose, Calif., pp. 108–111, 1999.

"Electron Gun with Astigmatic Flare-Reducing Beam Forming Region," U.S. Patent 4,234,814, Nov. 18, 1980.

Fink, D. G., et al.: "The Future of High-Definition Television," *SMPTE Journal*, SMPTE, White Plains, N.Y., vol. 89, February/March 1980.

Fink, D. G.: *Color Television Standards*, McGraw-Hill, New York, N.Y., 1986.

Fink, Donald, (ed.): *Television Engineering Handbook*, McGraw-Hill, New York, N.Y., 1957.

Fink, Donald, and Donald Christiansen (eds.): *Electronics Engineers Handbook*, 3rd ed., McGraw-Hill, New York, N.Y., 1989.

Fiore, J., and S. Kaplin: "A Second Generation Color Tube Providing More Than Twice the Brightness and Improved Contrast," *IEEE Trans. Consumer Electronics*, vol. CE-28, no. 1, pp. 65–73, February 1982.

Flechsig, W.: "CRT for the Production of Multicolored Pictures on a Luminescent Screen," French Patent 866,065, 1939.

Florence, J., and L. Yoder: "Display System Architectures for Digital Micromirror Device (DMD) Based Projectors," *Proc. SPIE*, SPIE, Bellingham, Wash., vol. 2650, Projection Displays II, pp. 193–208, 1996.

Fritz, Victor J.: "Full-Color Liquid Crystal Light Valve Projector for Shipboard Use," *Large Screen Projection Displays II*, William P. Bleha, Jr. (ed.), Proc. SPIE 1255, SPIE, Bellingham, Wash., pp. 59–68, 1990.

Fujio, T., J. Ishida, T. Komoto, and T. Nishizawa: "High-Definition Television Systems—Signal Standards and Transmission," *SMPTE Journal*, SMPTE, White Plains, N.Y., vol. 89, August 1980.

Gerhard-Multhaupt, R.: "Light Valve Technologies for HDTV Projection Displays: A Summary," *Proceedings of the ITS*, International Television Symposium, Montreux, Switzerland, 1991.

Glenn, W. E., C. E. Holton, G. J. Dixon, and P. J. Bos: "High-Efficiency Light Valve Projectors and High-Efficiency Laser Light Sources," *SMPTE Journal*, SMPTE, White Plains, N.Y., pp. 210–216, April 1997.

Glenn, William E.: "Large Screen Displays for Consumer and Theater Use," *Large Screen Projection Displays II*, William P. Bleha, Jr., (ed.), Proc. SPIE 1255, SPIE, Bellingham, Wash., pp. 36–43, 1990.

Glenn, William. E.: "Principles of Simultaneous Color Projection Using Fluid Deformation," *SMPTE Journal*, SMPTE, White Plains, N.Y., vol. 79, pg. 788, 1970.

Godfrey, R., et al.: "Development of the Permachrome Color Picture Tube," *IEEE Trans. Broadcast TV Receivers*, vol. BTR-14, no. 1, 1968.

Goede, Walter F: "Electronic Information Display Perspective," *SID Seminar Lecture Notes*, Society for Information Display, San Jose, Calif., vol. 1, pp. M-1/3–M1/49, May 17, 1999.

Good, W.: "Projection Television," *IEEE Trans.*, vol. CE-21, no. 3, pp. 206–212, August 1975.

Good, W.: "Recent Advances in the Single-Gun Color Television Light-Valve Projector," *Soc. Photo-Optical Instrumentation Engrs.*, vol. 59, 1975.

Gove, R. J., V. Markandey, S. Marshall, D. Doherty, G. Sextro, and M. DuVal: "High-Definition Display System Based on Digital Micromirror Device," International Workshop on HDTV (HDTV '94), International Institute for Communications, Turin, Italy (October 1994).

Gow, J., and R. Door: "Compatible Color Picture Presentation with the Single-Gun Tri Color Chromatron," *Proc. IRE*, vol. 42, no. 1, pp. 308–314, January 1954.

Gretag AG: "What You May Want to Know about the Technique of Eidophor," Regensdorf, Switzerland.

Grinberg, J. et al.: "Photoactivated Birefringent Liquid-Crystal Light Valve for Color Symbology Display," *IEEE Trans. Electron Devices*, vol. ED-22, no. 9, pp. 775–783, September 1975.

Hardy, A. C., and F. H. Perrin: *The Principles of Optics*, McGraw-Hill, New York, N.Y., 1932.

Hasker, J.: "Astigmatic Electron Gun for the Beam Indexing Color TV Display," *IEEE Trans. Electron Devices*, vol. ED-18, no. 9, pg. 703, September 1971.

Hayashi, M., N. Yamada, and B. Sastra: "Development of a 42-in. High-Definition Plasma-Addressed LCD," in *SID International Symposium Digest of Technical Papers*, Society for Information Display, San Jose, Calif., pp.280–284, 1999.

Herold, E.: "A History of Color TV Displays," *Proc. IEEE*, vol. 64, no. 9, pp. 1331–1337, September 1976.

Hockenbrock, Richard: "New Technology Advances for Brighter Color CRT Displays," *Display System Optics II*, Harry M. Assenheim (ed.), Proc. SPIE 1117, SPIE, Bellingham, Wash., pp. 219-226, 1989.

Hornbeck, Larry J.: "Digital Light Processing for High-Brightness, High-Resolution Applications," *Projection Displays III*, Electronic Imaging '97 Conference, SPIE, Bellingham, Wash., February 1997.

Hoskoshi, K., et al.: "A New Approach to a High Performance Electron Gun Design for Color Picture Tubes," 1980 IEEE Chicago Spring Conf. Consumer Electronics.

Howe, R., and B. Welham: "Developments in Plastic Optics for Projection Television Systems," *IEEE Trans.*, vol. CE-26, no. 1, pp. 44–53, February 1980.

Hu, C., Y. Yu and K. Wang: "Antiglare/Antistatic Coatings for Cathode Ray Tube Based on Polymer System," in *Display Technologies*, Shu-Hsia Chen and Shin-Tson Wu (eds)., Proc. SPIE 1815, SPIE, Bellingham, Wash., pp.42–48, 1992.

Hubel, David H.: *Eye, Brain and Vision*, Scientific American Library, New York, N.Y., 1988.

Hutter, Rudolph G. E.: "The Deflection of Electron Beams," *Advances in Image Pickup and Display*, B. Kazen (ed.), vol. 1, Academic Press, New York, pp. 212-215, 1974.

Ilcisen, K. J., et. al.: *Eurodisplay '96*, pg. 595, 1996.

Itah, N., et al.: "New Color Video Projection System with Glass Optics and Three Primary Color Tubes for Consumer Use," *IEEE Trans. Consumer Electronics*, vol. CE-25, no. 4, pp. 497–503, August 1979.

Johnson, A.: "Color Tubes for Data Display—A System Study," Philips ECG, Electronic Tube Division.

Judd, D. B.: "The 1931 C.I.E. Standard Observer and Coordinate System for Colorimetry," *Journal of the Optical Society of America*, vol. 23, 1933.

Kikuchi, M., et al.: "A New Coolant-Sealed CRT for Projection Color TV," *IEEE Trans.*, vol. CE-27, IEEE, New York, no. 3, pp. 478-485, August 1981.

Kingslake, Rudolf (ed.): *Applied Optics and Optical Engineering*, vol. 1, Chap. 6, Academic, New York, N.Y., 1965.

"Kodak Filters for Scientific and Technical Uses," Eastman Kodak Co., Rochester, N.Y.

Kurahashi, K., et al.: "An Outdoor Screen Color Display System," *SID Int. Symp. Digest 7*, Technical Papers, vol. XII, Society for Information Display, San Jose, Calif., pp. 132– 133, April 1981.

Lakatos, A. I., and R. F. Bergen: "Projection Display Using an Amorphous-Se-Type Ruticon Light Valve," *IEEE Trans. Electron Devices*, vol. ED-24, no. 7, pp. 930–934, July 1977.

Law, H.: "A Three-Gun Shadowmask Color Kinescope," *Proc. IRE*, vol. 39, pp. 1186–1194, October 1951.

Lim, G. S., et. al.: "New Driving Method for Improvement of Picture Quality in 40-in. AC PDP," in *Asia Display '98—Proceedings of the 18th International Display Research Conference*, Society for Information Display, San Jose, Calif., pp. 591–594, 1998.

Lucchesi, B., and M. Carpenter: "Pictures of Deflected Electron Spots from a Computer," *IEEE Trans. Consumer Electronics*, vol. CE-25, no. 4, pp. 468–474, 1979.

Luxenberg, H., and R. Kuehn: *Display Systems Engineering*, McGraw-Hill, New York, N.Y., 1968.

Maeda, M.: *Japan Display '83*, pg. 2, 1971.

Maseo, Imai, et. al.: "High-Brightness Liquid Crystal Light Valve Projector Using a New Polarization Converter," *Large Screen Projection Displays II*, William P. Bleha, Jr., (ed.), Proc. SPIE 1255, SPIE, Bellingham, Wash., pp. 52–58, 1990.

Masterson, W., and R. Barbin: "Designing Out the Problems of Wide-Angle Color TV Tube," *Electronics*, pp. 60–63, April 26, 1971.

Masuda, T., et. al: *Conference Record—International Display Resolution Conference*, pg. 357, 1994.

McKechnie, S.: Philips Laboratories (NA) report, 1981, unpublished.

Mears, N., "Method and Apparatus for Producing Perforated Metal Webs," U.S. Patent 2,762,149, 1956.

Mikoshiba, Shigeo: *Color Plasma Displays*, Society for Information Display, San Jose, Calif., pp M-4/3–M-4/68, 1999.

Mitsuhashi, Tetsuo: "HDTV and Large Screen Display," *Large-Screen Projection Displays II*, William P. Bleha, Jr., (ed.), Proc. SPIE 1255, SPIE, Bellingham, Wash., pp. 2– 12, 1990.

Mokhoff, N.: "A Step Toward Perfect Resolution," *IEEE Spectrum*, IEEE, New York, N.Y., vol. 18, no. 7, pp. 56–58, July 1981.

Morizono, M.: "Technological Trends in High Resolution Displays Including HDTV," *SID International Symposium Digest*, paper 3.1, Society for Information Display, San Jose, Calif., May 1990.

Morrell, A., et al.: *Color Picture Tubes*, Academic Press, New York, pp. 91-98, 1974.

Morrell, A.: "Color Picture Tube Design Trends," *Proc. SID*, Society for Information Display, San Jose, Calif., vol. 22, no. 1, pp. 3–9, 1981.

Morris, James E.: "Liquid-Crystal Displays," in *The Electrical Engineering Handbook*, Richard C. Dorf (ed.), CRC Press, Boca Raton, Fla., 1993.

Moss, H.: *Narrow Angle Electron Guns and Cathode Ray Tubes*, Academic, New York, N.Y., 1968.

Na, Young-Sun, et. al.: "A New Data Driver Circuit for Field Emission Display," in *Asia Display '98—Proceedings of the 18th International Display Research Conference*, Society for Information Display, San Jose, Calif., pp. 137–140, 1998.

Nasibov, A., et al.: "Electron-Beam Tube with a Laser Screen," *Sov. J. Quant. Electron.*, vol. 4, no. 3, pp. 296–300, September 1974.

Nishio, T., and K. Amemiya: "High-Luminance and High-Definition 50-in.-Diagonal Co-Planar Color PDPs with T-Shaped Electrodes," in *SID International Symposium Digest of Technical Papers*, Society for Information Display, San Jose, Calif., pp.268–272, 1999.

Oess, F.: "CRT Considerations for Raster Dot Alpha Numeric Presentations," *Proc. SID*, Society for Information Display, San Jose, Calif., vol. 20, no. 2, pp. 81–88, second quarter, 1979.

Ohkoshi, A., et al.: "A New 30V" Beam Index Color Cathode Ray Tube," *IEEE Trans. Consumer Electronics*, vol. CE-27, p. 433, August 1981.

Palac, K.: Method for Manufacturing a Color CRT Using Mask and Screen Masters, U.S. Patent 3,989,524, 1976.

Pease, Richard W.: "An Overview of Technology for Large Wall Screen Projection Using Lasers as a Light Source," *Large Screen Projection Displays II*, William P. Bleha, Jr., (ed.), Proc. SPIE 1255, SPIE, Bellingham, Wash., pp. 93–103, 1990.

Pfahnl, A.: "Aging of Electronic Phosphors in Cathode Ray Tubes," *Advances in Electron Tube Techniques*, Pergamon, New York, N.Y., pp. 204–208.

Phillips, Thomas E., et. al.: "1280 × 1024 Video Rate Laser-Addressed Liquid Crystal Light Valve Color Projection Display," *Optical Engineering*, Society of Photo-Optical Instrumentation Engineers, vol. 31, no. 11, pp. 2300–2312, November 1992.

Pitts, K., and N. Hurst: "How Much do People Prefer Widescreen (16 × 9) to Standard NTSC (4 × 3)?," *IEEE Transactions on Consumer Electronics*, vol. 35, no. 3, pp. 160–169, August 1989.

Pointer, R. M.: "The Gamut of Real Surface Colors," *Color Res. App.*, vol. 5, 1945.

Poorter, T., and F. W. deVrijer: "The Projection of Color Television Pictures," *SMPTE Journal*, SMPTE, White Plains, N.Y., vol. 68, pg. 141, 1959.

"Recommended Practice for Measurement of X-Radiation from Direct View TV Picture Tubes," Publication TEP 164, Electronics Industries Association, Washington, D.C., 1981.

Robbins, J., and D. Mackey: "Moire Pattern in Color TV," *IEEE Trans. Consumer Electronics*, vol. CE-28, no. 1, pp. 44–55, February 1982.

Robertson, A.: "Projection Television—1 Review of Practice," *Wireless World*, vol. 82, no. 1489, pp. 47–52, September 1976.

Robinder, R., D. Bates, P. Green: "A High Brightness Shadow Mask Color CRT for Cockpit Displays," *SID Digest*, Society for Information Display, vol. 14, pp. 72-73, 1983.

Rublack, W.: "In-Line Plural Beam CRT with an Aspherical Mask," U.S. Patent 3,435,668, 1969.

Sakamoto, Y.: and E. Miyazaki, *Japan Display '83*, pg. 30, 1983.

Sarma, Kalluri R.: "Active-Matrix LCDs," in *Seminar Lecture Notes*, Society for Information Display, San Jose, Calif., vol. 1, pp. M3/3–M3/45, 1999.

Say, D.: "Picture Tube Spot Analysis Using Direct Photography," *IEEE Trans. Consumer Electronics*, vol. CE-23, pp. 32–37, February 1977.

Say, D.: "The High Voltage Bipotential Approach to Enhanced Color Tube Performance," *IEEE Trans. Consumer Electronics*, vol. CE-24, no. 1, pg. 75, February 1978.

Schiecke, K.: "Projection Television: Correcting Distortions," *IEEE Spectrum*, IEEE, New York, N.Y., vol. 18, no. 11, pp. 40–45, November 1981.

Schwartz, J.: "Electron Beam Cathodoluminescent Panel Display," U.S. Patent 4,137,486.

Sears, F. W.: *Principles of Physics, III, Optics*, Addison-Wesley, Cambridge, Mass., 1946.

Sextro, G., I. Ballew, and J. Lwai: "High-Definition Projection System Using DMD Display Technology," *SID 95 Digest*, Society for Information Display, San Jose, Calif., pp. 70–73, 1995.

Sherr, S.: *Electronic Displays*, Wiley, New York, N.Y., 1979.

Sherr, S.: *Fundamentals of Display System Design*, Wiley-Interscience, New York, N.Y., 1970.

Sinclair, C.: "Small Flat Cathode Ray Tube," *SID Digest*, Society for Information Display, San Jose, Calif., pg. 138, 1981.

Stanley, T.: "Flat Cathode Ray Tube," U.S. Patent 4,031,427.

Swartz, J.: "Beam Index Tube Technology," *SID Proceedings*, Society for Information Display, San Jose, Calif., vol. 20, no. 2, p. 45, 1979.

Takeuchi, Kazuhiko, et. al.: "A 750-TV-Line Resolution Projector using 1.5 Megapixel a-Si TFT LC Modules," *SID 91 Digest*, Society for Information Display, San Jose, Calif., pp. 415–418, 1991.

Taneda, T., et al.: "A 1125-Scanning Line Laser Color TV Display," *SID 1973 Symp. Digest Technical Papers*, Society for Information Display, San Jose, Calif., vol. IV, pp. 86– 87, May 1973.

Tomioka, M., and Y. Hayshi: "Liquid Crystal Projection Display for HDTV," *Proceedings of the International Television Symposium*, ITS, Montreux, Switzerland, 1991.

Tong, Hua-Sou: "HDTV Display—A CRT Approach," *Display Technologies*, Shu-Hsia Chen and Shin-Tson Wu (eds.), Proc. SPIE 1815, SPIE, Bellingham, Wash., pp. 2–8, 1992.

Tsuruta, Masahiko, and Neil Neubert: "An Advanced High Resolution, High Brightness LCD Color Video Projector," *SMPTE Journal*, SMPTE, White Plains, N.Y., pp. 399– 403, June 1992.

Uba, T., K. Omae, R. Ashiya, and K. Saita: "16:9 Aspect Ratio 38V-High Resolution Trinitron for HDTV," *IEEE Transactions on Consumer Electronics*, vol. 34, no. 1., pp. 85–89, February 1988.

Van, M. W., and J. Von Esdonk: "A High Luminance High-Resolution Cathode-Ray Tube for Special Purposes," *IEEE Trans Electron Dev.*, IEEE, New York, N.Y., ED-30, pg. 193, 1983.

Wang, Q. H., J. B. Cheng, and Z. L. Lin: "A New YAG Phosphor Screen for Projection CRT," *Electron Lett.*, vol. 34, no. 14, pg. 1420, 1998.

Wang, Qionghua, Jianbo Cheng, Zulun Lin, and Gang Yang: "A High-Luminance and High-Resolution CRT for Projection HDTV Display," *Journal of the SID*, Society for Information Display, San Jose, Calif., vol. 7, no. 3, pg 183–186, 1999.

Wang, S., et al.: "Spectral and Spatial Distribution of X-Rays from Color Television Receivers," *Proc. Conf. Detection and Measurement of X-radiation from Color Television Receivers*, Washington, D.C., pp. 53–72, March 28–29, 1968.

Weber, Larry F.: "Plasma Displays," in *The Electrical Engineering Handbook*, Richard C. Dorf (ed.), CRC Press, Boca Raton, Fla., 1786–1798, 1993.

Wedding, Donald K., Sr.: "Large Area Full Color ac Plasma Display Monitor," *Large Screen Projection Displays II*, William P. Bleha, Jr., (ed.), Proc. SPIE 1255, SPIE, Bellingham, Wash., pp. 29–35, 1990.

Williams, Charles S., and Becklund, Orville A.: *Optics: A Short Course for Engineers and Scientists*, Wiley Interscience, New York, N.Y., 1972.

Wilson, J., and J. F. B. Hawkes: *Optoelectronics: An Introduction*, Prentice-Hall, London, pg. 145, 1989.

Woodhead, A., et al.: *1982 SID Digest*, Society for Information Display, San Jose, Calif., pg. 206, 1982.

"X-Radiation Measurement Procedures for Projection Tubes," TEPAC Publication 102, Electronic Industries Association, Washington, D. C.

Yamamoto, Y., Y. Nagaoka, Y. Nakajima, and T. Murao: "Super-compact Projection Lenses for Projection Television," *IEEE Transactions on Consumer Electronics*, IEEE, New York, N.Y., August 1986.

Yoshida, S., et al.: "25-V Inch 114-Degree Trinitron Color Picture Tube and Associated New Development," *Trans. BTR*, pp. 193-200, August 1974.

Yoshida, S., et al.: "A Wide Deflection Angle (114°) Trinitron Color Picture Tube," *IEEE Trans. Electron Devices*, vol. 19, no. 4, pp. 231–238, 1973.

Yoshida. S.: et al., "The Trinitron—A New Color Tube," *IEEE Trans. Consumer Electronics*, vol. CE-28, no. 1, pp. 56–64, February 1982.

Yoshizawa, T., S. Hatakeyama, A. Ueno, M. Tsukahara, K. Matsumi, K. Hirota: "A 61-in High-Definition Projection TV for the ATSC Standard," *SID 99 Digest*, Society for Information Display, San Jose, Calif., pp. 112–115, 1999.

Younse, J. M.: "Projection Display Systems Based on the Digital Micromirror Device (DMD)," *SPIE Conference on Microelectronic Structures and Microelectromechanical Devices for Optical Processing and Multimedia Applications*, Austin, Tex., SPIE Proceedings, SPIE, Bellingham, Wash., vol. 2641, pp. 64–75, Oct. 24, 1995.

6.3h Video Recording Systems

Ampex: *General Information, Volume 1*, Training Department, Ampex Corporation, Redwood City, Calif., 1983.

Anderson, D: "Fibre Channel-Arbitrated Loop: The Preferred Path to Higher I/O Performance, Flexibility in Design," Seagate Technology Paper #MN-24, Seagate, Scotts Valley, Calif., 1995.

Bate, G.: "Recent Developments in Magnetic Recording Materials," *J. Appl. Phys*. pg. 2447, 1981.

Bertram, H. N.: “Long Wavelength ac Bias Recording Theory,” *IEEE Trans. Magnetics*, vol. MAG-10, pp. 1039–1048, 1974.

Bozorth, Richard M.: *Ferromagnetism*, Van Nostrand, Princeton, N.J., 1961.

Chikazumi, Soshin: *Physics of Magnetism*, Wiley, New York, N.Y., 1964.

Epstein, Steve: “Video Recording Principles,” in *NAB Engineering Handbook*, 9th ed., Jerry C. Whitaker (ed.), National Association of Broadcasters, Washington, D.C., pp. 923–935, 1999.

Epstein, Steve: “Videotape Storage Systems,” in *The Electronics Handbook*, Jerry C. Whitaker (ed.), CRC Press, Boca Raton, Fla., pp. 1412–1433, 1996.

Felix, Michael O.: “Video Recording Systems,” in *Electronics Engineers’ Handbook*, 4th ed., Donald Christiansen (ed.), McGraw-Hill, New York, N.Y., pp. 24.81–24.92, 1996.

Fink, D. G., and D. Christiansen (eds.): *Electronic Engineers’ Handbook*, 2nd ed., McGraw-Hill, New York, N.Y., 1982.

Ginsburg, C. P.: *The Birth of Videotape Recording*, Ampex Corporation, Redwood City, Calif., 1981. Reprinted from notes of paper delivered to Society of Motion Picture and Television Engineers, October 5, 1957.

Goldberg, Thomas: “New Storage Technology,” *Proceedings of the Advanced Television Summit*, Intertec Publishing, Overland Park, Kan., 1996.

Grega, Joe: “Magnetic and Optical Recording Media,” in *NAB Engineering Handbook*, 9th ed., Jerry C. Whitaker (ed.), National Association of Broadcasters, Washington, D.C., pp. 893–906, 1999.

Hammer, P.: “The Birth of the VTR,” *Broadcast Engineering*, Intertec Publishing, Overland Park, Kan., vol. 28, no. 6, pp. 158–164, 1986.

Hawthorne, J. M., and C. J. Hefielinger: “Polyester Films,” in *Encyclopedia of Polymer Science and Technology*, N. M. Bikales (ed.), vol. 11, Wiley, New York, N.Y., pg. 42, 1969.

Heyn, T.: “The RAID Advantage,” Seagate Technology Paper, Seagate, Scotts Valley, Calif., 1995.

Jorgensen, F.: *The Complete Handbook of Magnetic Recording*, Tab Books, Blue Ridge Summit, Pa., 1980.

Kalil, F. (ed): *Magnetic Tape Recording for the Eighties*, NASA References Publication 1975, April 1982.

Kraus, John D.: *Electromagnetics*, McGraw-Hill, New York, N.Y., 1953.

Lehtinen, Rick, “Editing Systems,” *Broadcast Engineering*, Intertec Publishing, Overland Park. Kan., pp. 26–36, May 1996.

Lueck, L. B. (ed): *Symposium Proceedings Textbook*, Symposium on Magnetic Media Manufacturing Methods, Honolulu, May 25–27, 1983.

McConathy, Charles F.: “A Digital Video Disk Array Primer,” *SMPTE Journal*, SMPTE, New York, N.Y., pp. 220–223, April 1998.

McKnight, John G.: "Erasure of Magnetic Tape," *J. Audio Eng. Soc.*, Audio Engineering Society, New York, N.Y., vol. 11, no.3, pp. 223–232, 1963.

Mee, C. D., and E. D. Daniel: *Magnetic Recording Handbook*, McGraw-Hill, New York, N.Y., 1990.

Nylen, P., and E. Sunderland: *Modern Surface Coatings*, Interscience Publishers Division, Wiley, London, 1965.

Pear, C. B.: *Magnetic Recording in Science and Industry*, Reinhold, New York, N.Y., 1967.

Perry, R. H., and A. A. Nishimura: "Magnetic Tape," in *Encyclopedia of Chemical Technology*, 3d ed., Kirk Othmer (ed.), vol. 14, Wiley, New York, N.Y., pp. 732–753, 1981.

Plank, Bob: "Video Disk and Server Operation," *International Broadcast Engineer*, September 1995.

Robin, Michael, and Michel Poulin: "Multimedia and Television," in *Digital Television Fundamentals*, McGraw-Hill, New York, N.Y., pp. 455–488, 1997.

Roizen, J.: *Magnetic Video Recording Techniques*, Ampex Training Manual—General Information, vol. 1, Ampex Corporation, Redwood City, Calif., 1964.

Roters, Herbert C.: *Electromagnetic Devices*, Wiley, New York, N.Y., 1961.

Sanders, M.: "Technology Report: AST," *Video Systems*, Intertec Publishing, Overland Park, Kan., April 1980.

Sharrock, Michael P., and D. P. Stubs: "Perpendicular Magnetic Recording Technology: A Review," *SMPTE J.*, SMPTE, White Plains, N.Y., vol. 93, pp. 1127–1133, December 1984.

Smit, J., and H. P. J. Wijn: *Ferrite*, Wiley, New York, N.Y., 1959.

Tochihara, S.: "Magnetic Coatings and Their Applications in Japan," *Prog. Organic Coatings*, vol. 10, pp. 195–204, 1982.

Tyson, H: "Barracuda and Elite: Disk Drive Storage for Professional Audio/Video," Seagate Technology Paper #SV-25, Seagate, Scotts Valley, Calif., 1995.

Whitaker, Jerry C.: "Data Storage Systems," in *The Electronics Handbook*, Jerry C. Whitaker (ed.), CRC Press, Boca Raton, Fla., pp. 1445–1459, 1996.

Whitaker, Jerry C.: "Tape Recording Technology," *Broadcast Engineering*, Intertec Publishing, Overland Park, Kan., vol. 31, no. 11, pp. 78–108, 1989.

6.3i Video Production Standards, Equipment, and System Design

Ajemian, Ronald G.: "Fiber Optic Connector Considerations for Professional Audio," *Journal of the Audio Engineering Society*, Audio Engineering Society, New York, N.Y., June 1992.

ATSC: "Implementation Subcommittee Report on Findings," Draft Version 0.4, ATSC, Washington, D.C., September 21, 1998.

Baldwin, M. Jr.: "The Subjective Sharpness of Simulated Television Images," *Proceedings of the IRE*, vol. 28, July 1940.

Belton, J.: "The Development of the CinemaScope by Twentieth Century Fox," *SMPTE Journal*, vol. 97, SMPTE, White Plains, N.Y., September 1988.

Benson, K. B., and D. G. Fink: *HDTV: Advanced Television for the 1990s*, McGraw-Hill, New York, N.Y., 1990.

Course notes, "DTV Express," PBS/Harris, Alexandria, Va., 1998.

Crutchfield, E. B.: *NAB Engineering Handbook*, 8th ed., National Association of Broadcasters, Washington, D.C., 1992.

DeSantis, Gene, Jerry C. Whitaker, and C. Robert Paulson: *Interconnecting Electronic Systems*, CRC Press, Boca Raton, Fla., 1992.

Dick, Bradley: "Building Fiber-Optic Transmission Systems," *Broadcast Engineering*, Intertec Publishing, Overland Park, Kan., November 1991 and December 1991.

Fink, D. G.: "Perspectives on Television: The Role Played by the Two NTSCs in Preparing Television Service for the American Public," *Proceedings of the IEEE*, vol. 64, IEEE, New York, September 1976.

Fink, D. G: *Color Television Standards*, McGraw-Hill, New York, 1986.

Fink, D. G, et. al.: "The Future of High Definition Television," *SMPTE Journal*, vol. 9, SMPTE, White Plains, N.Y., February/March 1980.

Fujio, T., J. Ishida, T. Komoto, and T. Nishizawa: "High-Definition Television Systems—Signal Standards and Transmission," *SMPTE Journal*, vol. 89, SMPTE, White Plains, N.Y., August 1980.

Hamasaki, Kimio: "How to Handle Sound with Large Screen," *Proceedings of the ITS*, International Television Symposium, Montreux, Switzerland, 1991.

Hopkins, Robert: "What We've Learned from the DTV Experience," *Proceedings of Digital Television '98*, Intertec Publishing, Overland Park, Kan., 1998.

Holman, Tomlinson: "Psychoacoustics of Multi-Channel Sound Systems for Television," *Proceedings of HDTV World*, National Association of Broadcasters, Washington, D.C., 1992.

Holman, Tomlinson: "The Impact of Multi-Channel Sound on Conversion to ATV," *Perspectives on Wide Screen and HDTV Production*, National Association of Broadcasters, Washington, D.C., 1995.

Hubel, David H.: *Eye, Brain and Vision*, Scientific American Library, New York, 1988.

Judd, D. B.: "The 1931 C.I.E. Standard Observer and Coordinate System for Colorimetry," *Journal of the Optical Society of America*, vol. 23, 1933.

Keller, Thomas B.: "Proposal for Advanced HDTV Audio," *1991 HDTV World Conference Proceedings*, National Association of Broadcasters, Washington, D.C., April 1991.

Kelly, R. D., A. V. Bedbord, and M. Trainer: "Scanning Sequence and Repetition of Television Images," *Proceedings of the IRE*, vol. 24, April 1936.

Kelly, K. L.: "Color Designation of Lights," *Journal of the Optical Society of America*, vol. 33, 1943.

Lagadec, Roger, Ph.D.: "Audio for Television: Digital Sound in Production and Transmission," *Proceedings of the ITS*, International Television Symposium, Montreux, Switzerland, 1991.

Leathers, David: "Production Considerations for DTV," in *NAB Engineering Handbook*, 9th ed., Jerry C. Whitaker (ed.), National Association of Broadcasters, Washington, D.C., pp. 1067–1072, 1999.

Mendrala, Jim: "Mastering at 24P," *Broadcast Engineering*, Intertec Publishing, Overland Park, Kan., pp. 92–94, February 1999.

Miller, Howard: "Options in Advanced Television Broadcasting in North America," *Proceedings of the ITS*, International Television Symposium, Montreux, Switzerland, 1991.

Pearson, Eric: How to Specify and Choose Fiber-Optic Cables, Pearson Technologies, Acworth, GA, 1991.

Pitts, K. and N. Hurst: "How Much Do People Prefer Widescreen (16 × 9) to Standard NTSC (4 × 3)?," *IEEE Transactions on Consumer Electronics*, IEEE, New York, August 1989.

Pointer, R. M.: "The Gamut of Real Surface Colors, *Color Res. App.*, vol. 5, 1945.

Robin, Michael: "Digital Resolution," *Broadcast Engineering*, Intertec Publishing, Overland Park, Kan., pp. 44–48, April 1998.

Slamin, Brendan: "Sound for High Definition Television," *Proceedings of the ITS*, International Television Symposium, Montreux, Switzerland, 1991.

SMPTE Recommended Practice RP 199-1999, "Mapping of Pictures in Wide-Screen (16:9) Scanning Structure to Retain Original Aspect Ratio of the Work," SMPTE, White Plains, N.Y., 1999.

SMPTE Recommended Practice—Implementation of 24P, 25P, and 30P Segmented Frames for 1920 × 1080 Production Format, RP 211-2000, SMPTE, White Plains, N.Y., 2000.

"SMPTE Standard for Television—1280 × 720 Scanning, Analog and Digital Representation and Analog Interface," SMPTE 296M-2001, SMPTE, White Plains, N.Y., 1997.

"SMPTE Standard for Television—1920 × 1080 50 Hz Scanning and Interfaces," SMPTE 295M-1997, SMPTE, White Plains, N.Y., 1997.

"SMPTE Standard for Television—1920 × 1080 Scanning and Analog and Parallel Digital Interfaces for Multiple-Picture Rates," SMPTE 274-1998, SMPTE, White Plains, N.Y., 1998.

"SMPTE Standard for Television—720 × 483 Active Line at 59.94 Hz Progressive Scan Production—Digital Representation," SMPTE 293M-1996, SMPTE, White Plains, N.Y., 1996.

"SMPTE Standard for Television—720 × 483 Active Line at 59.94-Hz Progressive Scan Production Bit-Serial Interfaces," SMPTE 294M-1997, SMPTE, White Plains, N.Y., 1997.

"SMPTE Standard for Television—Composite Analog Video Signal NTSC for Studio Applications," SMPTE 170M-1999, SMPTE, White Plains, N.Y., 1999.

"SMPTE Standard for Television—Digital Representation and Bit-Parallel Interface—1125/60 High-Definition Production System," SMPTE 260M-1992, SMPTE, White Plains, N.Y., 1992.

"SMPTE Standard for Television—MPEG-2 4:2:2 Profile at High Level," SMPTE 308M-1998, SMPTE, White Plains, N.Y., 1998.

"SMPTE Standard for Television—Signal Parameters—1125-Line High-Definition Production Systems," SMPTE 240M-1995, SMPTE, White Plains, N.Y., 1995.

"SMPTE Standard for Television—Synchronous Serial Interface for MPEG-2 Digital Transport Stream," SMPTE 310M-1998, SMPTE, White Plains, N.Y., 1998.

SMPTE: "System Overview—Advanced System Control Architecture, S22.02, Revision 2.0," S22.02 Advanced System Control Architectures Working Group, SMPTE, White Plains, N.Y., March 27, 2000.

Suitable Sound Systems to Accompany High-Definition and Enhanced Television Systems: Report 1072. Recommendations and Reports to the CCIR, 1986. Broadcast Service—Sound. International Telecommunications Union, Geneva, 1986.

Thorpe, Larry: "The Great Debate: Interlaced Versus Progressive Scanning," *Proceedings of the Digital Television '97 Conference*, Intertec Publishing, Overland Park, Kan., December 1997.

Thorpe, Laurence: "A New Global HDTV Program Origination Standard: Implications for Production and Technology," *Proceedings of DTV99*, Intertec Publishing, Overland Park, Kan., 1999.

Thorpe, Laurence J.: "Applying High-Definition Television," *Television Engineering Handbook*, rev. ed., K. B. Benson and Jerry C. Whitaker (eds.), McGraw-Hill, New York, N.Y., pg. 23.4, 1991.

Torick, Emil L.: "HDTV: High Definition Video—Low Definition Audio?," *1991 HDTV World Conference Proceedings*, National Association of Broadcasters, Washington, D.C., April 1991.

Venkat, Giri, "Understanding ATSC Datacasting—A Driver for Digital Television," *Proceedings of the NAB Broadcast Engineering Conference*, National Association of Broadcasters, Washington, D.C., pp. 113–116, 1999.

Whitaker, Jerry C.: *The Communications Facility Design Handbook*, CRC Press, Boca Raton, Fla., 2000.

6.3j Film for Video Applications

Bauer, Richard W.: "Film for Television," *NAB Engineering Handbook*, 9th ed., Jerry C. Whitaker (ed.), National Association of Broadcasters, Washington, D.C., 1998.

Belton, J.: "The Development of the CinemaScope by Twentieth Century Fox," *SMPTE Journal*, vol. 97, SMPTE, White Plains, N.Y., September 1988.

Benson, K. B., and D. G. Fink: *HDTV: Advanced Television for the 1990s*, McGraw-Hill, New York, 1990.

Evans, R. N., W. T. Hanson, Jr., and W. L. Brewer: *Principles of Color Photography*, Wiley, New York, N.Y., 1953.

Fink, D. G, et. al.: "The Future of High Definition Television," *SMPTE Journal*, vol. 9, SMPTE, White Plains, N.Y., February/March 1980.

Fink, D. G: *Color Television Standards*, McGraw-Hill, New York, 1986.

Hubel, David H.: *Eye, Brain and Vision*, Scientific American Library, New York, 1988.

James, T. H., (ed.): *The Theory of the Photographic Process*, 4th ed., MacMillan, New York, N.Y., 1977.

James, T. H., and G. C. Higgins: *Fundamentals of Photographic Theory*, Morgan and Morgan, Inc., Dobbs Ferry, N.Y., 1968.

Judd, D. B.: "The 1931 C.I.E. Standard Observer and Coordinate System for Colorimetry," *Journal of the Optical Society of America*, vol. 23, 1933.

Kelly, K. L.: "Color Designation of Lights," *Journal of the Optical Society of America*, vol. 33, 1943.

Pointer, R. M.: "The Gamut of Real Surface Colors," *Color Res. App.*, vol. 5, 1945.

Woodlief, Thomas, Jr. (ed.): *SPSE Handbook of Photographic Science and Engineering*, Wiley, New York, N.Y., 1973.

Wyszecki, G., and W. S. Stiles: *Color Science*, Wiley, New York, N.Y., 1967.

6.3k Compression Technologies for Video and Audio

Arvind, R., et al.: "Images and Video Coding Standards," *AT&T Technical J.*, p. 86, 1993.

ATSC, "Guide to the Use of the ATSC Digital Television Standard," Advanced Television Systems Committee, Washington, D.C., doc. A/54, Oct. 4, 1995.

Bennett, Christopher: "Three MPEG Myths," *Proceedings of the 1996 NAB Broadcast Engineering Conference*, National Association of Broadcasters, Washington, D.C., pp. 129–136, 1996.

Bonomi, Mauro: "The Art and Science of Digital Video Compression," *NAB Broadcast Engineering Conference Proceedings*, National Association of Broadcasters, Washington, D.C., pp. 7–14, 1995.

Brandenburg, K., and Gerhard Stoll: "ISO-MPEG-1 Audio: A Generic Standard for Coding of High Quality Digital Audio," *92nd AES Convention Proceedings*, Audio Engineering Society, New York, N.Y., 1992, revised 1994.

Cugnini, Aldo G.: "MPEG-2 Bitstream Splicing," *Proceedings of the Digital Television '97 Conference*, Intertec Publishing, Overland Park, Kan., December 1997.

Dare, Peter: "The Future of Networking," *Broadcast Engineering*, Intertec Publishing, Overland Park, Kan., p. 36, April 1996.

DeWith, P. H. N.: "Motion-Adaptive Intraframe Transform Coding of Video Signals," *Philips J. Res.*, vol. 44, pp. 345–364, 1989.

Epstein, Steve: "Editing MPEG Bitstreams," *Broadcast Engineering*, Intertec Publishing, Overland Park, Kan., pp. 37–42, October 1997.

Fibush, David K.: "Testing MPEG-Compressed Signals," *Broadcast Engineering*, Overland Park, Kan., pp. 76–86, February 1996.

Freed, Ken: "Video Compression," *Broadcast Engineering*, Overland Park, Kan., pp. 46– 77, January 1997.

Gilge, M.: "Region-Oriented Transform Coding in Picture Communication," VDI-Verlag, Advancement Report, Series 10, 1990.

Herre, J., and B. Grill: "MPEG-4 Audio—Overview and Perspectives for the Broadcaster," *IBC 2000 Proceedings*, International Broadcast Convention, Amsterdam, September 2000.

IEEE Standard Dictionary of Electrical and Electronics Terms, ANSI/IEEE Standard 100-1984, Institute of Electrical and Electronics Engineers, New York, 1984.

"IEEE Standard Specifications for the Implementation of 8 × 8 Inverse Discrete Cosine Transform," std. 1180-1990, Dec. 6, 1990.

Jones, Ken: "The Television LAN," *Proceedings of the 1995 NAB Engineering Conference*, National Association of Broadcasters, Washington, D.C., p. 168, April 1995.

Lakhani, Gopal: "Video Compression Techniques and Standards," *The Electronics Handbook*, Jerry C. Whitaker (ed.), CRC Press, Boca Raton, Fla., pp. 1273–1282, 1996.

Lyman, Stephen: "A Multichannel Audio Infrastructure Based on Dolby E Coding," *Proceedings of the NAB Broadcast Engineering Conference*, National Association of Broadcasters, Washington, D.C., 1999.

Nelson, Lee J.: "Video Compression," *Broadcast Engineering*, Intertec Publishing, Overland Park, Kan., pp. 42–46, October 1995.

Netravali, A. N., and B. G. Haskell: *Digital Pictures, Representation, and Compression*, Plenum Press, 1988.

Robin, Michael, and Michel Poulin: *Digital Television Fundamentals*, McGraw-Hill, New York, N.Y., 1998.

Smith, Terry: "MPEG-2 Systems: A Tutorial Overview," *Transition to Digital Conference*, Broadcast Engineering, Overland Park, Kan., Nov. 21, 1996.

SMPTE Recommended Practice: RP 202-2000, *Video Alignment for MPEG-2 Coding*, SMPTE, White Plains, N.Y., 1999.

SMPTE Standard: SMPTE 308M, *MPEG-2 4:2:2 Profile at High Level*, SMPTE, White Plains, N.Y., 1998.

SMPTE Standard: SMPTE 319M-2000, *Transporting MPEG-2 Recoding Information Through 4:2:2 Component Digital Interfaces*, SMPTE, White Plains, N.Y., 2000.

SMPTE Standard: SMPTE 312M, *Splice Points for MPEG-2 Transport Streams*, SMPTE, White Plains, N.Y., 1999.

SMPTE Standard: SMPTE 327M-2000, *MPEG-2 Video Recoding Data Set*, SMPTE, White Plains, N.Y., 2000.

SMPTE Standard: SMPTE 328M-2000, *MPEG-2 Video Elementary Stream Editing Information*, SMPTE, White Plains, N.Y., 2000.

SMPTE Standard: SMPTE 329M-2000, *MPEG-2 Video Recoding Data Set—Compressed Stream Format*, SMPTE, White Plains, N.Y., 2000.

SMPTE Standard: SMPTE 351M-2000, *Transporting MPEG-2 Recoding Information through High-Definition Digital Interfaces*, SMPTE, White Plains, N.Y., 2000.

SMPTE Standard: SMPTE 353M-2000, *Transport of MPEG-2 Recoding Information as Ancillary Data Packets*, SMPTE, White Plains, N.Y., 2000.

Smyth, Stephen: "Digital Audio Data Compression," *Broadcast Engineering*, Intertec Publishing, Overland Park, Kan., February 1992.

Solari, Steve. J.: *Digital Video and Audio Compression*, McGraw-Hill, New York, N.Y., 1997.

Stallings, William: *ISDN and Broadband ISDN*, 2nd Ed., MacMillan, New York.

Symes, Peter D.: "Video Compression Systems," *NAB Engineering Handbook*, 9th Ed., Jerry C. Whitaker (ed.), National Association of Broadcasters, Washington, D.C., pp. 907–922, 1999.

Symes, Peter D.: *Video Compression*, McGraw-Hill, N.Y., 1998.

Taylor, P.: "Broadcast Quality and Compression," *Broadcast Engineering*, Intertec Publishing, Overland Park, Kan., p. 46, October 1995.

Terry, K. B., and S. B. Lyman: "Dolby E—A New Audio Distribution Format for Digital Broadcast Applications," *International Broadcasting Convention Proceedings*, IBC, London, England, pp. 204–209, September 1999.

Vernon, S., and T. Spath: "Carrying Multichannel Audio in a Stereo Production and Distribution Infrastructure," *Proceedings of IBC 2000*, International Broadcasting Convention, Amsterdam, September 2000.

Ward, Christopher, C. Pecota, X. Lee and G. Hughes: "Seamless Splicing for MPEG-2 Transport Stream Video Servers," *Proceedings, 33rd SMPTE Advanced Motion Imaging Conference*, SMPTE, White Plains, N.Y., 2000.

Whitaker, Jerry C., and Harold Winard (eds.): *The Information Age Dictionary*, Intertec Publishing/Bellcore, Overland Park, Kan., 1992.

Wylie, Fred: "Audio Compression Techniques," *The Electronics Handbook*, Jerry C. Whitaker (ed.), CRC Press, Boca Raton, Fla., pp. 1260–1272, 1996.

Wylie, Fred: "Audio Compression Technologies," *NAB Engineering Handbook*, 9th ed., Jerry C. Whitaker (ed.), National Association of Broadcasters, Washington, D.C., 1998.

6.31 Video Networking

ATSC, "Guide to the Use of the Digital Television Standard," Advanced Television Systems Committee, Washington, D.C., Doc. A/54, Oct. 4, 1995.

Craig, Donald: "Network Architectures: What does Isochronous Mean?," *IBC Daily News*, IBC, Amsterdam, September 1999.

Dahlgren, Michael W.: "Servicing Local Area Networks," *Broadcast Engineering*, Intertec Publishing, Overland Park, Kan., November 1989.

Fibush, David: *A Guide to Digital Television Systems and Measurement*, Tektronix, Beaverton, OR, 1994.

Gaggioni, H., M. Ueda, F. Saga, K. Tomita, and N. Kobayashi, "The Development of a High-Definition Serial Digital Interface," Sony Technical Paper, Sony Broadcast Group, San Jose, Calif., 1998.

Gallo and Hancock: *Networking Explained*, Digital Press, pp. 191–235, 1999.

Goldman, J: *Applied Data Communications: A Business Oriented Approach*, 2md ed., Wiley, New York, N.Y., 1998.

Goldman, J: *Local Area Networks: A Business Oriented Approach*, 2nd ed., Wiley, New York, N.Y., 2000.

Goldman, James E.: "Network Communication," in *The Electronics Handbook*, Jerry C. Whitaker (ed.), CRC Press, Boca Raton, Fla., 1996.

Held, G.: *Ethernet Networks: Design Implementation, Operation and Management*, Wiley, New York, N.Y., 1994.

Held, G.: *Internetworking LANs and WANs*, Wiley, New York, N.Y., 1993.

Held, G.: *Local Area Network Performance Issues and Answers*, Wiley, New York, N.Y., 1994.

Held, G.: *The Complete Modem Reference*, Wiley, New York, N.Y., 1994.

International Organization for Standardization: "Information Processing Systems—Open Systems Interconnection—Basic Reference Model," ISO 7498, 1984.

Legault, Alain, and Janet Matey: "Interconnectivity in the DTV Era—The Emergence of SDTI," *Proceedings of Digital Television '98*, Intertec Publishing, Overland Park, Kan., 1998.

Miller, Mark A.: "Servicing Local Area Networks," *Microservice Management*, Intertec Publishing, Overland Park, Kan., February 1990.

Miller, Mark A.: *LAN Troubleshooting Handbook*, M&T Books, Redwood City, Calif., 1990.

"Networking and Internet Broadcasting," Omneon Video Networks, Campbell, Calif, 1999.

"Networking and Production," Omneon Video Networks, Campbell, Calif., 1999.

Owen, Peter: "Gigabit Ethernet for Broadcast and Beyond," *Proceedings of DTV99*, Intertec Publishing, Overland Park, Kan., November 1999.

Piercy, John: "ATM Networked Video: Moving From Leased-Lines to Packetized Transmission," *Proceedings of the Transition to Digital Conference*, Intertec Publishing, Overland Park, Kan., 1996.

"SMPTE Recommended Practice—Error Detection Checkwords and Status Flags for Use in Bit-Serial Digital Interfaces for Television," RP 165-1994, SMPTE, White Plains, N.Y., 1994.

"SMPTE Recommended Practice—SDTI-CP MPEG Decoder Templates," RP 204, SMPTE, White Plains, N.Y., 1999.

"SMPTE Standard for Television—540 Mb/s Serial Digital Interface, SMPTE 344M-2000, SMPTE, White Plains, N.Y., 2000.

"SMPTE Standard for Television—540 Mb/s Serial Digital Interface: Source Image Format Mapping," SMPTE 347M, SMPTE, White Plains, N.Y., 2001.

"SMPTE Standard for Television—24-Bit Digital Audio Format for HDTV Bit-Serial Interface," SMPTE 299M-1997, SMPTE, White Plains, N.Y., 1997.

"SMPTE Standard for Television—Ancillary Data Packet and Space Formatting," SMPTE 291M-1998, SMPTE, White Plains, N.Y., 1998.

"SMPTE Standard for Television—Bit-Serial Digital Interface for High-Definition Television Systems," SMPTE 292M-1998, SMPTE, White Plains, N.Y., 1998.

"SMPTE Standard for Television—Element and Metadata Definitions for the SDTI-CP," SMPTE 331M-2000, SMPTE, White Plains, N.Y., 2000.

"SMPTE Standard for Television—Encapsulation of Data Packet Streams over SDTI (SDTI-PF)," SMPTE 332M-2000, SMPTE, White Plains, N.Y., 2000.

"SMPTE Standard for Television—General Exchange Format (GXF)," SMPTE 360M, SMPTE, White Plains, N.Y., 2001.

"SMPTE Standard for Television—High Data-Rate Serial Data Transport Interface (HD-SDTI)," SMPTE 348M-2000, SMPTE, White Plains, N.Y., 2000.

"SMPTE Standard for Television—Mapping of AES3 Data into MPEG-2 Transport Stream," SMPTE 302M-1998, SMPTE, White Plains, N.Y., 1998.

"SMPTE Standard for Television—SDTI Content Package Format (SDTI-CP)," SMPTE 326M-2000, SMPTE, White Plains, N.Y., 2000.

"SMPTE Standard for Television—Serial Data Transport Interface," SMPTE 305M-1998, SMPTE, White Plains, N.Y., 1998.

"SMPTE Standard for Television—Signals and Generic Data over High-Definition Interfaces," SMPTE 346M-2000, SMPTE, White Plains, N.Y., 2000.

"SMPTE Standard for Television—Transport of Alternate Source Image Formats through SMPTE 292M, SMPTE 349M, SMPTE, White Plains, N.Y., 2001.

"SMPTE Standard for Television—Vertical Ancillary Data Mapping for Bit-Serial Interface, SMPTE 334M-2000, SMPTE, White Plains, N.Y., 2000.

"SMPTE Standard for Television—Video Payload Identification for Digital Television Interfaces," SMPTE 352M, SMPTE, White Plains, N.Y., 2001.

"Technology Brief—Networking and Storage Strategies," Omneon Video Networks, Campbell, Calif., 1999.

Turow, Dan: "SDTI and the Evolution of Studio Interconnect," *International Broadcasting Convention Proceedings*, IBC, Amsterdam, September 1998.

Wilkinson, J. H., H. Sakamoto, and P. Horne: "SDDI as a Video Data Network Solution," *International Broadcasting Convention Proceedings*, IBC, Amsterdam, September 1997.

Wu, Tsong-Ho: "Network Switching Concepts," *The Electronics Handbook*, Jerry C. Whitaker (ed.), CRC Press, Boca Raton, Fla., p. 1513, 1996.

6.3m Digital Television Transmission Systems

Advanced Television Enhancement Forum Specification," Draft, Version 1.1r26 updated 2/2/99, ATVEF, Portland, Ore., 1999.

Allision, Arthur: "PSIP 101: What You Need to Know," *Broadcast Engineering*, Intertec Publishing, Overland Park, Kan., pp. 140–144, June 2001.

Arragon, J. P., J. Chatel, J. Raven, and R. Story: "Instrumentation for a Compatible HD-MAC Coding System Using DATV," *Conference Record*, International Broadcasting Conference, Brighton, Institution of Electrical Engineers, London, 1989.

ATSC: "Amendment No. 1 to ATSC Standard: Program and System Information Protocol for Terrestrial Broadcast and Cable," Doc. A/67, ATSC, Washington, D.C, December 17, 1999.

ATSC: "ATSC Data Broadcast Standard," Advanced Television Systems Committee, Washington, D.C., Doc. A/90, July 26, 2000.

ATSC: "ATSC Digital Television Standard," Advanced Television Systems Committee, Washington, D.C., Doc. A/53, September 16, 1995.

ATSC: "Conditional Access System for Terrestrial Broadcast," Advanced Television Systems Committee, Washington, D.C., Doc. A/70, July 1999.

ATSC: "Digital Audio Compression Standard (AC-3), Annex A: AC-3 Elementary Streams in an MPEG-2 Multiplex," Advanced Television Systems Committee, Washington, D.C., Doc. A/52, December 20, 1995.

ATSC: "Digital Audio Compression Standard (AC-3)," Advanced Television Systems Committee, Washington, D.C., Doc. A/52, Dec. 20, 1995.

ATSC: "Guide to the Use of the Digital Television Standard," Advanced Television Systems Committee, Washington, D.C., Doc. A/54, October 4, 1995.

ATSC: "Implementation of Data Broadcasting in a DTV Station," Advanced Television Systems Committee, Washington, D.C., Doc. IS/151, November 1999.

ATSC: "Performance Assessment of the ATSC Transmission System, Equipment, and Future Directions," ATSC Task Force on RF System Performance, Advanced Television Systems Committee, Washington, DC, revision: 1.0, April 12, 2001.

ATSC: "Program and System Information Protocol for Terrestrial Broadcast and Cable," Advanced Television Systems Committee, Washington, D.C., Doc. A/65, February 1998.

ATSC: "Technical Corrigendum No.1 to ATSC Standard: Program and System Information Protocol for Terrestrial Broadcast and Cable," Doc. A/66, ATSC, Washington, D.C., December 17, 1999.

ATTC: "Digital HDTV Grand Alliance System Record of Test Results," Advanced Television Test Center, Alexandria, Virginia, October 1995.

Basile, C.: "An HDTV MAC Format for FM Environments," International Conference on Consumer Electronics, IEEE, New York, June 1989.

Cadzow, James A.: *Discrete Time Systems*, Prentice-Hall, Inc., Englewood Cliffs, N.J., 1973.

Chairman, ITU-R Task Group 11/3, "Report of the Second Meeting of ITU-R Task Group 11/3, Geneva, Oct. 13-19, 1993," p. 40, Jan. 5, 1994.

Chernock, Richard: "Implementation Recommendations for Data Broadcast," *NAB Broadcast Engineering Conference Proceedings*, National Association of Broadcasters, Washington, D.C., pp. 315–322, April 2000.

Chini, A., Y. Wu, M. El-Tanany, and S. Mahmoud: "An OFDM-based Digital ATV Terrestrial Broadcasting System with a Filtered Decision Feedback Channel Estimator," *IEEE Trans. Broadcasting*, IEEE, New York, N.Y., vol. 44, no. 1, pp. 2–11, March 1998.

Chini, A., Y. Wu, M. El-Tanany, and S. Mahmoud: "Hardware Nonlinearities in Digital TV Broadcasting Using OFDM Modulation," *IEEE Trans. Broadcasting*, IEEE, New York, N.Y., vol. 44, no. 1, March 1998.

Clement, Pierre, and Eric Gourmelen: "Internet and Television Convergence: IP and MPEG-2 Implementation Issues," *Proceedings of the 33rd SMPTE Advanced Motion Imaging Conference*, SMPTE, White Plains, N.Y., February 1999.

"Conclusions of the Extraordinary Meeting of Study Group 11 on High-Definition Television," Doc. 11/410-E, International Radio Consultative Committee (CCIR), Geneva, Switzerland, June 1989.

Ehmer, R. H.: "Masking of Tones Vs. Noise Bands," *J. Acoust. Soc. Am.*, vol. 31, pp. 1253–1256, September 1959.

Ehmer, R. H.: "Masking Patterns of Tones," *J. Acoust. Soc. Am.*, vol. 31, pp. 1115–1120, August 1959.

Erez, Beth: "Protecting content in the Digital Home," *Proceedings of the 33rd SMPTE Advanced Motion Imaging Conference*, SMPTE, White Plains, N.Y., pp. 231–238, February 1999.

ETS-300-421, "Digital Broadcasting Systems for Television, Sound, and Data Services; Framing Structure, Channel Coding and Modulation for 11–12 GHz Satellite Services," DVB Project technical publication.

ETS-300-429, "Digital Broadcasting Systems for Television, Sound, and Data Services; Framing Structure, Channel Coding and Modulation for Cable Systems," DVB Project technical publication.

ETS-300-468, "Digital Broadcasting Systems for Television, Sound, and Data Services; Specification for Service Information (SI) in Digital Video Broadcasting (DVB) Systems," DVB Project technical publication.

ETS-300-472, "Digital Broadcasting Systems for Television, Sound, and Data Services; Specification for Carrying ITU-R System B Teletext in Digital Video Broadcasting (DVB) Bitstreams," DVB Project technical publication.

ETS-300-473, "Digital Broadcasting Systems for Television, Sound, and Data Services; Satellite Master Antenna Television (SMATV) Distribution Systems," DVB Project technical publication.

ETS 300-744: "Digital Broadcasting Systems for Television, Sound and Data Services: Framing Structure, Channel Coding and Modulation for Digital Terrestrial Television," ETS 300 744, 1997.

Eureka 95 HDTV Directorate, Progressing Towards the Real Dimension, Eureka 95 Communications Committee, Eindhoven, Netherlands, June 1991.

European Telecommunications Standards Institute: "Digital Video Broadcasting; Framing Structure, Channel Coding and Modulation for Digital Terrestrial Television (DVB-T)", March 1997.

FCC Office of Engineering and Technology: "DTV Report on COFDM and 8-VSB Performance," Federal Communications Commission, Washington, D.C., OET Report FCC/OET 99-2, September 30, 1999.

FCC Report and Order: "Closed Captioning Requirements for Digital Television Receivers," Federal Communications Commission, Washington, D.C., ET Docket 99-254 and MM Docket 95-176, adopted July 21, 2000.

Fibush, David K.: *A Guide to Digital Television Systems and Measurements*, Tektronix, Beaverton, Ore., 1997.

"HD-MAC Bandwidth Reduction Coding Principles," Draft Report AZ-11, International Radio Consultative Committee (CCIR), Geneva, Switzerland, January 1989.

Husak, Walt, et. al.: "On-channel Repeater for Digital Television Implementation and Field Testing," *Proceedings 1999 Broadcast Engineering Conference*, NAB'99, Las Vegas, National Association of Broadcasters, Washington, D.C., pp. 397–403, April 1999.

ITU Radiocommunication Study Groups, Special Rapporteur's Group: "Guide for the Use of Digital Television Terrestrial Broadcasting Systems Based on Performance Comparison of ATSC 8-VSB and DVB-T COFDM Transmission Systems," International Telecommunications Union, Geneva, Document 11A/65-E, May 11, 1999.

ITU-R Document TG11/3-2, "Outline of Work for Task Group 11/3, Digital Terrestrial Television Broadcasting," June 30, 1992.

ITU-R Recommendation BS-775, "Multi-channel Stereophonic Sound System with and Without Accompanying Picture."

ITU-R SG 11, Special Rapporteur—Region 1, "Protection Ratios and Reference Receivers for DTTB Frequency Planning," ITU-R Doc. 11C/46-E, March 18, 1999.

Jacklin, Martin: "The Multimedia Home Platform: On the Critical Path to Convergence," DVB Project technical publication, 1998.

Joint ERC/EBU: "Planning and Introduction of Terrestrial Digital Television (DVB-T) in Europe," Izmir, Dec. 1997.

Ligeti, A., and J. Zander: "Minimal Cost Coverage Planning for Single Frequency Networks", *IEEE Trans. Broadcasting*, IEEE, New York, N.Y., vol. 45, no. 1, March 1999.

Lucas, K.: "B-MAC: A Transmission Standard for Pay DBS," *SMPTE Journal*, SMPTE, White Plains, N.Y., November 1984.

Luetteke, Georg: "The DVB Multimedia Home Platform," DVB Project technical publication, 1998.

Mignone, V., and A. Morello: "CD3-OFDM: A Novel Demodulation Scheme for Fixed and Mobile Receivers," *IEEE Trans. Commu.*, IEEE, New York, N.Y., vol. 44, pp. 1144–1151, September 1996.

Moore, B. C. J., and B. R. Glasberg: "Formulae Describing Frequency Selectivity as a Function of Frequency and Level, and Their Use in Calculating Excitation Patterns," *Hearing Research*, vol. 28, pp. 209–225, 1987.

Morello, Alberto, et. al.: "Performance Assessment of a DVB-T Television System," *Proceedings of the International Television Symposium 1997*, Montreux, Switzerland, June 1997.

Muschallik, C.: "Improving an OFDM Reception Using an Adaptive Nyquist Windowing," *IEEE Trans. on Consumer Electronics,* no. 03, 1996.

Muschallik, C.: "Influence of RF Oscillators on an OFDM Signal", IEEE Trans. Consumer Electronics, IEEE, New York, N.Y., vol. 41, no. 3, pp. 592–603, August 1995.

NAB TV TechCheck: National Association of Broadcasters, Washington, D.C., February 1, 1999.

NAB TV TechCheck: National Association of Broadcasters, Washington, D.C., January 4, 1999.

NAB: "An Introduction to DTV Data Broadcasting," *NAB TV TechCheck*, National Association of Broadcasters, Washington, D.C., August 2, 1999.

NAB: "Digital TV Closed Captions," *NAB TV TechCheck*, National Association of Broadcasters, Washington, D.C., August 7, 2000.

NAB: "Navigation of DTV Data Broadcasting Services," *NAB TV TechCheck*, National Association of Broadcasters, Washington, D.C., November 1, 1999.

NAB: "Pay TV Services for DTV," *NAB TV TechCheck*, National Association of Broadcasters, Washington, D.C., October 4, 1999.

Pickford, N.: "Laboratory Testing of DTTB Modulation Systems," Laboratory Report 98/01, Australia Department of Communications and Arts, June 1998.

Pollet, T., M. van Bladel, and M. Moeneclaey. "BER Sensitivity of OFDM Systems to Carrier Frequency Offset and Wiener Phase Noise," *IEEE Trans. on Communications*, vol. 43, 1995.

Raven, J. G.: "High-Definition MAC: The Compatible Route to HDTV," *IEEE Transactions on Consumer Electronics*, vol. 34, pp. 61–63, IEEE, New York, February 1988.

Robertson, P., and S.Kaiser: "Analysis of the Effects of Phase-Noise in Orthogonal Frequency Division Multiplex (OFDM) Systems," ICC 1995, pp. 1652–1657, 1995.

Sabatier, J., D. Pommier, and M. Mathiue: "The D2-MAC-Packet System for All Transmission Channels," *SMPTE Journal*, SMPTE, White Plains, N.Y., November 1984.

Salter, J. E.: "Noise in a DVB-T System," BBC R&D Technical Note, R&D 0873(98), February 1998.

Sari, H., G. Karam, and I. Jeanclaude: "Channel Equalization and Carrier Synchronization in OFDM Systems," *IEEE Proc.*, 6th Tirrenia Workshop on Digital Communications, Tirrenia, Italy, pp. 191–202, September 1993.

Sariowan, H.: "Comparative Studies Of Data Broadcasting," *International Broadcasting Convention Proceedings*, IBC, London, England, pp. 115–119, 1999.

Schachlbauer, Horst: "European Perspective on Advanced Television for Terrestrial Broadcasting," *Proceedings of the ITS, International Television Symposium*, Montreux, Switzerland, 1991.

Sheth, Amit, and Wolfgang Klas (eds.): *Multimedia Data Management*. McGraw-Hill, New York, N.Y., 1996.

SMPTE Recommended Practice RP 203: "Real Time Opportunistic Data Flow Control in an MPEG-2 Transport Emission Multiplex," SMPTE, White Plains, N.Y., 1999.

SMPTE Standard: SMPTE 320M-1999, "Channel Assignments and Levels on Multichannel Audio Media," SMPTE, White Plains, N.Y., 1999.

SMPTE Standard: SMPTE 324M, "12-Channel Serial Interface for Digital Audio and Auxiliary Data," SMPTE, White Plains, N.Y., 1999.

SMPTE Standard: SMPTE 325M-1999, "Opportunistic Data Broadcast Flow Control," SMPTE, White Plains, N.Y., 1999.

SMPTE Standard: SMPTE 333M-1999, "DTV Closed-Caption Server to Encoder Interface," SMPTE, White Plains, N.Y., 1999.

Story, R.: "HDTV Motion-Adaptive Bandwidth Reduction Using DATV," BBC Research Department Report, RD 1986/5.

Story, R.: "Motion Compensated DATV Bandwidth Compression for HDTV," International Radio Consultative Committee (CCIR), Geneva, Switzerland, January 1989.

Stott, J. H.: "Explaining Some of the Magic of COFDM", *Proceedings of the International TV Symposium 1997*, Montreux, Switzerland, June 1997.

Stott, J. H: "The Effect of Phase Noise in COFDM", *EBU Technical Review*, Summer 1998.

Teichmann, Wolfgang: "HD-MAC Transmission on Cable," *Proceedings of the ITS, International Television Symposium*, Montreux, Switzerland, 1991.

Thomas, Gomer: "ATSC Datacasting—Opportunities and Challenges," *Proceedings of the 33rd SMPTE Advanced Motion Imaging Conference*, SMPTE, White Plains, N.Y., pp. 307–314, February 1999.

Todd, C., et. al.: "AC-3: Flexible Perceptual Coding for Audio Transmission and Storage," AES 96th Convention, Preprint 3796, Audio Engineering Society, New York, February 1994.

Tvede, Lars, Peter Pircher, and Jens Bodenkamp: *Data Broadcasting: The Technology and the Business*, John Wiley & Sons, New York, N.Y., 1999.

van Klinken, N., and W. Renirie: "Receiving DVB: Technical Challenges," *Proceedings of the International Broadcasting Convention*, IBC, Amsterdam, September 2000.

Vreeswijk, F., F. Fonsalas, T. Trew, C. Carey-Smith, and M. Haghiri: "HD-MAC Coding for High-Definition Television Signals," International Radio Consultative Committee (CCIR), Geneva, Switzerland, January 1989.

VSB/COFDM Project: "8-VSBCOFDM Comparison Report," National Association of Broadcasters and Maximum Service Television, Washington, D.C., December 2000.

Wu, Y., and M. El-Tanany: "OFDM System Performance Under Phase Noise Distortion and Frequency Selective Channels," *Proceedings of Int'l Workshop of HDTV 1997*, Montreux Switzerland, June 10–11, 1997.

Wu, Y., et. al.: "Canadian Digital Terrestrial Television System Technical Parameters," *IEEE Transactions on Broadcasting*, IEEE, New York, N.Y., to be published in 1999.

Wu, Y., M. Guillet, B. Ledoux, and B. Caron: "Results of Laboratory and Field Tests of a COFDM Modem for ATV Transmission over 6 MHz Channels," *SMPTE Journal*, SMPTE, White Plains, N.Y., vol. 107, February 1998.

Wugofski, T. W.: "A Presentation Engine for Broadcast Digital Television," *International Broadcasting Convention Proceedings*, IBC, London, England, pp. 451–456, 1999.

Zwicker, E.: "Subdivision of the Audible Frequency Range Into Critical Bands (Frequenzgruppen)," *J. Acoust. Soc. of Am.*, vol. 33, p. 248, February 1961.

6.3n Frequency Bands and Propagation

Bean, B. R., and E. J. Dutton: "Radio Meteorology," National Bureau of Standards Monograph 92, March 1, 1966.

Bullington, K.: "Radio Propagation at Frequencies above 30 Mc," *Proc. IRE*, pg. 1122, October 1947.

Bullington, K.: "Radio Propagation Variations at VHF and UHF," *Proc. IRE*, pg. 27, January 1950.

Burrows, C. R., and M. C. Gray: "The Effect of the Earth's Curvature on Groundwave Propagation," *Proc. IRE*, pg. 16, January 1941.

Collocott, T. C., A. B. Dobson, and W. R. Chambers (eds.): *Dictionary of Science & Technology*.

de Lisle, E. W.: "Computations of VHF and UHF Propagation for Radio Relay Applications," RCA, Report by International Division, New York, N.Y.

Dickson, F. H., J. J. Egli, J. W. Herbstreit, and G. S. Wickizer: "Large Reductions of VHF Transmission Loss and Fading by the Presence of a Mountain Obstacle in Beyond-Line-of-Sight Paths," *Proc. IRE*, vol. 41, no. 8, pg. 96, August 1953.

"Documents of the XVth Plenary Assembly," CCIR Report 238, vol. 5, Geneva, 1982.

"Documents of the XVth Plenary Assembly," CCIR Report 563, vol. 5, Geneva, 1982.

"Documents of the XVth Plenary Assembly," CCIR Report 881, vol. 5, Geneva, 1982.

Dougherty, H. T., and E. J. Dutton: "The Role of Elevated Ducting for Radio Service and Interference Fields," NTIA Report 81–69, March 1981.

Dye, D. W.: *Proc. Phys. Soc.*, Vol. 38, pp. 399–457, 1926.

Eckersley, T. L.: "Ultra-Short-Wave Refraction and Diffraction," *J. Inst. Elec. Engrs.*, pg. 286, March 1937.

Epstein, J., and D. Peterson: "An Experimental Study of Wave Propagation at 850 Mc," *Proc. IRE*, pg. 595, May 1953.

Fink, D. G., (ed.): *Television Engineering Handbook*, McGraw-Hill, New York, N.Y., 1957.

Frerking, M. E.: *Crystal Oscillator Design and Temperature Compensation*, Van Nostrand Reinhold, New York, N. Y., 1978.

Handbook of Physics, McGraw-Hill, New York, N.Y., 1958.

Hietala, Alexander W., and Duane C. Rabe: "Latched Accumulator Fractional-*N* Synthesis With Residual Error Reduction," United States Patent, Patent No. 5,093,632, March 3, 1992.

Judd, D. B., and G. Wyszecki: *Color in Business, Science and Industry*, 3rd ed., John Wiley and Sons, New York, N.Y.

Kaufman, Ed: *IES Illumination Handbook*, Illumination Engineering Society.

King, Nigel J. R.: "Phase Locked Loop Variable Frequency Generator," United States Patent, Patent No. 4,204,174, May 20, 1980.

Lapedes, D. N. (ed.): *The McGraw-Hill Encyclopedia of Science & Technology*, 2nd ed., McGraw-Hill, New York, N.Y.

Longley, A. G., and P. L. Rice: "Prediction of Tropospheric Radio Transmission over Irregular Terrain—A Computer Method," ESSA (Environmental Science Services Administration), U.S. Dept. of Commerce, Report ERL (Environment Research Laboratories) 79-ITS 67, July 1968.

McPetrie, J. S., and L. H. Ford: "An Experimental Investigation on the Propagation of Radio Waves over Bare Ridges in the Wavelength Range 10 cm to 10 m," J. *Inst. Elec. Engrs.*, pt. 3, vol. 93, pg. 527, 1946.

Megaw, E. C. S.: "Some Effects of Obstacles on the Propagation of Very Short Radio Waves," J. *Inst. Elec. Engrs.*, pt. 3, vol. 95, no. 34, pg. 97, March 1948.

National Bureau of Standards Circular 462, "Ionospheric Radio Propagation," June 1948.

NIST: *Manual of Regulations and Procedures for Federal Radio Frequency Management*, September 1995 edition, revisions for September 1996, January and May 1997, NTIA, Washington, D.C., 1997.

Norgard, John: "Electromagnetic Spectrum," *NAB Engineering Handbook*, 9th ed., Jerry C. Whitaker (ed.), National Association of Broadcasters, Washington, D.C., 1999.

Norgard, John: "Electromagnetic Spectrum," *The Electronics Handbook*, Jerry C. Whitaker (ed.), CRC Press, Boca Raton, Fla., 1996.

Norton, K. A.: "Ground Wave Intensity over a Finitely Conducting Spherical Earth," *Proc. IRE*, pg. 622, December 1941.

Norton, K. A.: "The Propagation of Radio Waves over a Finitely Conducting Spherical Earth," *Phil. Mag.*, June 1938.

"Radio Wave Propagation," Summary Technical Report of the Committee on Propagation of the National Defense Research Committee, Academic Press, New York, N.Y., 1949.

"Report of the Ad Hoc Committee, Federal Communications Commission," vol. 1, May 1949; vol. 2, July 1950.

Riley, Thomas A. D.: "Frequency Synthesizers Having Dividing Ratio Controlled Sigma-Delta Modulator," United States Patent, Patent No. 4,965,531, October 23, 1990.

Rohde, Ulrich L.: *Digital PLL Frequency Synthesizers*, Prentice-Hall, Englewood Cliffs, N.J., 1983.

Rohde, Ulrich L.: *Microwave and Wireless Synthesizers: Theory and Design*, John Wiley & Sons, New York, N.Y., pg. 209, 1997.

Selvidge, H.:"Diffraction Measurements at Ultra High Frequencies," *Proc. IRE*, pg. 10, January 1941.

Smith, E. E., and E. W. Davis: "Wind-induced Ions Thwart TV Reception," *IEEE Spectrum*, pp. 52—55, February 1981.

Stemson, A: *Photometry and Radiometry for Engineers*, John Wiley and Sons, New York, N.Y.

Tate, Jeffrey P., and Patricia F. Mead: "Crystal Oscillators," in *The Electronics Handbook*, Jerry C. Whitaker (ed.), CRC Press, Boca Raton, Fla., pp. 185–199, 1996.

The Cambridge Encyclopedia, Cambridge University Press, 1990.

The Columbia Encyclopedia, Columbia University Press, 1993.

"The Propagation of Radio Waves through the Standard Atmosphere," *Summary Technical Report of the Committee on Propagation*, vol. 3, National Defense Research Council, Washington, D.C., 1946, published by Academic Press, New York, N.Y.

van der Pol, Balth, and H. Bremmer: "The Diffraction of Electromagnetic Waves from an Electrical Point Source Round a Finitely Conducting Sphere, with Applications to Radiotelegraphy and to Theory of the Rainbow," pt. 1, *Phil. Mag.*, July, 1937; pt. 2, *Phil. Mag.*, November 1937.

Webster's New World Encyclopedia, Prentice Hall, 1992.

Wells, John Norman: "Frequency Synthesizers," European Patent, Patent No. 012579OB2, July 5, 1995.

Wyszecki, G., and W. S. Stiles: *Color Science, Concepts and Methods, Quantitative Data and Formulae*, 2nd ed., John Wiley and Sons, New York, N.Y.

6.3o Television Transmission Systems

ACATS, "ATV System Description: ATV-into-NTSC Co-channel Test #016," Grand Alliance Advisory Committee on Advanced Television, p. I-14-10, Dec. 7, 1994.

Aitken, S., D. Carr, G. Clayworth, R. Heppinstall, and A. Wheelhouse: "A New, Higher Power, IOT System for Analogue and Digital UHF Television Transmission," *Proceedings of the 1997 NAB Broadcast Engineering Conference*, National Association of Broadcasters, Washington, D.C., p. 531, 1997.

Andrew Corporation: "Broadcast Transmission Line Systems," Technical Bulletin 1063H, Orland Park, Ill., 1982.

Andrew Corporation: "Circular Waveguide: System Planning, Installation and Tuning," Technical Bulletin 1061H, Orland Park, Ill., 1980.

ATSC Standard: "Modulation And Coding Requirements For Digital TV (DTV) Applications Over Satellite," Doc. A/80, ATSC, Washington, D.C., July, 17, 1999.

ATSC, "Guide to the Use of the Digital Television Standard," Advanced Television Systems Committee, Washington, D.C., Doc. A/54, Oct. 4, 1995.

Ben-Dov, O., and C. Plummer: "Doubly Truncated Waveguide," *Broadcast Engineering*, Intertec Publishing, Overland Park, Kan., January 1989.

Benson, K. B., and J. C. Whitaker: *Television and Audio Handbook for Technicians and Engineers*, McGraw-Hill, New York, N.Y., 1989.

Cablewave Systems: "Rigid Coaxial Transmission Lines," Cablewave Systems Catalog 700, North Haven, Conn., 1989.

Cablewave Systems: "The Broadcaster's Guide to Transmission Line Systems," Technical Bulletin 21A, North Haven, Conn., 1976.

Carnt, P. S., and G. B. Townsend: *Colour Television—Volume 1: NTSC*; *Volume 2: PAL and SECAM*, ILIFFE Bookes Ltd. (Wireless World), London, 1969.

"CCIR Characteristics of Systems for Monochrome and Colour Television—Recommendations and Reports," Recommendations 470-1 (1974–1978) of the Fourteenth Plenary Assembly of CCIR in Kyoto, Japan, 1978.

CCIR Report 122-4, 1990.

Crutchfield, E. B. (ed.), *NAB Engineering Handbook*, 8th Ed., National Association of Broadcasters, Washington, D.C., 1992.

DeComier, Bill: "Inside FM Multiplexer Systems," *Broadcast Engineering*, Intertec Publishing, Overland Park, Kan., May 1988.

Ericksen, Dane E.: "A Review of IOT Performance," *Broadcast Engineering*, Intertec Publishing, Overland Park, Kan., pg. 36, July 1996.

Fink, D., and D. Christiansen (eds.), *Electronics Engineers' Handbook*, 2nd Ed., McGraw-Hill, New York, N.Y., 1982.

Fink, D., and D. Christiansen (eds.): *Electronics Engineers' Handbook*, 3rd Ed., McGraw-Hill, New York, N.Y., 1989.

Fink, Donald G. (ed.): *Color Television Standards*, McGraw-Hill, New York, N.Y., 1955.

Gilmore, A. S.: *Microwave Tubes*, Artech House, Dedham, Mass., pp. 196–200, 1986.

Harrison, Cecil: "Passive Filters," in *The Electronics Handbook*, Jerry C. Whitaker (ed.), CRC Press, Boca Raton, Fla., pp. 279–290, 1996.

Herbstreit, J. W., and J. Pouliquen: "International Standards for Color Television," *IEEE Spectrum*, IEEE, New York, N.Y., March 1967.

Heymans, Dennis: "Hot Switches and Combiners," *Broadcast Engineering*, Overland Park, Kan., December 1987.

Hirsch, C. J.: "Color Television Standards for Region 2," *IEEE Spectrum*, IEEE, New York, N.Y., February 1968.

Hulick, Timothy P.: "60 kW Diacrode UHF TV Transmitter Design, Performance and Field Report," *Proceedings of the 1996 NAB Broadcast Engineering Conference*, National Association of Broadcasters, Washington, D.C., p. 442, 1996.

Hulick, Timothy P.: "Very Simple Out-of-Band IMD Correctors for Adjacent Channel NTSC/DTV Transmitters," *Proceedings of the Digital Television '98 Conference*, Intertec Publishing, Overland Park, Kan., 1998.

Jordan, Edward C.: *Reference Data for Engineers: Radio, Electronics, Computer and Communications*, 7th ed., Howard W. Sams, Indianapolis, IN, 1985.

Krohe, Gary L.: "Using Circular Waveguide," *Broadcast Engineering*, Intertec Publishing, Overland Park, Kan., May 1986.

Ostroff, Nat S.: "A Unique Solution to the Design of an ATV Transmitter," *Proceedings of the 1996 NAB Broadcast Engineering Conference*, National Association of Broadcasters, Washington, D.C., p. 144, 1996.

Perelman, R., and T. Sullivan: "Selecting Flexible Coaxial Cable," *Broadcast Engineering*, Intertec Publishing, Overland Park, Kan., May 1988.

Plonka, Robert J.: "Planning Your Digital Television Transmission System," *Proceedings of the 1997 NAB Broadcast Engineering Conference*, National Association of Broadcasters, Washington, D.C., p. 89, 1997.

Priest, D. H., and M. B. Shrader: "The Klystrode—An Unusual Transmitting Tube with Potential for UHF-TV," *Proc. IEEE*, vol. 70, no. 11, pp. 1318–1325, November 1982.

Pritchard, D. H.: "U.S. Color Television Fundamentals—A Review," *SMPTE Journal*, SMPTE, White Plains, N.Y., vol. 86, pp. 819–828, November 1977.

Rhodes, Charles W.: "Terrestrial High-Definition Television," *The Electronics Handbook*, Jerry C. Whitaker (ed.), CRC Press, Boca Raton, Fla., pp. 1599–1610, 1996.

Roizen, J.: "Universal Color Television: An Electronic Fantasia," *IEEE Spectrum*, IEEE, New York, N.Y., March 1967.

Stenberg, James T.: "Using Super Power Isolators in the Broadcast Plant," *Proceedings of the Broadcast Engineering Conference*, Society of Broadcast Engineers, Indianapolis, IN, 1988.

Surette, Robert A.: "Combiners and Combining Networks," in *The Electronics Handbook*, Jerry C. Whitaker (ed.), CRC Press, Boca Raton, Fla., pp. 1368–1381, 1996.

Symons, R., M. Boyle, J. Cipolla, H. Schult, and R. True: "The Constant Efficiency Amplifier—A Progress Report," *Proceedings of the NAB Broadcast Engineering Conference*, National Association of Broadcasters, Washington, D.C., pp. 77–84, 1998.

Symons, Robert S.: "The Constant Efficiency Amplifier," *Proceedings of the NAB Broadcast Engineering Conference*, National Association of Broadcasters, Washington, D.C., pp. 523–530, 1997.

Tardy, Michel-Pierre: "The Experience of High-Power UHF Tetrodes," *Proceedings of the 1993 NAB Broadcast Engineering Conference*, National Association of Broadcasters, Washington, D.C., p. 261, 1993.

Terman, F. E.: *Radio Engineering*, 3rd ed., McGraw-Hill, New York, N.Y., 1947.

Vaughan, T., and E. Pivit: "High Power Isolator for UHF Television," *Proceedings of the NAB Engineering Conference*, National Association of Broadcasters, Washington, D.C., 1989.

Whitaker, Jerry C., G. DeSantis, and C. Paulson: *Interconnecting Electronic Systems*, CRC Press, Boca Raton, Fla., 1993.

Whitaker, Jerry C.: *Radio Frequency Transmission Systems: Design and Operation*, McGraw-Hill, New York, N.Y., 1990.

Whitaker, Jerry C.: "Microwave Power Tubes," *Power Vacuum Tubes Handbook*, Van Nostrand Reinhold, New York, p. 259, 1994.

6.3p Television Antenna Systems

Allnatt, J. W., and R. D. Prosser: "Subjective Quality of Television Pictures Impaired by Long Delayed Echoes," *Proc. IEEE*, vol. 112, no. 3, March 1965.

Ben-Dov, O.: "Measurement of Circularly Polarized Broadcast Antennas," *IEEE Trans. Broadcasting*, vol. BC-19, no. 1, pp. 28–32, March 1972.

Bendov, Oded: "Coverage Contour Optimization of HDTV and NTSC Antennas," *Proceedings of the 1996 NAB Broadcast Engineering Conference*, National Association of Broadcasters, Washington, D.C., p. 69, 1996.

Brawn, D. A., and B. F. Kellom: "Butterfly VHF Panel Antenna," *RCA Broadcast News*, vol. 138, pp. 8–12, March 1968.

Clark, R. N., and N. A. L. Davidson: "The V-Z Panel as a Side Mounted Antenna," *IEEE Trans. Broadcasting*, vol. BC-13, no. 1, pp. 3–136, January 1967.

DeVito, G. G., and L. Mania: "Improved Dipole Panel for Circular Polarization," *IEEE Trans. Broadcasting*, vol. BC-28, no. 2, pp. 65–72, June 1982.

DeVito, G: "Considerations on Antennas with no Null Radiation Pattern and Pre-established Maximum-Minimum Shifts in the Vertical Plane," *Alta Frequenza*, vol. XXXVIII, no.6, 1969.

Dudzinsky, S. J., Jr.: "Polarization Discrimination for Satellite Communications," *Proc. IEEE*, vol. 57, no. 12, pp. 2179–2180, December 1969.

Fisk, R. E., and J. A. Donovan: "A New CP Antenna for Television Broadcast Service," *IEEE Trans. Broadcasting*, vol. BC-22, no. 3, pp. 91–96, September 1976.

Fowler, A. D., and H. N. Christopher: "Effective Sum of Multiple Echoes in Television," J. *SMPTE*, SMPTE, White Plains, N. Y., vol. 58, June 1952.

Fumes, N., and K. N. Stokke: "Reflection Problems in Mountainous Areas: Tests with Circular Polarization for Television and VHF/FM Broadcasting in Norway," *EBU Review*, Technical Part, no. 184, pp. 266–271, December 1980.

Heymans, Dennis: "Channel Combining in an NTSC/ATV Environment," *Proceedings of the 1996 NAB Broadcast Engineering Conference*, National Association of Broadcasters, Washington, D.C., p. 165, 1996.

Hill, P. C. J.: "Measurements of Reradiation from Lattice Masts at VHF," *Proc. IEEE*, vol. III, no. 12, pp. 1957–1968, December 1964.

Hill, P. C. J.: "Methods for Shaping Vertical Pattern of VHF and UHF Transmitting Aerials," *Proc. IEEE*, vol. 116, no. 8, pp. 1325–1337, August 1969.

Johns, M. R., and M. A. Ralston: "The First Candelabra for Circularly Polarized Broadcast Antennas," *IEEE Trans. Broadcasting*, vol. BC-27, no. 4, pp. 77–82, December 1981.

Johnson, R. C, and H. Jasik: *Antenna Engineering Handbook*, 2d ed., McGraw-Hill, New York, N.Y., 1984.

Knight, P.: "Reradiation from Masts and Similar Objects at Radio Frequencies," *Proc. IEEE*, vol. 114, pp. 30–42, January 1967.

Kraus, J. D.: *Antennas*, McGraw-Hill, New York, N.Y., 1950.

Lessman, A. M.: "The Subjective Effect of Echoes in 525-Line Monochrome and NTSC Color Television and the Resulting Echo Time Weighting," *J. SMPTE*, SMPTE, White Plains, N.Y., vol. 1, December 1972.

Mertz, P.: "Influence of Echoes on Television Transmission," *J. SMPTE*, SMPTE, White Plains, N.Y., vol. 60, May 1953.

Moreno, T.: *Microwave Transmission Design Data*, Dover, New York, N.Y.

Perini, J., and M. H. Ideslis: "Radiation Pattern Synthesis for Broadcast Antennas," *IEEE Trans. Broadcasting*, vol. BC-18, no. 3, pg. 53, September 1972.

Perini, J.: "Echo Performance of TV Transmitting Systems," *IEEE Trans. Broadcasting*, vol. BC-16, no. 3, September 1970.

Perini, J.: "Improvement of Pattern Circularity of Panel Antenna Mounted on Large Towers," *IEEE Trans. Broadcasting*, vol. BC-14, no. 1, pp. 33–40, March 1968.

Plonka, Robert J.: "Can ATV Coverage Be Improved With Circular, Elliptical, or Vertical Polarized Antennas?" *Proceedings of the 1996 NAB Broadcast Engineering Conference*, National Association of Broadcasters, Washington, D.C., p. 155, 1996.

Praba, K.: "Computer-aided Design of Vertical Patterns for TV Antenna Arrays," *RCA Engineer*, vol. 18-4, January–February 1973.

Praba, K.: "R. F. Pulse Measurement Techniques and Picture Quality," *IEEE Trans. Broadcasting*, vol. BC-23, no. 1, pp. 12–17, March 1976.

"Predicting Characteristics of Multiple Antenna Arrays," *RCA Broadcast News*, vol. 97, pp. 63–68, October 1957.

Sargent, D. W.: "A New Technique for Measuring FM and TV Antenna Systems," *IEEE Trans. Broadcasting*, vol. BC-27, no. 4, December 1981.

Siukola, M. S.: "Size and Performance Trade Off Characteristics of Horizontally and Circularly Polarized TV Antennas," *IEEE Trans. Broadcasting*, vol. BC-23, no. 1, March 1976.

Siukola, M. S.: "The Traveling Wave VHF Television Transmitting Antenna," *IRE Trans. Broadcasting*, vol. BTR-3, no. 2, pp. 49-58, October 1957.

Siukola, M. S.: "TV Antenna Performance Evaluation with RF Pulse Techniques," *IEEE Trans. Broadcasting*, vol. BC-16, no. 3, September 1970.

Smith, Paul D.: "New Channel Combining Devices for DTV, *Proceedings of the 1997 NAB Broadcast Engineering Conference*, National Association of Broadcasters, Washington, D.C., p. 218, 1996.

"WBAL, WJZ and WMAR Build World's First Three-Antenna Candelabra," *RCA Broadcast News*, vol. 106, pp. 30–35, December 1959.

Wescott, H. H.: "A Closer Look at the Sutro Tower Antenna Systems," *RCA Broadcast News*, vol. 152, pp. 35–41, February 1944.

Whythe, D. J.: "Specification of the Impedance of Transmitting Aerials for Monochrome and Color Television Signals," Tech. Rep. E-115, BBC, London, 1968.

6.3q Television Receivers and Cable/Satellite Distribution Systems

A. DeVries et al, "Characteristics of Surface-Wave Integratable Filters (SWIFS)," *IEEE Trans.*, vol. BTR-17, no. 1, p. 16.

"Advanced Television Enhancement Forum Specification," Draft, Version 1.1r26 updated 2/2/99, ATVEF, Portland, Ore., 1999.

Altman, F. J., and W. Sichak: "A Simplified Diversity Communication System," *IRE Trans.*, vol. CS-4, March 1956.

Applebaum. S. P.: "Adaptive Arrays," *IEEE Trans.*, vol. AP-24, pg. 585, September 1976.

ATSC: "Guide to the Use of the ATSC Digital Television Standard," Advanced Television Systems Committee, Washington, D.C., Doc. A/54, October 4, 1995.

ATSC: "Performance Assessment of the ATSC Transmission System, Equipment, and Future Directions," ATSC Task Force on RF System Performance, Advanced Television Systems Committee, Washington, DC, revision: 1.0, April 12, 2001.

Baldwin, T. F., and D. S. McVoy: *Cable Communications*, Prentice-Hall, Englewood Cliffs, N.J., 1983.

Bendov, O.: "On the Validity of the Longley-Rice (50,90/10) Propagation Model for HDTV Coverage and Interference Analysis," *Proceedings of the Broadcast Engineering Conference*, National Association of Broadcasters, Washington, D.C., 1999.

Beverage, H. H., and H. O. Peterson: "Diversity Receiving System of RCA Communications, Inc., for Radiotelegraphy," *Proc. IRE*, vol. 19, pg. 531, April 1931.

Bonang, C., and C. Auvray-Kander: "Next Generation Broadband Networks for Interactive Services," in Proceedings of IBC 2000, International Broadcasting Convention, Amsterdam, 2000.

Brennan, D. G.: "Linear Diversity Combining Techniques," *Proc. IRE*, vol. 47, pg. 1075, June 1959.

"Cable Television Information Bulletin," Federal Communications Commission, Washington, D.C., June 2000.

Ciciora, Walter S.: "Cable Television," in *NAB Engineering Handbook*, 9th ed., Jerry C. Whitaker (ed.), National Association of Broadcasters, Washington, D.C., pp. 1339– 1363, 1999.

Ciciora, Walter, et. al.: "A Tutorial on Ghost Canceling in Television Systems," *IEEE Transactions on Consumer Electronics*, IEEE, New York, vol. CE-25, no. 1, pp. 9–44, February 1979.

Compton, R. T., Jr., R. J. Huff, W. G. Swarner, and A. A. Ksienski: "Adaptive Arrays for Communication Systems: An Overview of Research at the Ohio State University," *IEEE Trans.*, vol. AP-24, pg. 599, September 1976.

Cook, James H., Jr., Gary Springer, Jorge B. Vespoli: "Satellite Earth Stations," in *NAB Engineering Handbook*, Jerry C. Whitaker (ed.), National Association of Broadcasters, Washington, D.C., pp. 1285–1322, 1999.

Digital HDTV Grand Alliance System: *Record of Test Results*, October 1995.

Di Toro, M. J.: "Communications in Time-Frequency Spread Media," *Proc. IEEE*, vol. 56, October 1968.

Einolf, Charles: "DTV Receiver Performance in the Real World," *2000 Broadcast Engineering Conference Proceedings*, National Association of Broadcasters, Washington, D.C., pp. 478–482, 2000.

Elliot, R. S.: *Antenna Theory and Design*, Prentice-Hall, Englewood Cliffs, N.J., pg. 64, 1981.

Engelson, M., and J. Herbert: "Effective Characterization of CDMA Signals," *Microwave Journal*, pg. 90, January 1995.

FCC/ACATS SSWP2-1306: *Grand Alliance System Test Procedures*, May 18, 1994.

FCC Regulations, 47 CFR, 15.65, Washington, D.C.

Feinstein, J.: "Passive Microwave Components," *Electronic Engineers' Handbook*, D. Fink and D. Christiansen (eds.), Handbook, McGraw-Hill, New York, N.Y., 1982.

Fink, D. G., and D. Christiansen (eds.): *Electronic Engineer's Handbook*, 2nd ed., McGraw-Hill, New York, 1982.

Fockens, P., and C. G. Eilers: "Intercarrier Buzz Phenomena Analysis and Cures," *IEEE Trans. Consumer Electronics*, vol. CE-27, no. 3, pg. 381, August 1981.

Gibson, E. D.: "Automatic Equalization Using Time-Domain Equalizers," *Proc. IEEE*, vol. 53, pg. 1140, August 1965.

Grossner, N.: *Transformers for Electronic Circuits*, 2nd ed., McGraw-Hill, New York, pp. 344–358, 1983.

Hoff, L. E., and A. R. King: "Skywave Communication Techniques," Tech. Rep. 709, Naval Ocean Systems Center, San Diego, Calif, March 30,1981.

Hoffman, Gary A.: "IEEE 1394: The A/V Digital Interface of Choice," 1394 Technology Association Technical Brief, 1394 Technology Association, Santa Clara, Calif., 1999.

Howald, R.: "Understand the Mathematics of Phase Noise," *Microwaves & RF*, pg. 97, December 1993.

Hufford, G. A: "A Characterization of the Multipath in the HDTV Channel," *IEEE Trans. on Broadcasting*, vol. 38, no. 4, December 1992.

Hulst, G. D.: "Inverse Ionosphere," *IRE Trans.*, vol. CS-8, pg. 3, March 1960.

IEEE Guide for Surge Withstand Capability, (SWC) Tests, ANSI C37.90a-1974/IEEE Std. 472-1974, IEEE, New York, 1974.

Jasik, H., *Antenna Engineering Handbook*, McGraw-Hill, New York, Chapter 24, 1961.

Johnson, J. B:, "Thermal Agitation of Electricity in Conduction," *Phys. Rev.*, vol. 32, pg. 97, July 1928.

Kahn, L. R.: "Ratio Squarer," *Proc. IRE*, vol. 42, pg. 1704, November 1954.

Kase, C. A., and W. L. Pritchard: "Getting Set for Direct-Broadcast Satellites," *IEEE Spectrum*, vol. 18, no. 8, pp. 22–28, 1981.

Kraus, J. D.: *Antennas*, McGraw-Hill, New York, N. Y., Chapter 12, 1950.

Ledoux, B, P. Bouchard, S. Laflèche, Y. Wu, and B. Caron: "Performance of 8-VSB Digital Television Receivers," *Proceedings of the International Broadcasting Convention*, IBC, Amsterdam, September 2000.

Lo, Y. T.: "TV Receiving Antennas," in *Antenna Engineering Handbook*, H. Jasik (ed.), McGraw-Hill, New York, N.Y., pp. 24–25, 1961.

Lucky, R. W.: "Automatic Equalization for Digital Communication," *Bell Sys.Tech. J.*, vol. 44, pg. 547, April 1965.

Mack, C. L.: "Diversity Reception in UHF Long-Range Communications," *Proc. IRE*, vol. 43, October 1955.

Megawave Corp.: "Megawave/NAB Joint Technology Development Improved Antennas For NTSC Off-The-Air TV Reception," *NAB Engineering Conference Proceedings*, National Association of Broadcasters, Washington, D.C., 1996.

Miyazawa, H.: "Evaluation and Measurement of Airplane Flutter Interference," *IEEE Trans. on Broadcasting*, vol. 35, no. 4, pp 362–367, December 1989.

Monsen, P.: "Fading Channel Communications," *IEEE Commun..*, vol. 18, pg. 16, January 1980.

NAB TV TechCheck: "Canadians Perform Indoor DTV Reception Tests," National Association of Broadcasters, Washington, D.C., October 9, 2000..

NAB TV TechCheck: "CEA Establishes Definitions for Digital Television Products," National Association of Broadcasters, Washington, D.C., September 1, 2000.

NAB TV TechCheck: "Consumer Electronic Consortium Publishes Updated Specifications for Home Audio Video Interoperability," National Association of Broadcasters, Washington, D.C., May 21, 2001.

NAB TV TechCheck: "Digital Cable–DTV Receiver Compatibility Standard Announced," National Association of Broadcasters, Washington, D.C., November 15, 1999.

NAB TV TechCheck: "FCC Adopts Rules for Labeling of DTV Receivers," National Association of Broadcasters, Washington, D.C., September 25, 2000.

NAB TV TechCheck: "New Digital Receiver Technology Announced," National Association of Broadcasters, Washington, D.C., August 30, 1999.

Neal, C. B., and S. Goyal, "Frequency and Amplitude Phase Effects in Television Broadcast Systems," *IEEE Trans*., vol. CE-23, no. 3, pg. 241, August 1977.

Nyquist, H.: "Thermal Agitation of Electrical Charge in Conductors," *Phys. Rev.*, vol. 32, pg. 110, July 1928.

Peterson, O. H., H. H. Beverage, and J. B. Moore: "Diversity Telephone Receiving System of RCA Communications, Inc.," *Proc. IRE*, vol. 19, pg. 562, April 1931.

Pleasant, D.: "Practical Simulation of Bit Error Rates," *Applied Microwave and Wireless*, pg. 65, Spring 1995.

Plonka, Robert: "Can ATV Coverage be Improved with Circular, Elliptical or Vertical Polarized Antennas?," *NAB Engineering Conference Proceedings*, National Association of Broadcasters, Washington, D.C., 1996.

Price, R., and P. E. Green, Jr.: "A Communication Technique for Multipath Channels," *Proc. IRE*, vol. 46, ph. 555, March 1958.

Proakis, J. G.: "Advances in Equalization for Intersymbol Interference," in *Advances in Communication Systems: Theory & Application*, A. V. Balakrishnan and A. J. Viterbi (eds.), Academic Press, New York, N.Y., 1975.

Qureshi, S.: "Adaptive Equalization," *IEEE Commun*., vol. 20, pg. 9, March 1982.

Qureshi, Shahid U. H.: "Adaptive Equalization," *Proceedings of the IEEE*, IEEE, New York, vol. 73, no. 9, pp. 1349–1387, September 1985.

Radio Amateur's Handbook, American Radio Relay League, Newington, Conn., 1983.

"Receiver Planning Factors Applicable to All ATV Systems," Final Report of PS/WP3, Advanced Television Systems Committee, Washington, D.C., Dec. 1, 1994.

Riegler, R. L., and R. T. Compton, Jr.: "An Adaptive Array for Interference Rejection," *Proc. IEEE*, vol. 61, June 1973.

Rohde, Ulrich L.: "Key Components of Modern Receiver Design—Part 1," *QST*, pg. 29, May 1994.

Rohde, Ulrich L. Rohde and David P. Newkirk: *RF/Microwave Circuit Design for Wireless Applications*, John Wiley & Sons, New York, N.Y., 2000.

Rossweiler, G. C., F. Wallace, and C. Ottenhoff: "Analog versus Digital Null-Steering Controllers," *ICC '77 Conf. Rec.*, 1977.

Sgrignoli, Gary: "Preliminary DTV Field Test Results and Their Effects on VSB Receiver Design," ICEE '99.

"Standards Testing: Bit Error Rate," application note 3SW-8136-2, Tektronix, Beaverton, OR, July 1993.

"Television Receivers and Video Products," UL 1410, Sec. 71, Underwriters Laboratories, Inc., New York, 1981.

Ungerboeck, Gottfried: "Fractional Tap-Spacing Equalizer and Consequences for Clock Recovery in Data Modems," *IEEE Transactions on Communications*, IEEE, New York, vol. COM-24, no. 8, pp. 856–864, August 1976.

Using Vector Modulation Analysis in the Integration, Troubleshooting and Design of Digital RF Communications Systems, Product Note HP89400-8, Hewlett-Packard, Palo Alto, Calif., 1994.

Watson, R.: "Receiver Dynamic Range; Pt. 1, Guidelines for Receiver Analysis," *Microwaves & RF*, vol. 25, pg. 113, December 1986.

"Waveform Analysis: Noise and Jitter," application note 3SW8142-2, Tektronix, Beaverton, OR, March 1993.

Whitaker, Jerry C.: *Interactive TV Survival Guide*, McGraw-Hill, New York, N.Y., 2001.

Widrow, B., and J. M. McCool: "A Comparison of Adaptive Algorithms Based on the Methods of Steepest Descent and Random Search," *IEEE Trans.*, vol. AP-24, pg. 615, September 1976.

Widrow, B., P. E. Mantey, L. J. Griffiths, and B. B. Goode: "Adaptive Antenna Systems," *Proc. IEEE*, vol. 55, pg. 2143, December 1967.

Wilson, E.: "Evaluate the Distortion of Modular Cascades," *Microwaves*, vol. 20, March 1981.

Wugofski, T. W.: "A Presentation Engine for Broadcast Digital Television," *International Broadcasting Convention Proceedings*, IBC, London, England, pp. 451–456, 1999.

Yamada and Uematsu, "New Color TV with Composite SAW IF Filter Separating Sound and Picture Signals, " *IEEE Trans.*, vol. CE-28, no. 3, p. 193.

Broadcasting, vol. 35, no. 4, pp 362–367, December 1989.

Zborowski, R. W.: "Application of On-Channel Boosters to Fill Gaps in DTV Broadcast Coverage," *NAB Engineering Conference Proceedings*, National Association of Broadcasters, Washington, D.C., 2000.

6.3r Video Signal Measurement and Analysis

ANSI Standard T1.801.03-1996, "Digital Transport of One-Way Video Signals: Parameters for Objective Performance Assessment," ANSI, Washington, D.C., 1996.

ATSC, "Transmission Measurement and Compliance for Digital Television," Advanced Television Systems Committee, Washington, D.C., Doc. A/64, Nov. 17, 1997.

ATSC, "Transmission Measurement and Compliance for Digital Television," Advanced Television Systems Committee, Washington, D.C., Doc. A/64-Rev. ZA, May 30, 2000.

Bender, Walter, and Alan Blount: "The Role of Colorimetry and Context in Color Displays," *Human Vision, Visual Processing, and Digital Display III*, Bernice E. Rogowitz (ed.), Proc. SPIE 1666, SPIE, Bellingham, Wash., pp. 343–348, 1992.

Bentz, Carl, and Jerry C. Whitaker: "Video Transmission Measurements," in *Maintaining Electronic Systems*, CRC Press, Boca Raton, Fla., pp. 328–346, 1991.

Bentz, Carl: "Inside the Visual PA, Part 2," *Broadcast Engineering*, Intertec Publishing, Overland Park, Kan., November 1988.

Bentz, Carl: "Inside the Visual PA, Part 3," *Broadcast Engineering*, Intertec Publishing, Overland Park, Kan., December 1988.

Bishop, Donald M.: "Practical Applications of Picture Quality Measurements," *Proceedings of Digital Television '98*, Intertec Publishing, Overland Park, Kan., 1998.

Boston, J., and J. Kraenzel: "SDI Headroom and the Digital Cliff," *Broadcast Engineering*, Intertec Publishing, Overland Park, Kan., pg. 80, February 1997.

"Broadening the Applications of Zone Plate Generators," Application Note 20W7056, Tektronix, Beaverton, Oreg., 1992.

DTV Express Training Manual on Terrestrial DTV Broadcasting, Harris Corporation, Quincy, Ill., September 1998.

"Eye Diagrams and Sampling Oscilloscopes," *Hewlett-Packard Journal*, Hewlett-Packard, Palo Alto, Calif., pp. 8–9, December 1996.

Fibush, David K.: "Error Detection in Serial Digital Systems," *NAB Broadcast Engineering Conference Proceedings*, National Association of Broadcasters, Washington, D.C., pp. 346-354, 1993.

Fibush, David K.: "Picture Quality Measurements for Digital Television," *Proceedings of the Digital Television '97 Summit*, Intertec Publishing, Overland Park, Kan., December 1997.

Fibush, David K.: "Practical Application of Objective Picture Quality Measurements," *Proceedings IBC '97*, IEE, pp. 123–135, Sept. 16, 1997.

Finck, Konrad: "Digital Video Signal Analysis for Real-World Problems," in *NAB 1994 Broadcast Engineering Conference Proceedings*, National Association of Broadcasters, Washington, D.C., pg. 257, 1994.

Gloeggler, Peter: "Video Pickup Devices and Systems," in *NAB Engineering Handbook*," 9th Ed., Jerry C. Whitaker (ed.), National Association of Broadcasters, Washington, D.C., 1999.

Haines, Steve: "Serial Digital: The Networking Solution?," in *NAB 1994 Broadcast Engineering Conference Proceedings*, National Association of Broadcasters, Washington, D.C., pg. 270, 1994.

Hamada, T., S. Miyaji, and S. Matsumoto: "Picture Quality Assessment System by Three-Layered Bottom-Up Noise Weighting Considering Human Visual Perception," *SMPTE Journal*, SMPTE, White Plains, N.Y., pp. 20–26, January 1999.

MacAdam, D. L.: "Visual Sensitivities to Color Differences in Daylight," *J. Opt. Soc. Am.*, vol. 32, pp. 247–274, 1942.

Mertz, P.: "Television and the Scanning Process," *Proc. IRE*, vol. 29, pp. 529–537, October 1941.

Pank, Bob (ed.): *The Digital Fact Book*, 9th ed., Quantel Ltd, Newbury, England, 1998.

Quinn, S. F., and C. A. Siocos: "PLUGE Method of Adjusting Picture Monitors in Television Studios—A Technical Note," *SMPTE Journal*, SMPTE, White Plains, N.Y., vol. 76, pg. 925, September 1967.

Reed-Nickerson, Linc: "Understanding and Testing the 8-VSB Signal," *Broadcast Engineering*, Intertec Publishing, Overland Park, Kan., pp. 62–69, November 1997.

SMPTE Engineering Guideline EG 1-1990, "Alignment Color Bar Test Signal for Television Picture Monitors," SMPTE, White Plains, N.Y., 1990.

SMPTE Recommended Practice: RP 166-1995, "Critical Viewing Conditions for Evaluation of Color Television Pictures," SMPTE, White Plains, N.Y., 1995.

SMPTE Recommended Practice: RP 167-1995, "Alignment of NTSC Color Picture Monitors," SMPTE, White Plains, N.Y., 1995.

SMPTE Recommended Practice: SMPTE RP 192-1996, "Jitter Measurement Procedures in Bit-Serial Digital Interfaces," SMPTE, White Plains, N.Y., 1996.

SMPTE Standard: SMPTE 259M-1997, "Serial Digital Interface for 10-bit 4:2:2 Components and $4F_{sc}$ NTSC Composite Digital Signals," SMPTE, White Plains, N.Y., 1997.

SMPTE Standard: SMPTE 303M, "Color Reference Pattern," SMPTE, White Plains, N.Y., 1999.

Stremler, Ferrel G.: "Introduction to Communications Systems," Addison-Wesley Series in Electrical Engineering, Addison-Wesley, New York, December 1982.

Tannas, Lawrence E., Jr.: *Flat Panel Displays and CRTs*, Van Nostrand Reinhold, New York, pg. 18, 1985.

Uchida, Tadayuki, Yasuaki Nishida, and Yukihiro Nishida: "Picture Quality in Cascaded Video-Compression Systems for Digital Broadcasting," *SMPTE Journal*, SMPTE, White Plains, N.Y., pp. 27–38, January 1999.

Verona, Robert: "Comparison of CRT Display Measurement Techniques," *Helmet-Mounted Displays III*, Thomas M. Lippert (ed.), Proc. SPIE 1695, SPIE, Bellingham, Wash., pp. 117–127, 1992.

6.3s Standards and Practices

"Advanced Television Enhancement Forum Specification," Draft, Version 1.1r26 updated 2/2/99, ATVEF, Portland, Ore., 1999.

Appelquist, P.: "The HD-Divine Project: A Scandinavian Terrestrial HDTV System," *1993 NAB HDTV World Conference Proceedings*, National Association of Broadcasters, Washington, D.C., pg. 118, 1993.

ATSC Digital Television Standard, Doc. A/53, Advanced Television Systems Committee, Washington, D.C., 1996.

ATSC: "Comments of The Advanced Television Systems Committee, MM Docket No. 00-39," ATSC, Washington, D.C., May, 2000.

"ATV System Recommendation," *1993 NAB HDTV World Conference Proceedings*, National Association of Broadcasters, Washington, D.C., pp. 253–258, 1993.

Baron, Stanley: "International Standards for Digital Terrestrial Television Broadcast: How the ITU Achieved a Single-Decoder World," *Proceedings of the 1997 BEC*, National Association of Broadcasters, Washington, D.C., pp. 150–161, 1997.

Battison, John: "Making History," *Broadcast Engineering*, Intertec Publishing, Overland Park, Kan., June 1986.

Baumgartner, Fred, and Terrence Baun: "Broadcast Engineering Documentation," in *NAB Engineering Handbook*, 9th ed., Jerry C. Whitaker (ed.), National Association of Broadcasters, Washington, D.C., 1999.

Baumgartner, Fred, and Terrence Baun: "Engineering Documentation," in *The Electronics Handbook*, Jerry C. Whitaker (ed.), CRC Press, Boca Raton, Fla., 1996.

Benson, K. B., and D. G. Fink: *HDTV: Advanced Television for the 1990s*, McGraw-Hill, New York, N.Y., 1990.

Benson, K. B., and J. C. Whitaker (eds.): *Television and Audio Handbook for Engineers and Technicians*, McGraw-Hill, New York, N.Y., 1989.

Benson, K. B., and Jerry C. Whitaker (eds.): *Television Engineering Handbook*, rev. ed., McGraw-Hill, New York, N.Y., 1992.

CCIR Document PLEN/69-E (Rev. 1): "Minutes of the Third Plenary Meeting," pp. 2–4, May 29, 1990.

CCIR Report 801-3: "The Present State of High-Definition Television," pg. 37, June 1989.

CCIR Report 801-3: "The Present State of High-Definition Television," pg. 46, June 1989.

Code of Federal Regulations, 40, Part 761.

"Current Intelligence Bulletin #45," National Institute for Occupational Safety and Health, Division of Standards Development and Technology Transfer, February 24, 1986.

Delatore, J. P., E. M. Prell, and M. K. Vora: "Translating Customer Needs Into Product Specifications", Quality Progress, January 1989.

DeSantis, Gene: "Systems Engineering Concepts," in *NAB Engineering Handbook*, 9th ed., Jerry C. Whitaker (ed.), National Association of Broadcasters, Washington, D.C., 1999.

DeSantis, Gene: "Systems Engineering," in *The Electronics Handbook*, Jerry C. Whitaker (ed.), CRC Press, Boca Raton, Fla., 1996.

Digital Audio Compression (AC-3) Standard, Doc. A/52, Advanced Television Systems Committee, Washington, D.C., 1996.

"Dr. Vladimir K. Zworkin: 1889–1982," *Electronic Servicing and Technology*, Intertec Publishing, Overland Park, Kan., October 1982.

"Electrical Standards Reference Manual," U.S. Department of Labor, Washington, D.C.

Federal Communications Commission: Notice of Proposed Rule Making 00-83, FCC, Washington, D.C., March 8, 2000.

Finkelstein, L.: "Systems Theory", *IEE Proceedings*, vol. 135, Part A, no. 6, July 1988.

General Services Administration, Information Technology Service, National Communications System: *Glossary of Telecommunication Terms*, Technology and Standards Division, Federal Standard 1037C, General Services Administration, Washington, D.C., August, 7, 1996.

Gilder, George: "IBM-TV?," *Forbes*, Feb. 20, 1989.

Guide to the Use of the ATSC Digital Television Standard, Doc. A/54, Advanced Television Systems Committee, Washington, D.C., 1996.

Hammar, Willie: *Occupational Safety Management and Engineering*, Prentice Hall, New York, N.Y.

Hammett, William F.: "Meeting IEEE C95.1-1991 Requirements," *NAB 1993 Broadcast Engineering Conference Proceedings*, National Association of Broadcasters, Washington, D.C., pp. 471–476, April 1993.

Hoban, F. T., and W. M. Lawbaugh: *Readings In Systems Engineering*, NASA, Washington, D.C., 1993.

Hopkins, R.: "Advanced Television Systems," *IEEE Transactions on Consumer Electronics*, vol. 34, pp. 1–15, February 1988.

Krivocheev, Mark I., and S. N. Baron: "The First Twenty Years of HDTV: 1972–1992," *SMPTE Journal*, SMPTE, White Plains, N.Y., pg. 913, October 1993.

Lincoln, Donald: "TV in the Bay Area as Viewed from KPIX," *Broadcast Engineering*, Intertec Publishing, Overland Park, Kan., May 1979.

Markley, Donald: "Complying with RF Emission Standards," *Broadcast Engineering*, Intertec Publishing, Overland Park, Kan., May 1986.

McCroskey, Donald C.: "Standardization: History and Purpose," in *The Electronics Handbook*, Jerry C. Whitaker (ed.), CRC Press, Boca Raton, Fla., 1996.

McCroskey, Donald: "Setting Standards for the Future," *Broadcast Engineering*, Intertec Publishing, Overland Park, Kan., May 1989.

National Electrical Code, NFPA #70.

"Occupational Injuries and Illnesses in the United States by Industry," OSHA Bulletin 2278, U.S. Department of Labor, Washington, D.C, 1985.

OSHA, "Electrical Hazard Fact Sheets," U.S. Department of Labor, Washington, D.C, January 1987.

OSHA, "Handbook for Small Business," U.S. Department of Labor, Washington, D.C.

Pank, Bob (ed.): *The Digital Fact Book*, 9th ed., Quantel Ltd, Newbury, England, 1998.

Pfrimmer, Jack, "Identifying and Managing PCBs in Broadcast Facilities," *1987 NAB Engineering Conference Proceedings*, National Association of Broadcasters, Washington, D.C, 1987.

Program Guide for Digital Television, Doc. A/55, Advanced Television Systems Committee, Washington, D.C., 1996.

Reimers, U. H.: "The European Perspective for Digital Terrestrial Television, Part 1: Conclusions of the Working Group on Digital Terrestrial Television Broadcasting," *1993 NAB HDTV World Conference Proceedings*, National Association of Broadcasters, Washington, D.C., p. 117, 1993.

"Safety Precautions," Publication no. 3386A, Varian Associates, Palo Alto, Calif., March 1985.

Schow, Edison: "A Review of Television Systems and the Systems for Recording Television," *Sound and Video Contractor*, Intertec Publishing, Overland Park, Kan., May 1989.

Schreiber, W. F., A. B. Lippman, A. N. Netravali, E. H. Adelson, and D. H. Steelin: "Channel-Compatible 6-MHz HDTV Distribution Systems," *SMPTE Journal*, SMPTE, White Plains, N.Y., vol. 98, pp. 5-13, January 1989.

Schreiber, W. F., and A. B. Lippman: "Single-Channel HDTV Systems—Compatible and Noncompatible," Report ATRP-T-82, Advanced Television Research Program, MIT Media Laboratory, Cambridge, Mass., March 1988.

Schubin, Mark: "From Tiny Tubes to Giant Screens," *Video Review*, April 1989.

Shinners, S. M.: *A Guide to Systems Engineering and Management*, Lexington, 1976.

"Sinclair Seeks a Second Method to Transmit Digital-TV Signals," *Wall Street Journal*, Dow Jones, New York, N.Y., October 7, 1999.

Smith, Milford K., Jr.: "RF Radiation Compliance," *Proceedings of the Broadcast Engineering Conference*, Society of Broadcast Engineers, Indianapolis, IN, 1989.

SMPTE and EBU: "Task Force for Harmonized Standards for the Exchange of Program Material as Bitstreams," *SMPTE Journal*, SMPTE, White Plains, N.Y., pp. 605–815, July 1998.

System Engineering Management Guide, Defense Systems Management College, Virginia, 1983.

System Information for Digital Television, Doc. A/56, Advanced Television Systems Committee, Washington, D.C., 1996.

"Television Pioneering," *Broadcast Engineering*, Intertec Publishing, Overland Park, Kan., May 1979.

"Toxics Information Series," Office of Toxic Substances, July 1983.

Tuxal, J. G.: *Introductory System Engineering*, McGraw-Hill, New York, N.Y., 1972.

"Varian Associates: An Early History," Varian publication, Varian Associates, Palo Alto, Calif.

Whitaker, Jerry C.: *AC Power Systems*, 2nd Ed., CRC Press, Boca Raton, Fla., 1998.

Whitaker, Jerry C.: *Electronic Displays: Technology, Design, and Applications*, McGraw-Hill, New York, N.Y., 1994.

Whitaker, Jerry C.: G. DeSantis, and C. Paulson: *Interconnecting Electronic Systems*, CRC Press, Boca Raton, Fla., 1993.

Whitaker, Jerry C.: *Maintaining Electronic Systems*, CRC Press, Boca Raton, Fla. 1991.

Whitaker, Jerry C.: *Power Vacuum Tubes Handbook*, 2nd Ed., CRC Press, Boca Raton, Fla., 1999.

Whitaker, Jerry C.: *Radio Frequency Transmission Systems: Design and Operation*, McGraw-Hill, New York, N.Y., 1990.

Chapter

7

Index of Figures, Tables, and Subjects

7.1 Introduction

Because of the unique nature of the material contained in this book, the tradational subject index has been expanded to include a listing of figures and tables, organized by subject matter.

7.2 Index of Figures

7.2.a General Electronics

7.2.b The Physical Nature of Sound

7.2.c The Audio Spectra

7.2.d Architectural Acoustic Principles and Design Techniques

7.2.e Light, Vision, and Photometry

7.2.f Color Vision, Representation, and Reproduction

7.2.g Optical Components and Systems

7.2.h Digital Coding of Audio/Video Signals

7.2.i Microphone Devices and Systems

7.2.j Sound Reproduction Devices and Systems

7.2.k Electron Optics and Deflection

7.2.l Video Cameras

7.2.m Monochrome and Color Image Display

7.2.n Audio/Video Recording Systems

7.2.o Audio/Video Production Standards, Equipment, and Design

7.2.p Film for Video Applications

7.2.q Audio/Video Compression Systems

7.2.r Audio/Video Networking

7.2.s Digital Broadcast Transmission Systems

7.2.t Frequency Bands and Propagation

7.2.u Radio/Television Transmission Systems

7.2.v Radio/Television Transmitting Antennas

7.2.w Radio/Television Receivers and Cable/Satellite Systems

7.2.x Audio/Video Signal Measurement and Analysis

7.2.y Standards and Practices

7.3 Index of Tables

7.3.a General Electronics

7.3.b The Physical Nature of Sound

7.3.c Architectural Acoustic Principles and Design Techniques

7.3.d Light, Vision, and Photometry

7.3.e Color Vision, Representation, and Reproduction

7.3.f Digital Coding of Audio/Video Signals

7.3.g Sound Reproduction Devices and Systems

7.3.h Electron Optics and Deflection

7.3.i Video Cameras

7.3.j Monochrome and Color Display

7.3.k Audio/Video Recording Systems

7.3.l Audio/Video Production Standards, Equipment, and Design

7.3.m Film for Video Applications

7.3.n Audio/Video Compression Systems

7.3.o Audio/Video Networking

7.3.p Digital Broadcast Transmission Systems

7.3.q Frequency Bands and Propagation

7.3.r Radio/Television Transmission Systems

7.3.s Radio/Television Transmitting Antennas

7.3.t Radio/Television Receivers and Cable/Satellite Systems

7.3.u Audio/Video Signal Measurement and Analysis

7.3.v Standards and Practices

7.4 Subject Index

About the Author

Jerry C. Whitaker is Technical Director of the Advanced Television Systems Committee (ATSC), Washington, D.C. He was previously President of Technical Press, a consulting company based in the San Jose (CA) area. Mr. Whitaker has been involved in various aspects of the electronics industry for over 25 years, with specialization in communications. Current book titles include the following:

- Editor-in-chief, *Standard Handbook of Video and Television Engineering*, 3rd ed., McGraw-Hill, 2000
- Editor-in-chief, *Standard Handbook of Audio and Radio Engineering*, 2nd ed., McGraw-Hill, 2001
- *DTV Handbook*, 3rd ed., McGraw-Hill, 2001
- *Video Displays*, McGraw-Hill, 2000
- Editor, *Interactive Television Demystified*, McGraw-Hill, 2001
- Editor, *Video and Television Engineers' Field Manual*, McGraw-Hill, 2000
- *Radio Frequency Transmission Systems: Design and Operation*, McGraw-Hill, 1990
- *Maintaining Electronic Systems*, CRC Press, 1991
- Editor-in-chief, *The Electronics Handbook*, CRC Press, 1996
- *AC Power Systems Handbook*, 2nd ed., CRC Press, 1999
- *Power Vacuum Tubes Handbook*, 2nd ed., CRC Press, 1999
- *The Communications Facility Design Handbook*, CRC Press, 2000
- *The Resource Handbook of Electronics*, CRC Press, 2000
- Co-author, *Television and Audio Handbook for Engineers and Technicians*, McGraw-Hill, 1989
- Co-author, *Communications Receivers*, 3rd ed., McGraw-Hill, 2000
- Co-editor, *Information Age Dictionary*, Intertec Publishing/Bellcore, 1992

Mr. Whitaker has lectured extensively on the topic of electronic systems design, installation, and maintenance. He is the former editorial director and associate publisher of *Broadcast Engineering* and *Video Systems* magazines, and a former radio station chief engineer and television news producer.

Mr. Whitaker is a Fellow of the Society of Broadcast Engineers and an SBE-certified professional broadcast engineer. He is also a fellow of the Society of Motion Picture and Television Engineers, and a member of the Institute of Electrical and Electronics Engineers.

Mr. Whitaker has twice received a Jesse H. Neal Award *Certificate of Merit* from the Association of Business Publishers for editorial excellence. He has also been recognized as *Educator of the Year* by the Society of Broadcast Engineers.

Mr. Whitaker resides in Morgan Hill, California.